From Gondwana to the Ice Age

From Gondwana to the Ice Age

The geological development of New Zealand
over the last 100 million years

Malcolm Laird and John Bradshaw

CANTERBURY UNIVERSITY PRESS

First published in 2020 by
CANTERBURY UNIVERSITY PRESS
University of Canterbury
Private Bag 4800, Christchurch
NEW ZEALAND
www.canterbury.ac.nz/engage/cup

ISBN 978-1-927145-99-9

A catalogue record for this book is available from the
National Library of New Zealand.

Book design and layout: Smartwork Creative, www.smartworkcreative.co.nz
Index: Robin Briggs

Front cover images: (Left) Sea cliffs at Fox River mouth, West Coast, South Island: Cretaceous sedimentary
breccia formed during extension and metamorphic core-complex development. (J.D. Bradshaw)
(Right) Sea cliffs at Cape Kidnappers, Hawke's Bay, east coast North Island: early Pleistocene marginal
marine Kidnappers Group of the convergent margin suite. (B.D. Field, GNS Science)

Printed in China through Asia Pacific Offset

Published with assistance from
the Foundation for Research, Science and Technology (FRST) and GNS Science

Contents

FOREWORD BY PETER BARRETT **10**

INTRODUCTION **12**

1 OVERVIEW AND SYNTHESIS OF BASIN EVOLUTION **17**

1.1 Cretaceous **17**
1.1.1 Albian (Urutawan–early Ngaterian) 20
1.1.2 Early–middle Cenomanian (Ngaterian) 21
1.1.3 Late Coniacian–Maastrichtian (Piripauan–Haumurian) 21
1.2 Paleocene **23**
1.3 Eocene **25**
1.4 Oligocene **28**
1.5 Miocene **32**
1.6 Pliocene–Pleistocene **36**

2 STRUCTURAL FRAMEWORK AND PRE MID-CRETACEOUS HISTORY OF THE NEW ZEALAND REGION **39**

2.1 Introduction **39**
2.2 Structural outline **40**
2.3 Basement terranes **40**
2.3.1 Western Province terranes 40
2.3.1.1 Buller Terrane *41*
2.3.1.2 Takaka Terrane *41*
2.3.2 The Median Boundary 42
2.3.3 Eastern Province 43
2.3.3.1 Brook Street Terrane *43*
2.3.3.2 Murihiku Terrane *43*
2.3.3.3 Maitai Terrane *44*
2.3.3.4 Eastern Terranes group *44*
2.3.3.5 Caples Terrane *45*
2.3.3.6 Torlesse Terrane *45*

2.3.3.7 Kaweka Terrane *46*
2.3.3.8 Waipapa Composite Terrane *46*
2.3.3.9 Mount Camel suspect terrane *47*
2.3.3.10 Tectonic setting of the eastern group of terranes *47*
2.4 The Rangitata Orogeny **47**
2.5 Break-up tectonics **48**
2.6 The plate boundary debate **49**

3 OVERVIEW OF THE STRATIGRAPHY AND TECTONICS OF THE MID-CRETACEOUS TO RECENT SUCCESSION **51**

3.1 From convergent margin to extension, and the Albian Unconformity **51**
3.2 Stratigraphic hierarchy and synopsis of Cretaceous–Cenozoic geological history **52**
3.2.1 Introduction 52
3.2.2 The Cretaceous–Cenozoic megasequence 54
3.2.3 Tectono-sedimentary assemblages 54
3.2.3.1 Assemblage 1 *54*
3.2.3.2 Assemblage 2 *55*
3.2.3.3 Assemblage 3 *56*

4 ASSEMBLAGE 1 – THE EARLY EXTENSION PHASE: THE MID- AND EARLY LATE CRETACEOUS SUCCESSION **57**

4.1 Introduction **57**
4.1.1 Effect of basement rocks 57
4.1.2 Mid-Cretaceous erosion and unconformity 58
4.1.3 Problematic features of the Raukumara Peninsula 58
4.1.3.1 Koranga Formation *59*
4.1.3.2 'Basement' to 'cover' gradational relationships *60*
4.1.4 Problematic 'cover' strata of the Chatham Rise and the Canterbury Basin 61
4.1.5 Structural sub-belts of the East Coast Basin 62
4.2 The Clarence Series: Albian–Cenomanian (Urutawan–Ngaterian) successions **62**
4.2.1 Non-marine sedimentary successions of southern and western New Zealand 63
4.2.1.1 Western New Zealand *63*
4.2.1.2 Non-marine sedimentary successions of southern and eastern South Island basins *66*
4.2.2 Marine sedimentary successions of eastern New Zealand 69
4.2.2.1 Clarence Series sedimentary rocks of onshore Northland *69*
4.2.2.2 Clarence Series sedimentary rocks of the East Coast Basin *69*
4.2.2.3 Clarence Series sedimentary rocks of the Chatham Rise *76*
4.3 The Raukumara Series: Late Cenomanian–Late Coniacian (Arowhanan–Teratan) successions **77**
4.3.1 Non-marine successions of the South Island 77
4.3.2 Marine successions of Northland, the East Coast Basin and the Chatham Rise 78
4.3.2.1 Raukumara Series rocks of onshore Northland *78*
4.3.2.2 Raukumara Series rocks of the East Coast Basin *79*
4.3.2.3 Raukumara Series rocks of the Chatham Rise *82*
4.4 Igneous rocks of the early extension phase **82**
4.4.1 Clarence Series 82
4.4.1.1 Northland and East Coast allochthons *82*
4.4.1.2 Autochthonous silicic volcanics of Albian (Urutawan–Motuan) age *83*
4.4.1.3 Autochthonous igneous rocks of Cenomanian (late Ngaterian) age *83*
4.4.2 South Island Raukumara Series igneous rocks 84

5 LATE CRETACEOUS AND PALEOGENE TECTONIC SETTING: CHANGING EXTENSIONAL REGIMES 85

5.1 Introduction 85
5.2 Late Cretaceous–Early Paleocene 85
5.3 Late Paleocene–Early Miocene 86

6 PASSIVE MARGIN PHASE: THE LATEST CRETACEOUS AND PALEOGENE SUCCESSION 89

6.1 Introduction 89
6.2 Continental break-up and the onset of sea-floor spreading 89
6.2.1 Associated unconformities 90
6.2.2 The Waipounamu Erosion Surface 94
6.2.3 Relationship of the unconformities to break-up and sea-floor spreading 94
6.3 Late Cretaceous, Santonian (Piripauan) successions 95
6.3.1 Introduction 95
6.3.2 Description of Santonian (Piripauan) rocks 95
6.3.2.1 Northland Allochthon 95
6.3.2.2 East Coast Basin 95
6.3.2.3 Eastern basins of the South Island 97
6.4 Campanian–Paleocene (Haumurian–Teurian) succession: Sedimentation during sea-floor spreading (~84–~56 Ma) 98
6.4.1 Introduction 98
6.4.2 Maastrichtian–Paleocene (Haumurian–late Teurian) sediments of eastern New Zealand 99
6.4.2.1 Whangai facies 99
6.4.3 Other lithofacies of eastern New Zealand 103
6.4.3.1 Northland 103
6.4.3.2 Northeastern South Island 104
6.4.3.3 Chatham Islands and Chatham Rise 105
6.4.3.4 Southeastern South Island and offshore 106
6.4.4 The Waipawa ‘black shale’ facies (Late Paleocene/late Teurian: ~58–~56 Ma) 107
6.4.4.1 Eastern North Island 107
6.4.4.2 Eastern South Island 107
6.4.4.3 Western New Zealand 108
6.4.5 Haumurian–Teurian sediments of western New Zealand 108
6.4.5.1 The West Coast–Taranaki Rift 108
6.4.5.2 Sedimentary succession 108
6.5 Igneous rocks associated with break-up and sea-floor spreading 115
6.5.1 North Island 115
6.5.2 Eastern South Island and Chatham Islands 115
6.5.3 Western South Island 116
6.6 Latest Paleocene event (Latest Teurian: ~58–55.5 Ma): The cessation of Tasman Sea spreading 116
6.7 Latest Paleocene–Eocene successions (~56–~34 Ma) 118
6.7.2 Latest Paleocene–earliest Oligocene sediments of eastern New Zealand 118
6.7.2.1 Eastern North Island 118
6.7.2.2 Eastern South Island 118
6.7.2.3 Other sedimentary lithofacies of eastern South Island 120
6.7.3 Latest Paleocene–earliest Oligocene sediments of western New Zealand 123
6.7.3.1 Southland onshore basins 123

6.7.3.2 Solander offshore basin *127*
6.7.3.3 Westland basins *128*
6.7.3.4 Taranaki Basin *132*
6.7.3.5 Northland Basin *134*
6.7.3.6 Onshore North Island basins *136*
6.8 Late Paleocene–Eocene volcanism **138**
6.9 Eocene–Oligocene boundary event (~33.7 Ma) **139**
6.9.1 Origin of the event 140
6.10 Oligocene successions **141**
6.10.1 Introduction 141
6.10.2 Oligocene sediments of eastern New Zealand 141
6.10.3 Oligocene rocks of southern and western New Zealand 143
6.10.3.1 Eastern Southland *143*
6.10.3.2 Western Southland *144*
6.10.3.3 Offshore Southland basins *147*
6.10.3.4 West Coast South Island *148*
6.10.3.5 Taranaki Basin *149*
6.10.3.6 Northland Basin *151*
6.10.3.7 Onshore North Island basins *152*
6.10.4 Oligocene volcanism 154
6.10.4.1 Northland Ophiolite and Northland Allochthon *154*
6.10.4.2 Other Oligocene igneous rocks *155*

7 MID- TO LATE CENOZOIC TECTONISM: THE KAIKOURA OROGENY **156**

7.1 Introduction **156**
7.2 Tectonic architecture of the Kaikoura Orogeny **157**
7.3 The Hikurangi margin **159**
7.4 The Western margin of the Kaikoura Orogen **160**
7.5 Tectonic history **161**
7.5.1 Early–Middle Miocene (23–15 Ma) 161
7.5.2 ‘Far field’ Kaikoura events in New Zealand 163
7.5.3 Middle–Late Miocene (~13–11 Ma) 164
7.5.4 Latest Miocene–Pliocene 164

8 THE FINAL CONVERGENT MARGIN PHASE: THE NEOGENE ASSEMBLAGE **165**

8.1 Introduction **165**
8.2 Early to mid-Middle Miocene (Late Waitakian–Lillburnian) **166**
8.2.1 Northland and East Coast allochthons 166
8.2.2 East Coast Basin 168
8.2.3 Canterbury, Chatham Rise and Great South basins 174
8.2.4 Southland basins 176
8.2.4.1 Eastern Southland Basin *177*
8.2.4.2 Winton Basin *177*
8.2.4.3 Waiau Basin *178*
8.2.4.4 Te Anau Basin *179*
8.2.4.5 Solander and Balleny basins *180*
8.2.5 West Coast basins 182
8.2.6 Taranaki and western offshore Northland basins 184
8.2.7 Onshore Northland, South Auckland and King Country basins 186

8.3 Middle Miocene event **192**
8.4 Late Middle–Late Miocene **193**
8.4.1 East Coast Basin 193
8.4.2 Canterbury, Chatham Rise and Great South basins 198
8.4.3 Western Southland basins 200
8.4.3.1 Waiau Basin *200*
8.4.3.2 Te Anau Basin *202*
8.4.3.3 Solander and Balleny basins *202*
8.4.4 West Coast basins 203
8.4.5 Taranaki and western offshore Northland basins 205
8.4.6 Onshore Northland, South Auckland, King Country and Wanganui basins 209
8.5 Miocene–Pliocene boundary event (Kapitean–Opoitian boundary: ~5 Ma) **211**
8.6 Pliocene–Pleistocene **212**
8.6.1 East Coast Basin 212
8.6.2 Canterbury, Chatham Rise, Great South and Eastern Southland basins 215
8.6.3 Western Southland basins 218
8.6.4 West Coast 220
8.6.5 Taranaki and offshore west Northland basins 221
8.6.6 Onshore Northland and South Auckland basins 224
8.6.7 King Country and Wanganui basins 225
8.7 Neogene volcanism **227**
8.7.1 Eastern and southern New Zealand basins (Pacific plate) 227
8.7.2 Western and northern New Zealand basins (Australian plate) 229

9 EVENT STRATIGRAPHY **233**

9.1 Introduction **233**
9.2 Global events **233**
9.2.1 Cretaceous–Tertiary (K–T) boundary event (65 Ma) 233
9.2.2 Eustatic sea-level changes 235
9.3 Regional events **236**
9.3.1 Cenomanian tectono-magmatic event (late Ngaterian: ~97 Ma) 236
9.3.2 Late Cenomanian event (Ngaterian–Arowhanan boundary: ~95 Ma) 238
9.3.3 Late Campanian–early Maastrichtian event (late Haumurian: ~70 Ma) 238
9.3.3.1 Origin of the event *239*
9.3.4 Mid-Eocene event (Porangan–Bortonian: ~45–~40 Ma) 240
9.3.4.1 Origin of the event *240*
9.3.5 Intra-Oligocene unconformities: the Marshall Paraconformity 240
9.3.5.1 Origin of the unconformity(ies) *242*
9.3.6 Mid-Miocene event (~15–13 Ma) 242
9.3.7 Miocene–Pliocene boundary event (Kapitean–Opoitian boundary: ~5 Ma) 243
9.3.7.1 Origin of the event *244*

CONCLUDING REMARKS **244**

REFERENCES **245**

LOCATION INDEX **281**

GENERAL INDEX **285**

Foreword

Malcolm Laird's geological career spanned five decades, beginning in 1960 with a master's project at Auckland University College under Jack Grant-Mackie, mapping Mesozoic strata of the 26 square miles of hill country northeast of Taumarunui in the southern Waikato region. In those days the challenge was to recognise from sparse macrofossils in rare outcrops the recently defined stages and series of Marwick's new Hokonui System and the major structural features of the area. After graduating Malcolm took a position in the Greymouth office of the New Zealand Geological Survey, mapping the coal-bearing strata of the region, where the basic stratigraphy had already been established by Harold Wellman, Max Gage and Pat Suggate, with mapping based mainly on lithology, supported by a new microfossil-based chronology being developed by Marwick and Hornibrook.

In 1964 Malcolm was awarded a DSIR PhD scholarship, choosing to study at the University of Oxford under Harold Reading. Reading at this time was developing the old concept of facies analysis in a more rigorous way. He pursued this work through a series of graduate research projects on environments from mountains and plains to the deep sea, summarised in his classic *Sedimentary Environments and Facies* (1978). Malcolm's own D.Phil. project, submitted in 1969, analysed the sedimentology of Silurian clastic rocks in western Ireland. While waiting for the outcome he spent the summer in Finnmark examining the newly discovered late Precambrian glacial deposits, and the results from this work were later published in Norway.

Malcolm's choice of D.Phil. project was very much in line with his first trip to Antarctica in the summer of 1960–61. As geologist in a four-man dog-sledging party he had mapped Paleozoic strata in the Nimrod Glacier area in the central Transantarctic Mountains. There he discovered the first Archaeocyathid-bearing rocks in place on the continent, a find reported in *Nature* in 1962. The work of Malcolm and his team was crucial to establishing the paleontological link between South Africa, the Transantarctic Mountains and southern Australia. He went on to lead several further expeditions over the next two decades, mapping and interpreting strata of similar age both in New Zealand and in northern Victoria Land. The results are summarised in his review in Robert Tingey's *The Geology of Antarctica* (1991).

Though Malcolm plainly had a passion for the Antarctic and the Paleozoic, this was to be shared equally with the Cretaceous and Cenozoic strata of New Zealand, which became central to his career and contribution to New Zealand geology. After his return home in 1969 he was based in Christchurch, applying the facies approach to mapping with colleagues at the co-located New Zealand Geological Survey and University of Canterbury on Late Cretaceous and Cenozoic strata of east Canterbury and Marlborough.

In the following years he became a key figure in discussions that led in 1978 to the Cretaceous–Cenozoic Project, or CCP, to which he was appointed leader. He was also head of the Christchurch office of the New Zealand Geological Survey (and later GNS Science), and was widely admired for his gentlemanly demeanour at all times, and for his enthusiasm for debate in the field: passionate, even heated, but always respectful. The project produced a series of reports covering all of the major sedimentary basins of New Zealand, a well-organised and accessible body of knowledge on the geological history of the New Zealand region over the last 135 million years and the resources within the strata of this period.

Malcolm's gentle but firm leadership style, developed through the CCP experience, was later called on for his role as chief sedimentologist for the team drilling Antarctic Cenozoic strata off Cape Roberts in 1999. His peace-keeping efforts and geologically themed T-shirts during the core interpretation session each morning were legendary.

Malcolm took early retirement to stay in Christchurch after the closing of the GNS office there, having successfully negotiated funding from the New Zealand Foundation for Research, Science and Technology (FRST) to work on a book to provide an up-to-date synthesis of the work covered by the project that had been so much of his life for the previous 20 years. This has now been completed with the help of his friends and colleagues led by John Bradshaw, and it is a worthy record of this most comprehensive survey of New Zealand's Cretaceous–Cenozoic strata, and of the teamwork that produced it.

Peter Barrett
Emeritus Professor of Geology
Antarctic Research Centre
Victoria University of Wellington

Introduction

Scope of this publication

There have been a number of books and monographs about New Zealand geology, starting with Dr Ferdinand Hochstetter's *Geologie von Neu-Seeland*, published in 1864, and ending most recently with several publications by individual geologists in the 1970s and 1980s. Perhaps the most comprehensive of these was the two-volume *The Geology of New Zealand*, edited by Suggate *et al.* and published in 1978. All of these publications dealt with the full stratigraphic range of New Zealand geology, from Cambrian to Recent.

This volume is more restricted in its scope, dealing with geological events subsequent to a major reorganisation of tectonic plates in the New Zealand region in mid-Cretaceous times. This reorganisation resulted in a region-wide change from compression associated with convergent margin tectonics, to extension. It is a convenient event to start with for a variety of reasons, primarily because it marks the demarcation between the highly indurated, extensively deformed and commonly metamorphosed 'basement' rocks of New Zealand and the overlying less indurated and deformed 'cover strata'. It is also significant in terms of the economic resources of New Zealand's sedimentary rocks: the vast majority of the country's coal and petroleum resources lie within the mid-Cretaceous to Recent sedimentary succession.

The 'oil shocks' of the 1970s led to a renewed and urgent interest in New Zealand's sedimentary resources, which was reflected by a large quantity of new research and subsequent publication on the country's younger sedimentary geology. Spearheading this research was the New Zealand Geological Survey's Sedimentary Basin Studies programme, also known as the Cretaceous–Cenozoic Project (CCP). This project, instigated by Dr R.P. Suggate, then Director of the New Zealand Geological Survey, started in 1978. It was designed to investigate the sedimentary succession to a uniform standard, and provide a high-quality database which could serve as a springboard for future hydrocarbon exploration. A number of regional monographs, covering most of the onshore and adjacent offshore parts of the country were published under the auspices of the CCP before the project was wound down in the late 1990s without final completion.

This publication is the logical culmination of the project, and represents a synthesis of regional sedimentary information assembled from a variety of sources. It also includes any available information from the continental shelf and beyond.

The timing of this synthesis is highly appropriate. Not only had a huge volume of information been assembled during the existence of the CCP, but a surge of drilling and seismic acquisition by petroleum companies since the early 1970s has greatly extended our knowledge of the subsurface, both onshore and offshore. Similarly, knowledge of plate movements in the southwest Pacific has increased dramatically in the last two or three decades.

Our intention is to present for the first time a comprehensive history of sedimentary basin evolution in the New Zealand region for the last ~110 m.y. This has required integration of the timing of tectonic plate movements in the region with basin formation and development, assessment from igneous material of the tectonic environment, and study of the sedimentary processes involved, lithofacies analysis and fossil content in order to ascertain sedimentary environment. The study has been expanded to include large areas of the continental shelf surrounding New Zealand, many of the regional sedimentary basins having both onshore and offshore elements.

Sources of data

A major source of data has been the regional basin studies monographs published by the New Zealand Geological Survey and its successors, DSIR Geology and Geophysics and the Institute of Geological and Nuclear Sciences (GNS Science). The mid-Cretaceous and Cenozoic successions in New Zealand fall conveniently into 11 main basins, with other less well-studied basins lying further offshore (see Fig. 1.3). In all, eight monographs were published on these basins: East Coast, Chatham Rise, Canterbury, Great South Basin, Western Southland, West Coast, Taranaki and Northland. Three others, Clutha, Wanganui and King Country/South Auckland, were not completed. Studies of the Wanganui and King Country basins have subsequently been carried out by researchers from the University of Waikato, and the results published to date are incorporated in this volume.

Other major sources of information have been university masters and doctoral theses, and oil company reports in the public domain. In addition, many other publications, including geological maps and papers in scientific journals have provided material for this study. The results of drilling on the New Zealand continental shelf undertaken by the Deep Sea Drilling Project (DSDP) and by its successor the Ocean Drilling Program (ODP) have also been utilised where appropriate. We have carried out additional research and, in some instances, have found that reinterpretation in the light of new data has been required.

Stratigraphy

When the New Zealand Geological Survey's CCP began in 1978 lithostratigraphic subdivision and nomenclature were fully developed only in areas where detailed investigation for 1 : 63,360 scale maps had taken place, or which had been the subject of special study. New field examination stemming from the CCP resulted in a proliferation of new lithostratigraphic names, many of them specific to a particular sedimentary basin.

There are difficulties in resolving lithostratigraphy in a region characterised by ongoing and complex deformation and tectonic displacement, and rapid lateral facies change. In many cases the proliferation of member, formation and group names tends to obscure the larger picture of intra-basinal relationships. Nevertheless, although some areas in New Zealand are burdened with a complex stratigraphic nomenclature, attempts have been made to design a more region-wide stratigraphic framework. Nathan (1974) developed a stratigraphy based on groups for north Westland and Buller on the West Coast of the South Island which was recognisable throughout the region: Browne and Field (1985), Field and Browne (1986) and Andrews *et al.* (1987) were able to show that a number of separately named limestone units throughout the Canterbury–Marlborough region in the east of the South Island were facies variations of the Amuri Limestone.

More recently, the introduction of the principles of sequence stratigraphy has proved helpful. The identification of unconformity-bounded packages of strata (allostratigraphy) enables a more coherent view of genetically related strata. For example, Laird (1992) developed a basin-wide allostratigraphy for the Cretaceous rocks of Marlborough based on unconformities, and a similar exercise was carried out for the sedimentary succession in the Great South Basin (Beggs, 1993; Cook *et al.* 1999).

In this volume, we have tried to keep the use of local formation and member designations to a minimum, and refer to groups and lithofacies where possible. Where sequence stratigraphy has been applied, sequences or allo-units are used preferentially. Our emphasis throughout this study has been on chronostratigraphic correlation region-wide to emphasise lateral and vertical facies changes.

We have used an 'event stratigraphy' to subdivide the Cretaceous–Cenozoic record. The geological events utilised are, firstly: first-order major changes in plate boundaries and tectonic plate motion in the southwest Pacific region, causing fundamental changes in tectonic style throughout the New Zealand continental block, and secondly: second-order changes resulting from local deformation or changes in depositional environment brought about by factors such as rapid variations in global sea level. Most, but not all, of these events are marked by erosion surfaces or abrupt regional facies changes. They are discussed further in Chapters 3 and 9.

Dating and geological timescale

New Zealand has developed its own set of biostratographic stages, partly because of its endemic flora and fauna and partly to avoid problems caused by globally diachronous bio-events. The mid- and Late Cretaceous stages were originally defined solely on inoceramid species, but work over the last few years both on these bivalves and *Aucellina*, together with development of a dinoflagellate zonal system, has produced some substantial revisions to the correlation of New Zealand stages with the international timescale (Crampton *et al.* 2004a). The use of dinoflagellates as a zonal tool has proved particularly valuable for subdividing the last 20 m.y. of the Cretaceous, and correlating it with international stages (Crampton *et al.* 2004b). New isotopic dating has also been valuable for establishing international correlation (Crampton *et al.* 2004a). The Cretaceous–Tertiary (K–T) boundary is well exposed at many localities throughout New Zealand, and is defined in many cases not only by faunal and floral extinctions, but by geochemical anomalies, including a significant iridium spike (see Chapter 9). The Cenozoic portion of the geological timescale is subdivided principally using foraminifera, except for the Pliocene which is based on both foraminifera and pectens. Bio-events have in many cases been related to the geomagnetic polarity timescale of Cande and Kent (1995), and to isotopic dating. The Pliocene–Pleistocene segment of the timescale has been refined by using oxygen isotope ratios, paleomagnetism and tephrochronologic dating to supplement dating using biostratigraphy (Carter & Naish, 1998).

Malcolm G. Laird, 2015

The family view

This book represents the culmination of the career of Malcolm Laird. He was a geologist by training and commitment, and his work was published widely. After he took early retirement from GNS in the 1990s, his work on what had been the Cretaceous–Cenozoic Project (CCP), for which he had full responsibility until 1990, continued in the form of a proposed publication funded by the then FRST (New Zealand Foundation for Research, Science and Technology). In 1999 a contract was signed with Canterbury University Press for the finished product.

In the intervening years he worked steadily to get the work finished, with all the necessary additions of the most recent research; but he was frequently hindered by problems with diagrams, the changes in nomenclature of places and procedures, and his full involvement in the lives of his wife and three teenage daughters.

Over time, The Book became not merely an elephant in the study, but the Cretaceous dinosaur with an insatiable appetite. Then came the earthquakes, the need for the repair of the house, room by room, health problems requiring two separate surgeries, and family bereavements. These all compromised the conclusion of the project.

At the beginning of 2015, he was so exhausted that at the celebration of his 80th birthday in April, his long-time colleague, co-author and friend, John Bradshaw, was asked to join in a major effort to complete the work. Malcolm's unexpected death two months later led a group of Malcolm's colleagues to urge John to continue and to see it through to publication with their help.

As Malcolm's family, we express our deepest respect and grateful thanks for their time and effort.

Margaret Burrell, Tamsin, Islay and Fiona Laird, 2019

Addenda 2018

This book was conceived by Malcolm Laird as an accessible summary of all the data that had been acquired on stratigraphy and sedimentary basin development during the Cretaceous–Cenozoic Project. The text was mainly written before 2008 at which time Malcolm began to have some minor health problems that interfered with the work. In 2015 it was decided that we should revise the work when Malcolm returned from South America. Sadly, he died suddenly in 2015, but his colleagues felt that it was important that the work still be updated and published.

I am very much the junior author, having limited experience of Cenozoic rocks, and thus the text is largely as written by Malcolm. In collaboration with a number of advisors, however, the text has been edited and references to more-recent publications have been added. The stratigraphic figures have been simplified to make them legible at the page size. The original paleogeographic maps were drawn with the present-day configuration of the New Zealand continental crust and did not represent actual relative positions of the sedimentary basins at the time of deposition. Unfortunately, at the present time there are questions about the timing and extent of oroclinal bending and contrasting views on the age and dynamic history of the Alpine Fault plate boundary. With no consensus on basic paleogeography, drawing new series of paleogeographic maps of sedimentary basins, particularly for the Cretaceous and Paleogene, would be contentious and possibly of ephemeral value. The maps of King (2000) can be used as a basis for the Late Cenozoic. The geochronological ages of units have been brought into line with the Global Geochronological Scale and the subdivisions of New Zealand Geological Timescale as adopted by Raine *et al.* (2015).

Inevitably with a book of this scope there are references to an enormous number of localities where important sections occur. It is not feasible to produce a map with all these localities, but all can be found in the New Zealand 1 : 50,000 topographic map series, which is available in digital form from a number of suppliers. Alternatively, Google Earth or a good road atlas can be used. The 21 maps and notes of GNS Science's 1 : 250,000 Geological Map of New Zealand series (QMAP: www.gns.cri.nz) also provide a valuable guide to geological framework, context and location.

Not all geologists focus on sedimentology and sedimentary basins, but in New Zealand the tectonic drivers are strongly indicated and reflected by unusual and spectacular deposits, some of which are illustrated in the colour plates. Most of these places are well worth visiting by any student of the earth sciences. Rapid vertical and lateral facies change is the rule rather than the exception, and even a book of this length is not sufficient to cover the details.

Naming conventions used in this publication

The LINZ New Zealand Gazetteer has been used for place names, with the addition of 'Northland Peninsula', 'Taranaki Peninsula' and 'East Coast'. These three names are in common use in geological literature and are needed to distinguish onshore geology from offshore basins.

Stratigraphic names have followed the form in the New Zealand Stratigraphic Lexicon, except where a new name or revised status has appeared in recent publications.

Initial capitals have been used for 'Early', 'Middle' and 'Late' in sub-epochs of the Cenozoic. This allows more-flexible use of the same words in lower case. 'K–T boundary' rather than 'K–Pg boundary' has been retained from the original manuscript as it is still widely understood.

Acknowledgements

The book in its present form is not remotely 'all my own work'. I have enjoyed significant help from many people, in particular Brad Field, James Crampton, Greg Browne and other staff of GNS Science. Kari Bassett and Catherine Reid of the Department of Geological Sciences, University of Canterbury, helped to clarify the structure and content of Chapter 6. Catherine Montgomery and Katrina McCallum of Canterbury University Press were unfailingly helpful in leading me through the world of publication. I am grateful to copyeditor Janet Bray and proofreader Helen Eastward who together identified ambiguities and inconsistencies that my geological eye had missed, turning an inconsistent draft into a text with a consistent style. Residual errors and omissions are all mine. Thanks also to Robin Briggs for his thorough and comprehensive indexes.

Jarg Pettinga and Catherine Reid provided office space and support in the Department of Geological Sciences, University of Canterbury, during a period of upheaval and relocation. The technical staff were unfailingly helpful.

The Ministry for Research, Science and Technology funded the later stages of the research through the Foundation for Research, Science and Technology (FRST) and contributed to the cost of publication. GNS Science gave permission to reproduce maps, stratigraphic figures and photographs. Margaret Burrell encouraged me to press on with this project and gave me access to Malcolm's office, computers and hard drives.

John Bradshaw, 2019

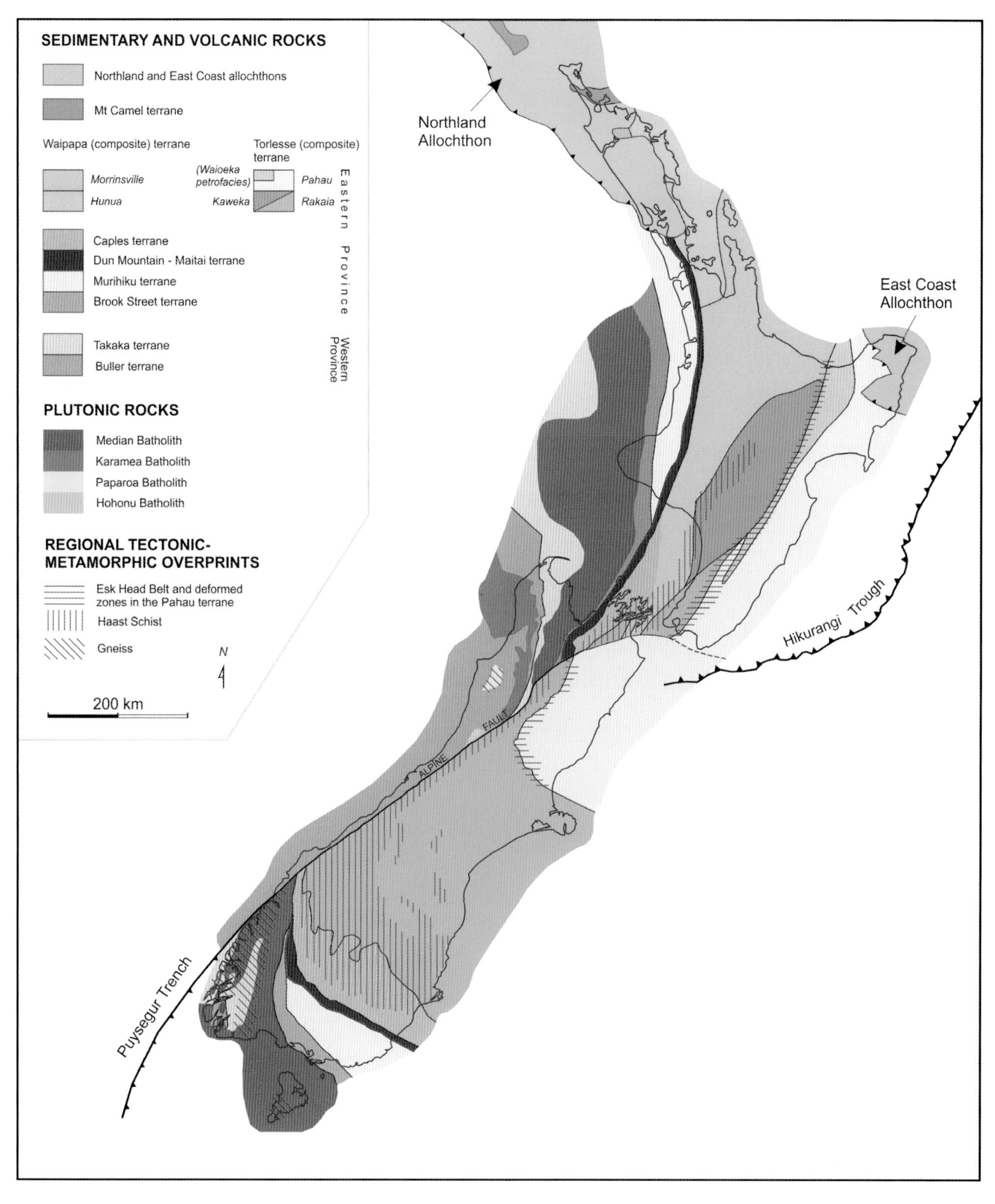

Plate 1. Map of the basement terranes recognised in New Zealand. (Source: GNS Science.)

(a)

(b)

Plate 2. Contrasted structure of Mesozoic basement sediments. (a) Triassic and Early Jurassic sediments of the Murihiku Terrane folded into the broad Southland Syncline: steep northern limb on the left and gently dipping southern limb on right. (b) View from Mount Sutton north to Aoraki Mount Cook. This image is entirely made up of Permian and Triassic sediments of the Rakaia Subterrane (Torlesse Terrane), and forms part of the Mesozoic accretionary prism. (Photos: J.D. Bradshaw.)

Plate 3. Within the Torlesse accretionary complex less-deformed lenses show either: (a) thickly bedded sandstone (near the head of the Adams River area, Westland), or (b) alternating thinly bedded sandstone and mudstone that commonly show graded bedding and the typical 'Bouma' sedimentary pattern (near the head of the Mathias River, Canterbury). (Photos: J.D. Bradshaw.)

(a)

(b)

Plate 4. (a) Representative section of the Esk Head Mélange made up of disrupted sandstone and mudstone derived from the Kaweka Terrane with blocks of white Triassic fossiliferous limestone and dark blocks of basaltic pillow lava (north slope of Esk Head, Hurunui River South Branch, Canterbury). (b) Early Cretaceous broken-formation within the Pahau Subterrane shows major bedding-parallel extension probably developed during the early stages of accretion (Penk River, Awatere Valley, Marlborough). (Photos: J.D. Bradshaw.)

(a)

(b)

Plate 5. (a) Fabrics developed during crustal extension in Cretaceous gneiss of the Paparoa Metamorphic Core Complex (Parsons Hill, Buller District). (b) Blocks of sandstone in mudstone forming an early Late Cretaceous olistostrome (Penk River, Awatere Valley, Marlborough). (Photos: J.D. Bradshaw.)

(a)

(b)

Plate 6. (a & b) Thickly bedded Hawks Crag Breccia on the south side of the Cretaceous Paparoa Metamorphic Core Complex (Fox River mouth, Buller District). (Photos: J.D. Bradshaw.)

(a)

(b)

Plate 7. (a) Korangan sandstone and conglomerate, dipping away from the camera, fills a channel cut in strongly deformed Pahau Terrane sediments of probable early Cretaceous age (Koranga, southwest of Matawai, Gisborne District). (b) The unconformity between steeply dipping Pahau Terrane sediments and sub-horizontal Te Wera Formation (with Malcolm Laird in the rain, Koranga area, southwest of Matawai, Gisborne District). (Photos: J.D. Bradshaw.)

Plate 8. (a) Gently dipping Albian sediments without complex deformation typical of cover sediment (Karekare Formation, Kokopumatara Stream, northeast of Matawai, Gisborne District). (b) Late Cretaceous Bluff Sandstone (Hapuku River, Marlborough). (Photos: J.D. Bradshaw.)

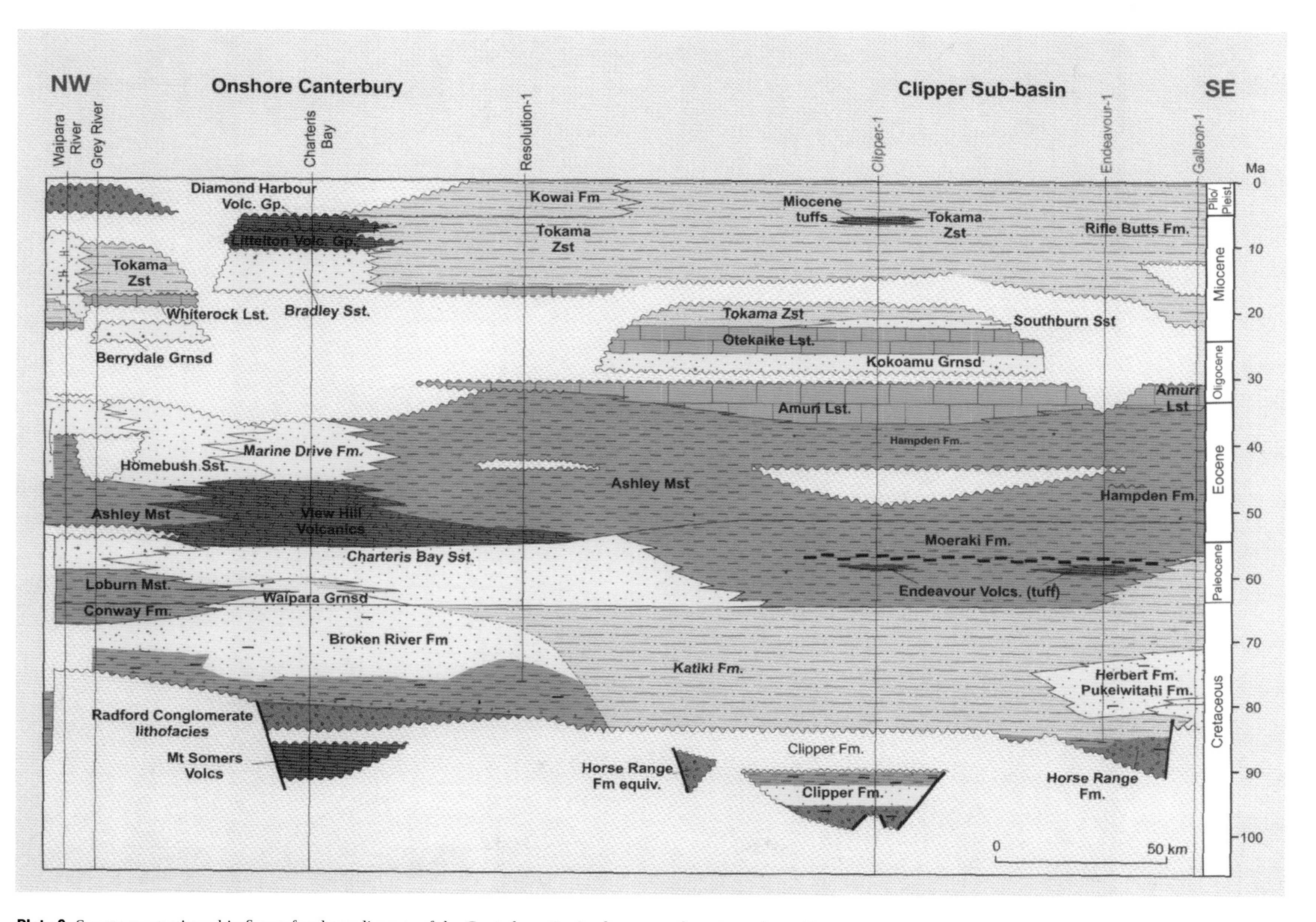

Plate 9. Summary stratigraphic figure for the sediments of the Canterbury Basin along a northwest–southeast line.
Fm, Formation; Gp, Group; Grnsd, Greensand; Sst, Sandstone; Lst, Limestone; Mst, Mudstone; Volc Gp, Volcanic Group; Volcs, Volcanics; Zst, Siltstone. (Source: GNS Science.)

(a)

(b)

Plate 10. (a) Part of the thick section of Brunner Coal Measures with channelised conglomerate and quartzose sandstone (Mount Rochfort, Buller Coalfield). (b) Four-metre seam overlain by quartzose sandstone in the Brunner Coal Measures (this view has since been destroyed during opencast mining) (Stockton Plateau, southwest Nelson). (Photos: J.D. Bradshaw.)

(a)

(b)

Plate 11. (a) Oligocene limestone on Flock Hill, in the Castle Hill tectonic basin, central Canterbury. The range behind is made of Triassic accretionary complex rocks. The limestones are interlayered with beds of sand- to pebble-size volcanic debris. (b) Pillow lavas with limestone filling the interstices (Waiareka Volcanics, Boatmans Harbour, Oamaru, north Otago). (Photos: J.D. Bradshaw.)

(a)

(b)

Plate 12. (a) Oligocene Spyglass Formation limestone with glauconitic interbeds resting with slight discordance on Amuri Limestone (shore platform, South Bay, Kaikoura). (b) Channelled and cross-bedded Oligocene Otekaike Limestone (Late Oligocene–Early Miocene) (Waihao Forks, south Canterbury). (Photos: J.D. Bradshaw.)

(a)

(b)

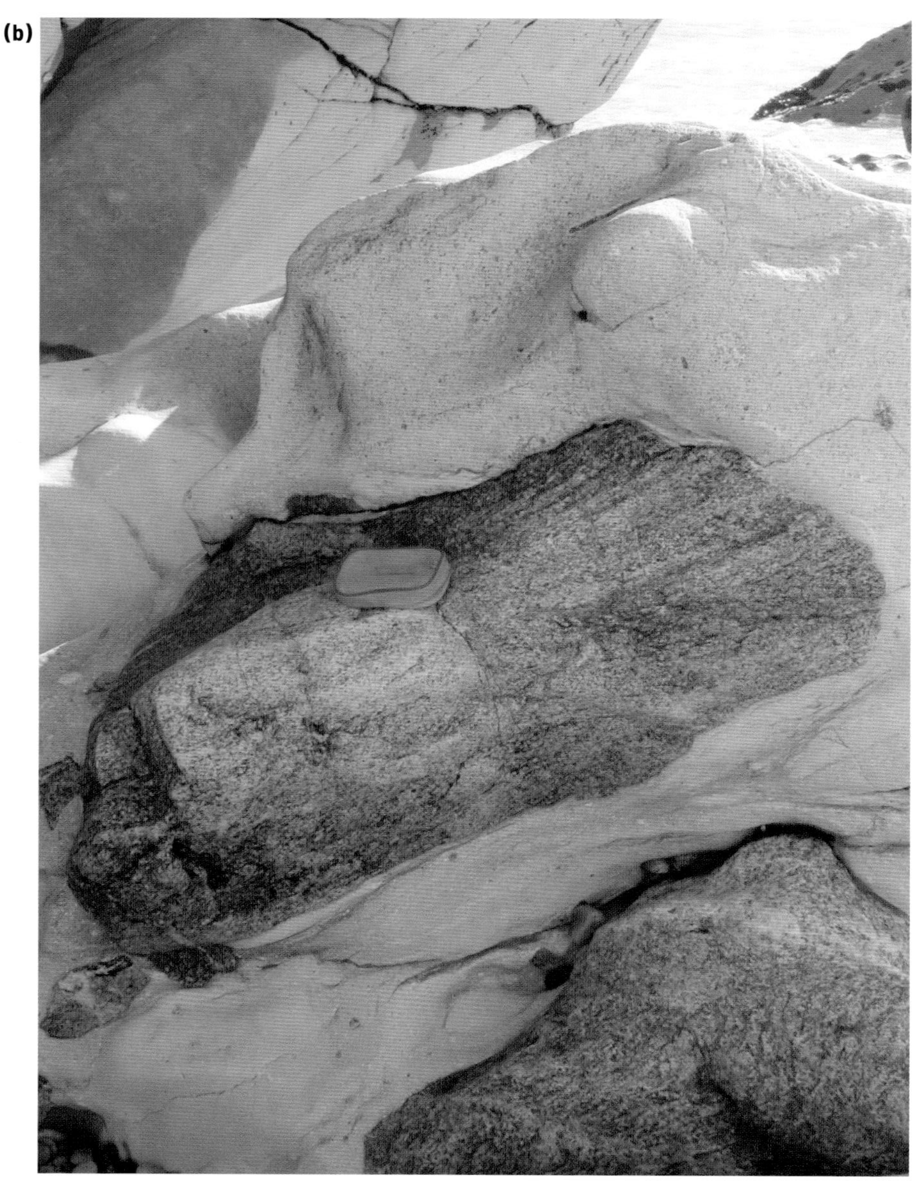

Plate 13. (a & b) Boulders of deformed granitoids in debris flows isolated within medium- to fine-grained limestone in Oligocene Karamea Limestone (Gentle Annie Point, Buller District). (Photos: J.D. Bradshaw.)

(a)

(b)

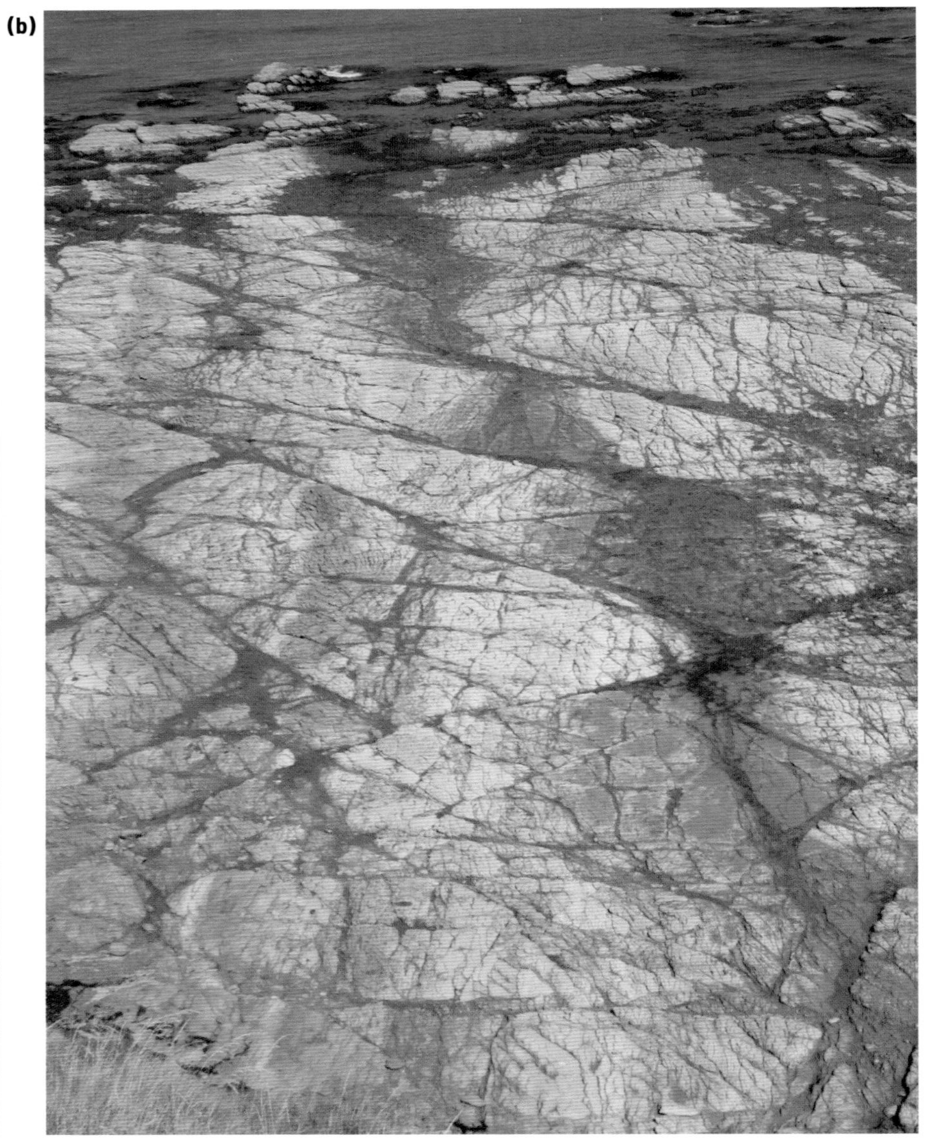

Plate 14. (a) Decimetre- and centimetre-bedded Early Miocene levee-overbank turbidites of the Whakataki Formation (north of Castlepoint, Wairarapa). (Source: GNS Science.) (b) Much faulted Early Miocene very fine sandstone to mudstone of the Waima Formation (shore platform, South Bay, Kaikoura). (Photo: J.D. Bradshaw.)

Plate 15. (a) Great Marlborough Conglomerate resting on Waiau Formation (Swale Stream, Clarence Valley, Marlborough). (b) Great Marlborough Conglomerate with a large block of Amuri Limestone in a conglomerate composed mainly of basement sandstone and Cretaceous volcanic boulders (Deadman Stream, coastal Marlborough). (Photos: J.D. Bradshaw.)

Plate 16. (a) Awapapa Limestone, one of the lenticular bioclastic coquinas of the Te Aute Limestone lithofacies. These lenses are of latest Pliocene to early Pleistocene age and are a widespread bluff-forming lithology (Awapapa, Poverty Bay). (b) Marginal marine Kidnappers Group sandstone with minor mudstone and white tuffaceous bands of early Pleistocene (early to mid-Castlecliffian) age (Cape Kidnappers, Hawke's Bay). (Photos: B.D. Field, GNS Science.)

CHAPTER 1

Overview and Synthesis of Basin Evolution

1.1 CRETACEOUS

The Cretaceous was a time of profound change in the New Zealand region. Throughout all but the later stages of the period New Zealand formed part of the paleo-Pacific-facing Gondwana continental margin and lay adjacent to Antarctica and Australia (Fig. 1.1). During Early Cretaceous times what was to become the New Zealand landmass lay in a position close to the South Pole (Veevers *et al.* 1991; Sutherland *et al.* 2001). The basement is divided into two provinces: the Western Province made up of mainly Paleozoic rocks, and the Eastern Province comprising rocks of Permian to Early Cretaceous age (see Plate 1). They are separated by the Median Tectonic Zone (see Chapter 2). The paleo-Pacific margin was characterised by convergent margin tectonics culminating in regional-scale deformation, the 'Rangitata Orogeny', in the late Early Cretaceous. Events associated with the orogeny in the Eastern Province include the folding of fore-arc basins and westward accretion and thrusting of basement terranes. In the Western Province, which was still an integral part of Gondwana, the Rangitata Orogeny was expressed by the widespread intrusion of granitoids of Early Cretaceous age, by local folding and faulting, and by final amalgamation of the Median Tectonic Zone

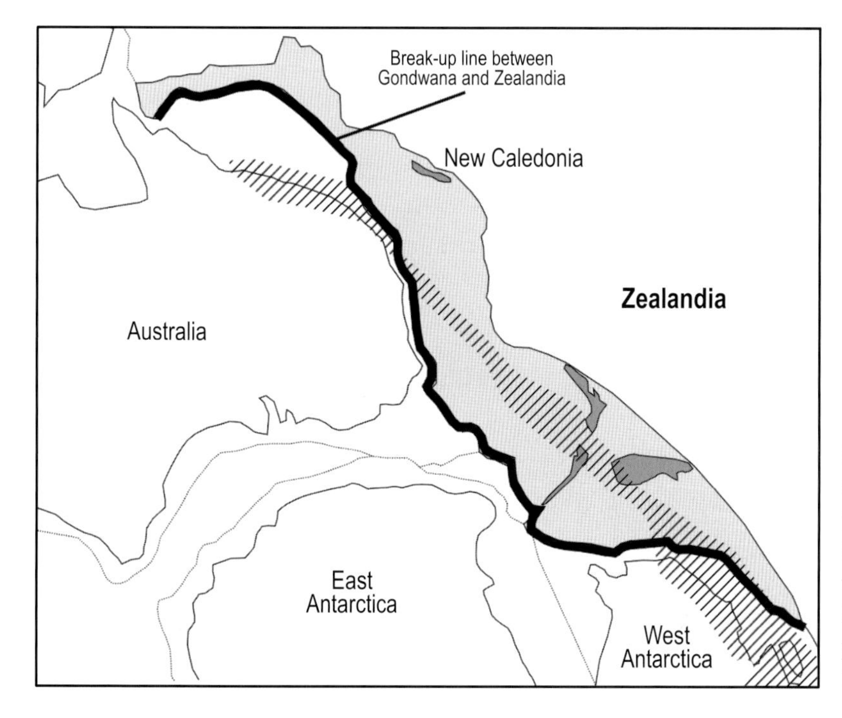

Fig. 1.1. Map showing the position of New Zealand within Zealandia on the Pacific margin of the Gondwana continent and its relationship to the future Australian, East Antarctic and West Antarctic plates. (Based on Sutherland, 1999, fig. 8.)

and the adjacent Takaka Terrane by intrusion of the Separation Point Batholith.

There is a well-established subdivision of series and stages within the New Zealand Late Cretaceous and Cenozoic and these are widely used in this book. Correlation with international schemes has varied over time but has become better constrained in recent years. The most recent version is presented in Figure 1.2.

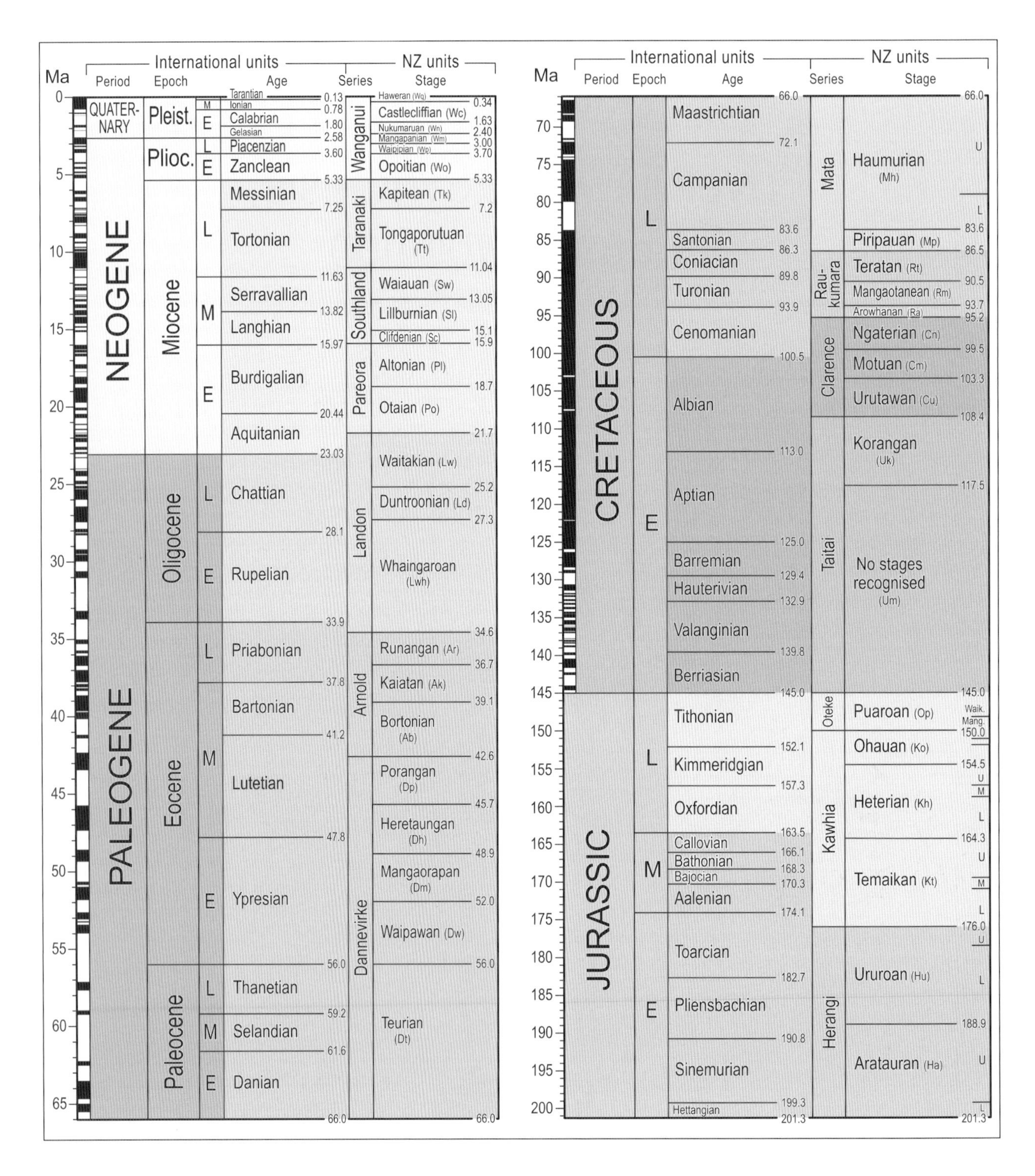

Fig. 1.2. The New Zealand geological time scale for the Jurassic to Recent, showing the correlation of local and international subdivisions and currently assigned ages. (From Raine *et al.* 2015, fig. 1.)

The following overview deals with the development of New Zealand region by region and sedimentary basin by sedimentary basin (Fig. 1.3), in a clockwise pattern from the north.

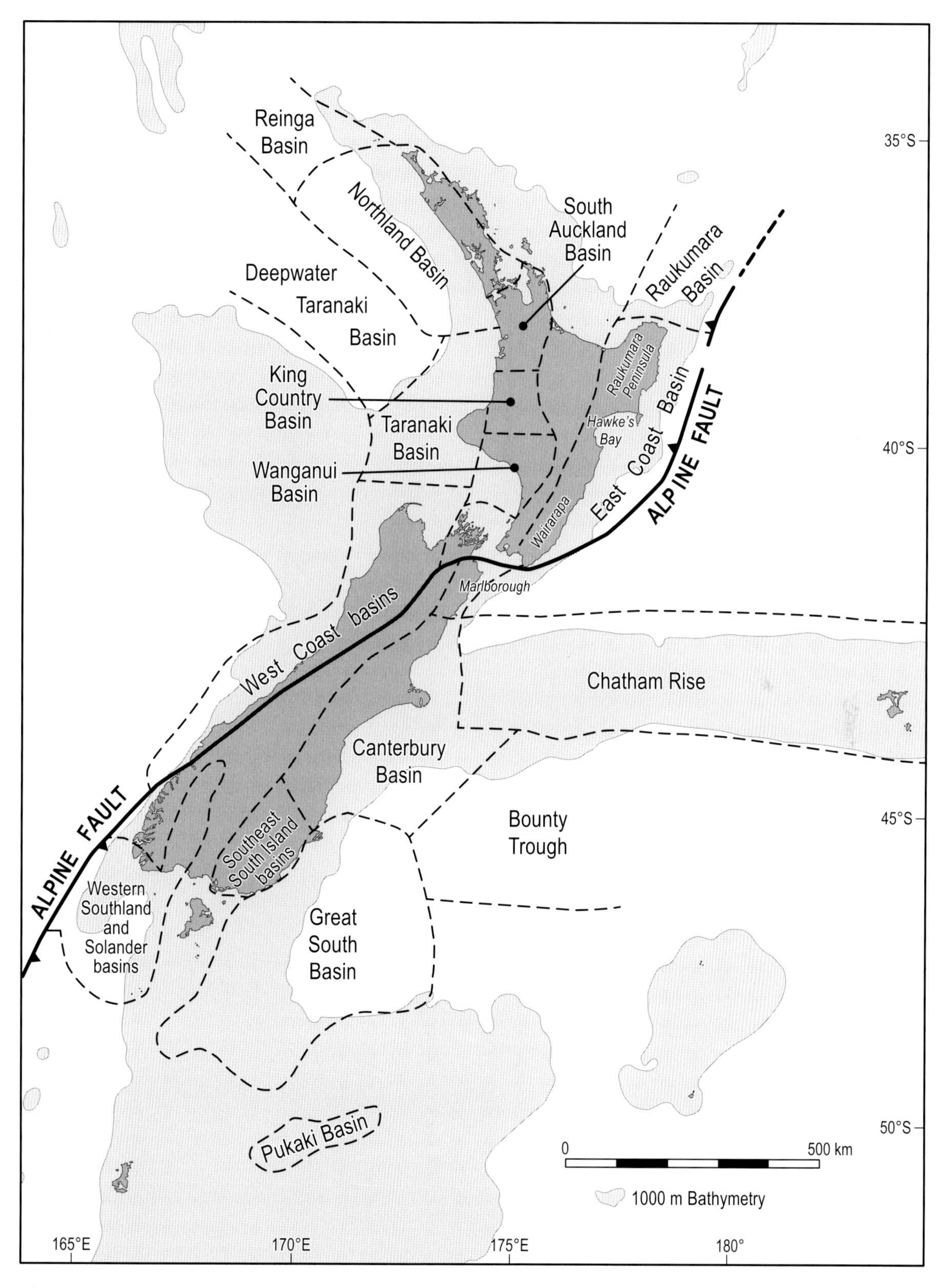

Fig. 1.3. Map of the New Zealand region, showing the main sedimentary basins onshore and offshore. (Source: GNS Science.)

1.1.1 Albian (Urutawan–early Ngaterian)

Convergent margin tectonics and associated subduction came to an end in southern and central New Zealand by ~105±5 Ma, an event probably caused by collision of the Hikurangi Plateau and possibly other plateau fragments with paleo-New Zealand at this time (Hoernle *et al.* 2006); only in the northeast (Raukumara Peninsula and Northland) is there a suggestion of continuing convergence after ~100 Ma. In the Western Province of the South Island, extension is indicated by the presence of metamorphic core complexes, and in both the Western and Eastern provinces by the formation of extensional grabens and half grabens, their inception dated in the northern Buller Terrane at ~102 Ma. On the western, Gondwana-facing, margin of the South Island the bounding faults of the developing grabens were oriented in what is currently west-northwest to northwest, parallel to the future Tasman Sea spreading centre, while on the paleo-Pacific margin the trends were more varied (Fig. 1.4). Palinspastic reconstructions indicate that the developing East Coast Basin and basins of the Chatham Rise were at that time adjacent and collinear (Sutherland *et al.* 2001; Crampton *et al.* 2003; and Fig. 1.4). The change in tectonic setting is also well-dated in the then adjacent West Antarctica, where there is an abrupt change from subduction-related magmatism to anorogenic 'A' type granitoids at ~102 Ma (Weaver *et al.* 1992).

As extension proceeded, much of the future New Zealand microcontinent, or 'Zealandia', in particular the central portion of both islands, the Western Province and the southeastern South Island including the Campbell Plateau and the Chatham Rise, remained above sea level and provided the source for thick non-marine successions in the newly developing marginal rift basins. In contrast, the thin continental crusts of the Pahau and related terranes underlying the developing East Coast Basin and

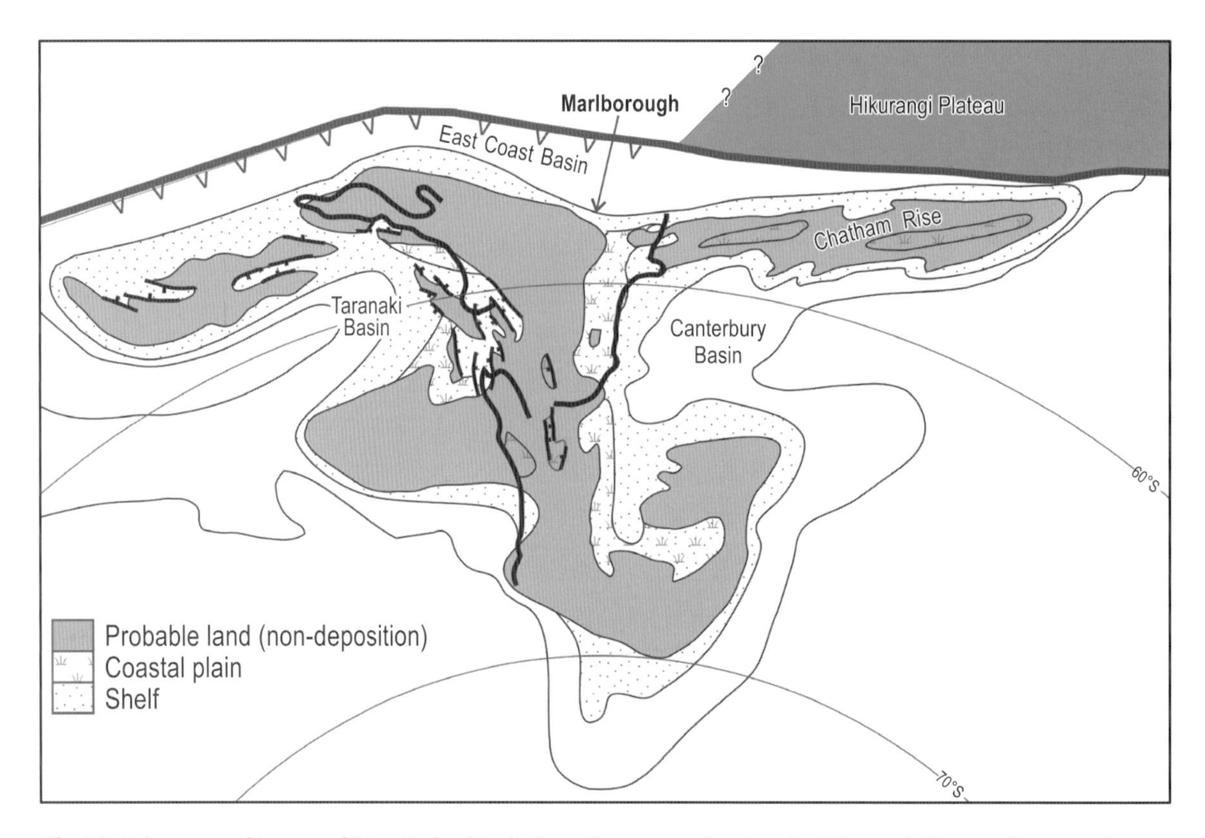

Fig. 1.4. Paleogeographic map of New Zealand in the Late Cretaceous, showing the Hikurangi Plateau adjacent to the Chatham Rise in the east. Evidence from subsequent research suggests that the Mesozoic subduction zone was still active to the northwest and that sedimentation in the East Coast and Northland regions may have been in a back-arc setting. Suggestions of large-scale Cretaceous strike-slip faulting and oroclinal bending are not included (see text). (Adapted from Crampton *et al.* 2003, fig. 5.)

eastern offshore Northland region were depressed below sea level, and commonly buried by bathyal marine deposits.

The change from convergent margin to extensional tectonics in the mid-Cretaceous (Albian) took place over a short space of time (probably <3 m.y.) in any one region, and is represented on land by an erosional unconformity. This was diachronous, ranging from early Albian (Urutawan) in western Raukumara Peninsula and parts of Marlborough, to latest Albian (Motuan: ~100 Ma) in eastern Raukumara Peninsula.

Non-marine deposits infilling the Albian–early Cenomanian (Urutawan–early Ngaterian) grabens and half grabens formed in southern New Zealand exhibit a commonly recurring motif: close to the active bounding fault locally derived breccia was deposited as talus or debris flows on alluvial fans, passing directly into fan deltas or via fluvial deposits into a lacustrine environment. Active faulting was the dominant control on subsidence and sediment supply, and this continued intermittently in some instances until the initiation of sea-floor spreading in the New Zealand region in late Santonian (late Piripauan to early Haumurian) times.

Throughout the East Coast Basin, graben fill consisted initially of deep-marine gravity-flow deposits forming large-scale fining-upwards units, the basal Albian deposits typically consisting of olistostromes or debris flows triggered by sporadic faulting along the basin margins. Shallow marine or shoreline deposits are preserved only in western Raukumara Peninsula, but the outcrop distribution and isopach pattern suggest that in southern Marlborough the Albian Basin closes to the present west and south and swings to the east to include the then collinear mid-Cretaceous half grabens of the Chatham Rise. The Albian succession in the Northland and East Coast allochthons consists almost entirely of turbidites sourced from the west and south, and submarine volcanics.

1.1.2 Early–middle Cenomanian (Ngaterian)

Early Cenomanian (Ngaterian: ~96–98 Ma) injection of intraplate igneous rocks in central New Zealand caused doming in Marlborough, resulting in shallowing and local uplift of the basin floor above sea level. This event included igneous activity in north and central Canterbury and on the Hikurangi Plateau, and igneous activity and uplift on the West Coast of the South Island and possibly also in southeastern Wairarapa. Igneous rocks of similar age on the Lord Howe Rise and in then-adjacent southeastern Australia and Antarctica may be related to this event (Hoernle *et al.* 2006). Tulloch *et al.* (2009b) review all the volcanism of this age in the context of Gondwana break-up.

In contrast to the non-marine basins, major faults bounding half grabens in the East Coast Basin became largely inactive by Cenomanian times, and, subsequently, accommodation was controlled by slow subsidence and episodic relative changes in sea level. In Marlborough a northeast-trending basin (current configuration: east–west after palinspastic reconstruction) persisted until latest Cretaceous times. Non-marine to shallow marine sediments filled the basin, with margin marine sandstone dominant near the basin margins in the south and west, and mudstone more abundant towards the northeast. Outer neritic to bathyal environments typify the East Coast Basin in Wairarapa and southern Hawke's Bay, with mudstone dominating the succession in the west and sandstone-rich turbidites in the east, with little change in lithology through Cenomanian and Coniacian times. Shoreline or shallow marine deposits are not preserved, and no western margin has been recognised. In the Raukumara Peninsula, sandstone and conglomerate of uncertain but possibly neritic environment occur in the west. In the central region, mudstone deposited at shelf to bathyal depths dominates, showing a generally deepening trend over time. To the east (oceanwards) the lower part of the mudstone interdigitates with and then overlies thick sandy turbidites of Cenomanian and older age, interpreted as forming part of a submarine fan.

1.1.3 Late Coniacian–Maastrichtian (Piripauan–Haumurian)

The paleogeography of New Zealand, including its continental shelf, had changed little up to Coniacian times. It was still attached to Gondwana, and was elevated above sea level, with the exception of eastern Northland, the East Coast Basin and the Chatham

Rise, which at that time formed a series of collinear marine basins facing northwards towards the paleo-Pacific (Crampton *et al.* 2003; and Fig. 1.4). This pattern changed in the late Coniacian (late Teratan–early Piripauan: ~87–86 Ma), with uplift and erosion in the East Coast Basin and the basins of the current southeastern South Island, probably driven by convective upwelling of the mantle just prior to sea-floor spreading (Laird & Bradshaw, 2004). The relative uplift caused a diachronous erosion surface ('break-up unconformity') to develop throughout the East Coast Basin, and a basinward facies shift in overlying sediments, with non-marine, shoreline or shallow marine facies overlying older bathyal deposits. This was immediately followed by an influx of coarse transgressive sands in the East Coast, Canterbury and Great South basins as subsidence began. The oldest of the transgressive sands overlying the break-up unconformity have been dated at ~87 Ma (late Coniacian/late Teratan) in eastern Marlborough. The timing of the first transgressive deposits may be linked to a change from rapid to slower cooling rates in the unroofing Paparoa Metamorphic Core Complex at ~90 Ma, which may correlate with the cessation of strong continental extension and the inception of sea-floor spreading in the Tasman Sea.

In the Canterbury and Great South basins the erosion surface was developed somewhat later, in late Coniacian times (early Piripauan: ~86 Ma), although there is evidence of marine influence in the Great South Basin below the erosion surface. Marine transgression encroached mainly from the present north and east, with paralic sands and coal swamps developed along the basin margin. Western areas of New Zealand were still mainly emergent at this time, although marine incursions may have begun in the New Caledonia Trough and the adjacent Lord Howe Rise and Norfolk Ridge.

A period of rapid subsidence beginning in late Santonian (late Piripauan to early Haumurian) times marked the onset of sea-floor spreading, and the separation of Zealandia from West Antarctica and Australia. A new widespread low-relief erosion surface, the Waipounamu Erosion Surface of LeMasurier and Landis (1996), was developed as transgression proceeded. The unconformity is strongly diachronous, and appears to be oldest (at ~85 Ma) in eastern Marlborough, where it has been most accurately dated (Crampton *et al.* 1999; Schiøler *et al.* 2002). Transgression accelerated and was widespread in what is now eastern New Zealand, resulting in a blanketing of the shallow transgressive sands by deeper-water muds and silts (the Whangai facies) accompanying a rapid rise in sea level, and a move of the shoreline further west over the present continental shelf. The first submergence of large tracts of the Chatham Rise and of onshore parts of the East Coast Basin, Canterbury Basin and Otago occurred, concurrent with local basaltic volcanism. Marine transgression also proceeded in east Northland.

The Campanian (early Haumurian) saw the beginning of subsidence and inundation of the present western continental shelf of New Zealand as sea-floor spreading began to form the Tasman Sea, and the Lord Howe Rise, Challenger Plateau and Norfolk Ridge became blanketed by shallow marine deposits. The first marine incursions also began in western New Zealand, in south Westland, although only a minor part of the area was affected. Most of the West Coast of the South Island and the western North Island remained emergent until Maastrichtian (latest Haumurian) times, when the first marine waters invaded parts of the Taranaki Basin and offshore portions of the Northland Basin. Again, most of the sediments were muds, reflecting derivation from a landmass of low relief. The central West Coast of the South Island and the western and central portion of the North Island from Wellington to Northland remained emergent.

At a time close to the Campanian–Maastrichtian boundary (late Haumurian: ~72 Ma) a major change in the stress regime in western and southern New Zealand resulted in local uplift, and in the development of new rift systems in the western Southland, West Coast and Taranaki basins with north-northeast orientations (current geography) nearly orthogonal to the mid-Cretaceous ones. A major new tectonic feature – the West Coast–Taranaki Rift – was associated with intraplate basaltic volcanism on the West Coast, but this did not extend north into the Taranaki Basin. While marine conditions continued to develop in the south, to the north the new fault-controlled basins filled with non-marine coal measures and local lacustrine deposits. By latest Cretaceous times marine incursions began in the Taranaki Basin and offshore portions of the Northland Basin, but the central West Coast and the currently onshore Western Southland basins remained emergent.

While rifting was occurring in the west, in the current east of the South Island some old faults were rejuvenated and new ones developed to form grabens and half grabens with mainly northwest or northeast orientations. The initial environment was paralic, typically with coal measures, which were flooded by marine transgression in the latest Cretaceous, with deposition of sands and muds. In Marlborough and southern Wairarapa, the tectonic response was uplift with a change to shallow marine sands inshore and siliceous, micritic limestone further offshore, followed by further deepening. Elsewhere, in the northern East Coast Basin, in eastern Northland and in the future Northland Allochthon offshore, and in the Canterbury Basin and Campbell Plateau, siliceous mudstones of the Whangai facies continued to blanket the region during the later Cretaceous and, in many cases, the Paleocene. Similar facies were also deposited on the Lord Howe Rise in the expanding Tasman Sea.

The final act in the Cretaceous was the event at the Cretaceous–Tertiary (K–T) boundary. The K–T boundary was originally identified in Western Europe as a level of major changes in sedimentation and faunas. It has subsequently been recognised as an event that resulted in the mass extinction of at least 50% of all species of plants and animals then in existence, including the dinosaurs and marine reptiles. It received worldwide public attention when Alvarez *et al*. (1980) suggested that the K–T boundary extinctions were caused by an asteroid impact. Critical supporting evidence for this theory is the presence of high concentrations of elements such as iridium, which is rare on earth but abundant in meteorites, in the thin layer of clay found at the K–T boundary in many sites worldwide. The K–T boundary is marked in Zealandia by mass extinction of planktic foraminifera and calcareous nannofossils. The boundary coincided with a decrease in carbonate productivity and an increase in siliceous productivity and siliceous microfossils, perhaps resulting from enhanced upwelling associated with Early Paleocene climatic deterioration. At several localities in the East Coast Basin where the K–T transition occurs within a shallow marine environment it is associated with an erosional unconformity and evidence for a fall in relative sea level.

1.2 PALEOCENE

Apart from localities where there was significant erosion across the K–T boundary, such as in parts of the East Coast Basin, in most regions there was little change in facies from latest Cretaceous (Maastrichtian) times into the Paleocene. Overall, deepening was coupled with region-wide shoreline retreat.

In eastern Northland, in the East Coast Basin from East Cape to central Wairarapa, in offshore south Canterbury and in the eastern and southern portion of the Great South Basin as far south as Campbell Island, deposition of Whangai facies mudstone in bathyal depths continued until the Late Paleocene. In Marlborough, deposition of muddy limestone continued from the latest Cretaceous into the Paleocene, and this facies spread to the north into southern Wairarapa in the Middle Paleocene, where it overlies Early Paleocene shallow-marine to shoreline sands. In the Canterbury Basin–Chatham Rise region, transgression continued and encroached on land, particularly in north Canterbury and on the Chatham Rise. This latter area shows progressive submergence as slow subsidence continued, and by the end of the Paleocene it was probably completely submerged.

Poor outcrop prevents adequate paleogeographic reconstruction of the Northland region during the Paleocene. The position of the shoreline in the North Island portion of the East Coast Basin is poorly known, but it probably lay further to the west than in the Late Cretaceous. Deepening from Late Cretaceous into the Paleocene is also evident in the sedimentary succession in Marlborough, and the shoreline almost certainly moved further to the west (present configuration).

In north Canterbury, transgression during the Paleocene encroached further inland than it did in the south. A northwest–southeast-trending (present configuration) emergent high (Hurunui High) separated deposition of marine mudstone in the south from a micritic limestone facies (Amuri Limestone) in the north, which was extending southwards from Marlborough (Field *et al.* 1997). A narrow seaway probably existed between the Hurunui High and the Chatham Rise. A low-lying landmass remained at the western margin of the region throughout Paleocene times, with a northeast-trending shoreline (present configuration). The western part of the Chatham Rise, extending to just north of Banks Peninsula, was still probably emergent, or nearly so, throughout most of the Paleocene, as it is apparently free of sediments of this age (Field *et al.* 1997). West and southwest of the rise mainly coarse-grained sediments were deposited in shallow seas flanking a northeast-trending low-lying landmass west of the present shoreline. In the southern Canterbury Basin paralic and nearshore environments were separated from bathyal settings by only a comparatively narrow neritic zone. Subsidence in the Clipper Basin continued, but mainly in the centre.

By the Paleocene, restricted marine environments extended over most of the Great South Basin, with marginal successions in the west and open sea to the northeast (Schiøler *et al.* 2010). Coal swamps extended no further southeast than the Tara-1–Toroa-1 area, which formed part of a coastal plain during the Early Paleocene, extending northeast to coastal Otago and including the Kaitangata area. Continued transgression resulted in coastal-plain coal measures giving way to nearshore glauconitic sandstones in the Late Paleocene. The central sub-basin had become a single depocentre where marine sediments accumulated. By the Late Paleocene the shoreline had moved to the west of Tara-1, but a narrow progradational zone persisted along the west margin of the basin. In the area east of Toroa-1, slope turbidite fans developed, and, further east, carbonate deposition resulted from a lack of clastic supply and restricted marine circulation (Cook *et al.* 1999).

In the Eastern Southland Basin and the western Southland basins the Paleocene coincides with a period of uplift and erosion, and no sediments are recorded. Most of the West Coast of the South Island was also exposed to subaerial erosion at this time, but marine transgression continued in the south Westland embayment from the Late Cretaceous into the Paleocene without apparent interruption. A general slowing down of tectonic activity is indicated, however, both by deposition of only a thin succession in the south, and by a marked increase in the quartz content of the non-marine Paleocene sediments in the Paparoa Trough and the Pakawau Basin in the north.

In the south of the Taranaki Basin the structurally controlled basin morphology of the Late Cretaceous continued into the Paleocene. The emergent portion of the northern West Coast extended into the area between the Pakawau and Manaia sub-basins almost as far as the present southern Taranaki coastline. These sub-basins were still actively subsiding until the Late Paleocene, and became filled with fluvial sediments. The paleogeography, however, was increasingly influenced by deepening of a marine basin to the northwest, and migration of the shoreline to the south and southwest. The fluvial systems of the Pakawau and Manaia sub-basins drained northwards into a shoreline complex of sand-rich sediments extending over present-day Taranaki west of the Taranaki Fault, and southwest to lie north of the Pakawau Basin. Sparse dinoflagellates in the Kupe South wells indicate minor marine incursions southwards as far as the Kupe South area.

To the north and northwest these shoreline sandy deposits pass gradationally into non-calcareous, dark-coloured, micaceous, carbonaceous mudstone extending over much of the Taranaki Basin. This mudstone was deposited in shelf and bathyal depths within the gradually deepening basin to the north and northwest of the coastal strand plain. By the end of the Paleocene a large shallow-marine embayment extended over most of the basin. This embayment was essentially land-locked, as eastern parts of the Challenger Plateau were still emergent. The floor of the bay shelved gently northwards, and quiet water deposition prevailed.

The position of the shoreline in the northeast is uncertain. In this area the coastal sand belt is abruptly truncated by the Taranaki Fault, but it presumably extended further to the northeast in Paleocene times. In the Te Ranga-1 well, which lies in the northeast of the Taranaki Basin adjacent to the Taranaki Fault, the Paleocene interval consists entirely of marine shelf siltstone, indicating that the paleo-shoreline is some distance east of this location. The shoreline may have extended east as far as the site of the southwestern Waikato Coalfield, where re-worked marine microfloras of Paleocene to Late Cretaceous age have been recorded. Normal faulting and differential subsidence ceased in the Late Paleocene, and the localised rift environment was replaced by more regional subsidence and sedimentation patterns. Subsidence rates began to wane, typifying the development of a passive margin basin. At this time the Taranaki area developed into a northwest-sloping continental margin along the inner reaches of the New Caledonia Basin.

North of the Taranaki Basin, west of Northland, similar gradually declining thermal subsidence continued in the Paleocene. In the Early Paleocene the shoreline lay at least 100 km west of the present coastline, as shown by the presence of non-marine gravels in the Waka Nui-1 petroleum well west of Kaipara Harbour. Later in the Paleocene, shallow marine conditions encroached eastwards into the present-day Northland Peninsula, although most of the peninsula was still above sea level. Off the present western shoreline, west of a probable coastal plain, open marine bathyal sedimentation became widespread. To the northeast, inundation and the subdued topography of the remaining landmass favoured accumulation of glauconite-rich greensand on a northeastern shelf (Isaac *et al.* 1994).

The Late Paleocene (~58–56 Ma) marked the end of sea-floor spreading in the Tasman Sea, and the beginning of a period of crustal cooling and passive continental margin subsidence. Some areas, such as the northern West Coast, southern Taranaki and much of the interior North Island, persisted as areas of emergent basement. Elsewhere, such as the Canterbury and Great South basins and the Chatham Rise, a prominent seismic horizon marks an erosion interval at this time.

Closely following the cessation of sea-floor spreading, much of the New Zealand region, but in particular eastern New Zealand, experienced widespread deposition of a distinctive organic-rich mud (Waipawa facies) at a water depth generally equivalent to the upper shelf or slope break. The appearance of this facies appears to have been coincident with a period of surface-water warming, and it is attributed to regional upwelling associated with changes in oceanic circulation. Throughout much of the offshore region, Paleocene deposition was brought to a close, following the deposition of organic-rich mud by sedimentation of deep-water, generally calcareous, muds. Notable variations in lithology occur in Marlborough and southern Wairarapa, where limestone continued to be deposited in the Late Paleocene. Limestone was also being deposited in a shelf environment over much of the site of the Chatham Islands, with marine tuff indicative of basaltic volcanism. Similar deposits also occurred in the western part of central Canterbury in latest Paleocene and Early Eocene times.

1.3 EOCENE

Early Eocene strata are absent over much of the present land area of New Zealand, suggesting that much of the axial portion of the landmass from Northland through to central and western Southland continued to be emergent and subject to erosion at this time. The central and much of the northern region of the West Coast of the South Island also remained emergent. However, in the western North Island and eastern New Zealand marine sedimentation continued with little disruption from latest Paleocene to Oligocene times.

In the East Coast Basin, subsidence continued throughout the Eocene, causing deepening depositional environments. Calcareous sediments dominated, with bathyal, slightly calcareous, muds being characteristic of the northern portion of the basin for most of the Eocene, while in southern Wairarapa and Marlborough highly calcareous muds (Amuri Limestone) dominated. In Marlborough, deposition in the Middle–Late Eocene was associated with basaltic intraplate volcanism.

The Canterbury region continued to have a low relief in the Eocene, and throughout this time continuing transgression resulted in much of the region being blanketed with outer neritic calcareous mudstone and micrite. At the beginning of the Eocene the region was still largely divided by the Chatham Rise, which was by then a submarine high or a very low-lying land area. Outer neritic to bathyal conditions were reached north and south of the rise, and the Hurunui High of north Canterbury was onlapped in the Early Eocene. In the northern part of the Canterbury Basin the Eocene was dominated by bathyal calcareous sediments (Amuri Limestone), which spread south from Marlborough. The base of the limestone is diachronous, becoming progressively younger towards the south, reflecting continuing transgression in the Eocene. In mid-north Canterbury the Amuri Limestone is confined to the Late Eocene, but further south the oldest members of the unit are Early Oligocene.

In north Canterbury the calcareous sediments forming the Amuri Limestone passed laterally southwards into outer-shelf to bathyal muds, which extended into central Canterbury from the Early to Late Eocene. Westwards, several kilometres inland of the present coastline, the muds passed gradationally into shoreline sands. Sub-alkaline volcanism, which began in the Late Paleocene in the central Canterbury region and in the Chatham Islands, continued into the Early Eocene. For the Chatham Rise in general, subsidence began in the Eocene, leading to deep submergence of large areas that had previously been exposed. The present northern edge of the rise marked the shelf edge and sloped to bathyal depths, with the Bounty Trough forming the rifted margin of the southern slope. As the rise subsided, outer-shelf to bathyal carbonate was deposited throughout the Eocene along its crest between the Chatham Islands Platform and the site of present-day Mernoo Bank.

In the offshore south Canterbury region, continued transgression resulted in steady deepening throughout the Eocene, from inner-shelf depths in the Early Eocene to outer-shelf in the Middle Eocene, resulting in widespread deposition of calcareous muds. By the Late Eocene, further deepening occurred, and deposition of carbonates (Amuri Limestone) became widespread. Volcanism at this time south of Oamaru resulted in interfingering of tuff and pillow lava with the carbonates at inner- to mid-shelf depths. In the west of the region the shoreline fluctuated in position during the Eocene and lay some tens of kilometres west and parallel to the present shoreline but in general advanced steadily northwestwards with ongoing transgression. It is marked by paralic sediments, including local coal measures. In southeastern Otago south of Dunedin, Eocene lithofacies are typical of shallow shelf depths in the Early Eocene, deepening to mid-shelf in Middle and Late Eocene times.

In the Great South Basin the change from Paleocene to Eocene sedimentation was marked by a significant increase in ocean influence. While Paleocene sediments were deposited in restricted, occasionally anoxic marine conditions, Eocene sediments were significantly more calcareous and contained faunas characteristic of open-ocean conditions. Low-lying land around the southern and eastern margins of the basin subsided below sea level at this time. In the central part of the basin the subsidence rate increased significantly during the Early Eocene. The basin was more exposed to ocean circulation towards the southeast, and here pelagic carbonate sediments, first deposited in the earliest Eocene, later extended progressively westwards.

By the Middle Eocene a new extensional plate boundary had propagated along the western margin of the Campbell Plateau and into southern New Zealand. This significantly affected drainage systems and sediment supply along the northwest margin of the basin, and may have been responsible for the observed change in sediment facies during the Middle Eocene. The formation of the Solander Basin and Solander Trough in the Middle Eocene is thought to have captured the supply of clastic sediment that was previously entering the Great South Basin along its northwest margin, causing a change from delta and submarine fan deposition to dominantly calcareous silt deposition in that basin. Relative sea level also rose rapidly in the Middle Eocene, resulting in marine transgression and landward migration of the shoreline of the Solander Basin. Basaltic volcanism associated with Late Eocene tectonism occurred along the northwest margin of the basin, and there is evidence for slight southeastward tilting. The Late Eocene also saw a sharp decline in sedimentation rates, with deposition dominated by marine carbonates.

No Early Eocene or early Middle Eocene sediments have been recorded throughout the Southland region, and it was presumably emergent. Marine transgression from the north began in the late Middle Eocene in the Te Anau Basin in fault-controlled sub-basins, possibly connected northwards to marine basins developing in the northern West Coast region that lay adjacent at that time. The drop in base-level in the late Middle–early Late Eocene caused fluvial systems with extensive coal swamps to extend throughout the Waiau Basin, through the Ohai Depression and into the Winton Basin in eastern Southland. From the beginning in Middle Eocene times, non-marine sedimentation was also dominant in the Balleny and Solander basins, although parts of the present offshore area may have been shallow marine.

Marine sedimentation continued into the early Late Eocene in the Te Anau Basin, when the sedimentation rate increased and a basin-wide fluvial complex developed. A major river system followed the trend of the present-day Eglinton and Hollyford valleys throughout the Late Eocene. Regional subsidence in the south during Late Eocene times resulted in the establishment of an extensive lake system over much of the Waiau Basin and Ohai Depression, and the retreat of the fluvial system to the west. It also resulted in a change to shallow marine conditions over a wider area involving most of the Balleny Basin. Initial Middle–Late Eocene deposition in the Solander Basin is inferred to have been in fault-controlled northeast-trending valleys, as fluvial wedges associated with scarps. The close of the Eocene saw continued subsidence and widespread marine transgression over most of the Southland basins, leaving only the Solander Ridge and the Hump Ridge–Stewart Island High emergent. In the Te Anau Basin the close of the Eocene saw the fluvial system replaced by a lagoon. Accelerated subsidence in the very latest Eocene–earliest Oligocene saw a marine incursion into the Te Anau Basin from the south.

On the West Coast the Early Eocene was a time of tectonic quiescence. The shoreline in south Westland was probably about the same as in the Paleocene. Onshore there is little sedimentary record, and most of the northern part of the region was a broad rolling subaerial plain.

The Middle Eocene was marked by the beginning of two processes that together controlled the subsequent pattern of Eocene sedimentation. The first was the start of a gradual marine transgression, which continued until Late Oligocene times and progressively inundated the whole of the land area exposed during the Early Eocene. The second was the formation or reactivation of small local fault-bounded basins, separated by areas of low-lying land or shallow sea. The largest subsiding area was the Paparoa Trough, which had expanded north and south from its latest Cretaceous–Paleocene extent, to form a narrow fault-controlled depression. By the end of the Eocene the central part of the West Coast region had become an archipelago of small islands with larger land areas to the north and south (Riordan *et al.* 2014).

In south Westland the sequence was already marine in the Middle Eocene, and steadily deepening. Elsewhere the transgression beginning in the Middle Eocene was marked by paralic coal measures (Brunner Coal Measures) that preceded the northwards-advancing shoreline. In the central West Coast the Paparoa Trough developed in the west, and by the end of the Middle Eocene the southern portion, to just north of Greymouth, had become submerged by transgressing seas. At the same time a thick fluvial succession was being deposited at the northern end, near Westport, and in other basins to the north and east in Middle and early Late Eocene time.

Marine transgression continued northwards through the Late Eocene, with the complete submergence of the axial area of the Paparoa Trough and probably much of the Reefton Basin. The area around Nelson City was also submerged by this time, although its connection with the marine area further south is uncertain. Swampy areas surrounded the shoreline, and fluvial sediment was deposited in a number of relatively small subsiding basins. Local uplift on the faulted margins of some basins provided a supply of coarse clastic material, but most of the land area was still relatively flat and low-lying and exposed to subaerial weathering. By the end of the Eocene much of the southern and central part of the region was submerged, with an irregular embayed shoreline surrounding low-lying land areas to the north and east.

By the Early Eocene in the Taranaki Basin to the north, fault control of local depocentres had entirely ceased. Early–Middle Eocene sedimentation is absent over virtually the entire southern part of the basin, while in the north, by contrast, the marine basin rapidly deepened as some former shelf areas subsided to bathyal depths. The deep marine facies extended into the offshore west Northland Basin, where the Eocene succession can be recognised on seismic reflection profiles as being of uniform thickness northwards from Taranaki Basin, continuing across the Reinga Basin and for at least 1000 km to the northwest to DSDP drillhole 206 in the New Caledonia Basin. West of the Kaipara Harbour there is a lobate thickening of the sequence similar to that described for the underlying Paleocene interval, and this is also inferred to be a delta lobe.

The subsidence in the northern Taranaki and west Northland basins appears to be a continuation of that which occurred in the Late Paleocene in the north, and the configuration of the Taranaki Basin suggests that there was northward tilting over the entire region during the Early Eocene. Subsidence ceased over a broad area in the south Taranaki Basin, and mild uplift may have occurred locally.

The growing marine embayment in the northern Taranaki Basin represented the southeastern end of the New Caledonia Basin, which extended into the emergent New Zealand and Challenger Plateau portion of Zealandia. The Taranaki passive margin was bounded to the east and south by land and to the southwest by the partly or completely emergent Challenger Plateau, and therefore evolved within an entirely intracontinental setting. To the south and east the marine embayment was bounded by a relatively narrow coastal plain along a fluctuating, northeast-trending shoreline. Shoreline and shelf depositional systems were very narrow in the northeast corner of the basin, probably reflecting influence by early movement on the Taranaki Fault. Ongoing subsidence resulted in expansion of the embayment, and Eocene deposition was characterised by coastal onlap and intermittent southward transgression of shallow marine deposits – a mirror image of the northward progression of paralic deposits accompanying northward marine transgression in the West Coast region to the south. However, the northwest Nelson region remained emergent, and formed a deeply weathered and low-lying land barrier between the West Coast and Taranaki marine embayments throughout the Late Eocene.

In the Late Eocene, tectonic style within the Taranaki Basin began to change in response to the progressive influence of the evolving Australian–Pacific plate boundary through New Zealand. Precursory deformation was marked by sediment accumulation within isolated faulted sub-basins in a probable extensional setting. At the same time, passive margin development continued uninterrupted over the remainder of the basin, eventually culminating in a regional hiatus in the Early Oligocene.

The central North Island from onshore Northland to the Wellington coast, like the central spine of the South Island, remained emergent throughout the early part of the Eocene. With the solitary exception of a small outcrop of Early Eocene coal measures caught up in the Northland Allochthon, there is no evidence of the deposition of sediments anywhere in this region in the Early–Middle Eocene.

Late Eocene regional subsidence in the North Taranaki and offshore west Northland basins also affected the adjacent onshore Northland, South Auckland and King Country basins. Faulted sub-basins also developed in the Northland and south Auckland onshore areas, controlling deposition in a relatively stable but gradually subsiding terrestrial to shallow marine environment, while the King Country and Wanganui basins remained emergent. This subsidence resulted first in non-marine sedimentation in inland alluvial-plain and coastal-plain settings, followed by gradual transition to marginal and shallow marine conditions spreading from the west in Northland, and southwards in the south Auckland region. The South Auckland and King Country basins were still separated from the marine northern Taranaki Basin by the emergent Herangi High. By the end of the Eocene, shallow marine conditions prevailed over most of the present Northland Peninsula, except for the extreme east, and were extending southwards via a re-entrant into the northern South Auckland Basin. Deep marine conditions prevailed to the northeast of the Northland Peninsula, lying north of an emergent high that occupied the Northland east coast and eastern offshore region south to the Coromandel Peninsula and linking with the northern end of the emergent portions of the South Auckland and northern Wanganui basins.

1.4 OLIGOCENE

The Oligocene, during which deposition was primarily of carbonates, corresponds to the period of maximum submergence of the New Zealand landmass resulting from region-wide crustal foundering and consequent marine inundation. Slow regional subsidence was overprinted by tectonism in part associated with initial development of significant motion on the Australian–Pacific plate boundary. In western Southland and on the West Coast of the South Island, local basins showed continued subsidence initiated in the Late Eocene, perhaps linked to the northward propagation of Emerald Basin spreading. At the same time, off the eastern margin of New Zealand, Pacific plate subduction was propagating southwards. Either of these influences could have resulted in local rapid Oligocene subsidence. The Oligocene was also host to diverse events affecting the New Zealand continental plateau, including the initiation of the Antarctic Circumpolar Current (ACC), and a postulated major global sea-level fall, as well as the mild broad tectonism of the plate boundary zone. Each of these processes resulted in erosion and widespread unconformity.

In the North Island portion of the East Coast Basin, Oligocene sedimentation was almost entirely dominated by massive calcareous muds, which form a thick blanket passing conformably upwards from the Eocene deposits. Paleodepths were generally bathyal. In contrast, in Marlborough much of the Oligocene is missing, or is limited to basaltic volcanics.

The Oligocene in the Canterbury Basin is notable for a basal erosional unconformity in the west, but apparently continuous Eocene–Oligocene sedimentation in the east. Rising relative sea level during the Early Oligocene led to the micritic Amuri Limestone reaching its maximum extent, forming a carbonate platform from Kaikoura to just northeast of Dunedin, and inland from the present coastline for ~50 km. This carbonate platform also extended along the whole length of the Chatham Rise. Submarine volcanism continued from the Late Eocene into the Early Oligocene in the Oamaru area. In the later part of the Early Oligocene, the area was swept by marine currents derived from the newly developed ACC, resulting, at least in part, in the development of the Marshall Paraconformity. The resulting erosion was accompanied by gentle deformation and, probably, a drop in sea level. The mid- to Late Oligocene was also marked by the development of the Endeavour High in the southern Canterbury Basin. This northeast-trending feature, lying along and offshore of the present coastline between the Oamaru area and Banks Peninsula, started developing in mid-Oligocene times and remained a high until the late Early Miocene. It was probably mainly submerged, but may have been emergent at times. The feature perhaps records reactivation of the mid- to Late Cretaceous Canterbury Bight High. In the Late Oligocene, volcanism occurred in the west in central and north Canterbury, while transgression continued. By the end of the Oligocene almost all of the Canterbury Basin was submerged, with the present onshore region drowned to shelf depths.

In the Great South Basin the base of the Oligocene is marked by a significant change in lithology and seismic facies over most of the basin. In central and eastern parts the sediments are carbonate-rich compared with the underlying deposits, although, where there are Eocene limestones, deposition appears to have been relatively continuous. As in the Canterbury Basin to the north a basal unconformity was developed near the western shelf edge, with progressive onlap above it indicating northwestward transgression. Further eastwards the basal contact of the Oligocene sediments appears conformable, but is still sharp, and represents a discontinuity. Because of a lack of seismic information the western shoreline of the Great South Basin during the Oligocene cannot be established in the south, and erosion in southeastern onshore Otago has made data sparse for this area. Early Oligocene sediments below the Marshall Paraconformity are lacking, and Late Oligocene deposits along the present Otago coastline consist of a carbonate platform deposited in shallow starved-shelf to inner-shelf environments. In the northwest, shallow-marine Late Oligocene deposits interfinger with non-marine sediments, delineating a fluctuating shoreline in the area, but information is lacking to the southwest.

Over most of northern and eastern Southland the basement is unconformably overlain, first by thin non-marine to estuarine beds, and then by shallow marine sandstone and limestone. The Winton Basin in the west and the then connected Waiau Basin of western Southland developed first, the non-marine beds being of latest Eocene to earliest Oligocene age, and the overlying marine beds ranging in age from Early to Late Oligocene. The basin underlying the eastern Southland Plains to the east developed later, the oldest sediments representing paralic condition of Late Oligocene age, passing up into Late Oligocene–Early Miocene marine beds.

At the end of the Eocene most of onshore western Southland was still above sea level, with fluvial and lacustrine conditions prevailing in most places. Region-wide subsidence in the earliest Oligocene imposed marine conditions over most of the region, although emergent hills remained adjacent to all the main basins until the Late Oligocene.

Parts of Fiordland were still emergent and supplying debris to depocentres in the Waiau Basin, where Early Oligocene transgression resulted in a marine embayment occupying the site of the Eocene Lake Orauea. It was initially bounded to the south-east by the Longwood High, which later in the Early Oligocene became submerged and covered by discontinuous carbonate banks in a shallow marine setting. Locally, in the eastern Waiau Basin, and along Hump Ridge at the southwestern margin of the basin, a shallow marine environment was rapidly replaced by submarine fan sediments and eventual submergence to bathyal depths during the Early Oligocene. Submarine fan breccias and sands related to local active faulting occupied slope- and base-of-slope settings along the entire northwestern side of the Waiau Basin by mid-Oligocene times, and merged eastwards into deep marine muds in the centre of the basin.

The Solander Ridge, the southern part of the Solander Basin, and a high between the Hautere and Parara sub-basins were emergent throughout the Oligocene. The regional Early Oligocene marine transgression produced shallow marine conditions across the Parara Sub-basin. In the Hautere Sub-basin Early Oligocene sediments are non-marine, and marine conditions were not established until the Late Oligocene.

Similar Early Oligocene transgression occurred over the Te Anau Basin, but here an additional factor was activity on the Moonlight and Hollyford fault systems, which resulted in the development of small fan deltas in local depocentres adjacent to and along the faults. These merged into a single large fan system, which extended southwards over the northern and central basin in the mid-Oligocene. On the southwest side of the basin the Murchison Mountains High remained emergent throughout the Early Oligocene, but in mid- to Late Oligocene times was inundated to become a shallow marine shelf. On the eastern side of the basin several small but very active submarine fans built north and west from the Takitimu Mountains block into the rapidly deepening basin. A deep trough followed the Moonlight Fault System along the northwest flank of the Takitimu block linking hitherto separated sub-basins from at least the mid-Oligocene (Turnbull *et al.* 1975).

The Waiau Basin had a major change in shape, depth and sedimentation pattern during the Late Oligocene. The Fiordland and Takitimu blocks subsided, the latter allowing deposition of muds to extend eastwards from the Waiau Basin, and possibly across the Longwood Range also. Another consequence was the deposition of a shallow-marine bioclastic calcareous facies along the western margin of both the Waiau and Te Anau basins, and also east of the Waiau Basin over the Southland Shelf (Winton Basin) in the Late Oligocene. This shelf environment prograded westwards into the eastern Waiau Basin, while bathyal conditions prevailed in the central Waiau Basin.

Much of the Solander Basin remained emergent during the Late Oligocene, although the Hautere Sub-basin became submerged at this time. The Waitutu Sub-basin also became wholly marine, and developed a series of small submarine fans prograding from the Hauroko Fault eastwards across the basin.

In the Te Anau Basin the Moonlight Fault System propagated northeastwards beyond Lake Wakatipu and established a shallow marine platform in that area in the Late Oligocene. Carbonate and terrigenous debris accumulated in local fault-controlled basins within the platform. In the latest Oligocene–earliest Miocene a major topographic boundary separated this platform from the basement terrane to the east, where Middle Miocene non-marine sediments rest on a deeply weathered erosion surface. By the Late Oligocene, in the main Te Anau Basin, the large mid-Oligocene fan system had either retreated to the north or its sediment supply had diminished, as only fan-fringe sands were being deposited in bathyal muds in the northern part of the basin. Bioclastic calcareous sediments accumulated on the basin margins, forming carbonate platforms.

In the western Balleny Basin, Early–mid-Oligocene sediments formed on a south-to-southwest-facing fault-controlled continental margin bounding the Oligocene 'Tasman Ocean'. Canyons incised into the narrow shelf fed numerous submarine fans lying on the edge of an abyssal plain that was floored mainly by Fiordland basement. By the mid-Oligocene the shoreline had migrated northwards over Fiordland and the Balleny Basin had subsided to bathyal depths. The western Balleny Basin was open ocean and received little terrigenous sediment. By the Late Oligocene only small remnants of earlier highs were still emergent in the Balleny Basin, although a small portion of the Fiordland block may have been still emergent. By analogy with the West Coast and western Southland basins that are presently onshore, these were probably fringed with shallow-marine bioclastic limestones forming a shelf or 'platform' facies.

About half the present land area of the West Coast region was submerged at the beginning of the Oligocene. A very irregular, deeply embayed coastline surrounded low-lying land, while local peat swamps close to the coast remained. The progressive marine transgression that had started in the Middle Eocene continued throughout the Oligocene and by the end of the Oligocene almost all, if not all, the land areas were covered by sea. In contrast to the small local basins that characterised Eocene sedimentation, most Oligocene subsidence was concentrated in the Paparoa Trough and Murchison Basin, which were dominated by deposits of calcareous muds or muddy limestone. The areas surrounding the basins were dominated by a platform facies characterised mainly by shallow-water bioclastic calcareous deposits. Sporadic uplift continued along major faults throughout the Oligocene, causing the localised occurrence of breccia and conglomerate.

By the earliest Oligocene the Taranaki Basin to the north was almost completely inundated. A shelf environment was established in the south and southeast, while basinal carbonate oozes were deposited in bathyal depths in the north and west. The eastern side of the Taranaki Fault remained emergent at this time. Early Oligocene strata are largely absent over the southern portion of the Taranaki Basin, marking a widespread unconformity. Final development of the Taranaki passive margin ceased in the Late Oligocene and a renewed phase of rapid subsidence began. This subsidence affected the entire basin, but was particularly pronounced in the east in areas adjacent to the Taranaki Fault, which appears to have been active at this time. In the far northwest the shelf had already subsided to bathyal depths during the Eocene passive margin phase, and here and in western offshore Northland, bathyal calcareous sediments accumulated. In central parts of the Taranaki Basin renewed subsidence began in the mid-Oligocene, and shelf areas plunged to bathyal water depths. In the south, rapid subsidence began slightly later, but submergence was only to shelf depths.

This period of accelerated subsidence in the Late Oligocene marked the onset of a new tectonic regime, ultimately related to the initiation of subduction in the northeast of New Zealand and the early development of the modern Australian–Pacific convergent plate boundary. Rapid subsidence began in Northland in the latest Oligocene, immediately prior to southwest-directed allochthon emplacement, the first indication of the plate boundary in that region. Here the Early Oligocene was a time of gradual subsidence of lowlands, and sediment starvation in basins. On onshore Northland and in the eastern offshore region, equivalent shallow-water shelf calcareous deposits, shallow equivalents of the bathyal western offshore sediments, were deposited over almost the entire area. This pattern indicates broad transgression over a subdued landscape, and by Late Oligocene time, much, if not all of the region had been inundated.

In the Oligocene in the south Auckland area, a marine transgression from the north inundated much of the northern alluvial plain. By the late Early Oligocene marine transgression had reached beyond Te Kuiti in the south, depositing, first, silts, and then calcareous sands and limestone. In the earliest Oligocene a western peninsula forming a northern extension of the proto-Herangi High delineated the western margin of the marine basin, but by the late Early Oligocene this peninsula became completely submerged north of Kawhia Harbour, and open marine shelf conditions existed over the region north of Te Kuiti.

North of Raglan Harbour Late Oligocene sedimentation consisted dominantly of calcareous silts or very fine sands, while to the south they were replaced by limestone that accumulated in an inner-shelf environment. This shallowed in the southwest in the King Country Basin near Awakino in response to the rising Herangi High, which must have been at least partially submerged during the Early Oligocene. Uplift and tilting of the high was coincident with the onset of rapid subsidence along the eastern Taranaki Basin margin directly west of the Herangi High in the early Late Oligocene, and it continued until the end of the Oligocene. This Late Oligocene phase of deformation developed in a mildly compressive regime, corresponding to a time of early plate boundary development.

1.5 MIOCENE

Changes in the style of sedimentation in the very early Miocene mark the early stages of a major phase of tectonism accompanying changes to the plate boundary system. It is shown by abrupt change from carbonate to siliciclastic sedimentation throughout the region, and the onset of New Zealand-wide uplift and regression. The beginning of the period is commonly marked by unconformities (see section 7.5.2).

In the East Coast region the change was marked by basal unconformities, flysch deposition, large-scale faulting and the emplacement of the East Coast (and Northland) allochthons. By the end of the earliest Miocene most of the East Coast region was mantled by muds and flysch. Depositional settings in the present onshore area, and probably also further east, were mainly bathyal, and this depositional pattern continued into the middle Early Miocene. By the late Early Miocene most of the region was still at bathyal depths, although paralic and shelf sediments resting on basement in the vicinity of Lake Waikaremoana indicate the proximity of a shoreline. By the end of the Early Miocene, and continuing into the early Middle Miocene, a large depocentre, containing a great thickness of sediment, had developed on the site of the present Raukumara Peninsula. Paleoenvironments generally showed deepening to the southeast in the north, with shelf environments in the west and bathyal muds associated with flysch in the east. Tectonism continued throughout. In Marlborough at this time vast quantities of gravel (Great Marlborough Conglomerate) were being shed onto a shallow marine shelf from rising highs of basement linked to the developing Marlborough Fault System (see Plate 15).

Middle Miocene paleogeography indicates general onlap in the north. To the west, basement was probably exposed and eroding, and late Middle Miocene and younger sediments rest directly on Cretaceous basement. A large depocentre formed in the north of the region, probably as a west-facing, arcuate sediment apron between a rising Raukumara Range and the present coast. A distinct shallowing in the south of the East Coast Basin resulted in few Middle Miocene sediments preserved in Marlborough. The Middle Miocene was characterised by dramatic periods of subsidence and uplift, indicating major faulting. These events produced a mix of bathyal flysch and muds along with paralic to neritic sands and calcareous sediments in much of the East Coast Basin. A shoreline was probably present in southern Wairarapa, where non-marine gravels were deposited in latest Middle Miocene times.

Widespread faulting influenced deposition during the Late Miocene, especially in the west of the region and in the east of the present onshore area. The seas transgressed over the previously exposed areas of southern Wairarapa, creating a shallow marine shelf, with local regression and exposure in latest Miocene times. Shallow marine gravels spread over much of northern Marlborough in the Late Miocene, the environment deepening and then shallowing back to shelf depths in latest Miocene times.

In the Canterbury Basin, Late Oligocene deposition of calcareous deposits and greensand, which continued into the Early Miocene, was followed by an influx of clastic sediment from the west in response to uplift along the site of the present Southern Alps. The sediments generally fine to the east, forming a prograding continental shelf wedge containing numerous channel structures. A shallow submarine high extending northwestwards from the Chatham Rise probably existed near Cheviot during most of the latest Oligocene and earliest Miocene, effectively separating areas of silts to the north from greensand, packstone and silts to the south. During the Early Miocene north Canterbury consisted of a shallow marine basin open to the southeast but bounded to the west by rising highs and to the north and east by eroding land areas. By the late Early Miocene the basin in the north Canterbury area had become considerably enlarged, with open, fully marine neritic conditions prevailing. In the far north and northeast, outer-shelf to bathyal conditions prevailed.

An isolated occurrence of limestone, the only record of Miocene deposition on the Chatham Islands, suggests that shallow marine conditions obtained there, as well as on the remainder of the Chatham Rise.

In the southern Canterbury Basin the Endeavour High appears to have remained a prominent submarine topographic feature from the Oligocene into the late Early Miocene, when it was fringed or capped with limestone. It probably provided a partial barrier to eastward sediment transport in late Early and early Middle Miocene times. The area between

Oamaru and Dunedin in the early Early Miocene was starved of sediment, suggesting that the Endeavour High may have extended as far south as Dunedin. Eruption of the Dunedin Volcanic Group probably started in middle Early Miocene times. By the late Early Miocene, deposition of non-marine to marginal-marine coal measures in the Kyeburn area and to the north defined a north-trending shoreline still far to the west of the present one.

By the Middle Miocene, uplift and erosion had stripped all Late Cretaceous and early Cenozoic sediments from parts of western inland Canterbury. Intraplate alkali and tholeiitic volcanism occurred in north Otago and central Canterbury between the late Early and latest Miocene. At Banks Peninsula there is no record of sedimentation at this time, and the first volcanics of the Governors Bay and Lyttelton volcanic groups mantled the Early Miocene erosion surface and dominated the area until the Pliocene. In the western part of central Canterbury, marginal marine and paralic sediments indicate that the shoreline was moving towards the basin. Further to the northeast, non-marine conglomerates and marine debris flows reflect active faulting and local uplift, and a generally southeastwards-prograding shoreline. The crest of the Chatham Rise was subject to widespread erosion at this time. Most of the southern half of the present onshore region was mantled by conglomerate, quartzose sands, clays and lignite in the Middle Miocene, reflecting continued transgression. In many places there is evidence of Late Miocene–Pliocene erosion having removed sediments of earlier Miocene age.

In the Dunedin area, Middle Miocene erosion and volcanism were followed by a minor transgression, giving rise to shallow marine sandstone and grainstone in the east. To the west, in Central Otago, continued erosion produced significant relief, leading to local accumulation of non-marine diatomite and oil shales. A period of volcanism, which began in the late Middle to early Late Miocene, extended to the Oamaru area.

Few data on the Miocene of the Great South Basin are available, as few samples were cored from the petroleum wells. Rapid Miocene to Quaternary shelf accretion along the northwest margin of the basin reflects an increase in clastic sediment supply with the change in tectonic regime, and coastal facies are inferred to lie just east of the present southeastern coastline of Otago and to continue to the southeast to the east of Stewart Island. The shelf area was probably narrow, however, as petroleum wells from Tara-1 south show bathyal sedimentation. In eastern parts of the basin the sediments consist of soft, foraminiferal oozes.

In Early Miocene times regional regression resulted in re-emergence above sea level of the onshore eastern Central Otago area. During the Early–Middle Miocene it became a non-marine depocentre for quartzose fluvial sediments and lake deposits. In Late Miocene times an increased pace of tectonism caused local faulting and uplift, resulting in the widespread deposition of immature sandstones and conglomerates. In eastern Southland the dominant Miocene environment is non-marine in the east, interfingering increasingly with marine facies to the west in the Winton Basin. In the Winton Basin, erosion has removed strata younger than Early Miocene, but the unit probably originally extended into the Middle Miocene. An Early Miocene shoreline probably lay close to the high between the eastern Southland Plains and the Winton Basin.

At the end of the Oligocene the present land area of western Southland was occupied by an elongate, northeast-trending basin, deepening to the west against a faulted Fiordland margin. It was bounded by a shallow marine platform west of the marginal fault systems, and by shallow-marine sand-dominated shelves to the east. The offshore area consisted of several depocentres separated by basement highs, including the Solander Ridge. The onshore Waitutu Sub-basin continued offshore to the southwest, separated by a carbonate platform from another deep marine basin in the Balleny area. In the Early Miocene, the onshore western shelf area abruptly subsided and the basin-wide Waicoe Mudstone facies extended westwards over eastern Fiordland. Uplift began northeast of the Te Anau Basin along the Livingstone Fault, and the Moonlight Sea of the Late Oligocene retreated rapidly westwards in the Early Miocene, persisting as a marine embayment over the Caples Terrane along the Moonlight Fault Zone. Uplift of the northern Te Anau Basin above sea level probably dates from Early Miocene times, cutting off the Oligocene marine connection with basins to the north. In the Waiau Basin a shelf edge and slope extended from the Takitimu Mountains to

Lake Hauroko. Deep-water Waicoe Mudstone facies, interspersed with submarine fan material, predominated west of the slope. To the east of the slope the shelf became shallower, accumulating bioclastic limestone.

In the offshore Solander Basin, which subsided in the middle Early Miocene, a shelf environment was predominant, except in the Waitutu Sub-basin, which was dominated by deep marine mudstone. Early Miocene sediments onlap the Hump Ridge–Stewart Island shelf from the south, and probably the Waiau Basin shallowed to the south onto an emergent Hump Ridge–Mid Bay High. General lack of coarse-grained Early Miocene sediment in both the Te Anau and Waiau basins suggests that much of Fiordland was submerged at this time, probably to outer-shelf depths.

The paleogeography of western Southland underwent only minor changes, such as increasing sand content and progressive shallowing into the Middle Miocene. Deep marine conditions continued in the western and northern Waiau Basin, and in the Burwood area of the southern Te Anau Basin. The shelf established in the Early Miocene continued to mark the eastern Te Anau Basin throughout the Middle Miocene. In the late Middle Miocene the southwestern side of the Te Anau Basin adjacent to the northern Monowai Sub-basin again subsided. Rapid shoaling then occurred until, in earliest Late Miocene times, marine sedimentation gave way to deposition of the terrestrial sediments of the Prospect Formation in both the Monowai Sub-basin and the Te Anau Basin.

The southwestern and eastern parts of the Solander Sub-basin were tectonically unstable in the Middle Miocene as a result of compressive movements on the Hauroko, Solander and Parara faults, as well as on the Hump Ridge–Stewart Island Thrust System. Movement on this thrust system also caused the uplift of Hump Ridge. In the adjacent Balleny Basin the Middle–Late Miocene sequence is concentrated in a southeast-trending trough that acted as a sediment conduit probably feeding into the western side of the Waitutu Sub-basin as a submarine valley fan system. The central part of the onshore Waitutu Sub-basin was still bathyal in the Middle Miocene.

At the beginning of the Late Miocene, further paleogeographic changes occurred throughout the western Southland region. Onshore they mainly affected the Te Anau and southern Waiau basins; in the Waitutu Sub-basin, sedimentation continued as before. The early Late Miocene saw maximum uplift of the Hump Ridge–Mid Bay High, with its resultant unconformity; and local uplift continued around large growing anticlines in the Solander Basin. Midway through the Late Miocene, Hump Ridge abruptly subsided and a marine incursion flooded the unconformity in the southern Waiau Basin, apart from the Mid Bay High. A sandy shallow marine-shelf facies was established in the Waiau Basin, although other areas were little affected.

In the Te Anau Basin, deposition of the conglomeratic Prospect Formation began in latest Middle Miocene time. This unit may represent diversion, north of the Te Anau Basin, of the major drainage system over eastern Fiordland that had fed the Monowai Formation in the Middle Miocene. The Burwood Sub-basin had, by now, effectively merged with the main Te Anau Basin. Deposition of Prospect Formation gravels continued throughout the Late Miocene, restricted in the southeast by the rising Takitimu Mountains. The Fiordland massif was also rising and shedding debris in the Te Anau Basin from the west, although Fiordland did not become a major gravel source until later, in the Pliocene.

In the West Coast region one of the most striking changes between the Late Oligocene, when virtually the whole of the area was submerged, and Early–early Middle Miocene paleogeography, is the emergence of large areas of land, including the whole of south Westland, most of the area immediately west of Nelson, and a narrow strip largely parallel to the Alpine Fault between the south end of Golden Bay and the Taramakau River. As a consequence of a newly developing pattern of uplift and subsidence the Early Miocene was a time of rapidly changing shorelines, and a complex embayed coastline developed in the west during this time. The Grey Valley Trough was a narrow, north-northeast-trending seaway with land on both sides, and, further north, the area between Inangahua and Punakaiki was submerged. Calcareous silts of a shelf environment was the dominant lithology except in the rapidly subsiding Murchison Basin, where turbidite sands were fed into the basin through channels from rising sources to the east and southwest. Much of northwest Nelson was emergent at this time, but the southern Taranaki Basin further north was submerged, with widespread deposition of calcareous muds. Relatively shallow-water depths prevailed over much of the region, with three distinctly deeper bathyal basins, which coincide broadly with the Grey Valley Trough, the Murchison Basin and the northernmost part of the region, including the southern end of the Taranaki Graben.

During the late Early Miocene the sea transgressed over much of the land previously exposed, so that by the end of this period northwest Nelson was again submerged by shallow seas, and the shoreline had moved southwards in the Greymouth area. Rising granitic areas along the eastern margin of the region continued to supply terrigenous detritus, but, in addition, a new source, outside the region to the east across the Alpine Fault, started to supply a very large amount of Permian volcaniclastic detritus to the Murchison Basin, where the rate of supply greatly exceeded the high rate of subsidence. The basin was rapidly infilled, and, by the end of the Early Miocene, braided rivers had started to prograde across it. Although much of the region was submerged during the late Early Miocene, there was progressive shallowing in northwest Nelson during the early Middle Miocene, so that much of the area was emergent by the end of this time. Local uplift along the Paparoa Range and the area to the west led to the formation of a mid-Miocene unconformity, and this area was probably also emergent, although low-lying. Offshore, west of the present coastline, marine-shelf sedimentation continued into the Late Miocene, but the only areas of marine sedimentation inland from the present coastline were in the Karamea area and in a narrow strip south of Greymouth. A rising source of Permian rocks to the northeast continued to supply fluvial gravel to the Murchison area. Regional paleogeography showed little change during the late Middle–Late Miocene, although the area west of the Paparoa Range started to subside and was submerged from the late Middle Miocene onwards.

The beginning of the Miocene in the Taranaki Basin is marked by a regional change from dominantly carbonate to dominantly clastic deposition, although platform carbonates were still being deposited in the south over northwest Nelson. A similar transition from carbonate- to terrigenous-dominated sedimentation is recorded in the neighbouring King Country Basin at about the same time, and the basins were probably connected in the north. Sedimentation was initially accompanied by rapid foreland subsidence in the east, immediately west of the active Taranaki Fault, with basin bathymetry and transgression reaching a maximum in the Early Miocene. By the end of the Early Miocene the foredeep along the basin's eastern margin was largely filled.

From late Middle to Late Miocene times the supply of coarse clastic sediment into the basin increased abruptly; the eastern foredeep filled and sediment started to spill over to the west. In general, Middle–Late Miocene strata form a northwestwards-thinning wedge that trends sub-parallel to the paleoslope. The continental margin offlap is depicted on seismic reflection profiles as a series of basinwards-stepping clinoforms, representing the successive positions of the coeval basin floor-slope-shelf. Collectively they illustrate a long-lived pattern of basin floor aggradation and slope progradation, and of shelf expansion since at least the beginning of the Middle Miocene. Submarine fan sandstones were deposited by turbidity currents and as fluidised mass flows at the base of the prograding slope, while muds and distal turbidites were being deposited further offshore to the northwest. These fan systems are distributed on a northeast trend through the centre of the basin, and were sourced from emerging landmasses in the south and east. The northwest Nelson area was undergoing uplift at this time. Volcanism began in the northern Taranaki Basin in the late Middle–Late Miocene, resulting in the deposition of a large thickness of volcaniclastics from sediment gravity flows in deep water. Fault-controlled uplift was confined to the east and south on individual thrust and inversion structures.

The Northland region saw the effects of the development of a plate boundary to the northeast of and parallel to the Northland Peninsula in the latest Oligocene–earliest Miocene. At this time rapid subsidence took place throughout the northern and all the western parts of the area. This subsidence was quickly followed by the emplacement of the Northland Allochthon from the northeast, i.e., from the Pacific side of the ancestral Northland landmass, which involved uplift of a Cretaceous and Paleogene passive-margin clastic wedge and Cretaceous–Paleocene ophiolites.

At this time most of northernmost Northland remained at bathyal depths. Minor non-marine to estuarine facies indicate local basin filling and perhaps uplift. Further south, rapid subsidence necessary to accommodate the nappes making up the allochthon was followed by similarly rapid uplift. Northeastern Northland was uplifted, and the allochthon and underlying sequences were folded, faulted and eroded. The nappe front reached south to Kaipara Harbour in the middle Early Miocene, and had advanced partway into the Waitemata Basin to the south a little later. Rapid subsidence in southern Northland is indicated by the abrupt transition from nearshore to a deep bathyal facies, including submarine fans in the Waitemata Group.

Beginning in the earliest Miocene, volcanoes east and west of the present site of the Northland Peninsula began erupting, and, by the middle Early Miocene, northwest-trending belts of calc-alkaline stratovolcanoes and volcanic complexes extended for at least 350 km parallel to, and on both flanks of, a now mainly emergent landmass. Volcanic activity continued into the late Early Miocene, ending by the close of the Early Miocene. Although the Northland volcanoes became extinct, andesitic eruptions continued at Great Barrier Island and northern Coromandel Peninsula. Southwest of the Northland region new volcanoes erupted; these were the precursors to the extensive Middle–Late Miocene andesitic volcanism in the Taranaki region. By this time most of Northland was land, except in the far north, where thick bathyal sediments were deposited in fault-bounded basins. Coromandel Peninsula was also dry land, and the Waitemata Basin became infilled and uplifted above sea level. Most of the region west of the present coast is inferred to have remained at deep bathyal depths.

The depositional history of offshore west Northland during the Middle–Late Miocene was one of passive filling of a basin which remained open to the west. The far north of the present onshore region subsided, and an extensive peninsula which extended northwest from the Kaitaia–Doubtless Bay area is inferred to have been wholly or partly submerged by Middle–Late Miocene marine transgression. Central and southern Northland are inferred to have been land. Gradual uplift of eastern Northland and subsidence west of the peninsula resulted in westward tilting of the land surface.

South of the Northland Peninsula the Early Miocene Waitemata Basin extended southwards into the south Auckland region, and sedimentation of limestone and sandstone continued in what was initially a shallow-water shelf environment. Rapid subsidence caused depression to bathyal depths late in the Early Miocene, with deposition of turbidites. South of Pirongia Mountain the sediments become muddier, and in the vicinity of Taumaranui they include an elongate body of turbidites, deposited in an outer-shelf to bathyal environment influenced by contemporaneous movement on the Ohura Fault. The Early Miocene basin pinches out to the west against the still elevated Patea–Tongaporutu High, and, in the south at the latitude of Stratford, against basement. The basin deposits thicken eastwards and terminate against a fault marking the eastern limit to the basin.

Erosion has removed Miocene sediments younger than Early Miocene in the north. South of Te Kuiti, however, Middle Miocene sandstones accumulated in a range of shelf environments. Reversal of movement on the Ohura Fault caused uplift to the east and restricted the early Middle Miocene basin to west of the fault, although in the late Middle Miocene transgression across the fault to the east of the fault occurred.

During Middle and Late Miocene times the southern portion of the Patea–Tongaporutu High became at least partly submerged in the west, allowing a connection between the adjacent King Country and Taranaki basins. However, the high still acted as a submerged western margin to the King Country Basin that, during the Late Miocene, received shallow-marine to paralic sediments.

In the Wanganui Basin to the south there was a prolonged period of erosion prior to subsidence and basin development in the Late Miocene. At this time sedimentation onlapped southwards onto the basement from the King Country Basin. The basin remained shallow, accumulating sediments in inner- to mid-shelf and paralic environments.

1.6 PLIOCENE–PLEISTOCENE

The beginning of the Pliocene Epoch at ~5 Ma was marked by accelerated uplift of the Southern Alps east of the Alpine Fault, and the outpouring of a huge volume of clastic sediments into most New Zealand basins, a response that has continued until the present day. This resulted in widespread shoreline regression, although large areas, such as the East Coast and Wanganui basins, continued to subside and remained substantially marine.

The East Coast region was almost entirely submerged at the beginning of the Pliocene. There was subsidence along the present axial ranges early in the Pliocene, and a marine strait formed in the southern Ruahine Range on the site of the present Manawatu Gorge (Trewick & Bland, 2012). Uplift separated marine sedimentation in the East Coast from that in the Wanganui Basin after about 2.3 Ma, and substantial uplift of the ranges has continued over the last million years. Plate boundary compression also began to uplift the coastal ranges through the Pliocene, resulting in conditions suitable for the accumulation of limestones of Te Aute lithofacies in the Hawke's Bay area. Continued compression saw almost the entire present onshore part of the region uplifted above sea level at or just before the end of the Nukumaruan (1.63 Ma). Most of Marlborough was emergent during the Pliocene–Pleistocene, apart from an area of shallow marine-shelf deposition south of Blenheim.

In the remainder of the South Island continued uplift associated with the rising Southern Alps and their extensions led to extensive erosion adjacent to the uplift, and to an influx of conglomerates in the Late Miocene–Early Pliocene. In the eastern South Island a conglomerate wedge advanced to the east to a position probably close to the present coastline, and interfingered with finer-grained shallow marine sediments. Reactivation of some of the major Cretaceous faults (e.g., Titri and Waihemo) might have started in the Pliocene, though main movements were probably Pleistocene. In Central Otago, widespread wedges of Late Miocene to Pleistocene non-marine conglomerate were formed in response to increased tectonism and uplift.

Most of the Chatham Rise remained a submarine platform, with little deposition apart from thin limestones. Major uplift occurred in latest Pliocene time, making the Chatham Islands largely emergent. In the Great South Basin a narrow progradational shelf-slope wedge developed in the west, while bathyal conditions persisted to the east. Pliocene–Pleistocene alkaline-to-subalkaline volcanic activity occurred on the Chatham Islands, in south Canterbury and on the Antipodes Islands.

In Early Pliocene (Opoitian) time, western Southland looked much as it does today, except for a more northerly shoreline in the Waiau Basin and a deep marine embayment reaching Lake Hauroko. An Early Pliocene connection between the Waiau and Te Anau basins, inherited from the Late Miocene, was probable.

In the Early Pliocene the southern Waiau Basin formed an embayment with fine-grained lithofacies and very-shallow-marine faunas. This suggests that it was sheltered from coarse clastic input and strong current influence, although tidal currents were significant and an open-ocean microfauna had access to the bay. The Mid Bay High barrier between Te Waewae Bay and the main Solander Basin, inherited from the Late Miocene, was probably not breached or submerged until the latest Pliocene. In the Te Anau Basin the Prospect Formation gravel plain still covered most of the basin. Rapid uplift did not begin around most of the basin until at least the mid-Pliocene, and uplift continued through the Pleistocene to the present day.

Further to the southwest in the offshore Solander and Balleny basins marine sedimentation continued from the Miocene, interrupted only by volcanic eruptions on Solander Island and its associated volcanic centres. By the Late Pliocene the sea had retreated from Te Waewae Bay, but it fluctuated throughout the Pleistocene from encroach to retreat, driven by the conflicting effects of continued uplift and southward tilting.

On the West Coast, erosion accompanying uplift of the Southern Alps led to the infill of the major basins by immature fluvial gravels (Rappahannock Group, Old Man Gravels and Moutere Gravels), and formation of widespread interbasinal alluvial gravel plains that covered much of the onshore part of the region. Continued compression drove rapid uplift of the West Coast mountain blocks during the Quaternary, leading to the formation of the present basin-and-range topography onshore.

In the Taranaki Basin the Western Stable Platform remained quiescent throughout the Pliocene–Pleistocene. Most parts of the basin experienced net subsidence, and the Northern and Central grabens and Toru Trough subsided dramatically. The

Pliocene–Pleistocene was also characterised by very high sediment input into the basin, and an accelerated northwestward progradation of the slope and shelf sedimentary wedge across the Western Stable Platform ensued.

Most of the Northland Peninsula is inferred to have been land during the Pliocene–Pleistocene, and clastic sediment was being shed from an uplifted landmass very similar to present-day Northland onto a western continental slope of irregular relief. The Kaipara and Waipoua basins received the greatest clastic sediment supply, with a large hinterland and narrow shelf.

In the south Auckland region a large shallow-marine embayment was present in the Pliocene south of the Manakau Harbour, extending to a few kilometres south of the Waikato River mouth. It was fed from the south and southeast by a major fluvial system that extended south almost to Te Kuiti, and east into the Bay of Plenty and the Hauraki Plain.

The Wanganui Basin and the southern part of the King Country Basin were mainly submerged during the Pliocene, the sea extending from central Taranaki to the flanks of the Ruahine Range. The main depocentre of Opoitian sedimentation was an elongate east–west shallow seaway that extended from southern Taranaki eastwards, becoming narrower and shallower across the area of the present axial ranges, and continuing into the Hawke's Bay region. Near the end of the Waipipian (~3.2 Ma) the basin was cut off from open water access, and water depths decreased from middle- to inner-shelf depths.

Uplift of the Ruahine Range commenced at the beginning of Nukumaruan times (~2.4 Ma), and the Wanganui Basin began to shallow, culminating in estuarine and terrestrial deposition in the uppermost Nukumaruan. The opening of Cook Strait is inferred to have commenced late in the Nukumaruan, although headward erosion from the south may have started before this time. Major uplift of the Tararua and Ruahine ranges took place in the mid-Pleistocene (Castlecliffian), leading to extensive gravel fans on the eastern side of the ranges, and abundant sediment supply to rivers in the Wanganui Basin, which steadily pushed the shoreline westwards.

Arc-related volcanism became widespread in the Quaternary of the northern North Island, with mainly basaltic volcanism in the Whangarei area, Auckland and the western south Auckland region. Andesitic activity in the Taupo Volcanic Zone started at ~2 Ma, with rhyolitic activity from ~1.6 Ma.

CHAPTER 2

Structural Framework and Pre Mid-Cretaceous History of the New Zealand Region

2.1 INTRODUCTION

Prior to latest Cretaceous times, New Zealand did not exist as a separate geological entity but formed part of the Gondwana continental margin adjacent to the Antarctic and Australian sectors (see Figs 1.1 and 2.1). During the Permian and most of the Mesozoic the margin was characterised by convergent margin tectonics and probably resembled parts of the present southwest Pacific margin, with more than one subduction zone and arc system. It represents over 200 m.y. of active margin history and, at normal plate tectonic rates, could accommodate major paleogeographic displacements. The outboard (now eastern) edge of the New Zealand region represented the youngest part of this convergent margin system, where subduction of the Phoenix plate continued until ~105±5 Ma. Convergence was then abruptly replaced by crustal extension, leading eventually to separation from Gondwana of New Zealand and its continental shelf (Chatham Rise, Campbell Plateau, Lord Howe Rise and Norfolk Ridge), collectively termed 'Zealandia' (Luyendyk, 1995; Mortimer, 2004).

Two models have been suggested for the end of subduction and the rapid change of tectonic environment. Initially, Bradshaw (1989) suggested that the oblique subduction of the Phoenix-Pacific Ridge eliminated the subduction zone and led to the attachment of the New Zealand continent to the receding Pacific plate. Subsequently, Luyendyk (1995) modified this interpretation, using a model based on the Neogene tectonics of California, and proposed that the ridge died as it neared the New Zealand margin, with a concomitant failure of the subduction system. Luyendyk (1995) did not discuss the additional complication caused by the presence of the Hikurangi Plateau (Fig. 2.1), a Cretaceous oceanic plateau (Mortimer & Parkinson, 1996). More-recent research suggests that this model is too simple and that an additional spreading ridge and plate may be involved (Sutherland & Hollis, 2001). With the loss of two active plate boundaries, the New Zealand continental crust was finally captured by the Pacific plate. Both models propose that the Pacific plate was moving away from the New Zealand margin at this time and consequently attachment led to crustal extension in New Zealand after the end of subduction. Final isolation of southern Zealandia, as part of the Pacific plate, was achieved no later than 84 Ma, as shown by the occurrence of anomaly 33r in the ocean floor of the Tasman Sea and the Southern Ocean. The ~20-m.y. period between the end of subduction and continental breakup is particularly important in terms of the initiation of sedimentary basins within the Zealandia crustal block, though there is a lack of consensus on tectonic events in this period (see section 2.6). There is widespread evidence that crustal extension and thinning continued after continental separation. Mortimer *et al.* (2006) have further constrained the extent and position of the Hikurangi Plateau and suggested additional tectonic elements (Fig. 2.1).

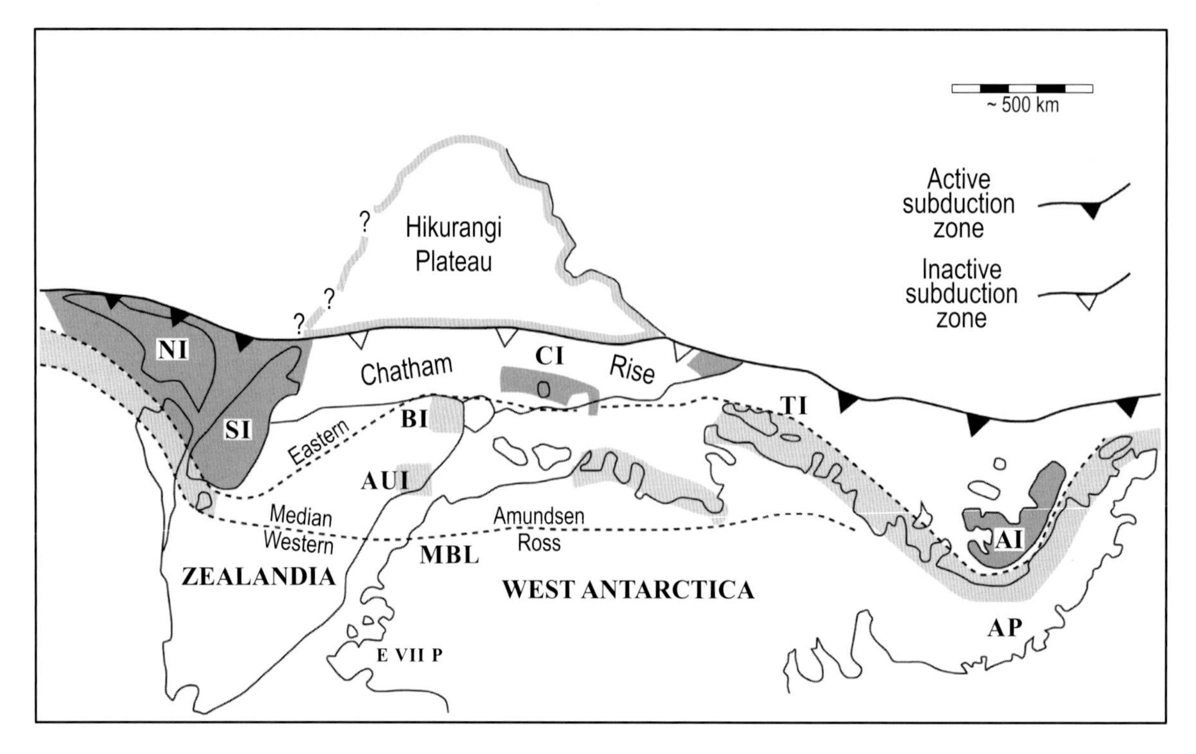

Fig. 2.1. Map of the southern margin of the Pacific between New Zealand and the Antarctic Peninsula. AI, Alexander Island; AP, Antarctic Peninsula; AUI, Auckland Island; BI, Bounty Island; CI, Chatham Islands; E VII P, Edward Seventh Peninsula; MBL, Marie Byrd Land; NI, North Island; SI, South Island; TI, Thurston Island. (Adapted from Mortimer *et al.* 2006, fig. 4.)

2.2 STRUCTURAL OUTLINE

In New Zealand, rocks older than mid-Cretaceous comprise a suite of igneous and sedimentary rocks that commonly show low- to high-grade metamorphism and are strongly deformed. Together they form the basement to the Late Cretaceous and Cenozoic sedimentary basins. Both offshore and onshore the basement rocks are disposed in curvilinear belts of contrasted character that are in tectonic contact. The basement is most easily described in terms of nine major and several minor tectono-stratigraphic terranes (Bishop *et al.* 1985; Bradshaw, 1989; Spörli & Ballance, 1989; Mortimer, 2004) that have been grouped into two 'superterranes' called the Western Province and the Eastern Province (see Plate 1). The provinces are separated by a Median Tectonic Zone or Median Batholith (Mortimer *et al.* 1999a,b), and the relative merits of these proposals are discussed by Scott (2013).

The 'superterranes' are not confined to New Zealand. The Western Province shows strong similarities to parts of Marie Byrd Land and northern Victoria Land in Antarctica and to the Lachlan Fold Belt of Australia (Cooper & Tulloch, 1992; Bradshaw, 2007; Bradshaw *et al.* 2009). Similarly, parts of the Median Tectonic Zone or Batholith appear to be developed on a much larger scale in Marie Byrd Land (Bradshaw *et al.* 1997). Representatives of the Eastern Province are restricted to New Zealand and possibly New Caledonia, although the accretionary complexes of the New England Orogen, the Antarctic Peninsula (Le May Group) and Patagonia (Duque de York Complex) show strong similarities.

2.3 BASEMENT TERRANES

2.3.1 Western Province terranes

The Western Province comprises two north–south-trending terranes (see Plate 1), the Buller and Takaka terranes (Cooper, 1989), which are separated by the Anatoki Fault (Jongens, 2006). They consist of early Paleozoic–Devonian sedimentary and metavolcanic rocks cut by granitoids of Late Devonian, mid-Carboniferous and Mesozoic ages.

2.3.1.1 Buller Terrane

The Buller Terrane, the westernmost terrane recognised in New Zealand, is far more extensive than the Takaka Terrane and probably extends under the Challenger Plateau to the west and the Campbell Plateau in the east. Paleozoic volcanic rocks have been intersected in the Tangaroa-1 petroleum well at almost 38° S on the Taranaki shelf (Mortimer *et al.* 1997) and have been interpreted as Western Province rocks. To the south the Western Province is known to underlie the southern half of the Campbell Plateau, where it has been encountered in petroleum exploration holes in the Great South Basin, and also on Campbell Island (Beggs *et al.* 1990; Cook *et al.* 1999; Adams *et al.* 2013c). Comparisons with Marie Byrd Land and northern Victoria Land (Antarctica) suggest a large volume of Buller Terrane-type rocks exist there too (Bradshaw *et al.* 1983, 1997; Bradshaw, 2007).

Two sedimentary facies are widely developed in the Buller Terrane: a monotonous turbidite succession, the Greenland Group and Golden Bay Group, of probable late Cambrian to Early Ordovician age; and a better differentiated succession of quartzose sandstone and graptolitic black shales of Early–Middle Ordovician age (Cooper, 1989; Rattenbury *et al.* 1998). The latter are seen only east of the Devonian Karamea Batholith where it overlies the Greenland Group. Geochronology data from slates (Adams *et al.* 1975; Adams, 2004) suggest initial metamorphism and cleavage formation in the early Silurian, with a maximum age of approximately 440 Ma. In tectonic contact with these older rocks, but almost certainly originally unconformable on them, is a shallow-marine to marginal-marine suite of quartzite, limestone and mudstone of Early Devonian age (Bradshaw, 1995). Greenland Group rocks of the Buller Terrane are also overlain by a small outlier of volcaniclastic Triassic sandstone cut by a Jurassic dolerite sill. The geochemical and isotopic composition of the dolerite closely resembles the Middle Jurassic Ferrar Dolerite of Antarctica (Mortimer *et al.* 1995).

The Buller Terrane is cut by granitoids of Late Devonian, mid-Carboniferous and Mesozoic ages (Muir *et al.* 1994, 1996). The Devonian granitoids, the Karamea Suite, show a range in composition that spans the ‘S’- and ‘I’-type fields (Muir *et al.* 1996). The mafic end of this range is similar in composition to the mafic intrusive rocks in the Takaka Terrane and was thought to indicate that the two were in close proximity by the Late Devonian, though this correlation now seems doubtful (Turnbull *et al.* 2013).

2.3.1.2 Takaka Terrane

The Takaka Terrane is diverse and comprises rocks of two discrete assemblages. The older one comprises Cambrian volcanic rocks, volcanogenic sandstones, quartzo-felspathic sandstone, limestone and conglomerate, and appears to be an arc-related suite of oceanic affinity (Münker & Cooper, 1995, 1999; Rattenbury *et al.* 1998). In tectonic contact with the arc suite are Cambrian quartzo-feldspathic turbidites of continental provenance. Basalts of back-arc affinity associated with the arc suite suggest a possible back-arc basin setting for the quartzo-feldspathic sandstones. Similar sandstones are the major component of a mélange that appears to intrude through and enclose the Cambrian formations in the area (Jongens *et al.* 2003). The arc assemblage in the north closely resembles contemporary rocks of the Bowers Terrane of northern Victoria Land (Bradshaw *et al.* 2009). The discovery of Cambrian granitic rocks in coastal Fiordland (Allibone *et al.* 2009b) further supports a link to the Ross–Delamerian Orogen of Australia–Antarctica, with granitic rocks a possible link to the Wilson Terrane.

The younger assemblage of rocks (late Cambrian to Early Devonian) has passive margin characteristics. It includes a thick sandstone–mudstone succession with major lenses of marble of Ordovician age and a Silurian section that is largely quartzite. This is overlain by over a kilometre of Early Devonian fine sandstone and mudstone (Rattenbury *et al.* 1998; Bradshaw, 2000). An unconformity reported below the Devonian strata (Willis, 1965) cannot be demonstrated and the structural history of the Devonian sediments is the same as that of the underlying rocks. The whole assemblage was strongly folded at the time of amalgamation of the Buller and Takaka terranes, probably in the Late Devonian (Jongens, 2006).

A small outlier of metamorphosed Permian and Triassic rocks unconformably overlies strongly deformed Ordovician rocks of the Takaka Terrane at Parapara Peak, west of Takaka (Campbell *et al.* 1998). Quartzo-feldspathic and calcareous schist intersected in the Moa-1B petroleum well on the Taranaki shelf to the north of the West Coast have been correlated with the Takaka Terrane (Mortimer *et al.* 1995).

Both the Buller and Takaka terranes are cut by Mesozoic granitoids, the most important of which is the Early Cretaceous Separation Point Batholith (Muir *et al.* 1995). This cuts both the Takaka Terrane and the magmatic rocks of the Median Tectonic Zone, stitching them together by 118 Ma (Bradshaw, 1993; Muir *et al.* 1994). Slightly younger granitoids (Rahu Suite) that are considered to be genetically related to the Separation Point Suite are widely developed in the Buller Terrane (Waight *et al.* 1997, 1998a).

Much of the high-grade and strongly deformed orthogneiss and paragneiss seen in the Buller Terrane and beneath the Takaka Terrane in Fiordland is metamorphic core complex rock separated by regional-scale detachment faults from lower-grade cover sequences (Gibson *et al.* 1988; Tulloch & Kimbrough, 1989). Cretaceous convergent margin magmatic rocks ranging in age from ~125 Ma to ~110 Ma (Tulloch & Kimbrough, 1989; Muir *et al.* 1994; Spell *et al.* 2000; Allibone *et al.* 2009a,b) are involved in the core complexes and overprinted by extensional deformation (see Chapter 4).

2.3.2 The Median Boundary

The boundary between the Western and Eastern provinces has been the subject of debate for nearly 50 years. Originally defined as the Median Tectonic Line between paired metamorphic belts by Landis and Coombs (1967), it was later renamed the 'Median Tectonic Zone' to embrace the complex of magmatic and sedimentary rocks that did not appear to relate to either province (Bradshaw, 1993). The Median Tectonic Zone comprises a narrow belt of basic to acid plutonic rocks and subsidiary units of volcanic and volcaniclastic sediments that range in age from Carboniferous to Early Cretaceous (Kimbrough *et al.* 1994). Many contacts are faulted, and at least three plutonic and two volcano-sedimentary assemblages have been identified within the Median Tectonic Zone (Williams, 1978; Bradshaw, 1993; Muir *et al.* 1998). Bishop *et al.* (1985) distinguished the volcano-sedimentary rocks as the Drumduan Terrane. Volcaniclastic sediments are typically plant bearing and are probably mainly subaerial (Williams & Smith, 1979; Johnston *et al.* 1987).

An alternative view of the boundary issue was proposed by Mortimer *et al.* (1999a,b), who place the whole assemblage in the Median Batholith, de-emphasising its tectonic significance (Mortimer, 2004). A wide cross-section in Stewart Island (see Plate 1) has been described (Allibone & Tulloch, 2004) and provides a clearer picture of the lithologies present. U–Pb dating (Kimbrough *et al.* 1994; Muir *et al.* 1994; Allibone *et al.* 2009a) shows that magmatism extended from the early Carboniferous to Early Cretaceous, with notable gaps in the Middle Permian and Early Jurassic. All the rocks are calc-alkaline. Some are said to resemble continental margin arc rocks of Andean type, but others (Late Jurassic–Early Cretaceous) appear to be geochemically more primitive (Muir *et al.* 1998; Tulloch & Kimbrough, 2003).

Other data, however, continue to point to tectonism at the boundary, such as the presence of deformation within granites (Allibone & Tulloch, 2008), the abrupt truncation of the stable-platform Ordovician–Devonian sedimentation at the edge of the narrow Takaka Terrane, and the paucity of the types of Carboniferous granites common in the correlative New England Fold-Belt (Tulloch *et al.* 2009a). Together these suggest that at times it was a zone of tectonic erosion. Scott (2013) provides a valuable review of the history and tectonics of this zone.

West of the North Island, Median Tectonic Zone rocks have been drilled in exploration well Tangaroa-1 (Mortimer *et al.* 1997), and dredge samples suggest further extension to the West Norfolk Ridge northwest of Northland (Mortimer *et al.* 1998, 2009). In Antarctica, arc magmatic rocks similar in age range to the Median Tectonic Zone are developed in a broad band (Amundsen Province) parallel to the coast of Marie Byrd Land and on Thurston Island (Bradshaw *et al.* 1996, 1998; Pankhurst *et al.* 1998; Mukasa & Dalziel, 2000). The pattern suggests a continuous, intermittently active, late Paleozoic–Mesozoic magmatic arc from Queensland to the Antarctic Peninsula.

2.3.3 Eastern Province

The Eastern Province is made up of terranes produced in convergent margin settings of Permian to mid-Cretaceous age, and includes intra-oceanic arcs, the possibility of a back-arc basin or fore-arc basins, and major accretionary complexes (Coombs *et al.* 1976, 1996; Bradshaw, 1989; Spörli & Ballance, 1989). The relationship between some of the terranes is still the subject of research and debate. There is a clear subdivision between a western group comprising the Brook Street, Murihiku and Maitai terranes, which appear to have a supra-subduction setting, and an eastern group comprising the Caples, Torlesse, Kaweka and Waipapa terranes, which are interpreted as being subduction-driven accretionary complexes.

2.3.3.1 Brook Street Terrane

The Brook Street Terrane comprises mainly Permian submarine volcanic arc rocks and related intra-arc volcanic sediments that locally exceed 14 km in thickness (Coombs *et al.* 1976; Houghton, 1981; Houghton & Landis, 1989). The arc rocks are overlain by 1 km of non-volcanic Upper Permian sandstone and limestone, and a minimum of 800 m of Jurassic sandstone and conglomerate rests unconformably on the Upper Permian (Landis *et al.* 1999). The conglomerate includes very large boulders of granitoids with U–Pb zircon ages in the Late Triassic–Early Jurassic, and Middle Jurassic plant fossils are present in the matrix. These rocks are in tectonic contact with the overlying Early Triassic Murihiku Terrane rocks along a major thrust (Letham Shear Zone) of probable Cretaceous age (Landis *et al.* 1999). The Upper Permian rocks suggest burial of an inactive arc margin during post-volcanic subsidence, and the unconformity below the conglomerate suggests pre-Jurassic folding and uplift.

The Brook Street Terrane lies on the western edge of the Eastern Province, and has a faulted contact with the Median Tectonic Zone / Median Batholith throughout its length. In Southland the Brook Street rocks are cut by plutons that appear to be related to the Median Tectonic Zone times (Mortimer *et al.* 1999a). The most recent geochronology suggests the plutons are of Permian age and indicates that the Brook Street Terrane became sutured to the Median Tectonic Zone by Late Permian times (McCoy-West *et al.* 2014).

Brook Street Terrane rocks have been identified west of the North Island (Mortimer *et al.* 1997) and in the West Norfolk Ridge (Mortimer *et al.* 1998).

2.3.3.2 Murihiku Terrane

The Murihiku Terrane consists almost entirely of a thick sequence of mainly Triassic and Jurassic volcanogenic sediments (Murihiku Supergroup) disposed in a major synclinorial structure (see Plate 2a) traceable intermittently for a distance of 450 km through the North and South Islands (Coombs *et al.* 1976; Suggate *et al.* 1978; Mortimer, 2004). Campbell *et al.* (2001, 2003) have suggested that Late Permian rocks (Kuriwao Group) exposed in an inlier east of Invercargill are continuous with Murihiku rocks and that the Murihiku Terrane includes these strata.

In the South Island, Triassic and Early Jurassic rocks have a minimum thickness of 9 km. A similar thickness of Late Triassic–Late Jurassic rocks is exposed in the coastal area southwest of Auckland in the North Island. Kamp and Liddell (2000) infer from fission track data that this was originally overlain by a further thickness of >3000 m of Cretaceous strata. Early Cretaceous rocks of Murihiku type have been intersected in a borehole in onshore Northland (Isaac *et al.* 1994). Murihiku rocks have also been found offshore (Mortimer *et al.* 2009). The only known basal contact is tectonic, and the upper contact is an unconformity below Cenozoic rocks so that the original total thickness of Murihiku rocks is unknown, but a minimum thickness of at least 12–15 km seems likely. Conglomeratic and marginal marine facies are developed more commonly in the western limb of the syncline and derivation is inferred from a Triassic–Jurassic volcanic arc lying to the southwest (Ballance & Campbell, 1993). The Murihiku rocks are generally considered to have formed in a fore-arc basin, although a back-arc setting has also been inferred (Coombs *et al.* 1976, 1996).

The sediments are mainly sandstone and mudstone of volcanic provenance and the present synclinal structure was probably developed sub-parallel to the original sedimentary basin margin, with conglomerate and marginal marine facies on the southwestern and western sides of the basin closest to the source. In the past it was suggested that the Brook Street Terrane was the dominant source of the detritus, but marked differences in composition and texture of volcanic clasts in conglomerate from those typical of the Brook Street Terrane, and widespread air-fall tuffs, indicate an active arc source in the Mesozoic (Ballance & Campbell, 1993; Bradshaw, 1994). Triassic granitoid clasts are also widespread (Graham & Korsch, 1990; Wandres *et al.* 2004b) and point to the uplift of plutonic rocks in the source only a little older than the depositional age.

A thick body (>5 km) of sedimentary rocks beneath the sea floor east of Stewart Island was correlated by Cook *et al.* (1999) with the Murihiku Supergroup. Although the sedimentary sequence appears similar to Murihiku rocks in seismic section, no petroleum drillholes penetrate the succession, and its correlation is still uncertain. The succession appears to lie across the boundaries of the Brook Street Terrane, Median Tectonic Zone, and potentially part of the Western Province, suggesting that the sediments were deposited in a basin subsequent to amalgamation of basement terranes and therefore not Murihiku Terrane. Recent research, however, suggests that the amalgamation of the Brook Street Terrane with parts of the Median Batholith was much earlier than previously thought (McCoy-West *et al.* 2014), making the Murihiku correlation a possibility. On the other hand, the identification of a core-complex type structure in southern Stewart Island suggests a possible Cretaceous age (Kula *et al.* 2007).

2.3.3.3 Maitai Terrane

The Maitai Terrane comprises Late Permian–Early Triassic sediments that rest unconformably on the Early Permian Dun Mountain Ophiolite (Kimbrough *et al.* 1992). The ophiolite contains the main elements of oceanic crust (pillow lavas, dike complex, mafic plutons and ultramafic tectonites (Coombs *et al.* 1976)), forming a disrupted and attenuated zone along the eastern margin of the terrane. Disruption and attenuation pre-date the deposition of the Permian sediments, and rapid variation in the thickness of the lower two formations suggests that extension continued to take place during the early phase of sedimentation (Adamson, 2008). Initial sedimentary breccia is succeeded by deep-water limestone, locally very thick, made up of fragments of the bivalve *Atomodesma*. The limestone is very variable in thickness and forms the lower part of a succession that is otherwise dominated by detrital sediment. The succession extends into the Middle Triassic, but is much finer grained than adjacent Murihiku rocks of the same age. The sediments exceed 4 km in thickness.

The western margin of the Terrane is very poorly exposed and appears to be delimited by faults.

The question of the relationship between the Maitai and Murihiku rocks is uncertain due to lack of exposure. The Maitai sediments appear more siliceous than the Murihiku rocks, but sandy arc-volcaniclastic detritus is still important and air-fall tuffs are common but generally thinner than in the Murihiku rocks. Until a genetic relationship can be demonstrated the retention of two discrete terranes is simpler.

The eastern boundary of the Maitai Terrane is also significant because it marks division between the supra-subduction group of terranes and the accretionary complexes. It is characterised by mélange (Craw, 1979; Rattenbury *et al.* 1998), or more sharply by the Livingstone Fault and related fractures. The mélange may represent a plate boundary zone between the crustal portion of the Dun Mountain Ophiolite, representing the basement of the upper plate, and geochemically diverse mafic rocks as relics of the subducted plate (Sivell & McCulloch, 2000).

2.3.3.4 Eastern Terranes group

The eastern group of terranes received little attention until the late 20th century despite the fact that they form more than half the basement of New Zealand. With the exception of the presence of gold mineralisation in schists in parts of Otago and Marlborough there was little incentive to tackle the problem of lithologically uniform, structurally complex and stratigraphically indivisible rocks commonly called the 'greywackes'. Widely scattered macrofossil localities showed the presence of Permian–Jurassic rocks, but no means of further subdivision seemed workable. During research to complete the first 1 : 250,000 maps of New Zealand, metamorphic zones based on mineralogy were mapped as the Haast Schist Group, and the lower-grade non-schistose rocks as the Torlesse Group (later Supergroup). Carter *et al.* (1974) suggested that these were not stratigraphic units, but metamorphic zones. At much the same time research showed that reasonably large coherent structures could be mapped (Bradshaw, 1972; Spörli & Lillie, 1974) and that in some cases the metamorphic zones cross-cut original stratigraphy (Andrews *et al.* 1974; Bishop, 1974b;

Bishop *et al.* 1976). The key to regional understanding finally appeared through a better appreciation of plate tectonic processes and a combination of detailed sedimentary petrology, geochemistry and isotope geochronology. New concepts and discoveries led to the development of a terrane-based terminology for the eastern zone (Bishop *et al.* 1985).

2.3.3.5 Caples Terrane

The Caples Terrane comprises regionally metamorphosed sedimentary and volcanic rocks between 5 and 7 km thick, and lies predominantly within the regionally extensive Haast Schist zone (Mortimer, 1993a). The sediments are of volcanic arc provenance, with local developments of mafic pillow lavas and cherts. In the western part of the terrane, subdivision into formations is possible and large-scale folds in bedding can be mapped (Bishop *et al.* 1976; Turnbull, 1979). To the northeast, strongly developed schistosity is the dominant feature, within which folds are widely mapped (Mortimer, 1993a). Sandstones are typically volcanic litharenites characterised by low Sr initial ratios at the time of Early Jurassic metamorphism (Graham & Mortimer, 1992). Permian- and Carboniferous-age conodonts in chert (Ford *et al.* 1999; Pound *et al.* 2014) are likely to relate to the oceanic substrate rather than to the deposition age of the bulk of the sediment, which is probably Late Permian–Triassic. Mélanges within the terrane include basalt, serpentinite and chert of probable ocean floor origin but do not appear to mark terrane boundaries. Coombs *et al.* (2000), however, suggest that the Caples Terrane may be composite and contain bodies that differ significantly from the above generalisations.

The Caples rocks pre-date Early Jurassic metamorphism. A pre-metamorphic suture between Caples Terrane and Torlesse Terrane can be mapped in Otago (Cox, 1991; Mortimer, 1993a; Turnbull, 2000) and in Marlborough (Mortimer, 1993b), based mainly on geochemical and isotopic composition.

2.3.3.6 Torlesse Terrane

The terms for subdivisions of the Torlesse rocks have varied between authors: some have adopted 'Torlesse Terrane' with several subterranes, while others prefer 'Torlesse Superterrane' containing several terranes. Here we adopt the first option. The Torlesse Terrane is very extensive and underlies much of the eastern South Island (see Plate 2b) and the North Island. In the South Island the terrane can be divided into the Rakaia (Permian–Triassic) and Pahau (Late Jurassic–Early Cretaceous) subterranes, separated by the Esk Head Mélange (Bradshaw, 1973, 1989; Bradshaw *et al.* 1981; Silberling *et al.* 1988) (see Plate 4). The Torlesse Terrane is mainly composed of marine strata dominated by thick successions of alternating sandstone and mudstone with rare conglomerates. Other minor constituents are chert, pillow lava and limestone; these three are considered to be ocean-floor relics. Numerous shear zones anastomose through the complex and fossils are very rare and no regional succession can be established.

The Torlesse Terrane grades into the Haast Schist to the southwest, but compared with the Caples Terrane the Rakaia Subterrane is much more quartzo-feldspathic and shows more-evolved isotopic compositions, with Sr initial ratios in Rakaia rocks significantly higher than those of Caples rocks at the time of Jurassic metamorphism. The Caples Terrane and the Rakaia Subterrane were amalgamated, deformed and metamorphosed in the Early Jurassic during terrane amalgamation (Graham & Mortimer, 1992). Further compressional and extensional deformations occurred in the Late Jurassic and Early Cretaceous so that the higher-grade rocks lie in a metamorphic core (Forster & Lister, 2003). Geochemical and isotopic evidence suggests that the 'Aspiring Lithological Association' of pelitic schist, metachert and greenschist that lies between typical Caples and Rakaia type schists in western Otago has affinities with the Rakaia Terrane (Mortimer & Roser, 1992), but, more recently, ages of detrital zircons in these rocks (Jugum *et al.* 2013) suggest they are younger than the adjacent Caples and Rakaia rocks and form a structural inlier of rocks related to the younger Waipapa Terrane (see below). Rocks of all three terranes are strongly deformed and schistose, and imply Cretaceous underplating and metamorphism.

Sandstones and mudstones in the Rakaia and Pahau subterranes (see Plate 3a,b) ranging in age from Permian to mid-Cretaceous appear very similar in hand specimen. Point counting (MacKinnon, 1983) and geochemistry (Roser & Korsch, 1999) show that there are differences with age: the Permian and late Mesozoic sandstones have more volcanic lithics, while the Late Triassic sandstones are more felsic. The uniformity of grain size and composition, and the lack of marker horizons, make them very difficult to map or subdivide. The uniformity over such a long period is difficult to explain, and their source is a matter of debate (Wandres *et al.* 2004a,b). Studies of the ages of detrital zircons (e.g., Wandres *et al.* 2004b; Adams *et al.* 2007, 2009) show that the youngest component is typically only slightly older than the age indicated by fossils, implying active replenishment of the source area by ongoing magmatism.

Detrital zircons in sandstone and igneous zircons from clasts in conglomerates show that Permian and Triassic magmatic arc rocks were the principal contributors to the Rakaia Subterrane (Wandres *et al.* 2004b). Linkage between the Rakaia Subterrane and the Pahau Subterrane is indicated by the abundance of sandstone clasts similar to older Rakaia-type sandstone in conglomerates within the Pahau Subterrane (Wandres & Bradshaw, 2005), supporting the contention that the two subterranes were in close proximity. On the other hand, initial Sr ratios in Pahau Subterrane rocks (Adams *et al.* 2009), differences in the abundance of volcaniclastic grains, and the presence of igneous pebbles of different composition and age (Wandres *et al.* 2004a; Wandres & Bradshaw, 2005) do not support derivation solely from older Rakaia sources, but indicate a significant contribution from a source similar to the Early Cretaceous components of the Median Batholith.

The two subterranes are separated by the Esk Head Mélange (see Plate 4a), a broad belt of highly disrupted rocks containing blocks of pillow lava, chert and limestone in an abundant matrix of sandstone and mudstone. In some places less-deformed patches show remnants of original bedding, particularly in sandstone-rich zones (Bradshaw, 1973; Silberling *et al.* 1988). A similar belt of mélanges crosses the North Island from the Rimutaka Range to Whakatane (Adams *et al.* 2011).

Fossils from Torlesse rock occur in two settings: those in beds of sandstone and mudstone, which form the bulk of the assemblage; and those occurring in lenses of chert and limestone in association with volcanic rocks. The former are predominately bivalves (Andrews *et al.* 1976). The second group occur in mélanges or as tectonic slices within younger rocks. They range in age from Carboniferous to Permian in the Rakaia Subterrane (Jenkins & Jenkins, 1971; Forsyth *et al.* 2006), Late Triassic in the Esk Head Mélange, and Triassic to Jurassic in Pahau Subterrane (Andrews *et al.* 1976; Silberling *et al.* 1988). The presence of Cretaceous sediments in the Pahau Subterrane is shown by plant microfossils and common Cretaceous detrital zircons (Adams *et al.* 2009). The Permian limestones with fusilinids and the overlying detrital sediments of possible equatorial origin (Cawood *et al.* 2002) represent a distinct exotic component in the Rakaia Subterrane. Fusilinids also occur as an exotic component in the Waipapa Terrane in Northland (Spörli *et al.* 2007 and references therein), and a similar equatorial origin is inferred.

Both the Rakaia Subterrane and the Pahau Subterrane continue into the North Island (see Plate 1). The whole terrane pattern in the North Island has, however, been considerably revised in recent years and is dominated by the Waipapa and Kaweka terranes.

2.3.3.7 Kaweka Terrane

Rocks in the central North Island, in part schistose, that were originally considered to be parts of the Torlesse Terrane have recently been shown to differ from typical Torlesse rocks in composition and age (Adams *et al.* 2011) and have been interpreted as a discrete terrane: the Kaweka Terrane. Interestingly, they appear to be intermediate between Waipapa and Torlesse in isotopic composition and also to bridge the gap in age between the Rakaia and Pahau subterranes. The terrane occupies a large area in the central North Island (see Plate 1) and the western boundary is not exposed. The eastern boundary of the Kaweka Terrane with Pahau Subterrane is a zone of mélange. Rocks of similar composition and detrital zircon populations occur in the South Island. They are found as components of the Esk Head Mélange and adjacent fault slices. The terrane boundaries are assumed rather than mapped, although many faults are present in the area. In effect the terrane appears to be the main contributor to the mélange. Adams *et al.* (2011) suggest that, as in the North Island, the Pahau Subterrane is almost entirely Early Cretaceous.

2.3.3.8 Waipapa Composite Terrane

The Waipapa Terrane (Spörli, 1978) lies west of the Taupo Volcanic Zone and is the basement unit of the west-central North Island. A small area of Waipapa Terrane mapped in central Northland is now considered to be Caples Terrane. The Terrane has two lithological types or facies. In the north the Hunua Facies comprises mainly thin-bedded volcaniclastic sandstones and mudstones that are complexly deformed with belts of mélange. Slices of chert and pillow basalt are widespread. This facies includes exotic Permian limestone and chert (Spörli & Balance, 1989; Spörli *et al.* 2007). The cherts are of particular interest, as they include a preserved fossiliferous Permian–Triassic boundary succession. These rocks are regarded as ocean floor relics. To the south the Morrinsville Facies is coarser and of simpler structure, lacking the cherts and basalts of ocean floor type. Rocks of Waipapa affinity occur within schists in Marlborough (Mortimer, 1993b) and in Otago (Jugum *et al.* 2013).

2.3.3.9 Mount Camel suspect terrane

This suspect terrane forms the basement in northern Northland, from the Karikari Peninsula northwards to the Three Kings Islands. Contacts with the adjacent Waipapa Terrane are everywhere concealed. Rocks of the Mount Camel suspect terrane belong to the Houhora Complex (Isaac *et al.* 1994), which includes the Rangiawhia Volcanics and the related Tokerau sedimentary facies. Deformational structures in the succession are associated with the emplacement of the Northland Allochthon in the Miocene (Toy & Sporli, 2008). The rocks are Early Cretaceous in age based on the bivalve *Inoceramus kapuus-ipuanus* of Albian (Urutawan–Motuan) age from Tokerau Facies near Mount Camel. U–Pb dating of zircons from siliceous volcaniclastics at Mount Camel gives ages in the range 104–131 Ma, which is compatible with the age indicated by the *Inoceramus*. Further research suggests an older period of deformation and accretion before the emplacement of the allochthon (Toy & Spörli, 2008). More-recent research confirms the Cretaceous age and a possible supra-subduction origin (Adams *et al.* 2013a).

2.3.3.10 Tectonic setting of the eastern group of terranes

All the continental crust of New Zealand east of the Livingstone Fault and related structures (Caples, Torlesse, Kaweka and Waipapa terranes) can be regarded as components of Permian–Mesozoic accretionary complexes. All are dominated by alternating sandstone–mudstone successions in which the main variable is the scale of the alternation. All other lithologies – chert, pillow lava, limestone and conglomerate – are very subsidiary, and all except conglomerate probably represent fault-bounded remnants of the oceanic substrate on which the detrital rocks were deposited, mainly by turbidity-current and mass-flow processes. Depositional settings are likely to have spanned the trench slope, trench and lower plate in a predominantly convergent margin setting. Some terranes and subterranes were fed from arcs dominated by active continental-margin volcanism (Caples and Waipapa terranes), while others (Torlesse and Kaweka terranes) were derived from arcs in which magmatic rocks were more evolved and plutonic rocks were more abundant. Periodic elevations of the accretionary complex to form a sedimentary outer-arc ridge lying at or above sea level may explain the local development of non-marine, shallow-marine and deltaic facies within the Rakaia Subterrane (Retallack, 1981) and in the Pahau Subterrane (Bassett & Orlowski, 2004). Similar Late Jurassic rocks occur in central Canterbury (Oliver *et al.* 1982) and are thus contemporaneous with Pahau Subterrane deposition but lie well to the southwest of the Esk Head Mélange. They lie in thrust contact below Rakaia Subterrane rocks, and their relationship is uncertain, but they are thought to record at least local emergence and erosion of the Rakaia Subterrane in Late Jurassic times (Cox & Barrell, 2007).

2.4 THE RANGITATA OROGENY

Historically, the Rangitata Orogeny was conceived as a major period of late Mesozoic deformation, metamorphism, plutonism and mountain building that consolidated the basement rocks of New Zealand (Suggate *et al.* 1978). Post-orogenic erosion was thought to have produced a widespread erosion surface that separated basement rocks from the younger Late Cretaceous–Cenozoic cover. More-recent research has compromised the simplicity of this interpretation and the term should be treated with caution. Given a convergent margin setting extending over ~150 m.y., complexity is not surprising, particularly in association with episodic accretion. Striking examples are the amalgamation deformation and metamorphism of the Caples Terrane and Rakaia Subterrane in the Jurassic and the accretion of the Brook Street Terrane in the Permian. Other events that were traditionally attributed to the orogeny are now assigned to crustal extension that preceded break-up. In parts of eastern New Zealand, however, the change in sedimentary character between basement and cover is slight, the difference in age is small and there is no clear evidence of subaerial exposure. There is typically a difference in structural style that can be correlated with the cessation of subduction-driven deformation at the trench, a process marked by abundant folding and faulting in soft sediment and pervasive bedding-parallel extension (see Plate 4b).

There were significant tectonic events in New Zealand during the mid-Cretaceous, shown by the folding of the Murihiku Terrane and its westward thrusting over the Brook Street Terrane on the Letham Shear Zone (Landis *et al.* 1999). There was a second phase of metamorphism on the Haast Schist and apparent under-plating of younger Waipapa Terrane rocks in Otago, Westland and Marlborough (Mortimer, 1993b; Jugum *et al.* 2013). Crustal thickening, possibly by under-plating, may also explain the development of the Western Fiordland Orthogneiss (Daczko *et al.* 2001). On a more local scale, thrusting of Rakaia Subterrane rocks over the Late Jurassic Clent Hills Group in Canterbury (Oliver *et al.* 1982), an event that pre-dates the extrusion of the mid-Cretaceous Mount Somers Volcanics at 95–99 Ma (Tappenden *et al.* 2002; Van der Meer *et al.* 2017), is a good example of Cretaceous deformation. These events collectively justify, at least for the present, the term 'Rangitata Orogeny'.

The major compressive events were completed in southern and central New Zealand by about 105±5 Ma (Bradshaw, 1989; Luyendyk, 1995; Bradshaw *et al.* 1996) and were replaced by crustal extension and the precursors to break-up. There are, however, indications that extension may have started between 120 and 100 Ma in western New Zealand before subduction terminated along the eastern margin (Waight *et al.* 1997; Spell *et al.* 2000; Forster & Lister, 2003) and may represent a type of back-arc extension with related magmatism. Slightly later in the eastern South Island, within-plate break-up related magmatic rocks occur, cutting folded Torlesse rocks (Pahau Subterrane), at least as early as 97 Ma (Weaver & Pankhurst, 1991; Baker & Seward, 1996).

On the other hand, in northern New Zealand (Raukumara Peninsula) there is evidence that sedimentation and compression related to subduction continued after 100 Ma (Ballance, 1993; Mazengarb & Harris, 1994). The suggestion that accretionary-complex rocks could be as young as ~100 Ma is supported by detrital zircon ages from this area that are slightly later than 100 Ma (Cawood *et al.* 1999).

The suggestion of Hoernle *et al.* (2005) that the Hikurangi Plateau collided only with the Chatham Rise portion of the New Zealand margin is relevant to the difference in time of subduction termination between southern and northern New Zealand. A portion of the Phoenix plate adjacent to northern New Zealand may have continued to subduct for several million years after subduction ceased in the south. The difference may reflect either the shape of the Hikurangi Plateau and/or the presence of a transform within the subducting plate.

2.5 BREAK-UP TECTONICS

In the mid-Cretaceous the future New Zealand continental fragment was roughly triangular and bounded by two types of continental margin (Fig. 2.1). Along the northeastern edge the old trench slope subsided progressively so that initially the overlying sediments closely resembled the underlying accretionary complex, lacking only the complex deformation. In the Late Cretaceous, new sedimentary basins and patterns developed (see Chapter 4). The other two sides of Zealandia are passive margins that were generated by break-up between West Antarctica to the southeast and Australia to the west. The break-up patterns were strongly asymmetrical. The margins in the Australian and Antarctica continents are topographically high, underlain by thick crust and bordered by narrow continental shelves. The New Zealand (Zealandia) margins, marked by extensive submarine continental plateaux and crustal thinning of the type proposed by Lister *et al.* (1991), seem to have affected the whole New Zealand region. In the Western Province, extension is indicated by the development of core complexes (Gibson *et al.* 1988; Tulloch & Kimbrough, 1989; Spell *et al.* 2000; Kula *et al.* 2009) (see Plate 5a) and the formation of extensional tectonic basins (see

Chapter 4). Similar extensional basins occur in the Eastern Province both onshore and offshore (see, e.g., Field *et al.* 1997).

In West Antarctica, subduction-related magmatism was abruptly replaced by anorogenic 'A' type granitoids in the Edward VII Peninsula at ~102 Ma (Weaver *et al.* 1992), and by syenites and alkali feldspar granites to the east in Marie Byrd Land. At about the same time a margin-parallel suite of tholeiitic dikes and layered gabbros at ~100 Ma was also developed (Storey *et al.* 1999). Core complex formation in Marie Byrd Land took place during the same interval (Richard *et al.* 1994).

In New Zealand, syenite, layered basic complexes, and related intraplate trachytes and alkali basalts appeared at ~97–96 Ma (Weaver & Pankhurst, 1991; Crampton *et al.* 2004b). Fission track ages (Baker & Seward, 1996) are consistent with uplift and cooling in the period 105–93 Ma in eastern New Zealand south of the latitude of Cook Strait. Cretaceous volcanism was widespread on the Tasman and Pacific margins as a precursor to break-up (Tulloch *et al.* 2009b).

Large-scale extension is also indicated by the Bounty Trough, a ~400 km-wide east–west depression south of the Chatham Rise (Davy, 1993). The trough appears to be floored by extremely thin continental crust, and extensional faults are well developed on the flanks. At the eastern end, where the Bounty Trough abuts the ocean floor of the southwest Pacific, Late Cretaceous magnetic anomalies run continuously across the end of the trough, indicating that the trough pre-dates break-up between New Zealand and Antarctica. Northwest of New Zealand the New Caledonia Basin does not align exactly with the Bounty Trough across a restored Alpine Fault, though the two may have originated at much the same time.

The New Caledonia Basin has a substantial fill of Late Cretaceous sediments that pre-date Tasman Sea opening (the Taranaki delta of Strogen *et al.* 2017), as does the Great South Basin (Cook *et al.* 1999), but the Bounty Trough has little sediment.

Both the Bounty Trough and the New Caledonia Basin are sub-parallel to the trend of the Mesozoic accretionary margin. On a palinspastic outline of Cretaceous Zealandia they are not aligned, however: the New Caledonia Basin lies inboard of the Median Batholith, while the Bounty Trough lies on the outboard side. The Tasman Sea margin has a similar orientation to the terrane boundaries in the north but not in the south. The line of continental separation between New Zealand and West Antarctica and the western margin of the Great South Basin are at approximately 45° to the accretionary margin and the terrane boundaries.

Pre-break-up paleomagnetic data place the future Zealandia close to the South Pole in the interval between 120 and 90 Ma (see, e.g., Sutherland *et al.* 2001). During the early part of the Cretaceous there may have been some northward migration of the Zealandia region due to 'roll-back' of the Phoenix plate margin. After attachment of Zealandia to the Pacific plate at ~105±5 Ma the northward migration became much more rapid and New Zealand approached middle latitudes by the end of the Cretaceous period. Luyendyk (1995) has drawn attention to the sympathy between the northward movement of New Zealand and the latitudinal changes of the Pacific plate. Larson *et al.* (1992) report an earlier southward migration of the Pacific plate towards the Gondwana margin in the Early Cretaceous followed by northward migration in the Late Cretaceous. This pattern correlates surprisingly well with the major events in the tectonic eastern Australia, New Zealand and West Antarctica.

Rey and Muller (2010), in a discussion of the fragmentation of continental margins, include a reinterpretation of the Pacific-Gondwana margin that may explain the complicated chronology outlined above.

2.6 THE PLATE BOUNDARY DEBATE

In recent years there has been much research relevant to the break-up phase, which indicates a complex Cretaceous tectonic history. A study of the Grebe Shear Zone (Scott *et al.* 2011) suggests that the outer part of the Median Batholith was uplifted at 128 Ma, followed by further syntectonic magmatism at 121 Ma and a cessation of deformation at 116 Ma. In the Paparoa Metamorphic Core Complex, Schulte *et al.* (2014) infer movement on the southern Pike River detachment zone started before 116 Ma and therefore before the intrusion of the Buckland Granite (110–109 Ma) that forms a large part of the core. The northern detachment fault is thought to have developed at the same time as the intrusion of the granite. A further thermal pulse is dated at ~75 Ma and regarded by Schulte *et al.* (2014) as related to the opening of the Tasman Sea, but this seems unlikely as the oldest Tasman Sea oceanic crust is nearly 10 m.y. older. Studies of widespread Cretaceous volcanic rocks extending from the Lord Howe Rise to eastern New Zealand suggest that they are a genetically related suite and show a progressive decline in crustal contribution to the magmas over time. They are thought to reflect crustal thinning over the period 101–82 Ma (Tulloch *et al.* 2009b). Plutonic magmatism (Waight *et al.* 1998a) also points to a changing tectonics setting prior to the arrival of the Hikurangi Plateau at the subduction zone and the end of subduction, placed at 110 and 105 Ma respectively by Davy (2014).

Despite the abundant evidence of crustal extension and basin formation in the interval between 105 and 85 Ma there is a growing body of evidence that suggests other tectonic events and a more complex setting (Fig. 2.2). These fall under four headings:

1. Evidence for Cretaceous oroclinal bending and lack of Miocene rotation (Turner *et al.* 2012; Mortimer, 2014)
2. Revised plate rotation data for this period that point to a very large sinistral strike-slip displacement across Zealandia along the line of the Alpine Fault (Lamb *et al.* 2016)
3. Evidence of late Cretaceous protoliths in the Alpine Schist close to the present Alpine Fault (Cooper & Ireland, 2013, 2015) that would require an active margin somewhere between the schists and the Median Batholith
4. Geochemical evidence for supra-subduction magmatism in the rocks of Late Cretaceous and Paleocene age, and deformation of similar age in the Northland Allochthon (Cluzel *et al.* 2010) that requires an extra plate and an extra plate boundary.

At the same time there is a large body of evidence consistent with major dextral motion on the Alpine Fault plate boundary in Miocene to Recent times, and for at least three decades the Alpine Fault has been regarded as essentially a Cenozoic structure. At present there is no consensus on these issues and the following chapters adopt the older convention of a purely Cenozoic Alpine Fault and orocline.

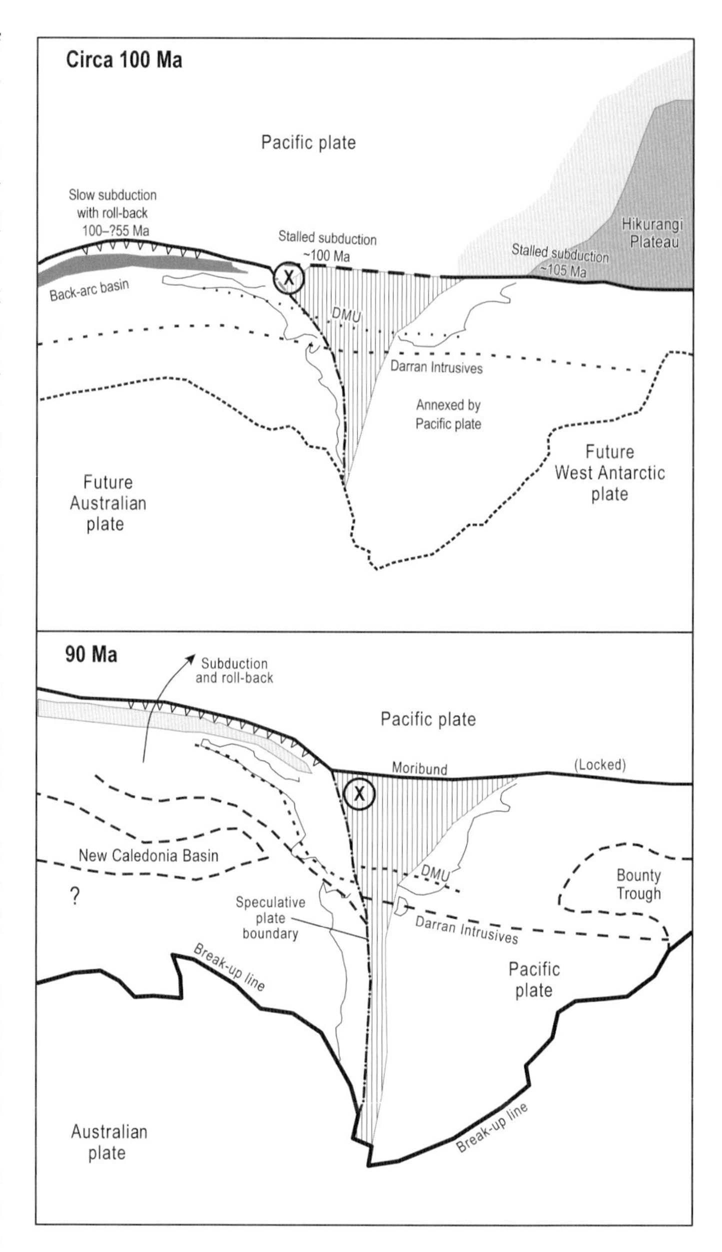

Fig. 2.2. Two maps showing possible tectonic configurations assuming the subduction stalled or ceased in the southeastern portion of the Pacific–Gondwana plate boundary but that subduction and Pacific plate roll-back continued in the northwest. The hachured area is the zone of strong late Cenozoic deformation that overprints and obscures Cretaceous deformation. X marks the limit of Late Cretaceous–Paleocene subduction. DMU, Dun Mountain Ultramafics.

CHAPTER 3

Overview of the Stratigraphy and Tectonics of the Mid-Cretaceous to Recent Succession

3.1 FROM CONVERGENT MARGIN TO EXTENSION, AND THE ALBIAN UNCONFORMITY

The change from subduction to extension is recorded throughout most of New Zealand by a major angular unconformity separating Paleozoic and older Mesozoic rocks from younger, generally less-deformed strata. The unconformity surface is almost certainly diachronous, ranging through the Albian Stage (113–100 Ma). In central Raukumara Peninsula and in parts of Marlborough the unconformity is overlain by sediments containing fossils of early to mid-Albian (Urutawan) age and, locally, possibly as old as late Aptian (Mazengarb & Speden, 2000; Crampton *et al.* 2004b; and Figs 3.1, 4.1a,b and 4.4). By contrast, in eastern Raukumara Peninsula U–Pb dating of zircons gives a probable youngest depositional age of ~100 Ma for basement rocks (Cawood *et al.* 1999; Adams *et al.* 2013a,b), indicating that subduction may have continued in the northeast until latest Albian times. At least locally, the change from convergent margin to extensional tectonics in the mid-Cretaceous took place over a short space of time (probably <3 m.y.). This is seen in parts of the East Coast Basin, notably the Wairarapa area, the unconformity separates basement and cover strata that both contain the index fossil of the late Albian Motuan Stage (Moore & Speden, 1984), inferred to have lasted ~3 m.y. (Cooper, 2004).

As outlined in Chapter 2, in the west and south of New Zealand the change is less easy to date. The younger granites were intruded during the early stages of crustal extension at ~110 Ma (Sagar & Palin, 2011; Schulte *et al.* 2014) and there is evidence that the granites themselves record a change from subduction to extension (Waight *et al.* 1998b). In Fiordland the relationship of granitoid dikes intruded into extensional shear zone suggests that the switch from a convergent to an extensional tectonic regime occurred between 111 and 108 Ma (Scott & Cooper, 2006). Tuffs and ignimbritic volcanic rocks in early extension basins in Otago have U–Pb zircon ages of 112 Ma (Tulloch *et al.* 2009b). Thus there appears to be an overlap between the commencement of extension in the southwest and the ending of subduction further to the northeast (Pacific) margin. The early extension could be viewed as a type of intra-arc or back-arc extension, particularly as it precedes continental break-up by more than 25 m.y.

In some areas of New Zealand, in particular the central and southwestern North Island, Cretaceous rocks are absent and may never have been deposited. Here strata of Eocene age or younger rest unconformably on basement rocks whose youngest age is Late Jurassic or earliest Cretaceous.

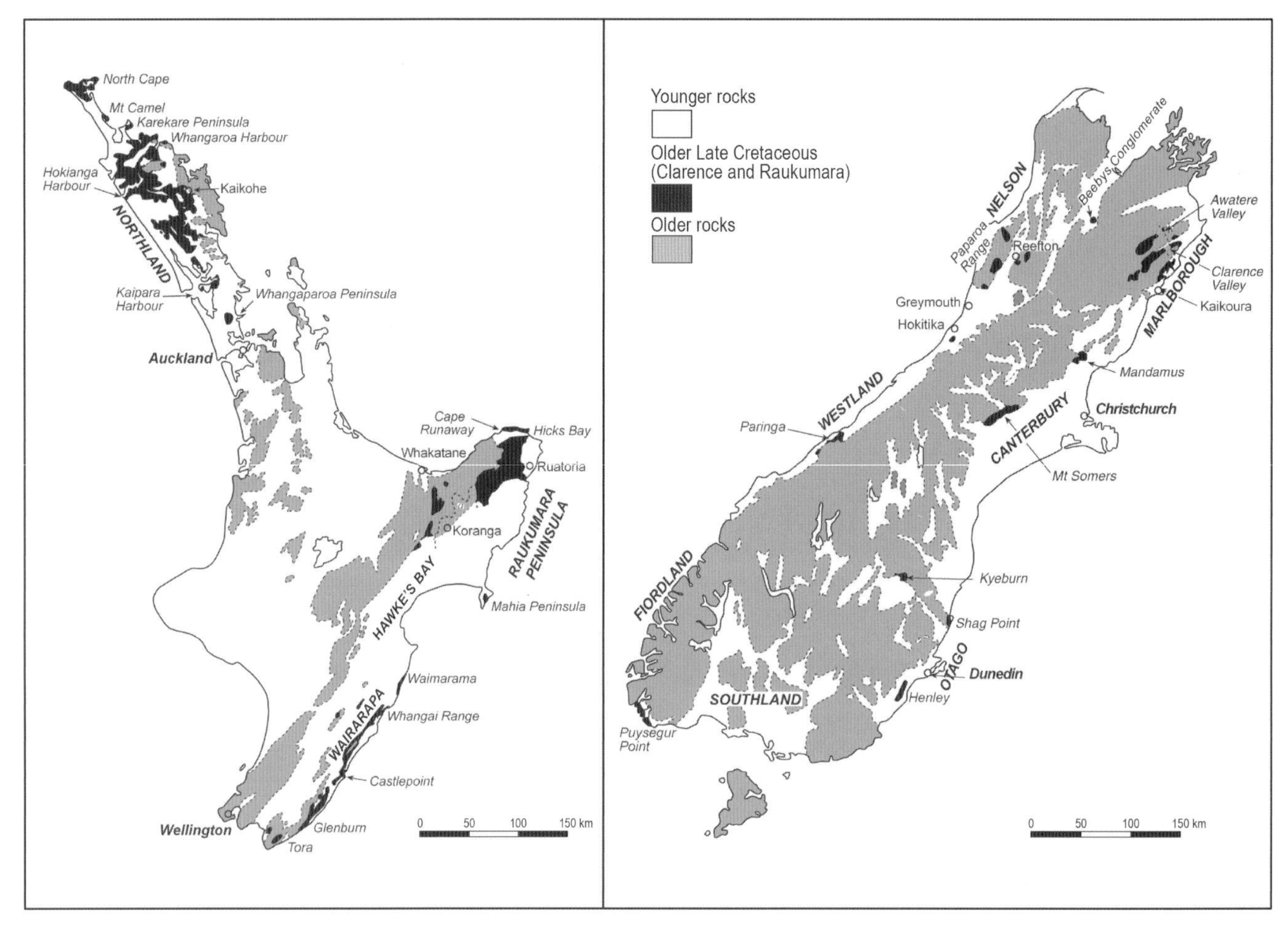

Fig. 3.1. Map of the onshore distribution of Late Cretaceous rocks in New Zealand. (Source: GNS Science.)

3.2 STRATIGRAPHIC HIERARCHY AND SYNOPSIS OF CRETACEOUS–CENOZOIC GEOLOGICAL HISTORY

3.2.1 Introduction

The filling of sedimentary basins commonly follows a similar pattern, recording a cycle of shoreward marine transgression, infilling and oceanward regression. These cycles are usually bounded above and below by unconformities. In some cases the cycles can be shown to relate to global eustatic sea-level changes driven by major climatic events. In other cases it is clear that eustatic changes are overwritten or completely obscured by local or regional tectonic events. Timespans of cycles vary through several orders of magnitude; those of over 50 m.y. are regarded as first-order events, with successively shorter second, third and so on cycles.

In Zealandia the characteristics of sedimentary basins and successions strongly reflect the tectonic evolution of the continent, and the sedimentary packages represent an 'event stratigraphy' with timespans that are determined by tectonic events. The stratigraphic development in the last 100 m.y. is a transgression–regression megasequence that can be divided into three assemblages representing the 'Rift', 'Drift' and 'Convergent Margin' phases of development (Fig. 3.2). Recently, Mortimer *et al.* (2014) have proposed a scheme of lithostratigraphic terms that cover the same interval. Assemblage 1 corresponds to the Momotu Supergroup, Assemblage 2 to the Haerenga and Waka supergroups, and Assemblage 3 to the Maui and Pakihi supergroups.

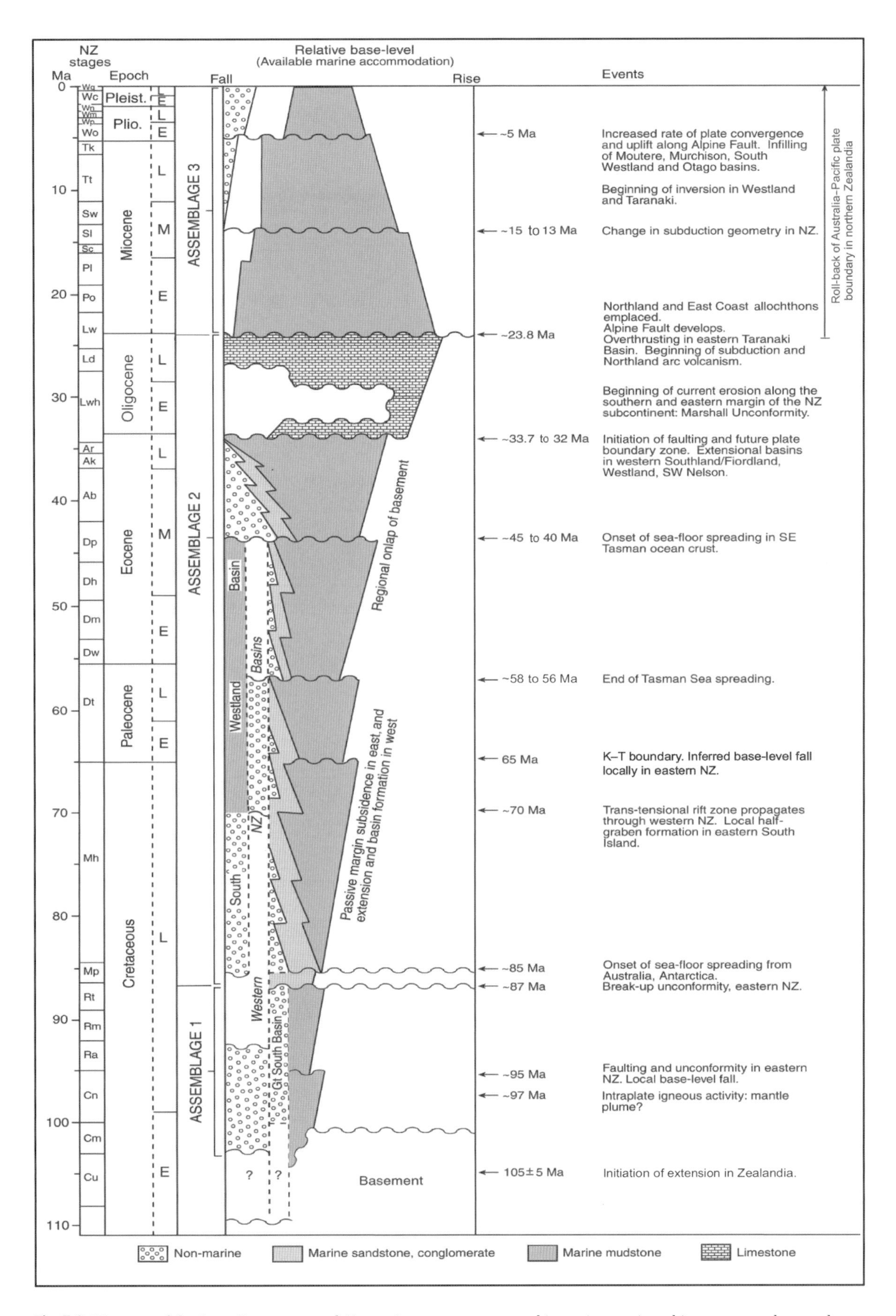

Fig. 3.2. Diagram of the Late Cretaceous and Cenozoic megasequence and its major stratigraphic sequences that result from the interaction of tectonic activity and sea-level fluctuations. (Source: GNS Science.)

3.2.2 The Cretaceous-Cenozoic megasequence

The post-basement Cretaceous–Cenozoic transgression–regression megasequence was named the 'Kaikoura Sequence' by Carter *et al.* (1974), and later the 'Kaikoura Synthem' (Carter, 1988a). The megasequence is separated from underlying basement rocks by a marked unconformity except where the deeply submerged outer margin of the Mesozoic accretionary zone remained below sea level during the Cretaceous. A summary figure (Fig. 3.2) shows the overall pattern and the three assemblages employed in this volume.

The oldest sediments overlying the basal megasequence boundary include dominantly coarse mainly marine deposits in the East Coast Basin, which is inferred to overlie thin 'crust' of the accretionary complex (see section 4.1.1), and, in most of the rest of the country, mainly coarse terrestrial deposits, commonly infilling half grabens formed in thicker crust (Laird, 1996b). The first major widespread marine transgression occurred in eastern and southern New Zealand at ~87 Ma, but in western New Zealand significant marine flooding began only in the latest Cretaceous (e.g., ~67–65 Ma in the Taranaki Basin).

A significant landward facies shift occurred close to the Eocene–Oligocene boundary in many parts of New Zealand. This is evidenced by the generally highly calcareous nature of Oligocene rocks, reflecting widespread submergence of former low-lying land areas and consequently a reduction in clastic detritus. The base of the carbonate sediments is a regional marine flooding surface, and is often marked by diachronous depositional hiatuses. In southeastern areas, such as the Great South Basin, the northern Canterbury Basin and Marlborough, the transition from clastic-dominated to carbonate-dominated sedimentation began much earlier, in keeping with the generally earlier onset of marine transgression in those areas. Nevertheless, the presence of Oligocene unconformities and diastems in these areas suggests that they too were affected by enhanced marine flooding in the Early Oligocene.

The carbonate-dominated interval throughout New Zealand contains a number of unconformities, disconformities or depositional hiatuses, the best documented of which is the widespread Marshall Paraconformity or Unconformity (Carter & Landis, 1972; Carter *et al.* 1982; Carter, 1985; Fulthorpe *et al.* 1996) of Early Oligocene (mid-Whaingaroan) age. Marine inundation of the paleo-New Zealand landmass was greatest in the early Waitakian (latest Oligocene). Carbonate deposition ends with an abrupt change to detrital sediments either concordantly or by a clearly marked unconformity that is also easily recognised on seismic reflection profiles. This surface is generally close to the Oligocene–Miocene boundary, but, like other key surfaces, it is diachronous.

The Neogene succession in most New Zealand basins is generally thick, clastic-dominated and regressive. This interval characteristically reflects the progressive infilling of previously formed and newly developing marine basins, and local progradation over the continental shelf: it corresponds to the highstand systems tract component of the first-order depositional cycle.

3.2.3 Tectono-sedimentary assemblages

3.2.3.1 Assemblage 1

Assemblage 1 comprises sediments deposited after the main phase of convergent margin tectonics in the mid-Cretaceous. The sediments generally reflect an extensional regime typified by continental rifting and graben formation. The assemblage is clearly separated in most areas from basement rocks by unconformity, and is terminated by region-wide transgression accompanying and following continental break-up and the separation, at ~85 Ma, of the New Zealand continental block (Zealandia) from the Australian–Antarctica segment of the Gondwana margin. Break-up on the Tasman and Campbell sectors may not have been synchronous (Kula *et al.* 2007).

With the exception of local channel-fill sediments of late Aptian (Korangan) age, which have been identified only in the Raukumara Peninsula, the oldest widespread supra-basement cover sediments in New Zealand are of Albian (Urutawan–Motuan) age. They crop out in the East Coast Basin, on the Chatham Islands, in north Otago, in western Southland and in the West Coast of the South Island. Strata of this age may also underlie latest Albian–early Cenomanian (Ngaterian) sediments drilled in the Great South and northern Taranaki basins, and may also form part of Tupou Complex within the Northland Allochthon (Isaac *et al.* 1994). In most places Albian strata pass upwards conformably into strata of early Cenomanian (Ngaterian) age. In eastern and southern New Zealand the mid- and Late Cretaceous succession is relatively complete. Significant tectonic change in the Late Cretaceous is indicated by two region-wide closely related unconformities: the earlier one, close to the late Coniacian–Santonian (Teratan–Piripauan) boundary (~86 Ma), represents the initial break-up unconformity. Erosion was widespread, extending in places to basement and followed by the deposition of transgressive marine sands of early to middle Santonian age. The later unconformity, also accompanied by major marine transgression over much of New Zealand, probably resulted from thermal subsidence as New Zealand started to move away from the thermal high at spreading ridges in the widening Tasman Sea and Southern Ocean. All of the Cretaceous succession above basement and below the older late Coniacian transgression has been incorporated into Assemblage 1. The younger unconformity is within Assemblage 2.

Early sedimentation, particularly in the South Island, was commonly within normal fault-controlled half grabens that evolved during a major period of crustal extension marked by core-complex formation that pre-dates continental break-up (Tulloch & Kimbrough, 1989; Laird, 1994, 1995). Reliable ages for these deposits are difficult to obtain, but the ~102 Ma (Muir *et al.* 1997) age from the base of the Pororari Group in the half graben northeast of the adjacent core-complex indicates a late Albian (late Motuan) age. Plant fossils in the group are consistent with this Albian age.

The early basin-fill successions of Assemblage 1 in the South Island generally comprise coarse-grained clastic rocks. The most common depositional settings were alluvial-fan, meandering-river and braided-river environments. Lacustrine, lake margin, and lacustrine fan delta settings are less common (Laird, 1996b). An age-equivalent unit in the offshore Taranaki and Northland basins (the Taniwha Formation) may have had a marginal marine influence (King & Thrasher, 1996; Strogen *et al.* 2017). In the Chatham Islands, strata of late Albian to early Cenomanian (late Motuan to Ngaterian) age were probably deposited in a fault-controlled marginal marine (estuarine-deltaic) environment (Campbell *et al.* 1993).

Mid-Cretaceous rocks of the East Coast of the North Island and northeastern South Island are also coarse-grained, and were deposited as marine olistostromes, debris flows and turbidites (Field *et al.* 1997). Thick, laterally extensive turbidite units occupying the eastern portion (Eastern Sub-belt) of the East Coast Basin of the North Island (Moore, 1988) may have been deposited in a large, semi-continuous, subsiding basin. In the Raukumara Peninsula, however, Mazengarb & Harris (1994) recognise compressional structures in rocks of this age, which they correlate with subduction-related deformation widely recognised in the Early Cretaceous rocks of Torlesse Terrane, and infer that subduction may have continued into the Late Cretaceous until the onset of sea-floor spreading (~85 Ma). Compressional features, however, may be caused by factors other than subduction, such as gravitational collapse of the accretionary complex after the cessation of mid-Cretaceous subduction (see also Knott, 2001). The presence of the regional Albian (Urutawan–Motuan) unconformity over most of the peninsula area also suggests that a major change in tectonic regime to one of extension occurred here in the mid-Cretaceous as in the remainder of New Zealand. There is, however, growing evidence of subduction continuing in the far north. This topic is discussed further in section 4.1.3 and in Chapter 2.

Although rifting continued into the early Late Cretaceous in some areas, notably the South Island, in other areas, such as in the Western Sub-belt of the North Island East Coast Basin, faulting died out in Cenomanian (late Ngaterian) times, and basins started to fill with fine-grained clastic deposits.

An abrupt end to the depositional assemblage occurred in late Coniacian–early Santonian (late Teratan–early Piripauan) times in eastern New Zealand, with erosion of a widespread unconformity (see above). Subsequent deposition of transgressive marine sands of dominantly early to middle Santonian age marks the beginning of Assemblage 2.

3.2.3.2 Assemblage 2

Assemblage 2 incorporates rocks deposited during continental break-up and the succeeding drift phase with related thermal subsidence in the Late Cretaceous–Paleogene. Differential subsidence is strong in some areas and associated with active normal faulting reflecting post-break-up extension. The assemblage starts with a diachronous erosion surface (see above). It is widely developed in eastern New

Zealand from East Cape to the Great South Basin. The age of the basal sediments ranges from late Coniacian to early Santonian (late Teratan to early Piripauan: ~87–~85.5 Ma) in the East Coast Basin, and is early Santonian (Piripauan) in the Great South Basin. The generally coarse-grained quartz-rich sediments represent a strong contrast to generally finer-grained deposits underlying the unconformity, and in the Great South Basin represent the first major marine transgression in the area.

A younger diachronous erosion surface is also widespread throughout New Zealand, and is particularly well developed in the east of the country, where it is overlain by fine-grained transgressive sediments ranging in age from middle Santonian to early Campanian (latest Piripauan to early Haumurian: ~85–~84 Ma). Western New Zealand remained emergent, with the first indications of marine encroachment occurring, in the latest Cretaceous, in south Westland and in parts of the Taranaki Basin–west Northland Basin areas.

Assemblage 2 includes post-rift graben-fill sediments of Late Cretaceous to Paleocene age in western basins, and mainly passive margin sediments in eastern basins. The western basins are thought to be linked to extension and northward propagation of a major transform in the Tasman spreading ridge (Laird, 1994; King & Thrasher, 1996). Crustal extension appears confined to a few sub-basins in southern Taranaki and Westland. In offshore Northland, graben-fill coal measures and overlapping marine sediments interpreted on seismic reflection profiles are inferred to be stratigraphic equivalents of similar strata in the Taranaki Basin. In areas between the downfaulted sub-basins in the west, erosion continued. In eastern New Zealand, latest Cretaceous sediments are dominated by fine-grained outer-shelf to bathyal deposits, typically mudstone or micritic limestone. These facies, showing a generally deepening bathyal trend, became dominant through Paleocene and Eocene times.

Sea-floor spreading in the Tasman Sea ceased in the Paleocene and led progressively to the formation or reactivation (see section 2.6) of the Australian–Pacific plate boundary through New Zealand. This event is marked in the Late Eocene and Oligocene times by the development of strongly subsiding fault-bounded extensional basins in western Southland and the West Coast.

Overall, transgression continued until the Oligocene, which corresponded to the period of maximum submergence of the New Zealand continent, and the deposition of dominantly carbonate rocks. In addition to carbonates, greensands are common in the Oligocene, indicating a considerable reduction or starvation of clastic supply caused by rapid marine transgression at this time. For much of New Zealand the Oligocene represents the culmination of passive margin development. Assemblage 2 was terminated by major influxes of clastic sediments accompanied by regression close to the Oligocene–Miocene boundary.

3.2.3.3 Assemblage 3

Assemblage 3 includes the rocks deposited during the Kaikoura Orogeny, a period of convergent margin tectonics that continues to the present day. The assemblage is separated from Assemblage 2 in most places by an unconformity and/or a profound change in sedimentary facies marked by an influx of primarily terrigenous sediment replacing the carbonate-dominated sediments of the upper part of Assemblage 2. The influx marks the beginning of a regional-scale regression driven by uplift and erosion coincident with the change to oblique convergence along the plate boundary in earliest Miocene (Waitakian–Otaian: ~22 Ma) times. At much the same time uplift near the continental margin in the far north led to the emplacement of the Northland and East Coast allochthons.

In most areas the basal portion of Assemblage 3 is fine-grained, although sandstones were deposited around the fringes of rising land areas in western Southland and Canterbury. In addition, alternating sandstone and siltstone or flysch-like sequences were deposited in the Murchison Basin, western Moutere Depression, Wairarapa and also western Southland. The first pulses of deep-water sandstone deposition began in eastern Taranaki in the middle of the Early Miocene (Otaian: ~20 Ma), but became more significant later and continued into early Middle Miocene (Altonian–Clifdenian: ~15 Ma) times. Similar influxes of sand are evident in the East Coast (Hawke's Bay and Wairarapa), although this is earlier and of earliest Miocene (Waitakian: ~22 Ma) age in places, and more variable both geographically and stratigraphically, as a result of local tectonic influence.

The beginning of the Pliocene (5.3 Ma) over much of the South Island was marked by the influx of a huge volume of coarse clastic sediments associated with accelerated uplift of the Southern Alps. Similar deposits in the eastern North Island developed in the Late Pliocene, in response to uplift along the axial ranges.

CHAPTER 4

Assemblage 1 – The Early Extension Phase: The Mid- and Early Late Cretaceous Succession

4.1 INTRODUCTION

This chapter examines the evolution of the sedimentary basins developed during the post-subduction to pre-drift period of New Zealand's geological history. Basin development depended on several factors, including the character of the basement, extensional tectonics and sea-level changes. The effects of sea-level changes are left to Chapter 9, but the other factors are introduced here.

4.1.1 Effect of basement rocks

The nature of the basement rocks, particularly their thickness and buoyancy, played an important part in the style of subsequent basin development and in the depositional environment of the overlying sediments. Most of the basement terranes of southern and western New Zealand, comprising the Western Province, the Median Tectonic Zone (or Median Batholith) (Mortimer *et al.* 2014), and the western portion of the Eastern Province, consisted of thickened crust, and crustal thicknesses range from 30–35 km for the Median Batholith west of the North Island (Stern & Davey, 1989) to 25–30 km for the Rakaia Subterrane in the eastern South Island (Davey *et al.* 1998; Mortimer *et al.* 2002). In these areas thick successions of non-marine rocks of mid-Cretaceous age were deposited in fault-bounded basins. Most of the thick continental crust, however, remained above sea level until continental break-up, when flooding occurred from the new western and south-eastern margins. In contrast, continental crust in the northeast, formed from the youngest and last accreted easternmost terranes (the Pahau Subterrane and correlatives), appears to average approximately 20 km thick (Eberhart-Phillips & Reyners, 1997). The thinner Pahau Subterrane underlies the entire East Coast Basin, and is the site of new mid- to Late Cretaceous basins with mainly deep marine deposits. The thinner crust in this region was less buoyant and was depressed below sea level when regional extension began. The newly named Kaweka Terrane is separated from overlying 'cover' rocks by an angular unconformity in north Canterbury.

It has been noted (Muir *et al.* 2000) that the central part of the Late Cretaceous–Cenozoic Taranaki Basin in the western North Island, referred to as the Taranaki Graben or Eastern Mobile Belt (King & Thrasher, 1996), appears to be underlain entirely by plutonic rocks of the Median Batholith. The development of the central basin is likely to have been strongly influenced by rejuvenation of faults within the Median Batholith that correspond to the Late Cretaceous–Cenozoic Cape Egmont Fault Zone and the Taranaki Fault.

Grobys *et al.* (2008) have calculated the thinning and concomitant areal expansion of the submerged continental crust around New Zealand.

4.1.2 Mid-Cretaceous erosion and unconformity

It is clear (see section 3.1) that the change from convergent margin to extensional tectonics is diachronous, particularly in the south and west where then mid-Cretaceous granites suggest a transition from a subduction to an extensional setting and there is evidence of granite intrusion contemporaneous with crustal extensions (Schulte *et al.* 2014). In parts of the East Coast Basin, notably the Wairarapa area (Moore & Speden, 1979, 1984; and Fig. 3.1), the unconformity separates basement and cover strata that both contain the index fossil of the Motuan Stage (mid- to late Albian), which is inferred to have a time range of ~4 m.y. (Raine *et al.* 2015). Overall, the evidence suggests the change took place in the Albian between 112 and 100 Ma. The sedimentary evidence and record of the change is discussed by region, below.

4.1.3 Problematic features of the Raukumara Peninsula

In the Raukumara Peninsula the division between basement and cover has long been a subject of debate. Two features, unique to the Raukumara Peninsula, have been described by earlier workers. These are: (1) the presence of relatively undeformed strata (Koranga Formation) of Aptian (Korangan) age in the western part of the region (Speden, 1975), resting with apparent unconformity on more-deformed

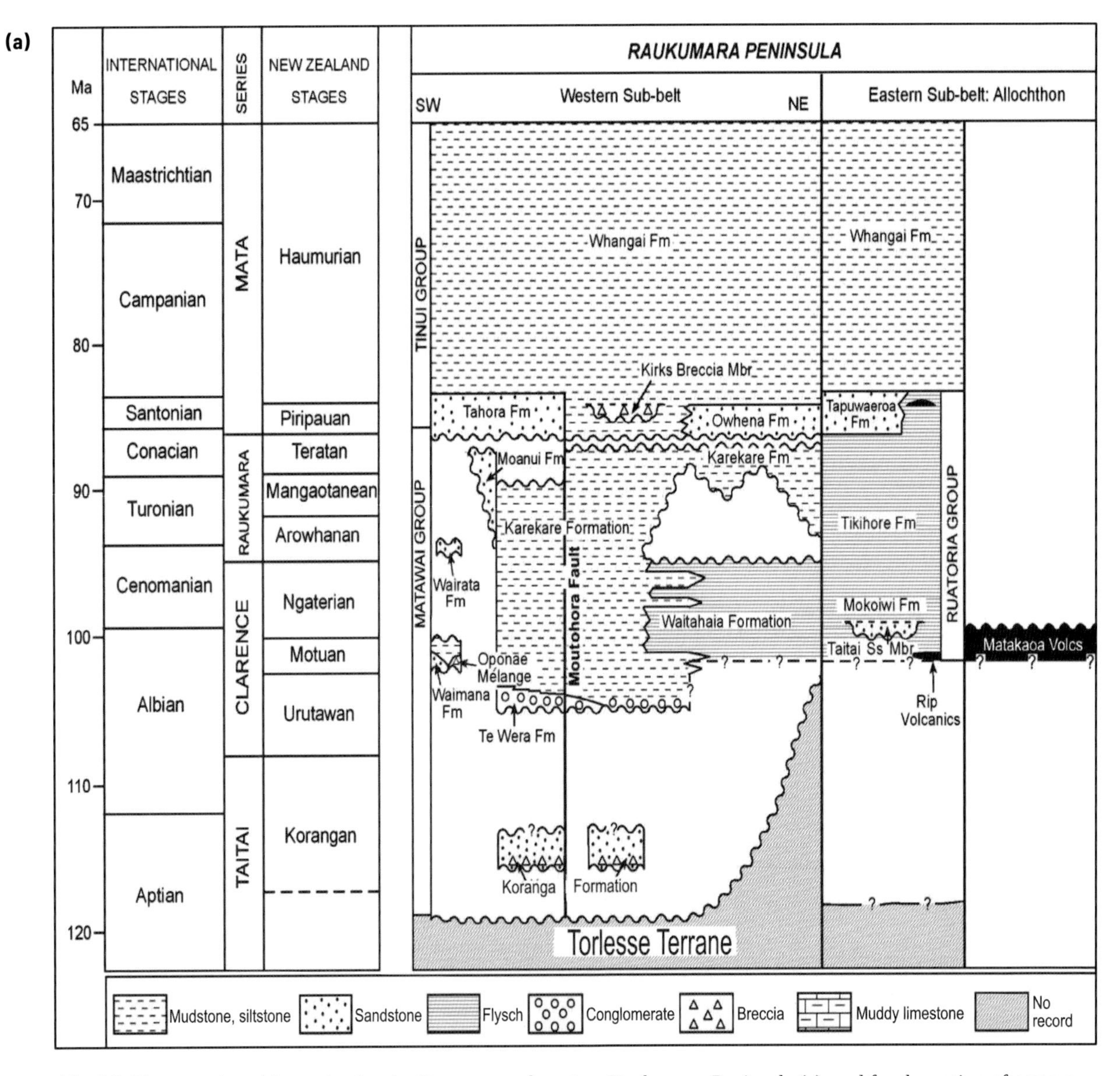

Fig. 4.1. Time-stratigraphic section for the Cretaceous of western Raukumara Peninsula (a), and for the region of eastern Raukumara Peninsula to Marlborough (b). Fm, Formation; Gp, Group; Mbr, Member; Ss, Sandstone; Volcs, Volcanics. (Adapted from Field *et al.* 1997.)

Torlesse Terrane (Pahau Subterrane, locally termed 'Urewera Group'), and in turn overlain unconformably by a fossiliferous shallow-marine cover strata, the Te Wera Formation (see Plate 7); and (2) inferred gradational relationships between 'basement' Urewera Group rocks and Urutawan cover strata in the middle portion of the region. The differences in stratigraphy from west to east across the Raukumara Peninsula led Mazengarb and Harris (1994) to the division of the Cretaceous rocks into five structural and stratigraphic 'domains', two of which are in the East Coast Allochthon. Our investigations have resulted in reinterpretation of the above points, and are discussed below.

4.1.3.1 Koranga Formation

Fossils of Aptian (Korangan) age, found in situ only in the Koranga area of western Raukumara Peninsula and in an outcrop at the Hawai River mouth in the Bay of Plenty, were originally inferred to be part of a continuous Early Cretaceous succession that passed essentially conformably up into Late Cretaceous strata (e.g., Wellman, 1959). Speden (1975) redefined the fossiliferous strata at Koranga as belonging to the Koranga Sandstone, and mapped an angular unconformity between it and underlying Torlesse Terrane (Urewera Group), and also between it and the overlying Te Wera Formation of early Albian (Urutawan) age (Fig. 4.1a; see Plate 7). Examination of the fossiliferous beds of

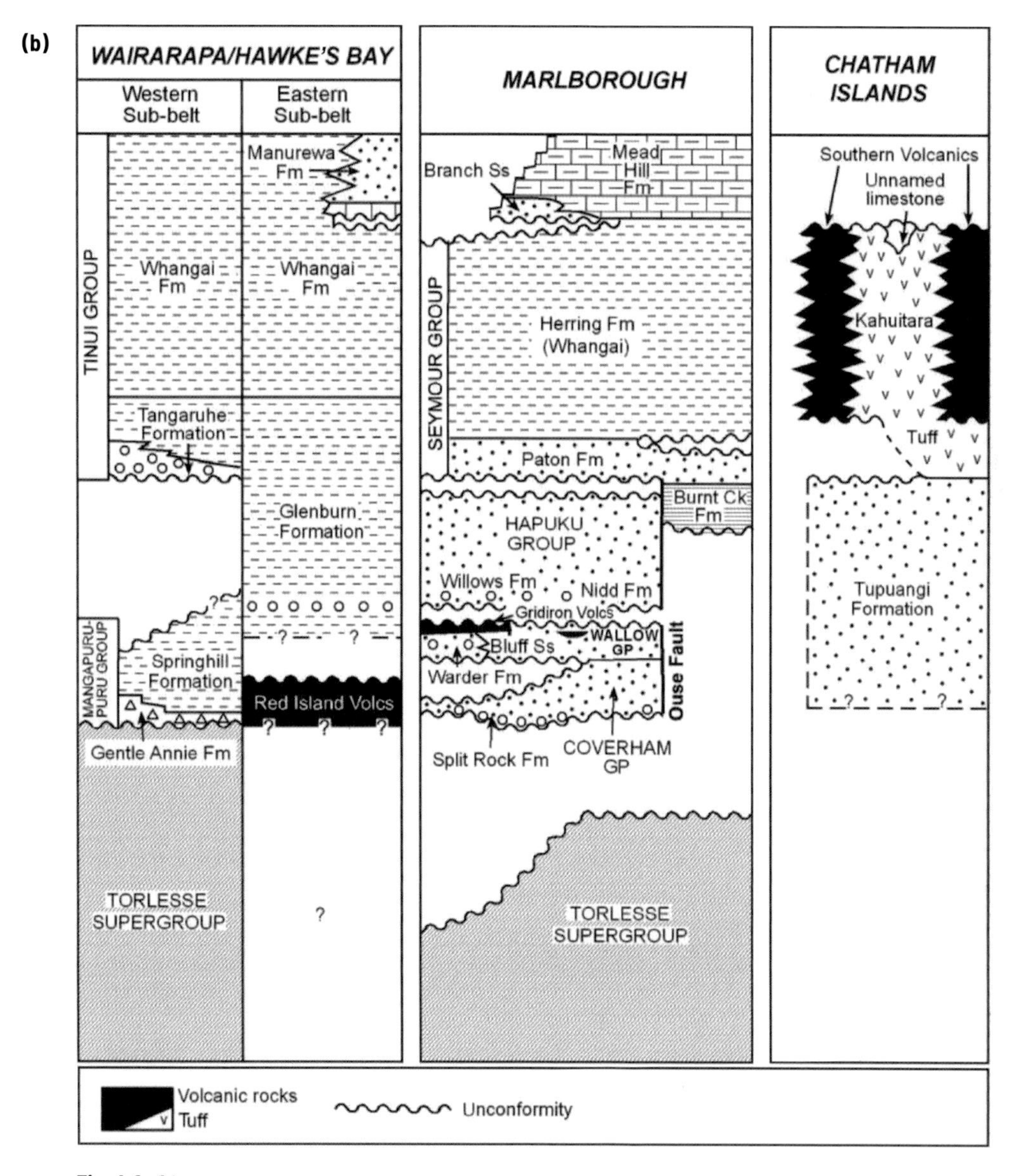

Fig. 4.1. (b)

Korangan age at Hawai River mouth has shown them to be channelled into the underlying Torlesse succession (Laird, 1996a). In the Matawai and Moutohora areas and at the type locality, fossiliferous and unfossiliferous strata mapped as Koranga Sandstone (Isaac, 1977) are also inferred to be channelled into Urewera Group strata (Laird, 1996a).

Mazengarb (1993) renamed the Koranga Sandstone as the 'Koranga Formation' at the type locality and in exposures to the north-northeast, as well as at the fossiliferous Korangan age rocks at Hawai River mouth. These comprise poorly sorted conglomerate in places showing slump features, locally with megaclasts up to 3.5 m long, and massive or graded-bedded sandstone. They are inferred to have been deposited by a series of sediment gravity flows, dominated by debris flows and turbidity currents, and are likely to have formed within a submarine canyon or submarine fan channel–lobe environment (Laird & Bradshaw, 1996). Fossils are inferred to have been redeposited from their shallow-water habitat into deeper water mainly by debris flows.

The stratigraphic affinities of the Koranga Formation have not yet been resolved. Its relatively simple structure contrasts strongly with the underlying Urewera Group into which it appears to be channelled. The lithofacies and fossiliferous nature also contrast strongly with the underlying rocks. All of these features suggest that the Koranga Formation is a unit distinct from and younger than the local Urewera Group basement.

Conglomerates in the Koranga Formation are predominantly made up of a mixture of Torlesse-derived sedimentary lithologies and fresh igneous rocks, indistinguishable from the components of conglomerates within the Urewera Group (Pahau Terrane). In addition, the magnetic susceptibility of Koranga Sandstone rocks is indistinguishable from that of the Torlesse Supergroup, but strikingly different from overlying Urutawan and younger rocks (Speden, 1975). This tends to suggest that the Koranga Sandstone is an integral part of the subduction-influenced basement rocks, rather than being part of the 'cover'. Some support is lent by the dominantly subduction-related signature of the fresh igneous pebbles in Korangan conglomerate (N. Mortimer, pers. comm. 1992), which is compatible with the convergent margin tectonic regime.

Where exposed the Koranga Formation is younger than the surrounding Urewera Group, and is channelled into substantially more deformed strata (see map in Speden, 1975). The critical issue is that the Aptian fauna of the Koranga Formation fauna is older than parts of the Urewera Group basement in the same region, although it is not known to be directly overlain by strata of the Urewera Group. Urewera Group rocks in northeastern Raukumara Peninsula are younger than the Koranga Formation, with fossils of late Albian to Cenomanian age (Mazengarb, 1993), and detrital zircons, possibly air-fall material, have ~100 Ma ages (Cawood *et al.* 1999; Adams *et al.* 2013a,b). It is suggested that the Koranga Formation represents deposition during a late and waning stage of subduction, in a submarine canyon or canyons channelled into rocks already showing strong subduction-related deformation.

4.1.3.2 'Basement' to 'cover' gradational relationships

In the middle portion of the Raukumara Peninsula, gradational relationships have been inferred between basement (Urewera Group) rocks and overlying strata of early Albian (Urutawan) age, in contrast to the unconformable relationship occurring in the west. Specifically, contacts in Kokopumatara Stream have been mapped and described as gradational (Mazengarb *et al.* 1991). A well-exposed contact in this stream re-examined by the authors (see Plate 7) and others, however, shows evidence of erosion and channelling, with indurated unfossiliferous Urewera Group rocks sharply overlain by notably less-indurated fossiliferous olistostromes of Albian (Urutawan to Motuan) age (Laird, 1996a; Laird & Bradshaw, 1996). So far, no unequivocal gradational contacts between the two units have been recorded, and they are considered to be separated by an unconformity in the central part of the region, as in the west (see Plates 7 and 8).

Overall, a discordant sedimentary contact separates Albian or younger rocks from older basement and it seems clear that the differences in stratigraphy from east to west that led Mazengarb (1993) and Mazengarb and Harris (1994) to divide the autochthonous Cretaceous rocks into three stratigraphic–structural domains are much less compelling than previously thought. In the remainder of this volume autochthonous Cretaceous rocks of the Raukumara Peninsula are treated as one structural entity.

In recent years this debate has been given an added twist by growing evidence from the allochthons that subduction did continue between East Cape and Northland, although the style changed from active accretion to volcanic arc and back-arc basin development (Toy & Spörli, 2008; Cluzel *et al*. 2010 and references therein).

4.1.4 Problematic 'cover' strata of the Chatham Rise and the Canterbury Basin

The oldest 'cover' strata cropping out in the Chatham Islands are rocks of late Albian (Motuan) and younger Cretaceous age (Fig. 4.2). They are exposed only on Pitt Island, and inferred from seismic profiling to occupy half grabens faulted into basement. No contacts are seen with older Torlesse rocks and Haast Schist that form the basement in the northern part of Chatham Island, and the relationship of the Pitt Island rocks to basement is unknown. Seismic reflection profiles from the eastern Chatham Rise (Wood & Ingham, 1989), however, reveal the presence of an additional unit, Seismic Unit IIA, between the offshore extensions of the Albian rocks and the basement (Wood *et al.* 1989). Seismic Unit IIA overlies the basement rocks (Seismic Unit I) with a contact of uncertain nature, and is separated by an angular unconformity from the overlying Albian graben-fill strata of Seismic Unit IIB of Pitt Island. Seismic Unit IIA appears on the seismic sections as a thick sequence of regular, moderately folded and faulted, parallel reflectors located between blocks of acoustic basement (Wood *et al.* 1989). The even parallel reflector configuration suggests a possible marine environment of deposition. A minimum stratigraphic thickness of 5–7 km is estimated.

Seismic Unit IIA must be no younger than mid-Cretaceous in age as it is evidently stratigraphically older than Albian rocks exposed on Pitt Island equated with Unit IIB. The graben-fill unit (Unit IIB) marks the beginning of the relatively little-deformed Cretaceous–Cenozoic sedimentary succession on the rise. Derived acoustic velocities of Unit IIA are more typical of Cretaceous sediments than of basement

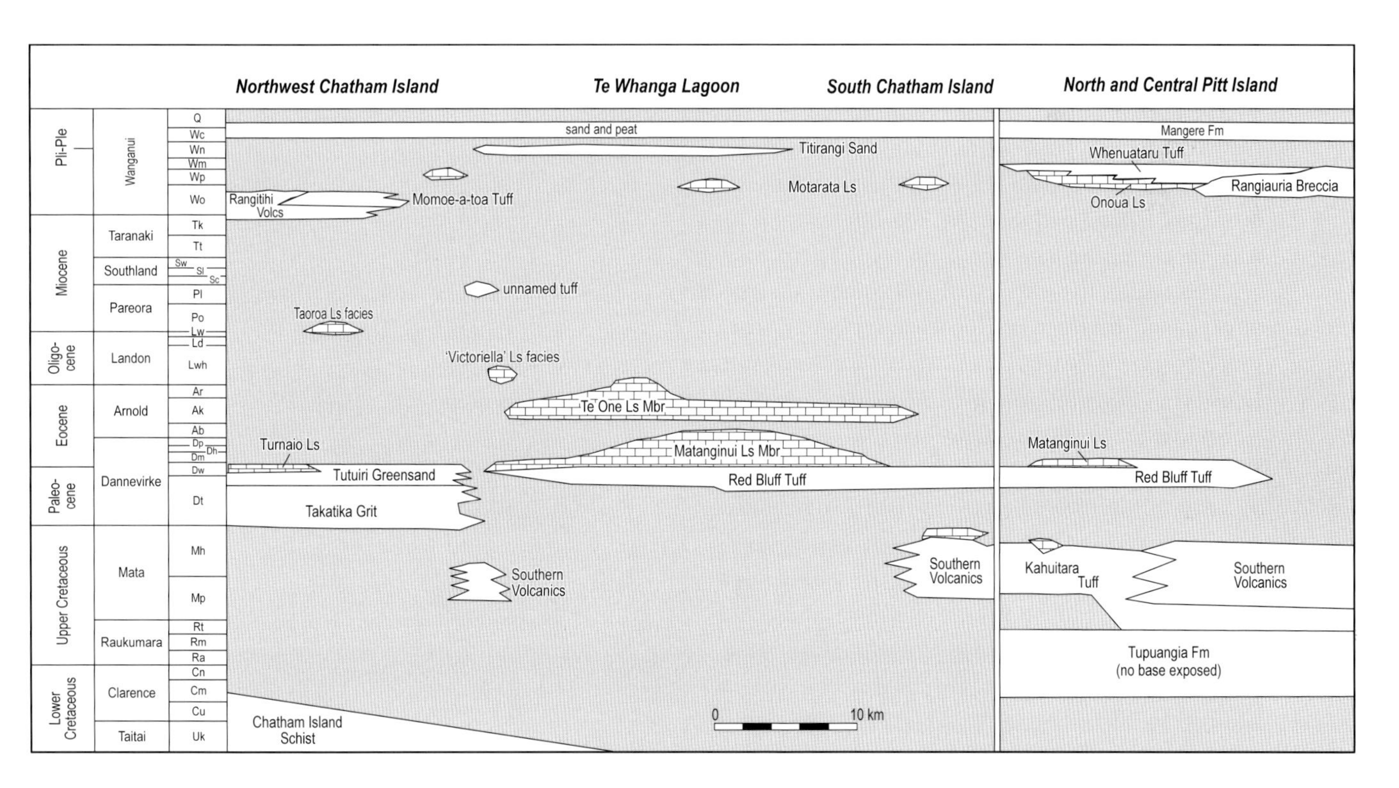

Fig. 4.2. Time-stratigraphic section of Cretaceous to Recent rocks of the Chatham Islands. The geological record for the Chatham Islands is very incomplete, although sedimentation may have been more continuous to the north and south. Fm, Formation; Ls, Limestone; Mbr, Member; Volcs, Volcanics. (Adapted from Campbell *et al.* 1993.)

rocks. The rocks appear to be less indurated than the outcropping basement, and it is concluded that the rocks are probably Late Jurassic or Early Cretaceous.

Strata giving a similar seismic profile unconformably underlie the Late Cretaceous–Cenozoic succession in south Canterbury, where they are inferred to correlate with the non-marine Jurassic Clent Hills Group, which crops out in the western Canterbury foothills.

4.1.5 Structural sub-belts of the East Coast Basin

As a result of Late Cenozoic deformation the East Coast Basin, comprising Raukumara Peninsula, Hawke's Bay–Wairarapa and Marlborough, forms part of one of the more structurally complex regions of New Zealand. Earlier geological studies have made clear that there are major structural differences and facies changes across this region, although a complex and incomplete stratigraphic framework made it difficult to recognise any pattern. It was not until Moore (1988) rationalised the stratigraphic nomenclature that it was possible to define structural divisions. Based on differences in lithofacies and structure Moore (1989a) recognised seven major, largely fault-bounded, structural blocks in the eastern North Island. These were in turn grouped into a Western Sub-belt and an Eastern Sub-belt, which are in tectonic contact and show marked differences (Fig. 4.3). In the Raukumara Peninsula the Eastern Sub-belt lies within the East Coast Allochthon, a series of nappes of Cretaceous–Oligocene rocks emplaced from the north-northeast (see section 7.2). Possible elements of the East Coast Allochthon have also been recognised in the Wairarapa (Delteil *et al.* 1996).

The sub-belt concept developed in the north may be valid in Marlborough as well (Laird & Schiøler, 2005). The Western Sub-belt is characterised by comparatively simple structure, variety in sedimentary facies but with common thick mudstone units, and relatively common, locally diachronous, unconformities, particularly in the west. By contrast, the Eastern Sub-belt is characterised by comparatively intense deformation, and essentially continuous deposition of mainly sandy flysch-like sediments from late Albian to Santonian (Motuan to Piripauan) times (Schiøler & Wilson, 1998). The major contrasts between the sub-belts occur in the Albian–Santonian (Urutawan–Piripauan) successions: differences are less apparent in the Campanian–Maastrichtian (Haumurian) and Paleocene.

4.2 THE CLARENCE SERIES: ALBIAN–CENOMANIAN (URUTAWAN–NGATERIAN) SUCCESSIONS

Rocks of Albian to Cenomanian (Urutawan to Ngaterian) age are the oldest sediments deposited on the erosion surface formed following the cessation of subduction-related deformation. They are represented in Northland, in the East Coast Basin, over much of the South Island, and in the northern Taranaki Basin (see Fig. 1.3). Rocks of this age are notably absent from the southern Taranaki Basin and central North Island. Thick non-marine successions are characteristic of the majority of South Island and western North Island mid-Cretaceous basins, while marine successions are typical of the East Coast Basin, and of onshore eastern Northland and the Northland Allochthon.

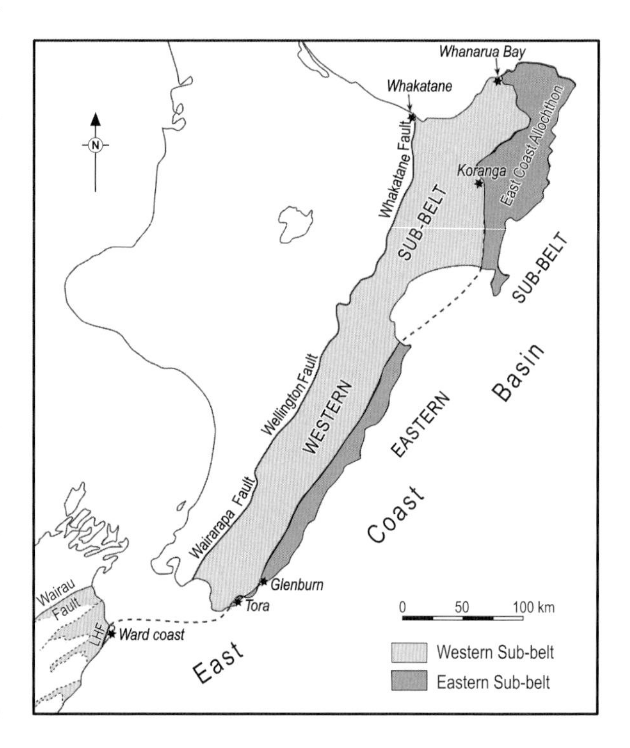

Fig. 4.3. Map of the distribution of the Western and Eastern sub-belts of Late Cretaceous marine sedimentation. LHF, London Hills Fault. (Source: GNS Science.)

4.2.1 Non-marine sedimentary successions of southern and western New Zealand

Thicknesses of up to several kilometres of non-marine sediments infill the basins formed as a result of early rifting in the New Zealand region. That these basins were commonly developed in half grabens or grabens is clearly shown on offshore seismic profiles and by the infill geometry of some outcropping successions. In many instances seismic reflection profiles indicate that the normal faults forming the basin margin decrease in dip downwards, and probably form listric faults at depth (e.g., Uruski, 1992). Half graben-fill is also often fan-shaped in section on seismic profiles, suggesting that active faulting continued during sedimentation (Turnbull *et al.* 1993). Offshore mid-Cretaceous graben-fill successions have so far been drilled only in the Great South Basin, where the oldest sediments penetrated are of Cenomanian (Ngaterian) age (Cook *et al.* 1999). However, the thickest parts of this succession have not been drilled, and sedimentation may well have started earlier.

Other than in the Chatham Islands (see section 4.1.4), non-marine half graben infill in the South Island exhibits a similar motif: close to the active bounding fault, locally derived breccia was deposited as talus or debris flows on alluvial fans, passing directly as fan deltas or via fluvial deposits into lacustrine environments (Laird, 1996b). Active faulting was the dominant control on subsidence and sediment supply.

4.2.1.1 Western New Zealand

Thick successions of mid-Albian to Coniacian (Motuan–Teratan) non-marine strata have been mapped on the West Coast of the South Island, and have also been recognised in a drillhole in the northern Taranaki Basin. Rocks of this age do not occur in the south Taranaki Basin, and are so far unknown in the offshore west Northland Basin.

On the West Coast of the South Island, outcrops of breccia collectively known as the 'Hawks Crag Breccia' (see Plate 6), a formation of late Albian–Cenomanian (Motuan–early Ngaterian) Pororari Group, are widespread over a northeast to southwest extent of about 320 km. The most extensive and diverse deposits are in the Paparoa Range of the northwestern South Island (Fig. 4.4a). Here, the Pororari Group relates to a north-northeast to south-southwest directed extensional event that formed the Paparoa Metamorphic Core Complex (Tulloch & Kimbrough, 1989). Extension was associated with a period of granitic intrusion and metamorphism in the mid-Cretaceous at ~110 Ma, as the tectonic regime was changing from convergence to extension (Waight *et al.* 1998b; Spell *et al.* 2000; Sagar & Palin, 2011; Schulte *et al.* 2014). The metamorphic core has $^{40}Ar/^{39}Ar$ cooling ages that indicate unroofing began before 110 Ma (Spell *et al.* 2000). Granitic mylonites are well developed below the detachment fault between the metamorphic core and the cover (see Plate 5a).

Stretching lineations found on foliation surfaces have a consistently south-southwest plunge in the southern Paparoa Range, but in the Buller Gorge area lineations are not well developed. Tulloch and Kimbrough (1989) proposed that the detachment faults on the northeast and southwest sides of the Paparoa Metamorphic Core Complex had opposite senses of shear. The structure within the two basins is unusual, with both successions dipping towards the core, a relationship that is not consistent with regional asymmetric extension. Schulte *et al.* (2014) suggest that the two detachments are of different ages, the southern one being significantly older. Concomitant brittle deformation formed extensional basins infilled with mid-Cretaceous Pororari Group sediments (Fig. 4.4b). U–Pb zircon geochronology of tuffaceous sediment from near the base of the Pororari Group in the Buller Gorge (northern basin) gives ages of close to 101 Ma (Muir *et al.* 1997) and is consistent with palynological data from the lower part of the Pororari Group that indicate a late-Albian age (Nathan *et al.* 1986). The duration of deposition of the Pororari Group in the region is unknown, as all exposures are truncated either by unconformities or faults. The youngest recorded palynological age is Cenomanian (Ngaterian to Arowhanan) (Raine, 1984).

The most diverse range of lithologies and sedimentary facies occurs in the southern Paparoa Range, in a southern basin that trends west-northwest to east-southeast, where over 2000 m of strata have been divided into three facies assemblages dominated by breccia (Hawks Crag Breccia; see Plate 6), massive or graded sandstone, and highly carbonaceous laminated mudstone (Laird, 1995). The three facies

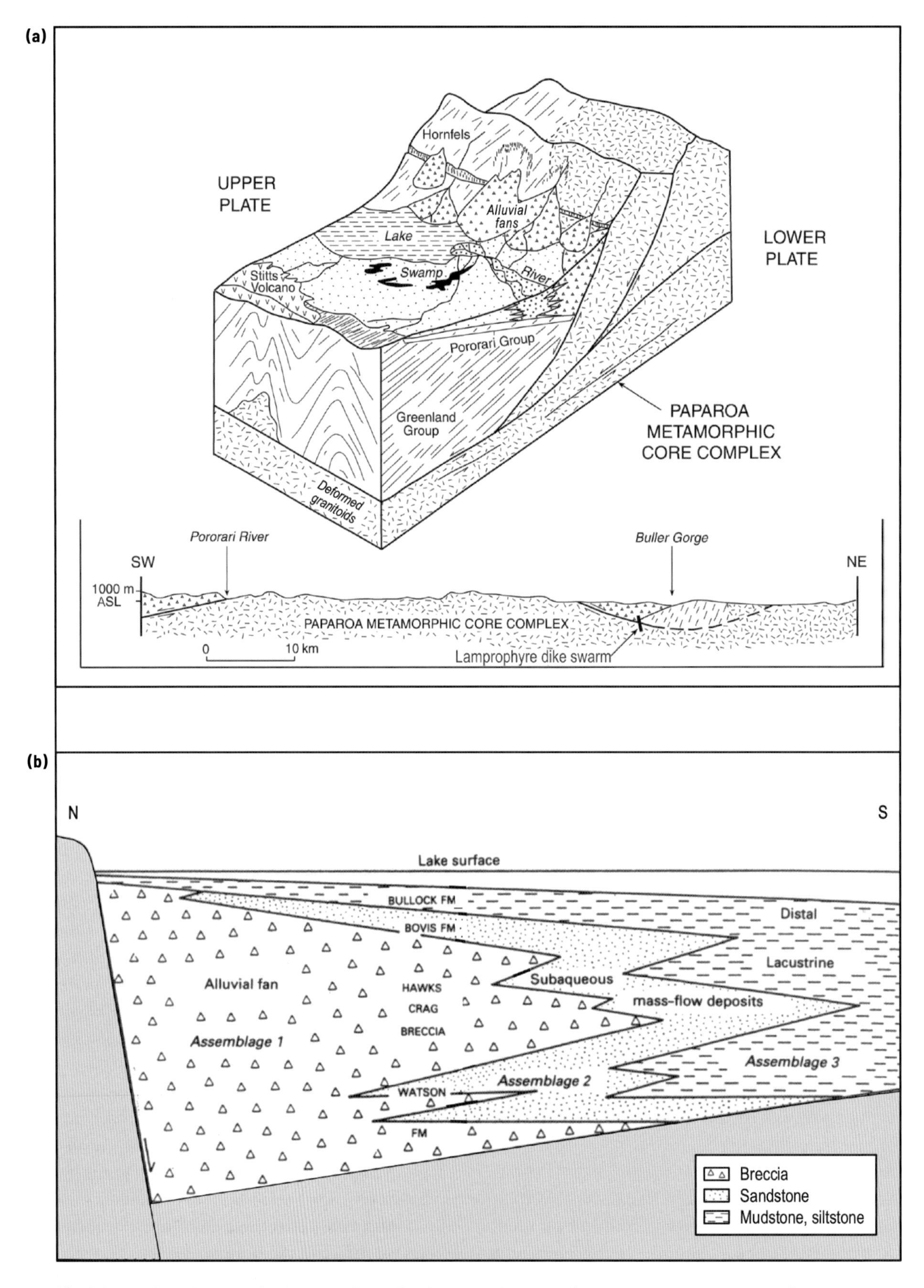

Fig. 4.4. (a) Block diagram of sedimentary basin development adjacent to the rising Cretaceous metamorphic core complex. ASL, above sea level. (From Laird & Bradshaw, 2004, fig. 4.) (b) Schematic section of the formations and facies relationships within the Pororari Group. FM, Formation. (Adapted from Laird, 1988.)

assemblages interfinger, and are considered to represent elements of a fan-delta system debouching into a lake (Fig. 4.4b). This fan-delta to lacustrine system is inferred to have infilled an actively subsiding half graben. The paleocurrent pattern and direction of thinning of units indicate sediment transport from the north-northeast and downthrow of the bounding fault to the south-southwest, consistent with the structural model.

In the Buller Gorge Basin the basal Pororari Group comprises 60 m of vitric tuffaceous rock, dark carbonaceous shale, and minor feldspathic sandstone of overlain concordantly at a sharp or erosive contact by up to 600 m of non-volcanogenic feldspathic sandstone, dark carbonaceous shale with local coal lenses, and conglomerate consisting of rounded clasts of basement rocks. These are thought to be deposits of floodplain with scattered shallow lakes and swamps, and are succeeded by clast and matrix-supported coarse Hawks Crag Breccia that may exceed 4500 m in thickness (Nathan, 1978c). As in the south, the breccia facies was probably deposited adjacent to active normal faults (see Plate 6).

Outcrops included in the Pororari Group also occur near Reefton and near Hokitika. Near Paringa, ~250 km to the southwest of the Paparoa Range, correlatives of the Pororari Group are represented by the Otumotu Formation (Nathan, 1977). The formation is at least 520 m thick, comprising a lower part of thick massive or imbricated, green and grey, angular-to-subrounded, boulder-to-pebble breccia and conglomerate that passes up into conglomerate interbedded with cross-bedded sandstone. These beds are inferred to be alluvial fan deposits dominated by debris flows in the lower part passing up into mixed finer debris flow, sheet flood and fluvial deposits (Adams, 1987). The gradationally overlying upper part of the succession is centimetre-bedded, moderately sorted cross-bedded medium and fine sandstones interbedded with laminated black carbonaceous mudstones with dolomite cement. These deposits were inferred to represent a change to semi-arid fluvial and lacustrine conditions (Adams, 1987). This upper portion has yielded pollen of Cenomanian to early Turonian (late Motuan to Arowhanan) age (Raine, 1984). Another possible correlative of the Pororari Group, the Beebys Conglomerate, lying approximately 100 km northeast of the Paparoa Range, has yielded a poorly constrained microfloral age in the range of late Cenomanian to early Santonian (Johnston, 1990), indicating that non-marine Pororari-like deposition may have extended locally well into the Late Cretaceous.

Deposits of Clarence Series age are absent from the south Taranaki Basin, but strata of Cenomanian (Ngaterian) age occupy a narrow strip in the northeast of the northern Taranaki Basin adjacent to and west of the Taranaki Fault. The only relevant petroleum exploration well, Te Ranga-1, was drilled 10 km south-southwest of Kawhia Harbour in the northeastern part of the Taranaki Basin (Higgs, 2007). Te Ranga-1 intersected a unit of coal measures (Taniwha Formation), comprising intercalated sandstone, siltstone, carbonaceous mudstone and coal (King & Thrasher, 1996). About 2400 m of undrilled strata are present below the well and above acoustic basement (Thrasher, 1992). The seismic expression of the Taniwha Formation shows that the unit continues as a narrow strip northwards from the vicinity of the Te Ranga well to just north of the Manakau Harbour before the trace becomes difficult to identify beneath Miocene volcanoes (Stagpoole, 1997). The unit is thickest against the Taranaki Fault, which truncates it, and it thins to the west.

The Deepwater Taranaki Basin, which extends offshore to the northwest from the north Taranaki continental shelf along the axis of the New Caledonia Basin, is inferred from seismic surveys to contain a northwestwards-prograding Cretaceous succession. It is interpreted as a delta system approximately 2000 m thick and 100 km long (Uruski & Baillie, 2002). The prograding sequence appears to underlie the latest Cretaceous marine-dominated section recognised in several wells at the shallower southeastern end of the New Caledonia Basin, and may be a correlative of mid- to Late Cretaceous successions in the South Island.

The presence of mid- to Late Cretaceous sediments in the offshore Northland Basin is still conjectural. Although such deposits had been inferred from seismic profiles to be present in half grabens (Gage & Kurata, 1996), the only petroleum well drilled for which information is available (Waka Nui-1) showed that Paleocene conglomerates rested unconformably on basement Murihiku Supergroup rocks of Middle Jurassic age (Milne & Quick, 1999).

4.2.1.2 Non-marine sedimentary successions of southern and eastern South Island basins

Non-marine equivalents of the Pororari Group also occur in southwest Fiordland and in offshore basins further to the south. Similar rocks occur onshore in eastern Otago and central Canterbury, and offshore in the Great South Basin. All sediments fill fault-controlled basins.

In Fiordland, Early Cretaceous sedimentary rocks of the Clarence Series are assigned to the Puysegur Group, which is exposed only along a 20 km-wide strip of the coast between Chalky Inlet and Gates Harbour (Turnbull *et al.* 1993) on the southwest extremity of Fiordland (see Fig. 1.3). The group is inferred from seismic evidence to be more extensive offshore in the Balleny Basin. Palynological dating shows that it is late Albian to early Cenomanian (Motuan to Ngaterian) in age. Outcrops of Puysegur Group rest unconformably on fresh to slightly weathered basement of granite and metasedimentary rocks, and are separated by a regional unconformity from overlying Cenozoic sequences (Turnbull *et al.* 1993). The lower part of the Puysegur Group consists of up to 200 m of interbedded granule-to-pebble conglomerates, cross-stratified sandstone, and variably carbonaceous mudstone containing thin bituminous coal lenses and rare rootlet horizons (Seek Cove Formation). Sedimentary features, including fining-upwards cycles, and the presence of plant fossils indicate deposition in a floodplain environment. Conglomerates and sandstones in the Seek Cove Formation are distinguished by their high proportion of volcanic detritus (70–90%), made up mainly of porphyritic rhyolite, andesite and possibly ignimbrite (Pocknall & Lindqvist, 1988). The volcanics have not yet been matched to a source but may well be part of the widespread Cretaceous post-subduction suite (see Tulloch *et al.* 2009b).

South of Preservation Inlet almost all of the exposed Cretaceous strata consist of a varied succession of lacustrine sediments (Windsor Formation) that is at least 1250 m thick. It is predominantly granite-derived, and is inferred to have formed in a lacustrine pro-delta fan and channel system. The contact with the underlying Seek Cove Formation is marked by a change from volcanogenic alluvial sediments to medium to thick parallel-bedded and graded-bedded feldspathic sandstone and sandstone with boulders. The abrupt change in composition suggests a change in provenance and unstable conditions. Paleocurrent data from the Puysegur Group are sparse and conflicting: cross-bedding in shallow-water strata near the top of the sequence indicates transport from southwest to northeast, while underlying pro-delta fan sediments contain slump folds suggesting a southwest-facing paleoslope, and sole markings indicate a south to southwest turbidity current flow.

Mid-Cretaceous sediments, reaching a maximum thickness of 3 km, are inferred to underlie portions of the offshore sector of the Balleny Basin (see Fig. 1.3). The Cretaceous succession overlies basement blocks that are normally faulted down to the west or northwest (Turnbull *et al.* 1993). Mid-Cretaceous rocks are also inferred to be present in the Hautere Sub-basin and the Waitutu Sub-basin within the adjacent Solander Basin (see Fig. 1.3). In the former sub-basin the thickness rarely exceeds 1 km, while in the Waitutu Sub-basin the thickness of the inferred Cretaceous sequence exceeds 2.5 km in places. The sediments fill a basin approximately 20 km wide, trending southwest and extending at least 130 km, beyond the region covered by seismic data.

In Otago, movement on the northwest–southeast-trending Waihemo Fault, active in mid- and Late Cretaceous times, produced at least two fault-angle depressions. The Kyeburn Basin differs from others examined, by being controlled by two orthogonal normal faults: the northwest–southeast-trending Waihemo Fault, and the orthogonal Danseys Pass Fault. The basin contains approximately 4000 m of proximal-alluvial-fan conglomerate and breccia, floodplain sandstone and lacustrine sediments (Kyeburn Formation) (Bishop & Laird, 1976). Paleocurrent directions and lithofacies variations show a close relationship to the basin-bounding faults. Silicic tuff horizons low in the Kyeburn succession have U–Pb ages of 112 Ma (Tulloch *et al.* 2009b), while pollen assemblages suggest a late Cenomanian or early Turonian (Arowhanan) age for the sediments (Bishop & Laird, 1976). No erosional break is evident within the succession, which appears to represent continuous deposition from Albian to early Turonian times. Pebble clasts of Torlesse sandstone are present only in the older half of the unit, being replaced progressively by schist clasts up-section. This indicates unroofing of the Haast Schist in this area by at least early Turonian times.

At the coastal end of the Waihemo Fault the Shag Point Group consists of similar non-marine conglomeratic deposits. Clasts are dominantly of schist, but the metamorphic grade of the schist increases up-sequence, suggesting significant uplift and unroofing on the southwest side of the fault (Mitchell *et al.* 2009). Pollen samples from the sediments (Raine, 1994), and U–Pb age determinations for an interbedded ignimbrite (Tulloch *et al.* 2009b), indicate that the Shag Point Group is coeval with the Kyeburn Formation.

Torlesse-derived conglomerate (Hewson Formation) of proximal alluvial braidplain to distal alluvial fan facies occurs in western Canterbury at Mount Somers. The base has not been seen, and only the top 65 m is exposed (Field *et al.* 1989). The sediments are overlain by the igneous rocks of the Mount Somers Volcanics Group (see section 4.4.1), which has been dated at between 94.5 Ma and 97 Ma (Tappenden *et al.* 2002), giving the sediments a minimum stratigraphic age of Cenomanian (Ngaterian).

The only offshore non-marine basin for which significant information is available is the Great South Basin (Fig. 4.5; see also Fig. 1.3 and Plate 9), and its northern extension into south Canterbury. The Great South Basin comprises a mainly northeast-trending system of grabens and half grabens that were initiated in the mid-Cretaceous (Beggs, 1993; Cook *et al.* 1999). Of eight petroleum exploration wells in the basin, four (possibly five) intersected a basal graben-fill succession – the Hoiho Group (Cook *et al.* 1999) of Cenomanian (Ngaterian) or older to early Santonian (Piripauan) age (Figs 4.5 and 4.6). All well intersections in the Hoiho Sequence comprise predominantly sandy coal measures, although in two wells close to bounding faults (Tara-1 and Hoiho-1C), conglomerate is also present. Regional seismic mapping shows that the coal-measure sandstones have a discontinuous and wedge-shaped geometry, suggesting fluvial valley fill, and adjacent conglomerates are associated with high-amplitude, steeply dipping reflectors interpreted as fault-scarp fans along the graben margins. The overlying late Cenomanian–Coniacian (Arowhanan–Teratan) non-marine sediments include low-amplitude, low-continuity or discontinuous reflectors interpreted as representing lacustrine facies (Cook *et al.* 1999). Lithofacies relationships show similarities to those in the onshore Kyeburn Formation.

The extension of the Great South Basin into south Canterbury includes two sub-basins: the Clipper Basin and the Caroline Basin. They contain sedimentary deposits, of probable late Cenomanian age, underlying a Late Cretaceous seismic horizon at the base of the terrestrial late Cenomanian–Coniacian (Arowhanan–Teratan) Clipper Formation (Field *et al.* 1989). In the northern part of the Clipper Basin these sediments are up to 1000 m thick, and, although not drilled, their seismic character suggests that they are coarse graben-fill deposits (Mound & Pratt, 1984), probably analogous to those of the Waihemo Fault depocentres. The oldest sediments in the Caroline Basin are inferred to be similar, and other occurrences are likely in smaller pockets scattered on the eastern part of the Canterbury Bight High and east of the Zapata High, although none of these basins have been drilled (Field *et al.* 1989).

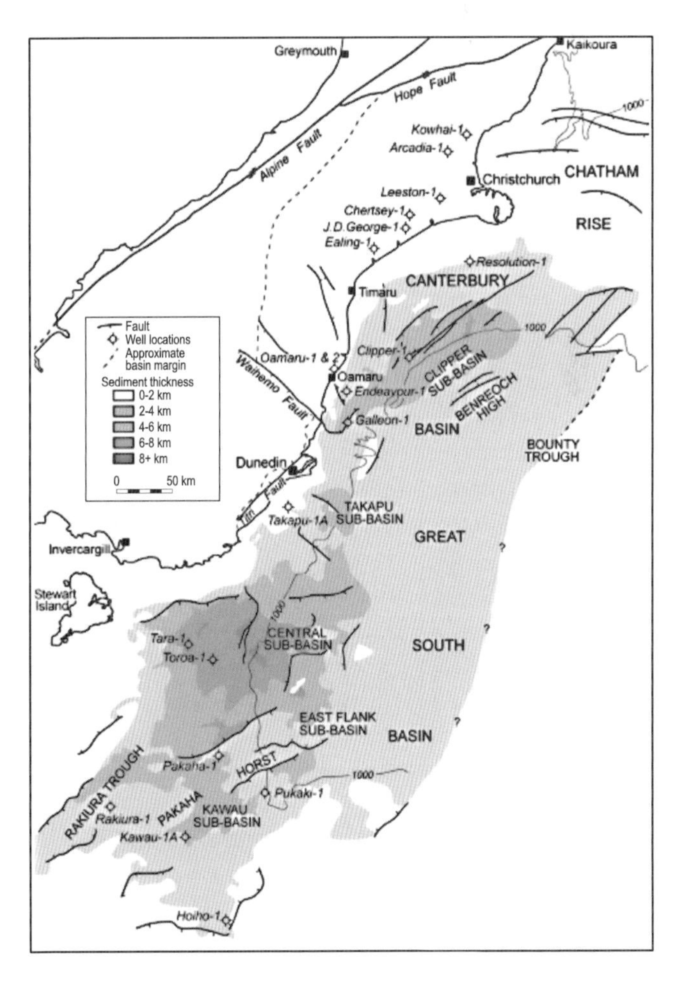

Fig. 4.5. Map of the Great South and Canterbury basins showing approximate sedimentary thicknesses and the locations of exploration drillholes. (Adapted from Cook *et al.* 1999.)

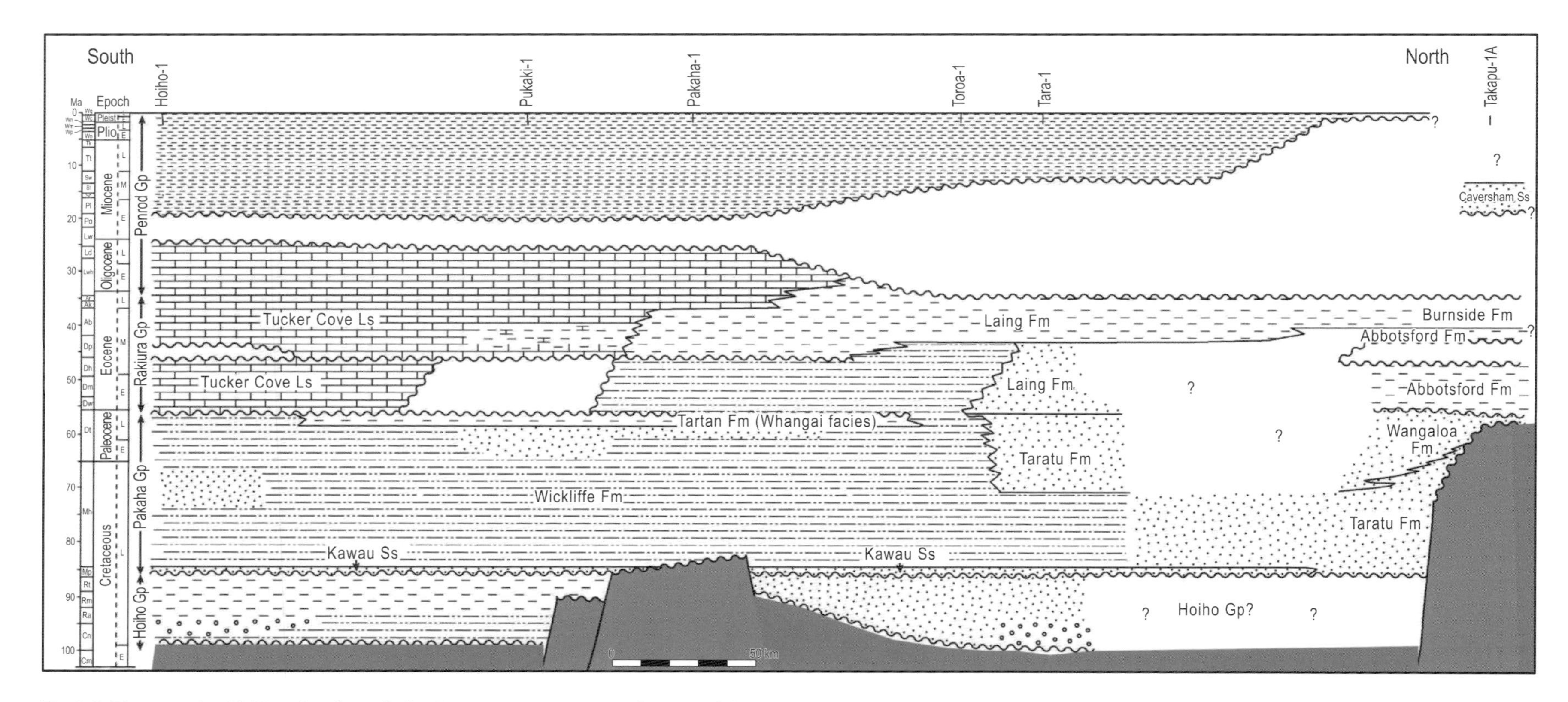

Fig. 4.6. Time-stratigraphic section through the Cretaceous–Cenozoic sediments of the Great South Basin. The formations are defined by drillhole data. Details can be found in section 4.2.1.2. Fm, Formation; Gp, Group; Ls, Limestone; Ss, Sandstone. (Adapted from Cook *et al.* 1999.)

4.2.2 Marine sedimentary successions of eastern New Zealand

4.2.2.1 Clarence Series sedimentary rocks of onshore Northland

Autochthonous successions

The only known area of mid-Cretaceous autochthonous rocks forming part of the cover strata onshore is at Whatuwhiwhi, Karikari Peninsula. Houhora Complex rocks of the Mount Camel Suspect Terrane are there unconformably overlain by an upwards-fining breccia to mudstone sequence 0–270 m thick (Whatuwhiwhi Formation) (Isaac *et al.* 1988). It consists of unbedded to poorly stratified basal breccia and conglomerate, derived mainly from the underlying Houhora Complex rocks, which pass upwards into intercalated breccia, conglomerate and sandstone. Conglomerate pebbles, cobbles and boulders are commonly well-rounded, suggesting fluvial transport. Stratigraphically higher sandstone and thin-bedded sandstone and mudstone are succeeded by massive mudstone. The rare occurrence of *Inoceramus* prisms indicates that the formation was deposited in a marine environment.

The breccia and conglomerate are interpreted as submarine rockfall and debris flow deposits, and sedimentary structures of the sandstone facies are consistent with turbidity current deposition. The upward transition from coarse-grained to fine-grained facies could reflect either decreased tectonism in the source area, or increased depth of deposition. No datable fossils have been recovered from the Whatuwhiwhi Formation. However, a U–Pb radiometric age of ~102 Ma was obtained for a silicic pyroclastic flow within the unit, indicating a late Albian (Motuan) age (Tulloch *et al.* 2009b; Adams *et al.* 2013b).

Allochthon rocks

During latest Oligocene to earliest Miocene times, a series of nappes was emplaced from the northeast and now overlies much of onshore and offshore Northland, forming the Northland Allochthon (see section 8.2.1). Two of the nappes, the Tupou Complex and the Tangihua Complex, include sediments containing fossils of probable mid-Cretaceous age. The sediments are inferred to have been originally deposited under marine conditions to the northeast of the present coastline (Isaac *et al.* 1994). The Tangihua Complex is dominated by volcanic rocks, and is discussed in section 4.4.1 (see Figs 4.7, 7.1 and 7.2).

Rocks of the Tupou Complex have been recognised only from the Whangaroa Harbour area, and remain poorly known. They are indurated, commonly veined, jointed and sheared. The major lithologies – unstratified to poorly stratified pebble conglomerate and pebbly sandstone – are in beds up to at least 3 m thick. Apparently massive, structureless, fine- to coarse-grained lithic sandstone, in beds or composite beds metres to tens of metres thick, and alternating centimetre- to decimetre-bedded sandstone and mudstone, are commonly intercalated with conglomerates.

Tupou conglomerates and sandstones are inferred to have been deposited by sediment gravity-flow mechanisms into a deep marine (outer-shelf to bathyal) basin. Conglomerate clasts are mainly sandstone, mudstone, vein quartz and siliceous volcanics, suggesting derivation from a landmass of Waipapa Terrane and Mount Camel Suspect Terrane rocks. A single collection of *Aucellina euglypha* from a concretion gives a minimum late Albian (Motuan) age for the Tupou Complex. The complex is overlain unconformably by the Late Cretaceous (Santonian/Piripauan) Mangakahia Complex near Whangaroa. The similarity of lithofacies and ages of the Tupou Complex and Mount Camel Suspect Terrane rocks suggests a possible relationship.

Although now well-separated geographically the East Coast Allochthon is a continuation of the same tectonic belt (Cluzel *et al.* 2010). It too contains Cretaceous sediments and these show strong similarities to those of the Eastern Sub-belt of the East Coast Basin (see below).

4.2.2.2 Clarence Series sedimentary rocks of the East Coast Basin

The Cretaceous 'cover' strata of the East Coast region are present in three geographically discrete areas – Raukumara Peninsula, southern Hawke's Bay and Wairarapa, and Marlborough – that are now separated by areas with no Cretaceous outcrop. They have been subdivided into two distinct parts: the Western and Eastern Sub-belts (Moore, 1988; and section 4.1.5).

Western Sub-belt: basal coarse facies

A common mid-Cretaceous stratigraphy for the Western Sub-belt can be recognised from the Raukumara Peninsula to Marlborough, although there are local differences. Mid- to late Albian (Urutawan) strata have been definitely recognised only in the Raukumara Peninsula and in the Coverham area of the Clarence Valley, Marlborough (Crampton *et al.* 2004b), although the original distribution was likely to have been greater (Figs 4.1b and 4.3).

In the Raukumara Peninsula, mid- to late Albian (Urutawan) strata rest with marked unconformity on basement Torlesse rocks including the late Aptian Koranga Formation (see discussion in section 4.1.3). The basal sediments are coarsest in the southwest. In the Koranga area the lowest 67 m of cover strata consist of conglomerate, breccia and coarse- to medium-grained sandstone (Te Wera Formation) (Speden, 1975). Here, the unconformity surface is irregular, with a relief of at least 1 m. A basal coarse facies, which consists of breccia–conglomerate comprising angular blocks of indurated sandstone up to 0.5 m in diameter and well-rounded pebbles up to 1 cm in diameter, passes up gradationally into a well-bedded succession of fossiliferous metre-bedded coarse- to medium-grained sandstones. Most sandstone beds are sharp-based, and show poor grading with common parallel lamination, and a gradational passage upwards into overlying mudstone. Other beds include isolated, sometimes vertically elongated, rafts of mudstone at their base. Channelling is common throughout the sandstone succession, with local channel complexes reaching 2 m in depth. One kilometre to the east of the type locality, siltstone fills a channel at least 15 m deep that is cut into the sandstone succession (Speden, 1975). The basal breccia–conglomerate resting on the unconformity surface is inferred to represent debris flow deposits, and is overlain by thick graded proximal turbidites. Although the succession has previously been interpreted as a transgressive inshore environment (Speden, 1975; Mazengarb & Speden, 2000), the characteristics listed above, including the common presence of channelling, suggest deposition in the lower portion of a submarine canyon or the upper portion of a submarine fan, in a moderately deep water environment (Laird & Bradshaw, 1996). The macrofossils present are dominated by epifaunal bivalves, which include *Maccoyella* n. sp., *Aucellina* cf. *gryphaeoides*, and *Pseudolimea* cf. *echinata*, and are probably transported. They indicate a mid- to late Albian (Urutawan) age for the basal sediments.

Conglomerate clasts in the Te Wera Formation indicate a dominantly acid to intermediate igneous terrain as source, with a secondary source of basement Torlesse rocks. The presence of partly altered volcanic glass shards in conglomerate horizons (Challis, in Speden, 1975) and of tuff horizons (Mazengarb *et al.* 1991) and young detrital zircons (Adams *et al.* 2013b) support the existence of a volcanic arc in the region. Paleocurrent determinations are variable, but generally between north and east.

To the northwest and north of Koranga, geographically isolated coarse basal facies potentially equivalent to the Urutawan Te Wera Formation include the Waimana Sandstone and Oponae Mélange. The Waimana Sandstone caps the range between the Waimana and Waiotahi rivers, south-southeast of Whakatane (Speden, 1975). The poorly exposed unit, which reaches a thickness of 120 m, consists of fine- to medium-grained sandstone with some interbedded siltstone, lenses of conglomerate and breccia, especially near the base, and scattered fossils. Fine parallel lamination is relatively common, and large-scale cross-bedding also occurs near the base. Deposition at shallow depth on the inner shelf was inferred. No diagnostic macrofossils have been collected from the Waimana Sandstone, but dinoflagellates indicate that the age is late Albian (Field *et al.* 1997).

Oponae Mélange (Fig. 4.1a), known only from near Oponae, in the Waioeka Gorge north of Koranga, is interpreted as an olistostrome unconformably overlying basement (Mazengarb, 1993). Thickness of the olistostrome exceeds 200 m near Oponae, but the thickness varies laterally, and the formation is absent in some localities. It consists of angular blocks and rounded pebbles of sandstone, mudstone, coal, bedded chert, marble, basalt and other igneous pebbles enclosed in a sheared and folded mudstone or very fine-grained sandstone matrix. Macrofossils are present, but rare. Outsize blocks of chert, up to 50 m thick and possibly over 100 m long, occur near the inferred base of the mélange (Feary, 1979). The origin of the Oponae olistostrome has not been determined, but the presence of scattered marine macrofossils indicates a marine environment probably similar to the basal

Te Wera Formation. The age is, at least in part, late Albian (Motuan).

In the Koranga and Oponae areas the basal coarse deposits are overlain, and locally interfinger, with fine-grained sediments of the Karekare Formation (Mazengarb & Speden, 2000). To the northeast, Karekare Formation mudstone, locally with sandstone bands, directly overlies Torlesse rocks. The basal or near-basal sediments commonly consist of mudstone with slump folding, other soft-sediment deformation and sedimentary breccia. At Motu Falls ~1400 m of siltstone-dominated Karekare Formation of late Albian (Urutawan and Motuan) age is present, with tuff layers in the upper 300 m (Motuan portion) of the succession. Zircons from a 5-m-thick tuff with well-preserved glass shards gave a date of 101.6±0.2 Ma (Crampton *et al.* 2004b; Tulloch *et al.* 2009b). The easternmost exposures of basement contact occur in Kokopumatara Stream, 50 km northeast of Koranga (see Plate 7) where it rests unconformably on basement (see discussion in section 4.1.3.2 and Laird & Bradshaw, 1996). The basal rocks are an olistostrome, which is up to 90 m thick and dominated by clasts of siltstone and very fine-grained sandstone, some showing soft-sediment deformation. The age of the basal sediments is probably Urutawan, and inferred to be a lateral equivalent of the Te Wera Formation. Prominent intercalations of thinly bedded sandstone alternating with mudstone higher in the succession are inferred to represent interfingering elements of a Waitahaia Formation that occurs in an extensive block to the northeast. The Waitahaia Formation (Mazengarb & Speden, 2000) is 1400 m thick and comprises a succession of alternating graded-bedded sandstones and mudstones, with minor occurrences of slump packets, conglomerate and tuff. The Waitahaia Formation is interpreted as a major submarine fan complex dominated by turbidite deposits. A sedimentary contact with basement has not been seen, but the age of the exposed sediments is late Albian to Cenomanian (Motuan to Ngaterian). Numerous flute casts in the Waitahaia Formation indicate paleocurrent flows to the north and northwest, and locally towards the west (Mazengarb, 1993).

In the Wairarapa–southern Hawke's Bay area coarse basal deposits (Gentle Annie Formation, Mangapurupuru Group (Fig. 4.1b) in the Western Sub-belt crop out sporadically from the White Rocks–Tora area in the south to the Whangai Range in the north. They consist of massive olistostromes with minor intervals of moderately fossiliferous, well-bedded sandstone–mudstone and massive mudstone (Crampton, 1989). The olistostromes include very poorly sorted pebbly sandy mudstone, pebbly silty sandstone, muddy sandy matrix-supported conglomerate, and minor clast-supported conglomerate. Clasts are angular to rounded, and comprise mainly indurated sandstone, with minor mudstone and igneous material. Most clasts are pebble-sized or smaller, though blocks up to 10 m across are common, and megaclasts range up to 100 m in length (Crampton, 1997; Field *et al.* 1997). Intervals of centimetre- to metre-bedded sandstone and mudstone, and massive mudstone, are generally less than 50 m thick, and are laterally discontinuous. Thin tuffaceous beds occur locally. Thickness of the Gentle Annie Formation varies throughout the region, but reaches 700 m in the White Rocks–Tora area in the south. It is overlain with apparent conformity by a succession consisting dominantly of mudstone: the Springhill Formation. A late Albian (Motuan) age is indicated from a number of fossil localities (Crampton, 1989). There is little direct evidence for depth of deposition. Fossils of shallow-water aspect are present, but are likely to have been transported to deeper water by mass-flow mechanisms. The Gentle Annie olistostromes are inferred to have been deposited by a number of discrete flows adjacent to a fault-controlled basin margin (Crampton, 1989). Sedimentary clasts were probably derived largely from Torlesse rocks (Pahau Subterrane) and from coeval Mangapurupuru Group sediments.

In Marlborough, rocks of mid- to late Albian (Urutawan) age (Fig. 4.1b), have so far been recognised with certainty only at Coverham in the Clarence Valley (Crampton *et al.* 2004b). The fossil *Mytiloides ipuanus* is known to occur at several other localities, however, and deposits of this age may prove to be more extensive than currently mapped. At Coverham the Urutawan strata coincide approximately with a ~200-m-thick informal unit (Champagne Formation of Ritchie, 1986), exposed in Ouse Stream. The rocks consist of basal thick conglomerate beds overlain by packets of fossiliferous sandstone and mudstone showing strong structural and soft-sediment deformation. They rest with marked angular unconformity on

Pahau Subterrane basement, and are in turn overlain unconformably by over 500 m of silty mudstone and subordinate centimetre to decimetre sandstone of the Split Rock Formation of late Albian (Motuan) age (Crampton, 1998). This formation is unusually fine-grained. In the middle Awatere River and its tributary Penk River, to the north of Coverham, alternating sandstones and mudstones, previously mapped as basement Pahau Subterrane, show soft sediment deformation similar to that of the 'Champagne Formation' (see Plate 4b), and contain fossils with affinities to *M. ipuanus* (Laird & Bradshaw, 1996). They are tentatively correlated with the Urutawan Champagne Formation and are overlain unconformably by sediments of Motuan age.

Elsewhere in Marlborough the basal deposits are variable in lithology, but in almost all cases are coarse-grained, in many instances consisting of clast- or matrix-supported conglomerate containing shelly fossils overlying channelled Torlesse Terrane (Laird, 1992). Some of these channels are between 500 m and 800 m deep and may be submarine canyons (Montague, 1981).

Throughout the region, basal coarse-grained deposits (Coverham Group, Split Rock Formation and correlatives) fine upwards, first into turbidite successions, commonly hundreds of metres thick, and then into massive or laminated mudstones. In the Awatere Valley a succession of synsedimentary faults, 2–5 km apart, defines a series of distinct half grabens (Laird & Lewis, 1986). These half grabens are filled with coarse, late Albian–early Cenomanian (Motuan–early Ngaterian) marine sediments, locally in excess of 600 m in thickness, of mainly sediment gravity flow origin. In one instance the base of an inferred half graben infill consists of an olistostrome 240 m thick and including clasts up to 20 m in diameter, overlying channelled Torlesse rocks. It may represent the fill of a submarine canyon (see Plate 4b).

Depth and paleoenvironment for much of the basal coarse successions in Marlborough are uncertain, mainly because of their redeposited nature. In the case of the thick olistostrome that may have been emplaced in a single depositional episode, bathyal depths appear to have been reached at least locally. By contrast, in the adjacent basin, common ripple marks and some possible wave-induced ripples suggest deposition above at least storm wave base (Laird, 1992). Possible shallow marine deposits have so far been recognised only in a drillhole in the Hapuku Valley, north of Kaikoura (Laird, 1982). To the south of this locality no Motuan rocks are known and a basin margin is inferred. The position of the western margin is unknown, and the original relative position of the sections described above is likely to have been very different in the Cretaceous (Crampton *et al.* 2003).

A common characteristic of all the late Albian–early Cenomanian (Motuan–early Ngaterian) successions in Marlborough is the marked fining-upwards trend throughout. All successions start with olistostromes, debris flows or other coarse deposits on an unconformity, and pass upwards gradationally through turbidites into mudstone, and mudstone becomes a widespread blanketing lithology (see Fig. 4.1 and next section).

As a result of late Cenozoic dislocation, paleocurrent data and isopachs from Marlborough appear inconsistent. They do, however, contribute to a palinspastic reconstruction of the Marlborough region that suggests a basin opening towards the northeast (present coordinates), as discussed by Crampton *et al.* (2003).

Western Sub-belt: mudstone-dominated facies

Throughout the Western Sub-belt of the East Coast Basin, coarse-grained sediments of Albian (Urutawan–Motuan) age commonly pass gradationally but rapidly upwards into mudstone of early Cenomanian (latest Motuan and early Ngaterian) age. The mudstones may be massive or finely laminated, and locally include thin centimetre-bedded, very fine-grained sandstone layers, inferred to represent distal turbidites. This final mudstone phase is interpreted to reflect a cessation of tectonic activity on the faulted margins followed by a quiet infilling of the basins by sediment from suspension. Alternatively, it may result from a relative rise in sea level.

In the Raukumara Peninsula area the mudstone succession (Karekare Formation) conformably overlies the Te Wera Formation, Waimana Sandstone and Oponae Mélange (Fig. 4.1a). Locally, in the Oponae area, mudstone rests unconformably on basement where the Oponae Mélange is absent. The mudstone succession appears to be continuous, without significant stratigraphic breaks, and ranges in age from Albian to Santonian (Mazengarb & Speden, 2000), reaching a thickness in excess of 2600 m. The dominant lithology is massive or laminated poorly bedded mudstone. Thin packets of alternating sandstone and mudstone, often graded and showing Bouma sequences, are well developed, particularly towards the east in the upper part of the Kokopumatara Stream. Minor developments of coloured mudstone, tuff and conglomerate also occur. Macrofossils, mostly *Inoceramus* species, are locally abundant. Further east the Karekare Formation overlies and onlaps the Waitahaia Formation turbidite fan. Paleocurrent directions determined from turbidites within the mudstones suggest flow to the north or northeast (Mazengarb, 1993).

Heavy mineral analysis suggests that the mudstone is mainly Torlesse-derived, with a minor primary igneous source (Smale & Laird, 1995). Foraminiferal evidence suggests that the depositional environment ranged from shallow-shelf depths through to bathyal.

In the Hawke's Bay to Wairarapa area the basal coarse facies of the Gentle Annie Formation grades vertically and laterally into a succession (Springhill Formation; Fig 4.1b) consisting dominantly of massive mudstone, with lesser alternating sandstone–mudstone and siltstone. Conglomerate, thick-bedded sandstone and tuff occur locally at the base. Estimates of thickness range from 750 m in the Whangai Range area to between 650 m and approximately 2200 m in the Tinui area west of Castlepoint. The Springhill Formation is largely of late Albian to early Cenomanian (Motuan–Ngaterian) age. Interbedded glauconitic grit, sandstone and mudstone near Tinui, of late Cenomanian to early Turonian (Arowhanan) age, represent the only known exception (Crampton, 1997).

The few paleocurrent directions gathered from the Springhill Formation indicate west-flowing to north-flowing sediment transport (Crampton, 1997). Heavy mineral data are consistent with derivation of the sediments from an indurated sedimentary provenance, such as the underlying Pahau Subterrane rocks.

There is little direct evidence for depth of deposition for the mudstone of the Springhill Formation. Fossils of shallow-water aspect are present, but may have been transported by debris flows to deeper water. Crampton (1997) noted that the formation displays a broadly fining-upwards trend, and coarse-grained turbidites are confined largely to lower parts of the unit, while upper parts are composed of massive or thick-bedded mudstone with only minor proportions of fine-grained turbidites. He inferred that lower parts of the unit were deposited at outermost-shelf or slope depths, while the upper parts record probable shallowing to mid- or outer-shelf depths.

In Marlborough also, the basal coarse facies passes upwards with rapid transition into mudstone-dominated successions (Swale Siltstone and correlatives) (Field *et al.* 1997) of mainly early Cenomanian (early Ngaterian) age. Mudstones are massive or laminated, with scattered thin packets of alternating sandstone and mudstone. Thickness of the mudstone facies reaches 700 m at Coverham, and possibly up to 1000 m in the upper Awatere Valley (Montague, 1981).

Paleocurrent data are sparse, but flutes on the base of intercalated turbidites in mudstones in Mororimu Stream (southwest of Clarence River mouth) show flow from the southeast towards the northwest. Depth of deposition is likely to be similar (outer-shelf to bathyal) to that of the correlative and similar Springhill and Karekare formations in the North Island.

Western Sub-belt: Cenomanian (Ngaterian) non-marine to shallow marine facies

In western and southern Marlborough the fill of the late Albian–Cenomanian half grabens was followed by deposition of marine and non-marine conglomerates, sandstones and mudstones, and by the eruption of extensive alkaline basaltic lavas and related dikes that heralded a new depositional cycle represented by the Wallow Group. In southern Marlborough these younger sediments and volcanics, of Cenomanian (late Ngaterian) age, rest with angular unconformity both on older Albian graben-fill deposits of the Coverham Group and on Torlesse rocks in the west. They are, however, conformable with sediments of the Coverham Group in the east (Laird, 1992). Heavy mineral studies indicate that significant compaction of the Coverham Group took place before the deposition of the Wallow Group and this is thought to indicate a tectonic event that also produced local unconformity between the groups (Smale, 1993; Smale & Laird, 1995; and see Fig. 4.1).

The basal Warder Formation of the Wallow Group comprises sandstone, minor conglomerate and carbonaceous horizons and occurs in the extreme west and southwest of the Clarence Valley. Locally, the sediments are arranged in a succession of fining-upwards rhythms, inferred to represent the deposits of a meandering stream (Browne & Reay, 1993). The non-marine beds pass laterally to the northeast into metre-bedded sandstones with marine shelly fossils (Bluff Sandstone Formation). Horizons of conglomerate occur locally, particularly in the west and southwest, while thin units of alternating sandstone and mudstone occur with increasing frequency and thickness towards the northeast. The metre-bedded sandstone is typically fine- to very fine-grained, and indistinctly parallel-laminated. Pyrite nodules are locally common, and there are rare shelly and burrowed horizons. Hummocky cross-stratification has been recognised in southern outcrops, inferred to be proximal to the paleo-shoreline (Crampton, 1988). Thickening upward successions, totalling up to 600 m and consisting of basal massive mudstone passing upwards into laminated mudstone then into alternating sandstone and then into massive sandstone typical of the Bluff Sandstone, occur southwest of Muzzle Stream in the Clarence Valley, and in the Hapuku River (see Plate 8). They are interpreted as representing part of a prograding highstand systems tract. The group thins towards the east and northeast, and becomes dominated by mudstone, commonly with intercalated packets of graded-bedded sandstone with common sole structures interpreted as turbidites.

The locally developed unconformable surface separating the Wallow Group from older rocks has been interpreted as a globally important sequence boundary (Wilgus *et al.* 1988), resulting from a relative sea-level fall and seaward facies shift. This particular unconformity has, however, not been recognised outside Marlborough, and relative sea-level fall is inferred to have resulted from uplift of the Albian marine Split Rock Formation as a result of intrusion of, and doming by, magmas of the Tapuaenuku Igneous Complex in the Marlborough area at ~96 Ma (Baker & Seward, 1996; and see section 4.4.1.3).

The clastic sediments, both marine and non-marine, reach a maximum thickness of 600 m. In most localities in the Clarence Valley they pass upwards into, or are interbedded with, alkaline basalts and tuffs (Gridiron Volcanics Formation, in the Clarence Valley; Lookout Volcanics Formation, in the Awatere Valley; see Plate 8 and section 4.4.1.3). In the Awatere Valley a slight angular unconformity separates Bluff Sandstone from flows of the Lookout Volcanics Formation. Dikes and sills of similar composition to the lavas intrude the underlying Cretaceous strata.

Paleocurrent data are sparse in the Wallow Group, and are restricted mainly to the Warder Formation and to distal turbidites of the Bluff Sandstone. Trough cross-beds in the Warder Formation indicate a dominantly easterly flow direction, which is generally consistent with the orientation of carbonised logs and branches. To the northeast, in Bluff Stream, orientation of symmetrical ripple crests is north-northwest, perhaps implying a north-northwest–south-southeast directed coastline, which is consistent with determinations from Seymour Stream. Flutes and other sole structures from Bluff Sandstone at Coverham suggest paleocurrents directed towards the north or north-northeast, and, in Mororimu Stream (see Plate 8b), one of the most easterly outcrops available, grooves and flutes indicate a flow towards the northwest. This swing in paleocurrent direction from east-directed through north and then northwest-directed suggests that the basin is likely to close to the south, with a north or north-northwest trending basin margin to the west, and then a northeast–southwest directed margin off the present shoreline to the east. This is in accord with paleo-facies trends, which suggest a similar basin configuration (note the significant Cenozoic rotation and displacement of areas implicit in Crampton *et al.* 2003).

Eastern Sub-belt

Cretaceous 'cover' strata of the Eastern Sub-belt of the North Island are characterised by a thick, flysch-like facies, ranging in age from Cenomanian to Santonian (Ngaterian to Piripauan), which extends from East Cape to Glenburn in southern Wairarapa (Fig. 4.1b).

In the Raukumara Peninsula, Cretaceous rocks of the Eastern Sub-belt lie within the East Coast Allochthon. In the northeast the nappes comprise Early Cretaceous to ?Eocene submarine basaltic lava, brecciated pillow lava, hyaloclastic breccia, and tuff, assigned to the Matakaoa Volcanics (Cluzel *et al.* 2010; and see section 4.4.1). The sedimentary portion of the allochthon (Ruatoria Group) (Mazengarb *et al.* 1991) is characterised by the predominance of alternating sandstone and mudstone. The component formations are typically present in fault-bounded, structurally complex outcrops.

Mokoiwi Formation (Speden, 1976), the oldest unit, occurs principally in the Tapuaeroa Valley and has a faulted basal contact. The thickness cannot be determined due to its intense internal deformation; a structural thickness of 900–1200 m has been inferred (Speden, 1976). The Mokoiwi Formation is a tectono-stratigraphic unit rather than lithostratigraphic formation (Field *et al.* 1997). The unit is typically mudstone dominated with varying proportions of thin, fine- to medium-grained sandstone beds. There are beds of calcareous siltstone, slump horizons, and local lenses of conglomerate up to 30 m thick. In the northern Tapuaeroa Valley the Mokoiwi Formation includes a band of mafic lavas, associated tuffs and tuffaceous sediments, known as the Rip Volcanics (see section 4.4.1). Also within the Mokoiwi Formation are lenticular sandstone masses, mapped as Taitai Sandstone Member, that are typically massive, poorly sorted fine- to coarse-grained sandstone, with minor conglomerate and breccia (Speden, 1976). The conglomerate includes subangular to subrounded sedimentary clasts, some of which are of intraformational origin, bituminous coal and a mixture of acidic to intermediate igneous rocks (Mazengarb & Speden, 2000). The Taitai Sandstone Member forms the crest of Mounts Wharekia and Aorangi, and of several other massifs in the vicinity, probably including Mount Taitai (Mazengarb & Speden, 2000).

An age range of late Albian to Cenomanian (Motuan to Ngaterian) is suggested for the Mokoiwi Formation by macrofossils and dinoflagellates (Speden, 1976; Wilson, 1976). The sandstones and mudstones are inferred to have been deposited on a moderately deep-water turbidite fan complex, the Taitai Sandstone Member, representing channel- or canyon-fill within the fan system (Mazengarb & Speden, 2000).

The younger Tikihore Formation (Black, 1980; Mazengarb, 1993; Mazengarb & Harris, 1994), which is less indurated and deformed than the Mokoiwi Formation, extends over much of the northeastern portion of the Raukumara Peninsula, and is also present at Mahia Peninsula. Its base is commonly fault-bounded and the original relationship with the older Mokoiwi Formation is uncertain. However, locally there is apparent gradation, and the Mokoiwi and Tikihore formations may have been part of an originally continuous sedimentary succession (Mazengarb & Speden, 2000). The predominant lithofacies is centimetre- to metre-bedded alternating sandstone and mudstone. Sandstone beds are generally fine-grained and normally graded, with common sole structures including flute casts and trace fossils. Internal sedimentary structures are well developed, and include cross-stratification and convolute lamination. Incomplete Bouma sequences are common, and sandstones are probably turbidites. Fossils of various inoceramid bivalve species occur throughout and confirm that the age of the succession ranges from Cenomanian to Santonian (Ngaterian to Piripauan). Like the Mokoiwi Formation, the Tikihore Formation is inferred to have accumulated on a submarine fan (Mazengarb & Speden, 2000).

In the Eastern Sub-belt of Hawke's Bay–Wairarapa, the oldest outcrops of cover strata are of Cenomanian (Ngaterian) age and occur at Glenburn on the southern Wairarapa coast (Fig. 4.1b). No major stratigraphic breaks occur in the Cretaceous succession of the Eastern Sub-belt, and Cenomanian–Santonian (Ngaterian–Piripauan) strata are all included in the Glenburn Formation (Crampton, 1997).

The base of the Glenburn Formation is not exposed, although Albian (Urutawan–Motuan) volcanics exposed at Red Island and Hinemahanga Rocks on the coast south of Waimarama in southern Hawke's Bay may represent basement to the formation (Crampton, 1997). The volcanics have faulted contacts with the sediments and their stratigraphic relationship has not been established. Geochemical data suggest that the Red Island and Hinemahanga basalts are low-K tholeiites erupted in an island arc setting.

At Glenburn, less than 100 m of the middle Cenomanian (late Ngaterian) portion of the Glenburn Formation is exposed. Here it consists of fossiliferous, mudstone-dominated sediments, locally with intercalations of centimetre- to decimetre-bedded graded sandstones, some with Bouma sequences. The Clarence Series beds are separated from the overlying Raukumara Series rocks by a conglomerate unit up to 12 m thick. The mudstone-dominated units are inferred to represent hemipelagic deposits, and the thin, graded sandstones are interpreted as turbidites.

No Eastern Sub-belt rocks of Clarence Series age have been recognised in Marlborough.

4.2.2.3 Clarence Series sedimentary rocks of the Chatham Rise

The most southerly mid-Cretaceous marine sediments occur on Chatham Island and are associated with marginal marine to non-marine strata. Between 300 and 400 m of late Albian to Coniacian (Motuan to Teratan) -aged sediments of the Tupuangi Formation (Waihere Bay Group; Fig. 4.7) are exposed on Pitt Island, with the base unseen (Campbell *et al.* 1988, 1993). Seismic refraction profiling on the Chatham Rise suggests that the section is underlain by at least a further 560 m of sediments (Wood *et al.* 1989). The outcropping mid-Cretaceous sediments are correlated with the infill of east–west trending grabens identified from seismic studies and occupying the crest of the Chatham Rise. These were probably associated with the initiation of a major rift structure to the south, now occupied by the Bounty Trough (Wood *et al.* 1989).

The oldest part of the exposed sequence (late Albian/Motuan) consists of 80 m of poorly sorted, commonly channelled and cross-bedded sandstone and conglomerate. Cross-bedding suggests southwest- and south-directed flow and is essentially unimodal. The microflora consists entirely of terrestrial species, and a fluvial environment of deposition is inferred. This succession is overlain by 60 m of latest Albian (latest Motuan–early Ngaterian) shallow or marginal marine sediments, locally densely burrowed and with poorly preserved marine bivalves. The lower portion shows strong tidal influence, with channelling up to 2 m deep. Clay drapes and flaser and linsen structures become abundant higher in the succession, and cross-bedding azimuths are bimodal with northeast–southwest distribution. Coaly horizons become common and burrows rare in the upper part of the unit, with rootlet horizons and a coalified tree in apparent growth position. Thick beds of fine sandstone with abundant round clay clasts become common. The occurrence of dinoflagellate cysts confirms marine influence and this part of the succession is interpreted to have accumulated in a coastal marsh setting.

The succeeding 240 m of section, ranging in age from latest Albian (early Ngaterian) to Turonian–Coniacian (Mangaotanean–Teratan), shows an increasing marine influence towards the top. It is characterised by depositional cycles 1.6–6 m thick, each consisting commonly of a basal clay unit, passing upwards into wavy-laminated very fine sandstone interbedded with mudstone, cross-bedded or parallel-bedded fine sandstone, muddy sand with rootlets and pyrite nodules, and lignite up to 1 m thick. The sands are commonly carbonaceous, and tree trunks are seen in growth position. Cross-bedding measurements indicate a south- and westward-flowing current system. At the top of the succession cross-bedded fine to medium sand is interbedded with tuffaceous layers over an interval of 2.7 m before the formation passes conformably upwards into massive tuff (Kahuitara Tuff), of Santonian to early Campanian (Piripauan to early Haumurian) age, deposited in a shallow marine environment (Campbell *et al.* 1993).

Most of the Tupuangi Formation is inferred to have accumulated in the terrestrial/marginal marine part of a delta system. Terrestrial environments probably dominated, but abundant dinoflagellates suggest greater marine influence increases towards the top of the formation. Lateral equivalents of the Tupuangi Formation, but of a deeper water siltstone facies, are found as blocks incorporated in volcanic breccia on The Sisters islands, 20 km north of Chatham Island and 100 km north of Pitt Island (Campbell *et al.* 1993).

Detrital composition indicates that the source rocks for the unit consisted of indurated sandstone and shale (possibly Torlesse rocks and schist and Seismic Unit IIA), granite and granodiorite plutons, rhyolite and probably a minor source of high-grade metamorphic rocks.

The offshore equivalent of the late Albian–Coniacian (Motuan–Teratan) Tupuangi Formation is likely to be Seismic Unit IIB, which occurs along the crest of the Chatham Rise, conspicuously preserved in east–west trending grabens (Wood *et al.* 1989). The reflectors diverge towards the main bounding faults of the grabens, indicating that deposition was contemporaneous with faulting. The sediment accumulation is over 1500 m in several of the grabens and over 2500 m in one of them: the South Mernoo graben. No Cretaceous samples have been dredged from the rise, and no wells have been drilled.

4.3 THE RAUKUMARA SERIES: LATE CENOMANIAN–LATE CONIACIAN (AROWHANAN–TERATAN) SUCCESSIONS

Raukumara Series rocks are nearly everywhere co-extensive with Clarence Series strata. Unlike the Clarence Series rocks most Raukumara Series strata, with a few exceptions, were developed in gently subsiding basins, rather than in half grabens. Like the Clarence Series rocks, non-marine successions are characteristic of the majority of South Island and western North Island basins, while marine successions are typical of the East Coast Basin and the Chatham Rise.

4.3.1 Non-marine successions of the South Island

Many non-marine successions of the South Island appear to represent essentially continuous deposition from Clarence Series through Raukumara Series, without the prominent break at the Clarence–Raukumara boundary that is common in the marine sediments of the East Coast Basin. As noted earlier (see section 4.2.1.2), there is continuity of deposition on Pitt Island (Chatham Islands) and at Kyeburn in north Otago. In the Canterbury Basin and coastal Otago, and in the Great South Basin, however, a break in deposition is evident at or close to the Clarence–Raukumara boundary.

In the offshore Canterbury Clipper Basin the top of the Clarence Series half graben-fill deposits is marked by a prominent mid-Cretaceous seismic horizon, recording a down-sequence increase in acoustic velocity. The overlying succession (Clipper Formation), in which a petroleum prospecting well has been drilled, comprises 480 m of mainly non-marine deposits of probable Raukumara Series age. It consists of four units that make a fining-upwards sequence: a pebbly lower sandstone unit at the base, an overlying sandstone unit, a sandstone–mudstone interbedded unit, and a fine-grained carbonaceous coal measures unit at the top (Hawkes *et al.* 1985). Depositional environments range from alluvial fan near the base of the formation to a fluvio-deltaic setting near the top. The presence of marine dinoflagellates near the top of the interbedded unit suggests a paralic setting for the upper parts of the formation.

The mid-Cretaceous seismic horizon continues south into the Great South Basin (Cook *et al.* 1999), although a stratigraphic break has not been recognised in drillholes here, and the Clarence and Raukumara Series strata are combined into the non-marine Hoiho Group (see section 4.2.1.2). Onshore, however, 550 m of coarse non-marine sediments (Shag Point Group) occur on the Otago coast at Shag Point and at Taieri River mouth. The bulk of the succession consists of conglomerate that passes upwards into 100 m of finer-grained coal measures. The conglomerate has yielded a palynoflora of Raukumara Series age and the gradationally overlying coal measures are of Mata Series age (Raine, 1994). Elsewhere (see section 4.2.1.2), and particularly south of the Waihemo Fault, rocks of Clarence Series age are present at the base of the Shag Point Group.

The correlative Henley Breccia occurs in the coastal area south of Dunedin, around the lower Taieri River, where it rests unconformably on a moderately weathered surface cut on schist. It consists of a ~1200 m unit of breccia and poorly sorted conglomerate, with very subordinate sandstone and siltstone. Clasts generally range in size from 2 to 30 cm, but locally angular blocks of schist up to 3 m occur. The clasts consist of schistose sandstone, vein quartz and schist derived from the adjacent Caples Terrane. The Henley Breccia is thought to be the deposits of streams and debris flows draining from the west that accumulated in a half graben formed on the southeastern side of the active northeast-trending Titri Fault. Clasts are commonly coated with specular haematite and the breccia is usually stained a distinctive pinkish-red colour by hydrated iron oxide minerals in the matrix. Rare microfloras and angiosperm leaves occurring low in the succession give a poorly constrained age of late Cenomanian to late Coniacian (Arowhanan to early Piripauan) (Bishop, 1994).

Sediments of Raukumara age have not been identified elsewhere in Otago or Southland, except as the upper portion of the Kyeburn Formation (see section 4.2.1.2). On the West Coast of the South Island the upper part of the Pororari Group has been lost to erosion, although possible correlative sections near Paringa (south Westland) and to the north near Nelson extend into the Raukumara Series (see section 4.2.1.1). No sediments of Raukumara age are known from either the Taranaki or the offshore west Northland basins. Nevertheless, Raukumara-aged strata are likely to be present in at least parts of the west Northland Basin, as coal measures of Turonian (Mangaotanean) age have been dredged from the West Norfolk Ridge, a westward extension of the offshore west Northland Basin (Herzer *et al.* 1999).

4.3.2 Marine successions of Northland, the East Coast Basin and the Chatham Rise

4.3.2.1 Raukumara Series rocks of onshore Northland

Northland Allochthon

No rocks of Raukumara age are known from the autochthon although they are present in the allochthon. Two of the four complexes that make up the Northland Allochthon, the Tangihua Complex and the Mangakahia Complex (Isaac *et al.* 1994), either contain fossils of Raukumara Series age or were likely to have accumulated during Raukumara times. The Tangihua Complex (see section 4.2.2.1) probably developed from mid-Cretaceous to Paleocene times (Isaac *et al.* 1994). Mangakahia Complex rocks are present over large areas of Northland as far south as Kaipara Harbour, with smaller areas as far south as Silverdale, west of the Whangaparaoa Peninsula. They are typically present in discrete nappes. Structural complexity makes it difficult to resolve original stratigraphic relationships, and the Mangakahia rocks are best described in terms of the five major lithofacies (Fig. 4.7).

The Motukaraka facies comprises centimetre- to decimetre-bedded, alternating, fine to very fine sandstone and mudstone. The facies crops out mainly at Hokianga, but small areas are present at Kaipara Harbour. Typically, sandstone beds are 5–30 cm thick, with sharp bases bearing, in some cases, flute and groove casts, and with well-developed, complete or partial Bouma sequences. Mudstone interbeds are typically massive, in places parallel-laminated on a millimetre scale. No accurate estimate of thickness is possible, but a minimum of several hundred metres is likely. Dinoflagellates are mainly of Turonian (Arowhanan to Mangaotanean) age, but Cenomanian (Ngaterian) *Inoceramus* shells occur locally.

The Punakitere Sandstone facies and Awapoko facies have similar characteristics, and comprise quartzo-feldspathic, often micaceous, sandstones cropping out mainly in the Hokianga, Whangaroa and Kaipara harbours. Sandstone beds of this facies are decimetre- to metre-bedded (up to 3.9 m), moderately sorted and medium- to fine-grained. Most beds have sharp basal contacts and normal grading, and are either massive or show vague centimetre-scale parallel lamination. A small proportion show Bouma ABC(D) intervals. Trough cross-bedding is locally present. Some beds have basal decimetre-thick granule or fine pebble conglomerate that locally includes a transported and abraded shelly fauna. In places the shells form thin (up to 0.5 m) beds of bioclastic limestone. Conglomerate clasts are subangular to rounded with a maximum diameter of 10 cm, and consist

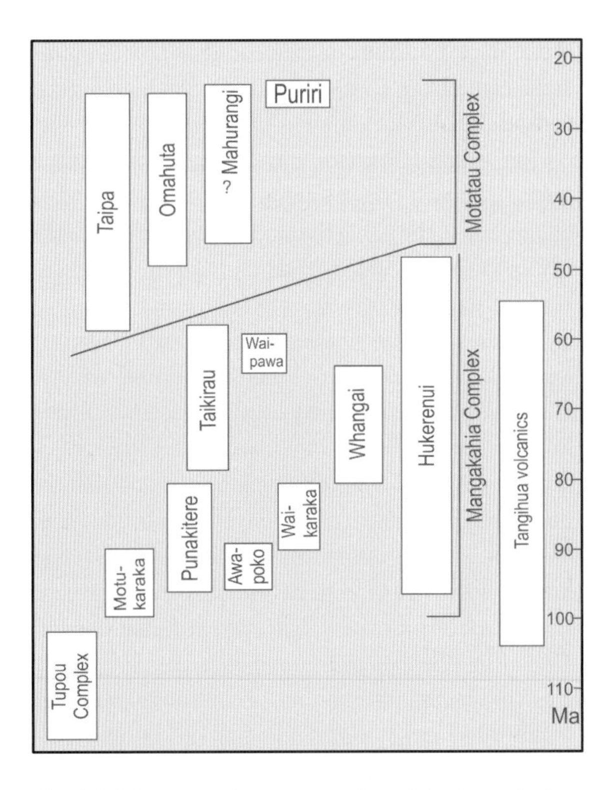

Fig. 4.7. Diagrammatic representation of the chronologically overlapping Cretaceous components of the Northland Allochthon. The white rectangles represent the age range of identified component rock bodies within the allochthon. All are in tectonic contact. (Adapted from Isaac, 1996.)

mainly of dark-grey siliceous volcanics, and locally derived sandstone and mudstone. In Whangaroa Harbour the Punakitere facies unconformably overlies an irregular eroded surface cut on Tupou Complex rocks (Brook & Hayward, 1989). Both facies have Late Cretaceous dinoflagellates ranging in age from Turonian–Santonian (Arowhanan–Piripauan) (Wilson, 1991). Inoceramids are abundant in this facies, although mainly as loose material, and cover the same age range.

Contacts between these facies and other lithofacies are either obscured or obviously tectonic, except in one instance, where Punakitere sandstone facies appears to overlie Whangai Formation siliceous mudstone concordantly, with some interbedding at the contact (Brook *et al.* 1988). The original stratigraphic thickness of the facies is estimated to have been 1–3 km.

The Waikaraka Mudstone is a facies recognised only near Kaikohe. It consists of sphaeroidally weathering siltstone intercalated with metre-bedded, graded granule- to fine-sandstone beds, commonly with abundant carbonised plant fragments. A Coniacian (Teratan) age for the facies is confirmed from inoceramids, which are abundant in concretions and in thin beds of broken shells.

The name ‘Hukerenui Mudstone’ is applied to mudstones widespread over northern Northland that are red, brown, green or grey, typically non-calcareous and commonly highly sheared. The mudstones are present both as discrete units within the Northland Allochthon, and as thin layers at tectonic contacts between lithofacies. The unit commonly overlies the basal décollement of the allochthon. Foraminifera from Hukerenui Mudstones give ages ranging from Turonian (Arowhanan) to Middle Eocene.

4.3.2.2 Raukumara Series rocks of the East Coast Basin

Western Sub-belt

In the Raukumara Peninsula, rocks of Raukumara Series age occur in both the east and west resting unconformably on older rocks, while in the middle part of the region, sedimentation appears to be continuous from Clarence times. In the west, near Oponae, rocks of Raukumara age (Wairata Sandstone) rest unconformably on early Cenomanian (Ngaterian) mudstone (Karekare Formation), and in places on basement. The formation is ~250 m thick, and consists dominantly of thick-bedded, very fine-grained, moderately sorted sandstone, with minor interbedded mudstone and intraformational breccia (Mazengarb, 1993). Scattered macrofossils indicate that the base is late Cenomanian to early Turonian (Arowhanan) in age. The upper age limit has not been established. The environment of deposition is uncertain, but Mazengarb (1993) considered it to be shallow marine.

West of Koranga the Moanui Formation has a very restricted distribution between two faults that appear to have been active in Raukumara times (Moore, 1978a,b). This unit, which rests unconformably on basement, is up to 500 m thick, and consists of a broadly fining-upwards succession, starting with coarse basal conglomerate passing gradationally upwards into fine- to medium-grained sandstone. Mudstone dominates in the upper part of the formation. The conglomerate consists of sub- to well-rounded pebbles and cobbles of indurated sandstone, with minor volcanic rock types, concretions and pebbly sandstone and siltstone clasts. Large subangular blocks of Torlesse-type (Urewera Group) sandstone exceeding 2 m in diameter are present in some conglomerates. Overlying and interbedded sandstone beds are often graded, show parallel- and cross-lamination and contain abundant fine plant debris (Moore, 1978a). Poorly bedded, concretionary, fossiliferous mudstone dominates the upper part of the formation. Intraformational slump horizons occur high in the sequence.

The base of the Moanui Formation is diachronous and varies from Turonian to Coniacian (Arowhanan to Teratan) in age, while the youngest part of the formation is Coniacian (Teratan). The environment of deposition is uncertain. Mazengarb (1993) suggested a shallow marine environment, but the presence of outsize angular clasts in the basal conglomerate, and of graded-bedded sandstone and slumping higher in the sequence, may indicate deposition of sediment gravity-flow deposits and slumping in deeper water as a result of active faulting (cf. Burnt Creek Formation, Marlborough: see below). The unit was derived from both Urewera Group basement rocks and an acid to intermediate igneous source.

East of Koranga, in the central portion of the Raukumara Peninsula, Raukumara Series strata are composed of Karekare Formation mudstone, continuous with that of Clarence Series age, or rest conformably or unconformably on turbidites of the late Albian–early Cenomanian Waitahaia Formation. Lithologies are generally similar to those in the older portions of the Karekare Formation. Slump structures occur locally, in one instance up to 44 m thick, but are uncommon. A distinctive feature of the Raukumara Series portion is the occurrence of red and green mudstone horizons. The origin of the colouring is uncertain (Mazengarb, 1993). Evidence for the depositional environment of the Karekare Formation is best recorded from Mangaotane Stream (central Raukumara Peninsula). Here there is an essentially complete succession through the unit, which rests gradationally on the late Cenomanian (late Ngaterian) Waitahaia Formation. Microfauna and microflora indicate that the Waitahaia Formation and the lower portion of the Karekare Formation were deposited in a restricted, deep bathyal or deeper environment, remote from land, while later sediments were probably deposited in shallower, bathyal to outer-shelf depths (Crampton *et al.* 2001).

In the lower Mata River area, 15 km southwest of Ruatoria, Raukumara Series strata, previously mapped as belonging to the Tikihore Formation of the East Coast Allochthon (see below under *Eastern Sub-belt*), are now regarded as autochthonous because they lie structurally below the basal thrust of the allochthon (Mazengarb & Speden, 2000). The succession is similar to the allochthonous Tikihore Formation, consisting dominantly of alternating sandstones and mudstones, with a thickness exceeding 1000 m. Sandstone beds are typically fine-grained with Bouma sequences, and have common directional sole structures. The section comprises five sequences, each defined by a channelled surface succeeded by a slumped horizon. These are inferred to have formed as a result of rapid relative sea-level falls followed by gradual relative rises (Laird *et al.* 1998). Inoceramids are common throughout the succession, giving an age range of the fauna of late Cenomanian–Santonian (Arowhanan–Piripauan). No nearshore features have been described, and the succession is inferred to have been deposited below storm wave base, probably on a ramp in outer neritic to upper bathyal depths. Sole marks indicate sediment transport towards the west and northwest during Turonian (Arowhanan and Mangaotanean) times, and to the north in Coniacian (Teratan) times (Laird, 1991; Laird *et al.* 1998).

Over western and central Marlborough an erosional break and major facies change occurs in the late Cenomanian (close to the Ngaterian–Arowhanan boundary). There is no angular discordance, but there is a basinward facies shift above the unconformity, and the break is inferred to reflect a drop in relative sea level (Laird, 1992). The overlying succession (Hapuku Group) (Laird *et al.* 1994) consists of a late Cenomanian–Coniacian (Arowhanan–Teratan) condensed sequence, 0–98 m thick, of non-marine to mainly shallow-marine clastic sediments dominated by burrowed glauconitic sandstone.

At the base of the succession in the extreme southwest of the Clarence Valley there are a few metres of non-marine beds, dominated by coarse conglomerate and woody material. These beds pass rapidly upwards and laterally into fossiliferous shallow-marine sandstones and sandy mudstones. The basal deposits in most cases rest on an eroded surface of basaltic rocks or Wallow Group sandstone with a relief of up to 1 m. The infill of erosional scours includes small pebbles of quartz and basalt, and fragments of *Inoceramus* and belemnite guards. Large *Ophiomorpha* burrows, and pyrite nodules, are common. In one instance a sharp-based micritic limestone with prominent cone-in-cone structure marks the base of the unit.

The Hapuku Group is characterised by its highly burrowed nature. In addition to large horizontal *Ophiomorpha* burrows up to several decimetres long, common small millimetre- to centimetre-scale vertical and horizontal burrows occur. In the west and south the bulk of the sediments consist of indistinctly bedded purple-brown siltstone and very fine-grained muddy sandstone. By contrast, glauconite is common in the east (e.g., at Wharekiri and Mororimu streams) where bedding is more distinct, with some graded beds and large-scale cross-bedding at several horizons.

Rocks of Raukumara age are often poorly exposed, which may account for the paucity of previous description. Apart from one isolated occurrence (J. Crampton, pers. comm. 2002), they have not been recognised in the Awatere Valley. They appear to be absent in outcrops in the middle Clarence Valley, where they either were not deposited or, more probably, were removed by erosion prior to deposition of the overlying sediments of the Seymour Group. Overall, the highly burrowed glauconitic nature of many of the deposits, and the local presence of cross-bedding, suggest that the group was deposited in a mainly shallow marine environment. No directions of sediment transport have been recorded. However, the presence of non-marine strata in the extreme southwest of the outcrop area, and the apparently increasingly more-offshore basinal nature towards the northeast, suggest deposition in a basin with axis oriented towards the northeast (present coordinates). Key macrofossils indicate an age range from late Cenomanian to late Coniacian (Arowhanan to Teratan) for the bulk of the sequence. Dinoflagellates confirm the dominantly Raukumara ages. Lenticular outcrops of pillow basalts occur interbedded with strata of probable Coniacian (Teratan) age in the Coverham area (Crampton & Laird, 1997; and see section 4.4.2).

The only exception to the stratigraphic relationships of Raukumara-aged sediments outlined above is found east of a major structure, the Ouse Fault, which occurs in northeastern Marlborough. This is a zone of structural complexity, which has been mapped northeast from the junction of Ouse Stream and Clarence River, through the eastern Coverham area and into the upper Kekerengu Valley (Ritchie, 1986), beyond which it has not been traced. In contrast to western Marlborough, Cretaceous rocks older than Raukumara Series are absent east of the Ouse Fault. The Raukumara sediments (Burnt Creek Formation) rest on basement Torlesse rocks with angular unconformity. They comprise a fining-upwards succession of conglomerate and pebbly mudstone, interbedded sandstone and mudstone couplets, siltstone and horizons of chaotic contorted beds (Crampton & Laird, 1997). The coarser lithologies, which are abundant towards the base of the unit and include irregular blocks of basement rocks up to 3 m in diameter, are inferred to have been emplaced by debris flows, and are succeeded by turbidites.

There are few environmental indicators, although the common occurrence of various types of sediment gravity-flow deposits coupled with slumping in the Burnt Creek succession suggests deposition on or at the base of a slope and, possibly, coeval tectonic activity. Microfloras suggest deposition some distance from land, and an outer neritic or upper bathyal environment seems probable.

The age of the base of the Burnt Creek Formation varies from Turonian to Coniacian (Mangaotanean to Teratan), and is diachronous. The youngest part of the formation is Coniacian (Teratan). Late Cenomanian–early Turonian (Arowhanan) fossils in clasts at the base of the formation indicate that older sediments were deposited in the region and subsequently eroded. Isopachs and slump-fold and paleocurrent data indicate deposition within a small east–west- or northeast–southwest-trending asymmetric basin with a north-facing paleoslope (Crampton & Laird, 1997). With regard to content and stratigraphic relationships the Burnt Creek Formation is strikingly similar to the contemporaneous Moanui Formation of the Raukumara Peninsula, which is also separated from adjacent areas of contrasted stratigraphy by faults, inferred to have been active during Raukumara times.

Eastern Sub-belt

In the East Coast Allochthon of the Raukumara Peninsula, alternating sandstone and mudstone of the Tikihore Formation was continuous from early Cenomanian (Ngaterian) into Raukumara Series times without a noticeable break in sedimentation. Sedimentation was similar to that in the Cenomanian, i.e., sandy turbidites. Paleocurrent directions indicate flow mainly to the north or northwest. Glauconite appears and increases in abundance from about the Coniacian (Teratan) onwards (Mazengarb, 1993). Inoceramids occur throughout and indicate an age range up to Santonian (Piripauan).

Raukumara Series rocks in the Eastern Sub-belt of the southern Hawke's Bay–Wairarapa area are all included in the Glenburn Formation, and outcrop occurs from Glenburn in the south to just south of Hawke Bay in the north. The unit comprises mudstone, with conglomerate and fossiliferous sandstone-dominated alternating sandstone and mudstone lenses.

The oldest part of the Glenburn Formation occurs only in the Glenburn area in southern Wairarapa, where it is in sedimentary contact with early Cenomanian (Ngaterian)-aged sediments (Crampton, 1997; and see section 4.2.2.2). The contact is sharp and concordant, being marked by a clast-supported conglomerate unit up to 12 m thick, which is deeply loaded into the underlying mudstone-dominated facies. The overlying late Cenomanian (Arowhanan) succession is dominated by decimetre- to metre-bedded sandstone, commonly with graded bedding. Within 100 m of the contact, nested shallow channel systems infilled with thick-bedded sandstone occur, some beds showing possible hummocky stratification (authors' observation). The abrupt change at the series boundary from probably hemipelagic mudstone to a succession of coarse facies suggests either a possible tectonic event or a relative drop in sea level.

Younger Raukumara sediments are exposed mainly in northern Wairarapa, and are dominated by alternating sandstone and mudstone with minor conglomerate and mudstone units. Conglomerate beds commonly display rapid variations in thickness, and often pinch out laterally over a few tens of metres. The sandstones are generally well sorted, moderately to highly carbonaceous, and fine- to very fine-grained, and are commonly normally graded and display partial Bouma sequences. Convolute and slumped horizons are abundant. Concretionary layers and concretions up to 2 m across are common in all lithologies, and pyrite is abundant locally. Scattered sections composed largely of mudstone are typically burrowed (including *Zoophycos*), are moderately to highly fossiliferous and include thin sandstone layers (Crampton, 1997).

Sparse paleocurrent indicators in the Glenburn Formation of the Wairarapa suggest that currents flowed mainly towards the east and southeast (Crampton, 1997). Factor analysis of heavy mineral suites suggests that the Glenburn Formation may have been recycled, in part, from the Springhill Formation of the Western Sub-belt (Smale, 1993). Conglomerate and thick sandstone beds were probably deposited by a variety of mechanisms, including high-concentration turbidity currents to debris flows. Thinner, graded sandstones are interpreted as turbidites, and mudstone-dominated units are inferred to represent hemipelagic sedimentation and dilute turbidites. Crampton (1997) inferred that these rocks were deposited on one or more submarine fans at outer-shelf or upper-bathyal depths.

No Eastern Sub-belt rocks of Raukumara Series age have been recognised in Marlborough.

4.3.2.3 Raukumara Series rocks of the Chatham Rise

Sediments of Raukumara age have so far been recorded only on Pitt Island, where they occur in the southern part of Waihere Bay and Tupuangi Beach. They have a gradational relationship with Clarence Series age sediments, and, in turn, pass gradationally upwards into Santonian (Piripauan) and younger age rocks (Campbell *et al.* 1993). The sediments are dominated by parallel or cross-bedded fine sandstone with some clay layers and lignite horizons. These beds show an increasing marine aspect, with tidal influence up-section, an interpretation supported by micropaleontology (Fig. 4.2).

4.4 IGNEOUS ROCKS OF THE EARLY EXTENSION PHASE

4.4.1 Clarence Series

Igneous rocks of Albian (Urutawan to early Ngaterian) age include rhyolitic tuffs and rare flows, and, in the allochthonous rocks of Northland and Raukumara Peninsula, supra-subduction basaltic lavas (Cluzel *et al.* 2010). Igneous rocks of Cenomanian age are found predominantly in Marlborough and Canterbury, and include mainly intraplate alkaline basalts.

4.4.1.1 Northland and East Coast allochthons

The Tangihua Complex (Brook *et al.* 1988) of the Northland Allochthon is a suite of submarine basaltic lavas, with basalt, dolerite, gabbro, and dioritic subvolcanic intrusives, tuff and tuff breccia. Pelagic mudstone and micritic limestone are minor constituents. The complex crops out as numerous discrete masses, which range in area from a few square metres to over 450 km^2. Most are in northern and central Northland. Tangihua massifs have been identified in the offshore west Northland Basin from geophysical surveys and dredge samples (Isaac *et al.* 1994).

Rock types within the complex are dominated by basaltic pillow lavas, flows, breccias and hyaloclastites with lesser sheeted dikes. The volcanic rocks cover a wide range of compositions, ages and tectonic affinities that reflect a complex succession of tectonic events within and at the edge of the Australian plate (Nicholson *et al.* 2000, 2007; Cluzel *et al.* 2010). The North Cape mass includes the only thick and laterally extensive plutonic sequence, which consists of serpentinised peridotite overlain by gabbro hosting microgabbro and microdiorite sills. The lower part of the gabbro shows well-developed mineralogical layering and banding, with microgabbro and microdiorite sills becoming increasingly common upwards to form a sheeted intrusion complex at the top of the present exposure.

Geochemical and petrological studies have shown that the Tangihua Complex has a wide variety of magmatic affinities, ranging through tholeiitic, calc-alkaline to alkaline (see references in Nicholson *et al.* 2000). Nicholson *et al.* (2000) interpreted the geochemistry as showing that the Tangihua basalts originated in an arc to back-arc setting. Fossils are rare in the Tangihua Complex, but sparse fragments of *Inoceramus* and ?*Aucellina* in sedimentary inclusions suggest a late Early Cretaceous age for the lower part of the succession (Isaac *et al.* 1994).

Igneous rocks forming part of the East Coast Allochthon in eastern Raukumara Peninsula, are assigned to the Matakaoa Volcanics and are correlatives of the Tangihua igneous rocks and part of the same complex (Cluzel *et al.* 2010). There is evidence for strong deformation overprinted by further deformation. Emplacement from the northeast is indicated. The sediments of the Northland nappes include rocks of deeper-water and more distal character than those of East Cape, which show similarities to those of the Eastern Sub-belt. Matakaoa igneous rocks form the northern tip of the Raukumara Peninsula between Cape Runaway and Hicks Bay. A separate Pukeamaru massif extends inland southwest of Hicks Bay, with a small outlier at Te Kiwikiwi Hill (Mazengarb & Speden, 2000). The volcanics have intercalated sedimentary lenses composed of pink to light-brown limestone and tuffaceous limestone, red-brown tuffaceous mudstone and minor chert (Spörli & Aita, 1992). Fossils from intercalated sediments have yielded late Early Cretaceous (Albian), Late Cretaceous, and Late Paleocene to Early Eocene ages (Strong, 1976, 1980; Spörli & Aita, 1992). Spörli and Aita (1992) suggested that the Matakaoa Volcanics are possibly up to 10 km thick, although due to structural complexity a true thickness is difficult to establish.

Geochemical studies of the Matakaoa Volcanics show that they are very similar to the Tangihua volcanics and formed in a similar supra-subduction basin (Nicholson *et al.* 2000; Cluzel *et al.* 2010).

In the northern Tapuaeroa Valley (eastern Raukumara Peninsula) the mid- to late Albian (Motuan) Mokoiwi Formation includes a somewhat discontinuous stratiform band of mafic lavas, intermediate and silicic tuffs, and tuffaceous sediments known as the Rip Volcanics (Pirajno, 1979). The unit is up to 600 m thick and can be followed laterally for 11 km. The basalts are of back-arc affinity, and similar in geochemistry to the Matakaoa Volcanics (Mortimer, N., in Mazengarb & Speden, 2000; S.D. Weaver, pers. comm. 2003).

4.4.1.2 Autochthonous silicic volcanics of Albian (Urutawan–Motuan) age

In the autochthonous Western Sub-belt of the East Coast Basin, silicic tuffs with glass shards occur in the Urutawan Te Wera Formation west of Koranga in the Raukumara Peninsula (Speden, 1975), and in the Motuan-aged segment of the succession at the Motu Falls section to the north (Crampton *et al.* 2004b; Tulloch *et al.* 2009b). A U–Pb age of 101.6±0.2 Ma, obtained from a 5-m-thick tuff bed from this latter section, is in close agreement with the Buller Gorge tuff ages (see section 4.4.1.3). Tuffs in a similar stratigraphic position also occur to the east of this locality in Kokopumatara Stream (Mazengarb *et al.* 1991). In eastern Wairarapa, similar thin, white, probably siliceous tuffs occur low in the Albian Springhill Formation west of Castlepoint (Moore & Speden, 1984). These are part of a suite of Cretaceous volcanic rocks reviewed by Tulloch *et al.* (2009b) that show an evolution in chemical composition over time. Coupled with detrital zircon data (Adams *et al.* 2013b and references therein) they suggest a change over time from volcanic-arc type composition to distinctly alkaline within-plate character.

4.4.1.3 Autochthonous igneous rocks of Cenomanian (late Ngaterian) age

The Cenomanian volcanics of Marlborough (Gridiron and Lookout formations) consist mainly of subaerial basalt to trachybasalt lava flows, chemically mildly alkaline, consistent with magma generation in an extensional tectonic regime (Warner, 1990). Dikes and sills of similar composition to the lavas intrude the underlying Cretaceous strata. The thickest exposed section of volcanic rocks in the Awatere Valley exceeds 700 m. In the Clarence Valley, flows reach a maximum cumulative thickness of 170 m in Seymour Stream, and thin and eventually disappear in the northeast and southwest. The easternmost occurrence is an isolated outcrop of pillow basalt at the top of the Bluff Sandstone in Swale Stream, Coverham. A U–Pb age of ~96 Ma has been obtained from the plutonic Tapuaenuku Igneous Complex in the Inland Kaikoura Range, lying between the Awatere and Clarence valleys, which is probably genetically related (Baker & Seward, 1996). A comparable $^{40}Ar/^{39}Ar$ age of 96.1±0.6 Ma has been obtained for a basalt flow in Seymour Stream, central Clarence Valley (Crampton *et al.* 2004b).

Eighty kilometres to the south, in north Canterbury, the Mandamus Igneous Complex consists of geochemically similar basalt to trachybasalt volcanic rocks, and alkali gabbro, syenite and lamprophyric dikes. The complex has an Rb–Sr age of 97.0±0.5 Ma, and shows intraplate geochemistry consistent with extensional environment (Weaver & Pankhurst, 1991). Ages comparable to the above have also been obtained from andesites and garnet-bearing rhyolites of the Mount Somers Volcanics Group of central Canterbury, which extends from Banks Peninsula to the alpine foothills west of Mount Somers (Tappenden, 2003). The group may be very widespread, as it has also been recognised beneath the Canterbury Plains in the JD George-1 petroleum drillhole, and offshore magnetic data suggest it may also be present to the east and south of Banks Peninsula. The geochemistry suggests high mantle heat flow and resultant crustal anatexis, and is compatible with an environment of crustal extension (Barley, 1987).

The above igneous province may extend as far north as Cape Palliser in the southeastern North Island, where dike rocks geochemically similar to the Tapuaenuku Igneous Complex occur (Challis, 1960). The dike rocks show considerable alteration, and have proved difficult to date. A clast from the fault-bounded Buttress Point Conglomerate of the southern West Coast of the South Island, which consists predominantly of alkaline andesitic clasts, has yielded a U–Pb zircon age of 96.9±1.6 Ma (Phillips *et al.* 2005). The source of the conglomerate is unknown, but the magmatism is contemporaneous with others in the Cretaceous volcanic suite.

4.4.2 South Island Raukumara Series igneous rocks

Lenticular outcrops of pillow basalts occur interbedded with strata of probable Coniacian (Teratan) age in the Coverham area of Marlborough (Crampton & Laird, 1997; and see section 4.3.2.2). Their geochemistry clearly indicates an intraplate, alkaline basalt affinity, and is very similar to older Cenomanian (Ngaterian) volcanics in Marlborough.

CHAPTER 5

Late Cretaceous and Paleogene Tectonic Setting: Changing Extensional Regimes

5.1 INTRODUCTION

Late Cretaceous and Cenozoic tectonics in the New Zealand region can be conveniently divided into four phases:

1. Mid- to Late Cretaceous extension regime prior to continental break-up
2. Late Cretaceous–Paleocene opening of the Tasman Sea and Southern Ocean
3. Paleocene–Oligocene formation of the Tasman ocean crust, with a poorly defined, mainly extensional plate boundary zone through New Zealand
4. Miocene to Recent development of a well-defined plate boundary (Alpine Fault and precursors) linking a spreading ridge, and later a subduction zone, south of New Zealand, with subduction zones lying to the north.

The first phase has been discussed in Chapters 2–4, and the last phase corresponds to the 'Kaikoura Orogeny', as discussed in Chapter 7. The unifying character of the tectonic events associated with the Late Cretaceous–Paleogene is the eastward movement of the Australian plate margin in a series of jumps, annexing parts of the Pacific plate.

5.2 LATE CRETACEOUS–EARLY PALEOCENE

The situation leading to crustal extension has already been discussed (Chapters 2 and 3). During this period a number of mainly small non-marine basins formed in the onshore South Island (Pororari, Henley, Kyeburn, etc. basins), but much larger ones developed in the offshore areas (Great South and New Caledonia basins). In the southeast, the oceanic Hikurangi Plateau lay along the inactive northern margin of the Chatham Rise but subduction continued west of the edge of the Hikurangi Plateau in the East Cape to Northland region into the Late Cretaceous–Paleocene (Chapter 4). Details of the relationship between ocean and continental crust are difficult to determine in the region because of the strong overprint of deformation related to the active Cenozoic plate boundary.

The quasi-oceanic Bounty Trough is an early expression of Cretaceous extension and pre-dates separation of Zealandia from Antarctica (Chapter 2). Much of the Bounty Trough appears to be starved of sediment, although at its western end thick sediments infill Cretaceous half grabens (Davy, 1993). In western New Zealand, elements of the Taranaki Basin also originated in the Late Cretaceous and continued to develop after continental separation with further active faulting (King & Thrasher, 1996; Strogen *et al.* 2017).

The opening of the Tasman Sea and the South Pacific portion of the Southern Ocean in the Late Cretaceous isolated New Zealand from Australia and Antarctica, and formed two passive margins, both characterised by extensive crustal thinning producing large continental plateaux and rises (Grobys *et al.* 2008). After separation, extensional faulting did not stop entirely and there are a number of basins where ongoing tectonic control is important, such as the Ohai, Paparoa and Pakawau basins. After the cessation of subduction in East Cape and Marlborough, further unconformities were developed within Late Cretaceous marine rocks, indicating ongoing faulting and tilting.

Originally Zealandia was considered to be entirely within the Pacific plate, with spreading ridges forming the plate boundaries between the Pacific, Antarctic and Australian plates. An important reinterpretation of the Late Cretaceous and Paleocene igneous rocks and related sediments of the Northland and East Cape allochthons as being of back-arc basin character (Nicholson *et al.* 2000; Toy & Spörli, 2008; Cluzel *et al.* 2010) means that there was a subduction zone northeast of New Zealand, probably dipping to the southwest. This has important implication for Zealandia as a whole. It is generally accepted that southern Zealandia, after the cessation of mid-Cretaceous subduction, became part of the Pacific plate separated by the Tasman Sea spreading ridge from the Australian plate. Since a plate cannot subduct under itself, the situation in the far north poses a problem. Either the subducting plate is not the Pacific plate, or the overriding plate is the Australian plate or possibly a micro-plate involving crust east of the New Caledonia Basin. The simplest solution is that it is the Australian plate, but this would require the presence of a plate boundary through Zealandia as early as the Late Cretaceous. Reconsideration of Cretaceous rotation poles does point to a poorly defined plate boundary through Zealandia in the Cretaceous (see Fig. 2.2). Lamb *et al.* (2016) also advocate a plate boundary through New Zealand based on revised plate rotation data. Their analysis further suggests a major sinistral displacement within Zealandia.

Although it seems doubtful whether the New Caledonia Basin actually contained a spreading ridge, it is possible that active extension continued into the Late Cretaceous, with a 'Lord Howe Rise block' (or micro-plate) moving northwards more slowly than the rest of the Pacific plate. Within New Zealand this differential movement may have caused the West Coast–Taranaki Rift (Laird, 1994; King & Thrasher, 1996), and the northeast-trending faults that control the Paparoa and Pakawau basins. These northeast-trending faults may represent reactivated transfer faults generated during mid-Cretaceous northeastward extension (Laird, 1994). There is also a change in orientation of magnetic anomalies at this time (King & Thrasher, 1996, their fig. 2.8). King and Thrasher (1996) refer to this period as the syn-rift–drift phase of basin development, although actual rifting from Australia had taken place much earlier.

Elsewhere in New Zealand, after separation, extensional faulting did not stop entirely and in a number of other basins ongoing tectonic control is important (e.g., Canterbury Basin). Unconformities are also developed within Cretaceous marine rocks along the inactive Pacific margin in the East Coast Basin, where there is evidence of synsedimentary faulting and block rotation that may relate to stress partitioning and adjustment between the older Cretaceous subducted slab and the accretionary complex.

5.3 LATE PALEOCENE–EARLY MIOCENE

The spreading rate appears to have decreased on the Pacific–Antarctic Ridge in the Paleocene, and spreading in the Tasman ceased entirely after anomaly 24 at ~53 Ma (Gaina *et al.* 1998). Southwest of New Zealand the central Tasman plate boundary was soon replaced by a new spreading ridge almost perpendicular to the first. The new spreading ridge produced a series of east–northeast-trending magnetic lineations in the 'southeast Tasman oceanic crust' (STOC) of Wood *et al.* (1996). These authors identify anomalies 18–11 (40–30 Ma) in this set. Further to the southwest Weissel *et al.* (1977) identified anomaly 22 (52 Ma), although this has recently been reinterpreted by Lawver *et al.* (1992) as anomaly 24. If this revision is correct, it implies an extremely rapid relocation of plate boundaries and a short-lived ridge–ridge–ridge triple junction. The new plate boundary is parallel to (and may have developed along the line of) transforms in the older Tasman crust. Crust identified by Cande *et al.* (1995, 2000) near the Iselin Bank on the Antarctic margin supports the notion of the new spreading ridge developing along an old transform in the Tasman Sea ocean crust. The new pattern represents a linking of the Southeast Indian Ridge with the Phoenix-Pacific Ridge and correlates with increased northward movement of Australia.

With the end of Tasman Sea spreading the eastern boundary of the Australian plate moved abruptly eastwards; the location of the Australian–Pacific plate boundary during the Late Paleocene is uncertain. The northeast end of STOC abuts the Resolution Ridge in a series of faulted and rotated blocks of oceanic and continental crust, which are probably detached parts of the original western margin of the Campbell Plateau (Wood *et al.* 1996; Lamarche *et al.* 1997). A further set of younger anomalies trending east–southeast has been detected in the northeast part of STOC, pointing to a further change in relative plate movement after 30 Ma (Early Oligocene). These data (Wood *et al.* 1996) have clarified the plate boundary changes that followed demise of spreading in the central Tasman Sea and point to an abrupt change in the direction of relative movement of the Pacific and Australian plates at about 50 Ma. If part of Zealandia was already within a separate plate at that time, significant change in motion is indicated (see end of section 5.2). The direction and extent of the STOC lineation set and the inclusion of New Zealand continent crustal material in the Resolution Ridge suggest that a new plate boundary zone cut northeastwards through the New Zealand continent from the tip of STOC. Exactly where this new plate boundary was and what form it took are crucial to understanding Eocene–Oligocene tectonic development of New Zealand, but unfortunately the nature and position are poorly constrained. Westward thrusting of the Pacific plate and regional shortening in the Neogene mean that the early plate boundary probably does not correspond to the present surface trace of the Alpine Fault in the South Island. Lebrun *et al.* (2000) suggest that the Moonlight Fault could be an early manifestation of the plate boundary. Alternatively it could well be below the present edge of the Pacific plate.

The margin of eastern Australia and the southwestern margin of the Challenger Plateau and Lord Howe Rise have been passive margins since Cretaceous break-up. With the cessation of central Tasman Sea spreading this crust became part of the Australian plate. The Resolution Ridge as the southeastern limit of continental crust in the enlarged Australian plate must retain its original relationship to the Challenger Plateau and the western South Island. Similarly there is no reason to suppose that the Campbell Plateau and eastern South Island have not been part of the Pacific plate since the mid-Cretaceous. Therefore, the movement that took place between the plates, expressed as northwesterly directed extension to the southeast of the Resolution Ridge, must have been expressed in a similar way within New Zealand in a zone that commences near Fiordland. Strands of the boundary may have passed along both margins of the Fiordland block, and these include the Moonlight Fault and Picton Fault in the Mesozoic schist belt. In the North Island the boundary probably lies within a wedge-shaped zone of major Neogene crustal strain and may explain the present juxtaposition of contrasted belts (Eastern and Western sub-belts; Chapter 4) of Late Cretaceous to Paleocene geology (Moore, 1988). In the Eocene–Oligocene the northeastern limit of the Australian plate is not well defined but must include northwestern New Zealand, the New Caledonia Basin and the Norfolk ridge.

Commonly, interpretations concerned mainly with regional-scale plate tectonics have portrayed Cretaceous–Oligocene plate displacement as having taken place along a line equivalent to the Alpine Fault (e.g., Weissel *et al.* 1977; Walcott, 1984a; Lawver *et al.* 1992; Lamb *et al.* 2016). Those familiar with New Zealand geology, however, have found little support for the existence of the Alpine Fault (*sensu stricto*) before the Oligocene, and Cooper *et al.* (1987) concluded fault inception to be as late as Late Oligocene–Early Miocene, based on ages of dikes near the present Alpine Fault. Kamp (1986) also considered that the Alpine Fault could be no older than Early Miocene and proposed an earlier structural boundary, the Challenger Rift System, extending from south of Fiordland to the Northland Peninsula. The rift zone was thought to be 100–200 km wide and 1200 km long and characterised by interconnected fault-bounded basins and half grabens with 2–4 km of Eocene–Oligocene sediments. The zone was interpreted as a developing plate boundary zone that almost reached the stage of break-up and sea-floor spreading (Kamp, 1986). This interpretation fits well with the western South Island, where extension and basin formation is clearly evident in the Late Eocene–Early Oligocene. Evidence cited by Kamp for the rift in the western North Island is, however, not compelling. In Taranaki, active faulting associated with basin formation appears to be Late Cretaceous–Paleocene (related to a Lord Howe plate margin) with only minor Oligocene faulting near the southern limit of the basin (King & Thrasher, 1996). The Challenger Rift therefore appears to terminate in the area of the Taranaki Bight, a situation consistent with Paleogene rotation poles lying in central New Zealand (Fig. 5.1) (see also Strogen *et al.* 2017).

The anomalies recognised by Wood *et al.* (1996) record, assuming they are half of an originally symmetrical set, crustal extension of 200 km with a spreading rate of 15 mm/yr. This is higher than the rate identified by Sutherland (1995), and Wood *et al.* (1996) suggest that this is because the stage pole for 40–30 Ma was more remote than that predicted by Sutherland. The poles are derived only from analysis of data from oceanic crust and deliberately do not involve data from the geology of the continental crust. Reconstructed configurations for the Late Eocene show the position of the extensional basins of Nelson and Westland lying to the southwest of Fiordland and the contemporaneous extensional basins of western Southland (Sutherland, 1995, fig. 4; 1999, fig. 3). A more northerly position (conventional) for the West Coast basins in the Eocene would point to a Late Eocene pole position to the north of that proposed by Sutherland (1995), somewhere in the South Taranaki Bight a little to the north of his error eclipse for ~40 Ma, and near the southern limit of the Taranaki Basin. Coupled with counterclockwise rotation of the Pacific plate, such a pole would accord with the rapid extension seen in the ocean floor and with more-limited extension in southern New Zealand (western Southland, Westland and Nelson). North of the pole, weakly developed strike slip or transpression would be expected. Demonstration of Eocene–Oligocene displacement between East and West Antarctica (Cande *et al.* 2000; Cande & Stock, 2004) may significantly influence reconstructions.

More-recent work on the relationship between East and West Antarctica is summarised by Lamb *et al.* (2016). They propose a sinistral Cretaceous proto-Alpine Fault, a proposition that is consistent with earlier strictly rigid plate reconstructions (e.g., Lawver *et al.* 1992). The exact position of this fault is uncertain (see discussion of Cretaceous tectonics in section 2.6).

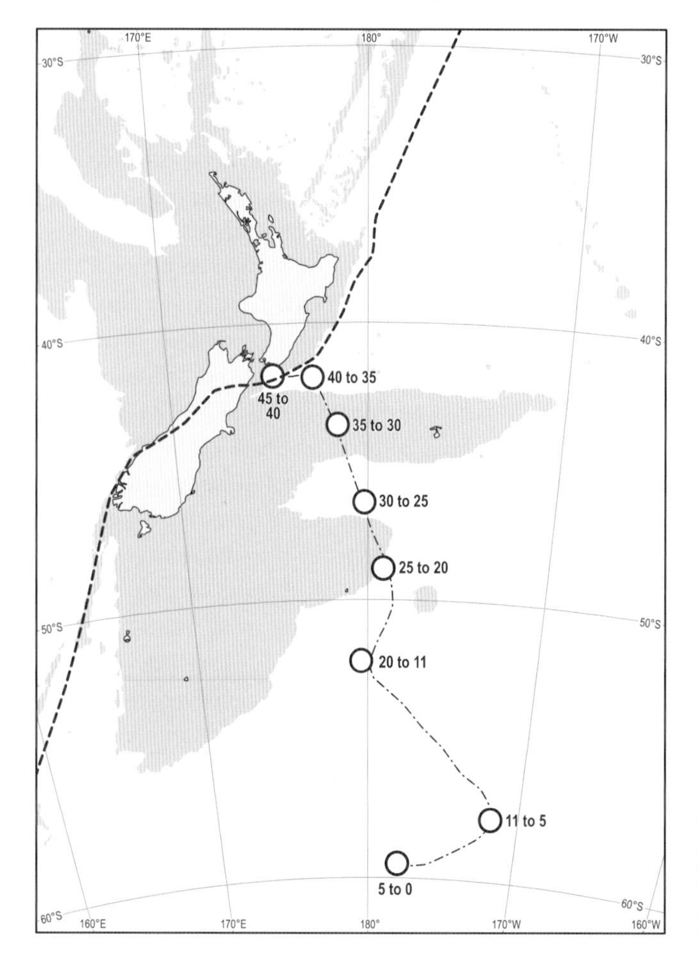

Fig. 5.1. Stage poles for the relative motion between the Australian and Pacific plates between 45 Ma (Middle Eocene) and the Present. (Based on Sutherland, 1995, fig. 5.)

CHAPTER 6

Passive Margin Phase: The Latest Cretaceous and Paleogene Succession

6.1 INTRODUCTION

The Late Cretaceous to Paleogene phase has traditionally been thought of as representing the passive margin during the 'drift' phase of the break-up of Gondwana. During this time, sea-floor spreading commenced in the Tasman Sea, separating New Zealand from the rest of Gondwana. The eastern margin in particular records progressive transgression over the previously active margin. However, the supposedly 'passive' subsidence is interspersed with further rifting, particularly along the western margin of New Zealand. These Late Cretaceous rifts are host to many of the economic basins, such as the petroleum-producing Taranaki Basin and the correlative coal-bearing basins. Rifting in the Taranaki and West Coast basins ceases at roughly the same time that sea-floor spreading ceases in the Tasman Sea, suggesting that these events are tectonically related. Subsequently, rifting starts in the south with the opening of the Emerald Basin. Between episodes of rifting around the country are periods of more-passive subsidence and an overall marine transgression, culminating in nearly all of Zealandia being submerged in the Oligocene.

6.2 CONTINENTAL BREAK-UP AND THE ONSET OF SEA-FLOOR SPREADING

Rifting and half graben formation accompanied by largely coarse-grained deposits was the dominant tectono-sedimentary theme introduced in the mid-Cretaceous, and it continued over much of the New Zealand region until Raukumara times. Locally (e.g., in the East Coast Basin), half graben formation had largely ceased during Cenomanian (Ngaterian) times and it gave way generally to slow subsidence, with sedimentation influenced broadly by relative sea-level changes, and by some localised tectonic events. By the late Coniacian (late Teratan: ~87 Ma) much of the New Zealand region had been reduced to an exposed landmass of low relief, with sedimentation, mainly fine-grained, largely restricted to the East Coast, Canterbury and Great South basins. The future Zealandia continent still formed an integral part of East Gondwana (Australo-Antarctica).

This pattern changed in the late Coniacian (late Teratan), and widespread transgression, accelerating in the Santonian (Piripauan–early Haumurian), took place in the Zealandia region. This rapid change in base level coincided with the onset of break-up and sea-floor spreading, and the drift of New Zealand away from East Gondwana. It heralded a long-drawn-out period of slow passive margin subsidence, interrupted by intermittent tectonic events, and general marine transgression throughout most of New Zealand, reaching its peak in Oligocene times.

6.2.1 Associated unconformities

Break-up and the beginning of sea-floor spreading are represented in New Zealand by the occurrence of two distinct and conspicuously diachronous erosional breaks in the sedimentary succession. The older break, of late Coniacian to early Santonian (late Teratan to early Piripauan: ~87–85 Ma) age, has been identified only in eastern New Zealand; the younger break, with a maximum age of late Santonian (close to the Piripauan–Haumurian boundary: ~85 Ma), is a major feature recognised throughout much of New Zealand (Crampton *et al.* 1999; Laird & Bradshaw, 2004).

The sediments immediately overlying the older break in most cases comprise thick sandstone units forming the base of fining-upwards transgressive successions, which contrast with the underlying mainly mudstone-dominated strata. The younger break is overlain by notably fine-grained sediments.

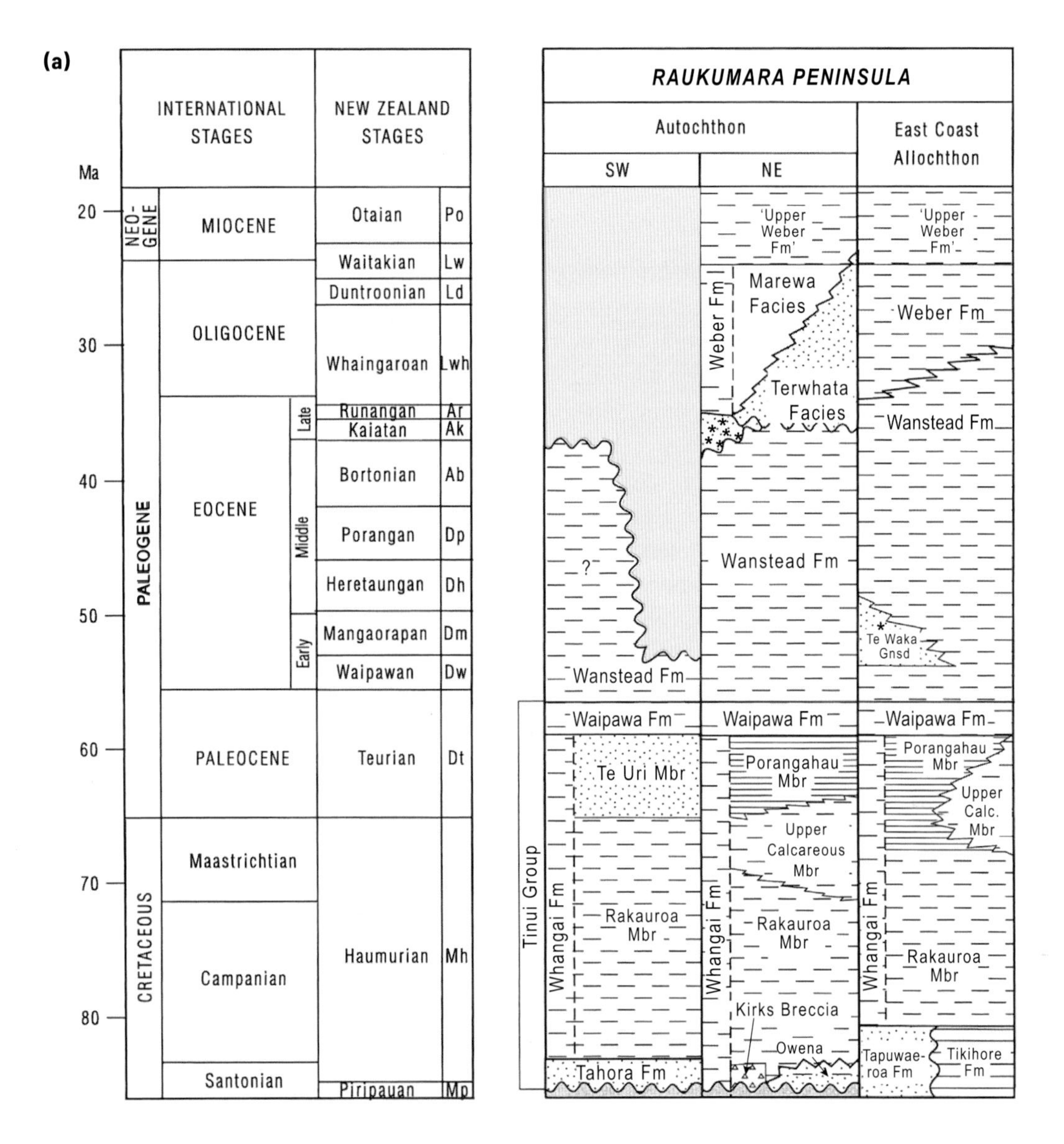

Fig. 6.1. (a & b) Time-stratigraphic section of the Paleogene rocks of the East Coast Basin from the Raukumara Peninsula to Marlborough. Fm, Formation; Gnsd, Greensand; Ls, Limestone; Mbr, Member; Siltst, Siltstone; Ss, Sandstone. (Adapted from Field *et al.* 1997.)

In the East Coast Basin, strata of Coniacian (Teratan) and older age forming part of the Western Sub-belt in the North Island, and southwestern outcrops in Marlborough, are overlain with slight to moderate angular discordance by sandstone-dominated deposits of late Coniacian to Santonian (late Teratan to Piripauan) age (Field *et al.* 1997; and Figs 4.1 and 6.1a). In the Raukumara Peninsula and Marlborough, inshore to shallow-marine transgressive deposits unconformably overlie dominantly muddy strata deposited in a shelf to bathyal environment, suggesting uplift and/or a relative drop in sea level (Laird, 1992; Mazengarb, 1993). By contrast, in the Western Sub-belt of the Hawke's Bay and Wairarapa regions an angular unconformity of similar age separates fine-grained probable shelf to bathyal sediments of late Albian to Cenomanian (Motuan to Ngaterian) age from overlying bathyal coarse-grained glauconitic debris flow deposits and turbidites of late Coniacian (late Teratan) age (Crampton, 1997; and Fig. 6.1b). Although there is no equivalent unconformity in the Eastern Sub-belt of the East Coast Basin, there is a

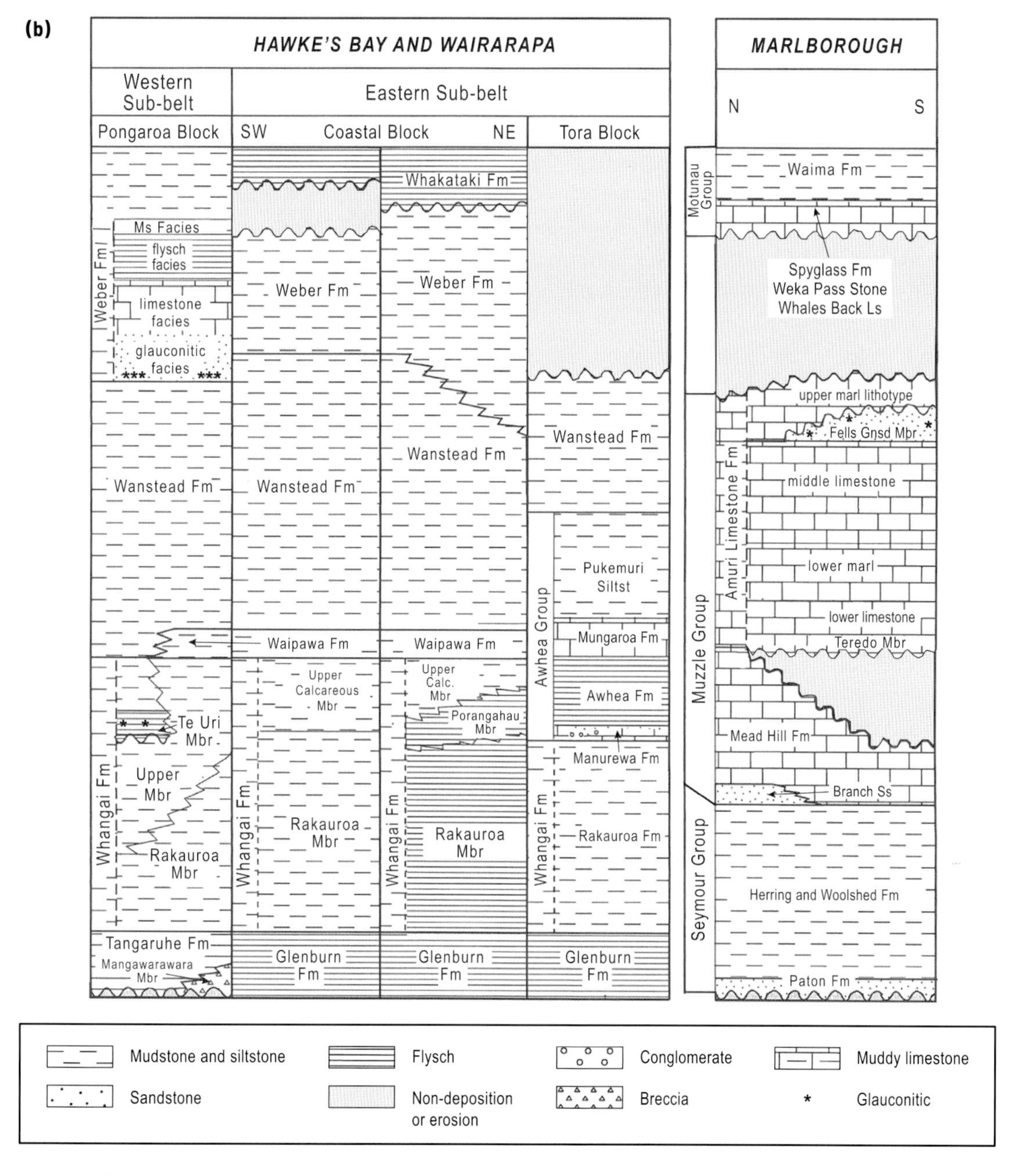

Fig. 6.1. (b)

rapid transition to thick, mainly turbidite sandstones in late Coniacian to early Santonian (late Teratan to Piripauan) times. In northeast Marlborough the late Coniacian unconformity is replaced by a rapid transition upwards into glauconitic sandstone, in part of turbidite character (Laird, 1992).

Further south in the South Island, most of the Canterbury region and Chatham Rise was subaerial in late Coniacian (late Teratan) times, although, in the south, onlap by marine sediments was proceeding. In the Canterbury Basin (see Plate 9) the basal late Coniacian sediments rest unconformably on older sediments or on basement (Field *et al.* 1989; and Fig. 6.2). This break in sedimentation is represented offshore by a prominent seismic reflector that extends southwards into the adjacent Great South Basin. There it overlies a regional transgressive surface that separates mainly terrestrial strata from the overlying dominantly marine sediments, here dated as early Santonian (Cook *et al.* 1999). Along the northwest margin of the basin, however, terrestrial depositional environments persisted and recognition of a break in sedimentation is difficult.

A second, major facies disjunction separates the late Coniacian–Santonian (late Teratan–Piripauan) mainly transgressive sandstones in eastern New Zealand from overlying mudstone-dominated sediments (referred to as the 'Whangai facies'). Outboard, in eastern areas of the East Coast Basin, it is recognisable as a rapid transition from sandy to muddy sediments. Near the basin margin in the west the facies change is in many cases associated with a locally strongly diachronous unconformity. The most detailed regional studies have been carried out in Marlborough, where the unconformity youngs towards the western basin margin by about 3–4 Ma over a distance of 10 km, suggesting slow subsidence of the continental margin and consequent transgression (Crampton *et al.* 1999; Schiøler *et al.* 2002). The oldest age on this unconformity has been dated with moderate accuracy in northeast Marlborough, where it is located within the Santonian just below the C34–C33 magnetochron boundary, at ~85 Ma (close to the Piripauan–Haumurian boundary) (Crampton *et al.* 1999).

In the Canterbury and Great South basins (Fig. 6.2) similar facies of equivalent age occur. In the Great South Basin light grey shales and clays unconformably overlie transgressive late Coniacian–early Santonian sands. Shorewards, the succession passes

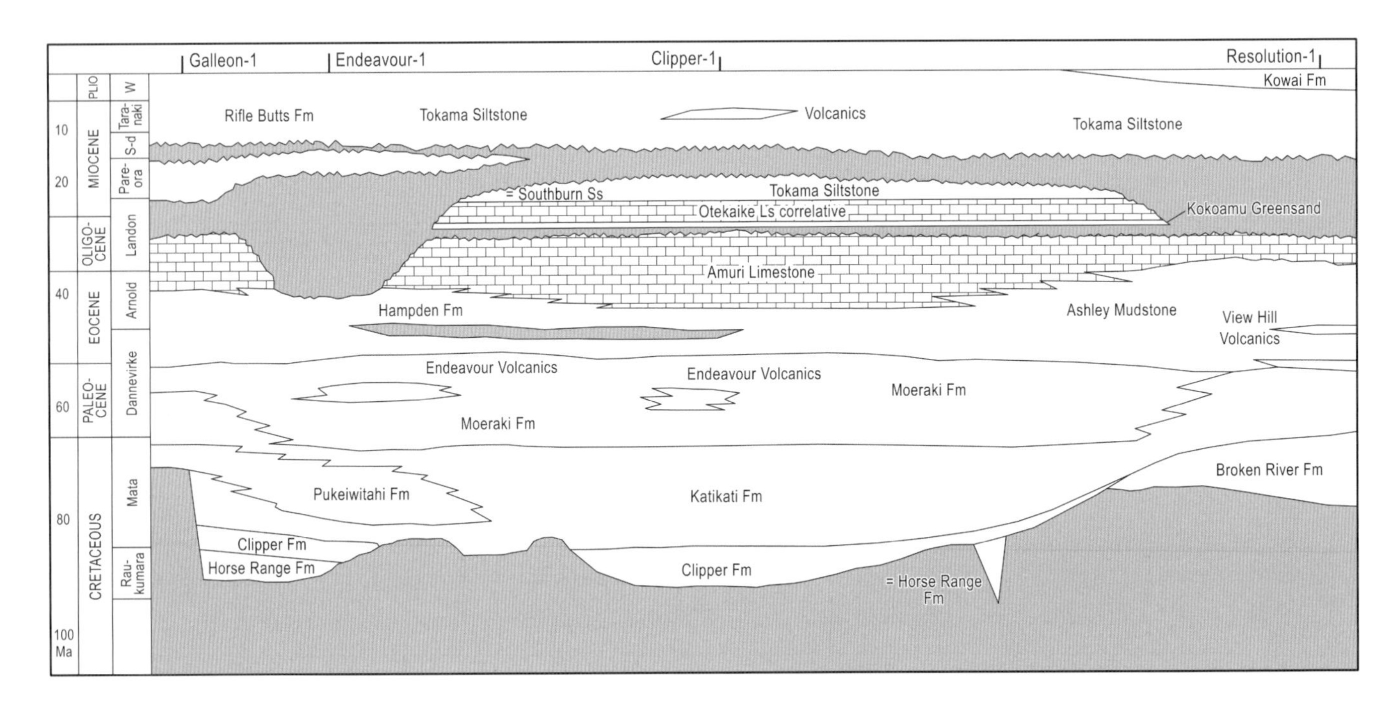

Fig. 6.2. Time-stratigraphic section from northeast to southwest through the offshore part of the Canterbury Basin. Most Cretaceous to Eocene units are defined from drillhole data. See sections 6.2.1 and 8.2.3. Fm, Formation; Ls, Limestone; Ss, Sandstone. (Adapted from Field *et al.* 1989.)

laterally into sandy shoreline facies and then coal measures (Cook *et al.* 1999).

In western New Zealand, shoreline and shallow-marine sands of poorly constrained Campanian to Maastrichtian (Haumurian) age (Fig. 6.3) resting on coal measures are the first indication of a transgression that occupied a marine embayment in the southern part of the West Coast Basin (Nathan *et al.* 1986). To the north, in the Taranaki Basin (Fig. 6.3), a major south-directed transgression beginning in the Maastrichtian caused the sea to flood across coastal floodplains, depositing shallow marine siltstones and coastal sandstones (Wizevich, 1994; King & Thrasher, 1996; Browne *et al.* 2008; Strogen *et al.* 2017).

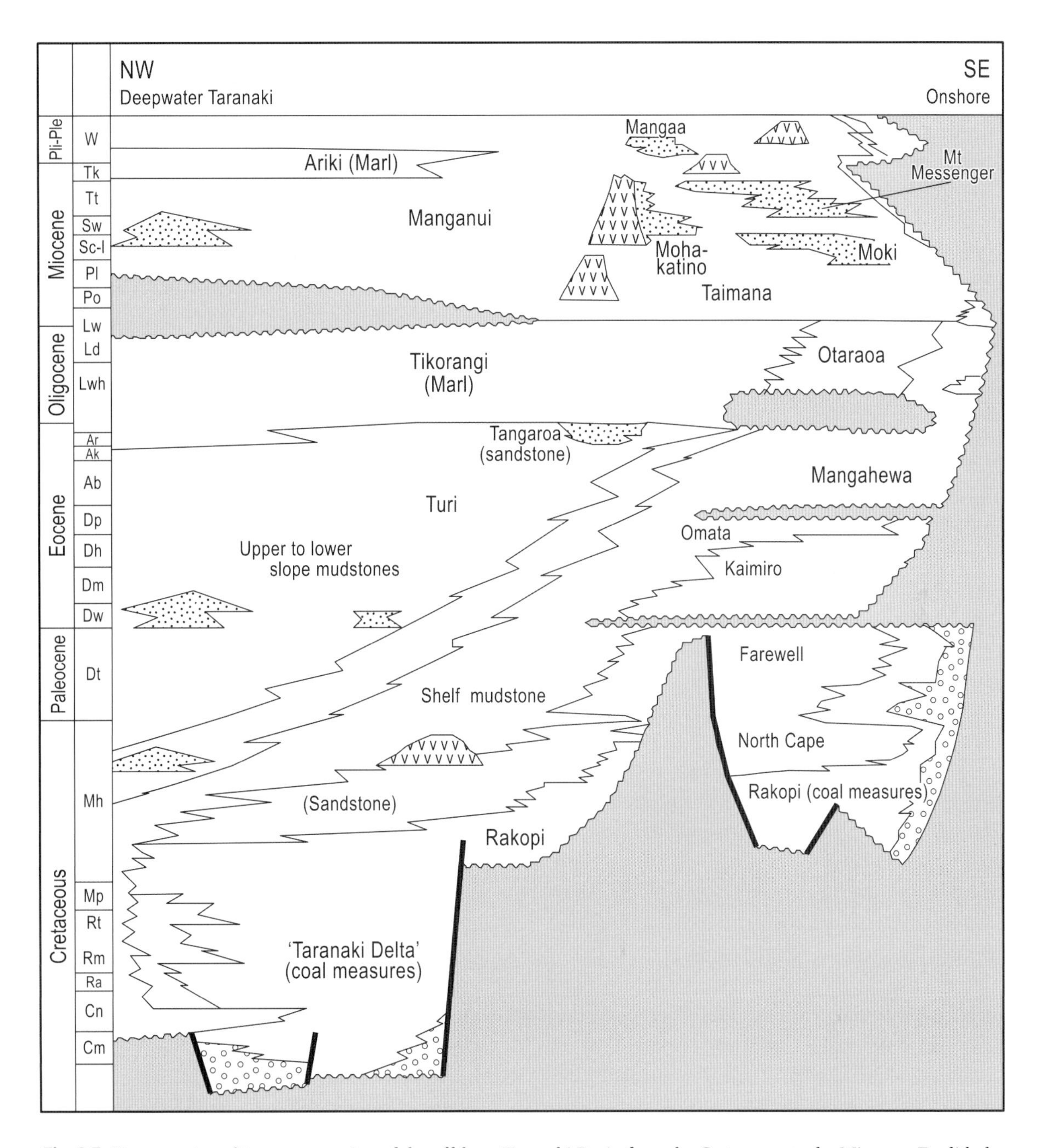

Fig. 6.3. Time-stratigraphic reconstruction of the offshore Taranaki Basin from the Cretaceous to the Miocene. For lithologies of named formations, see sections 4.2.1, 6.7.3.4, 6.10.3.5 and 8.2.6. (Based on Strogen *et al.* 2017, fig. 3.)

6.2.2 The Waipounamu Erosion Surface

Over much of onshore New Zealand, low relief erosion surfaces are well preserved over areas totalling more than 100,000 km^2. They are conspicuous in upland areas of the southern South Island, particularly those in Otago, and are veneered in many areas by fluvial gravel and in others by thin sequences of shallow marine deposits (Bishop, 1994; Landis *et al.* 2008). In addition, similar surfaces also bevel basement rock at scattered localities on the northern part of the South Island (Grindley, 1980), on the North Island (Korsch & Wellman, 1988), on the Chatham Islands (Campbell *et al.* 1993), and on Campbell Island, southern Campbell Plateau (Beggs, 1978). Offshore seismic profiles and scattered drillholes indicate the presence of low-relief, Late Cretaceous erosion surfaces (planar unconformities) beneath the Chatham Rise (Wood *et al.* 1989), Challenger Plateau (Nathan *et al.* 1986), Campbell Plateau (Houtz *et al.* 1967), Otago continental shelf (Carter, 1988b) and Canterbury region (Field *et al.* 1989). These data can be interpreted as indicating that an original erosion surface, including currently offshore localities, may have extended over an area as great as 106 km^2. The surface, however, appears to be a composite of features produced by fluvial erosion, followed by marine transgression and wave planation that took place during and after the separation of New Zealand from Australo-Antarctica. The term 'peneplain' is therefore inappropriate, and the name 'Waipounamu Erosion Surface' – using the Māori name for the South Island – has been suggested by LeMasurier and Landis (1996).

The Waipounamu Erosion Surface is clearly diachronous and has a long history in the South Island. The basal deposits on the surface are Late Cretaceous on the coast and become progressively younger when traced inland, and Paleocene, Eocene and Oligocene sedimentary rocks rest transgressively on the surface between the coast and Central Otago.

Crampton *et al.* (1999) suggested that the prominent unconformity in the late Santonian (near Piripauan–Haumurian boundary) represents the basinward expression of the Waipounamu Erosion Surface, although the surface may be the result of landward merging of both this and the older late Coniacian (near Teratan–Piripauan boundary) unconformity (Laird & Bradshaw, 2004).

6.2.3 Relationship of the unconformities to break-up and sea-floor spreading

Gaina *et al.* (1998) considered that sea-floor spreading in the Tasman Sea began before chron 34y (at ~84 Ma). The oldest reliably identified sea floor southeast of New Zealand was formed during chron 33r, which converts to an age of 83–79 Ma (Sutherland, 1999). However, a linear positive magnetic anomaly is present inboard of anomaly 33, and, while its correlation is uncertain, it may be part of anomaly 34 (118–83 Ma) and represent the earliest oceanic crust formed between southeastern New Zealand and Marie Byrd Land, Antarctica (Sutherland, 1999). Thus evidence from both the Tasman Sea and the southern Pacific Ocean suggests that sea-floor spreading around the Zealandia region may have started at or before ~84 Ma. Larter *et al.* (2002) suggest that the Chatham Rise may have started to separate from West Antarctica as early as 90 Ma, with the separation of the Campbell Plateau starting later, during chron 33r (83–79 Ma).

Convective upwelling of the mantle just prior to sea-floor spreading would involve secondary heating and thermal expansion, leading to a period of uplift and possible emergence at, or just prior to, break-up. Erosion during the uplift pulse causes a break-up unconformity, which we infer is represented by the older unconformity beneath transgressive coarse clastic sediments in the Raukumara Peninsula, Marlborough, the southern Canterbury Basin and the Great South Basin. Flooding is likely to have occurred first in the eastern, Pacific-facing, coastal areas, resulting in the late Coniacian (Teratan) transgression beginning at ~87 Ma.

The widespread nature of the subsequent Santonian (Piripauan–Haumurian boundary) unconformity or facies disjunction, and the rapid deepening of sedimentary environments above it, suggest rapid subsidence of the New Zealand continental block. That marine environments began encroaching on the New Zealand landmass from the southwest and northwest, as well as from the east, suggests that separation had occurred between New Zealand and Australo-Antarctica, and that sea-floor spreading and associated passive margin subsidence had begun. This inference is supported by the U–Pb zircon dates of 83±5 Ma (Tulloch *et al.* 1994) and 81.7±1.8 Ma (Waight *et al.* 1997) on the

West Coast A-type French Creek Granite, inferred to have been intruded shortly after the time sea-floor spreading began. Van der Meer *et al.* (2013) provide revised ages for related lamprophyre dikes. A-type granites are considered to be typical of rift-related non-orogenic settings, as are many alkaline basalt suites. The age of the beginning of post-separation subsidence and major widespread transgression (~85 Ma) coincides closely with these isotopic dates and the first magnetic anomaly (chron 34y at ~84 Ma) in oceanic crust of the Tasman Sea (Gaina *et al.* 1998).

6.3 LATE CRETACEOUS, SANTONIAN (PIRIPAUAN) SUCCESSIONS

6.3.1 Introduction

In the southeast of the South Island, subsidence of the Campbell Plateau and Chatham Rise began in response to the beginning of break-up, resulting in marine transgression beginning in late Coniacian (late Teratan) times. The shoreline advanced westwards in the Great South and Canterbury basins, and northwards over the Chatham Rise from the then open Bounty Trough, although the sedimentary basins were still dominated by non-marine deposits.

By contrast, regression preceded transgression in parts of the East Coast Basin. Although bathyal conditions continued in the Wairarapa area (Fig. 6.1b), in other parts of the East Coast Basin coarse erosion-based non-marine to shallow-marine sandstone units prograded to the northwest over older deeper-marine units, beginning in late Coniacian (late Teratan) times. This regression is also evident in the sediments of the East Coast Allochthon, but not in the Northland Allochthon, where bathyal conditions continued to prevail. The Northland Peninsula and areas to the west were probably still emergent, with local non-marine deposition possible west of the present shoreline. There are no known occurrences in outcrop or drillholes of strata of Piripauan age in western New Zealand from western Southland to Northland.

6.3.2 Description of Santonian (Piripauan) rocks

6.3.2.1 Northland Allochthon

Scattered macro- and microfaunas in the Mangakahia Complex (see section 4.3.2.1) indicate that Piripauan strata are present in the Punakitere Sandstone at Kaipara and Whangaroa harbours, and in float boulders from Awapoko facies sandstone. Foraminifera from the Hukerenui Mudstone yield wide-ranging ages, including Piripauan (Isaac *et al.* 1994). No clear stratigraphic relationships between the Santonian fossiliferous beds and surrounding strata could be determined.

6.3.2.2 East Coast Basin

There is a strong contrast between the sedimentary records of the Piripauan strata of the Western and Eastern sub-belts of the East Coast Basin. In the Western Sub-belt, Torlesse basement and Albian–Coniacian (Urutawan–Teratan) strata are overlain unconformably by formations of the shallow-marine to bathyal Tinui Group; while in the Eastern Sub-belt, turbidites predominate. In both Western and Eastern sub-belts the Piripauan deposits are overlain by mudstones of the Whangai Formation (Fig. 6.1b).

Western Sub-belt

In the Raukumara Peninsula the basal Tinui Group deposits (Fig. 6.1a) are dominated by massive to well-bedded sandstones (Tahora Formation; Isaac *et al.* 1991), which are exposed in a narrow strip in the west. The Tahora Formation has two main and one minor lithofacies. The westernmost lithofacies (Maungataniwha Sandstone) is a massive to poorly bedded, fine-grained quartzose sandstone, up to 500 m thick. It has common shellbeds and lenses, and is locally burrow-mottled with rare, steeply inclined *Ophiomorpha* burrows (Crampton & Moore, 1990). Sedimentary structures include probable hummocky cross-stratification, asymmetric ripples, and flaser and linsen bedding. It is considered to represent beach to shallow inshore environments (Crampton & Moore, 1990), although a fluvial environment for the basal portion of the westernmost outcrops cannot be ruled out (Browne & Reay, 1993). To the north-east the Maungataniwha Sandstone passes laterally into and is complexly interbedded with the Mutuera

Member, a glauconitic, poorly bedded bioturbated siltstone. It is less than 50 m thick and is considered to be a more offshore deposit. A locally restricted unit (Houpapa Member) consists of well-bedded alternating sandstone and mudstone; beds are often graded and with sole marks, and are interpreted as channel-fill turbidites.

A lateral equivalent, the Owhena Formation, extends in a narrow belt to the northeast of the Tahora Formation, reaching a maximum thickness of 100 m in the southwest, and thinning and finally disappearing to the northeast. It consists of a fining-upwards unit of sandstone, siltstone and mudstone (Phillips, 1985), which rests unconformably on Matawai Group strata, and passes upwards conformably into the Whangai Formation. The lower part of the Owhena Formation consists of a flysch-like facies, and is considered to have been deposited in deeper water than the Tahora Formation.

Paleocurrent indicators are relatively rare in the Tahora Formation. Small-scale channels are oriented southwest to northeast, and sediment transport from the southwest is indicated. No paleocurrent data are available for the Owhena Formation, but the general deepening environment to the northeast and orientation of channels in the Tahora Formation suggest a source and paleo-shoreline to the southwest. Diverse macrofaunas, microfaunas and microfloras indicate an age range from late Coniacian to late Santonian (Piripauan) for the Tahora and Owhena formations (Phillips, 1985; Isaac *et al.* 1991; Schiøler *et al.* 2002).

In the Western Sub-belt of the Wairarapa–Hawke's Bay region the Tangaruhe Formation (Fig. 6.1b), which unconformably overlies Springhill Formation, shows a laterally variable thickness, with a maximum of 280 m (Crampton, 1997). The lower part of the formation is characterised by relatively coarse alternating sandstone/mudstone and granular–pebbly sandstone or pebbly mudstone. The sandstone component of the upper Tangaruhe Formation is typically centimetre-bedded, commonly graded with Bouma sequences and fine- to very fine-grained. Glauconite makes up 1% to 30% of these beds. Slightly glauconitic mudstone dominates the upper part of the unit. Sandstone dikes of the same composition are abundant. Locally the formation consists of massive beds of poorly to very poorly sorted, richly glauconitic (up to 75%), granular–pebbly sandy mudstone to muddy sandstone (Mangawarawara Member) inferred to have been deposited by debris flows (Crampton, 1997). The presence of debris flows and turbidites, and abrupt lateral variations in facies and thickness in the lower part of the Tangaruhe Formation, are evidence for significant local relief during deposition, high rates of sedimentation and possibly synsedimentary tectonic activity. Much of the upper part of the formation is intensely bioturbated, a characteristic of accumulation in a low energy, tectonically quiescent environment. Deposition in an anoxic basin at outer neritic or upper bathyal depths seems probable (Crampton, 1997).

Paleocurrent directions recorded for the Tangaruhe Formation are consistently from the southwest. The presence of the coarsest facies (Mangawarawara Member) in the west also supports derivation from this direction. Preliminary factor analysis of heavy mineral samples suggests that the Tangaruhe Formation may have been reworked from the underlying Springhill Formation (Smale in Crampton, 1997). The age of the Tangaruhe Formation (Crampton, 1997), based on pollen, dinoflagellates, foraminifera, and rare macrofossils, is late Coniacian–late Santonian (late Teratan–early Haumurian) and it is thus an age equivalent of the Tahora and Owhena formations of the Raukumara Peninsula.

Western Sub-belt, South Island

The age-equivalent deposits in Marlborough, the Paton Formation (Webb, 1971), consists dominantly of slightly to moderately glauconitic, fine- to medium-grained sandstone (Fig. 6.1b). The Paton Formation is unconformable on older rocks, except in the east, where it rests with apparent conformity on the Raukumara-aged Burnt Creek Formation (Crampton & Laird, 1997). Basal scours up to one or two metres deep are locally eroded into the underlying strata and infilled with pebbly or gritty coarse to medium-grained sandstone (Laird, 1992). A common characteristic of the Paton Formation is its strongly burrowed character, including large *Ophiomorpha*. Locally the basal contact is concordant but sharp, and is underlain by a weathered zone with scattered pebbles of Torlesse Sandstone, blebs of glauconite and small rounded phosphatic clasts, indicating a probable hiatus (Laird, 1992). To the south of Kaikoura, Paton greensand passes into massive or cross-bedded, locally shelly non-glauconitic sandstone (Okarahia Sandstone) suggesting a transition into a shallow-marine to possibly non-marine environment, indicating the proximity of a shoreline (Warren & Speden, 1978). In the middle Clarence Valley, also, greensand gives way in the most southwesterly exposures to grey muddy sandstone with nearshore characteristics (Schiøler *et al.* 2002). The Piripauan shoreline is likely to closely parallel that for the Ngaterian, with the marine basin closing to the west and south, and opening to the northeast on the current geographic configuration (Laird, 1992; Crampton *et al.* 2003). This is supported by lithofacies variations. In the northeast of the outcrop area, near Kekerengu, bedding becomes more distinct, and includes decimetre-thick massive or graded-bedded sandstone beds with rare sole marks. Isopachs on the Paton Formation (Laird, 1992; Crampton *et al.* 2003) show a thickening of the sequence towards the northeast, reaching a maximum of 320 m at Kekerengu River.

A shallow marine environment is inferred for the bulk of the basal glauconitic sandstones, passing into nearshore or non-marine to the south and southwest. To the northeast there is a rapid change to deeper marine shown by the presence of probable turbidites, and minimum paleo-water depths of 100–150 m are indicated from foraminiferal studies (Strong *et al.* 1995). An overall deepening trend up-section is evident in the best-studied sections through the Paton Formation in eastern Marlborough, showing continuing transgression through time (Schiøler *et al.* 2002).

Eastern Sub-belt

Rocks of the East Coast Allochthon make up the Eastern Sub-belt in the Raukumara Peninsula (Fig. 6.1a) and sedimentation of turbidites was continuous from the Coniacian into the Santonian (Teratan into the Piripauan), but in the late Coniacian (late Teratan) there is a rapid transition into thick-bedded more proximal sandstone: the Tapuwaeroa Formation of Mazengarb and Speden (2000). Individual sandstone beds are up to 6 m thick, include a high proportion of carbonaceous material and are commonly graded with Bouma sequences. Thickness of the succession reaches 1150 m. Paleocurrent data indicate a prevalent northwest to north-northwest flow direction (Mazengarb, 1993). Age of the Tapuwaeroa Formation (Fig. 6.1a) ranges from late Coniacian to Campanian (late Teratan to early Haumurian).

In the Eastern Sub-belt of the Wairarapa–southern Hawke's Bay region, deposition of turbidites of the Glenburn Formation (see Chapter 4) continued from Clarence Series times through to the late Santonian. No Eastern Sub-belt Santonian strata have so far been recognised in Marlborough.

6.3.2.3 Eastern basins of the South Island

On the Chatham Islands, Santonian strata are probably represented by the lower part of the Kahuitara Tuff, which passes upwards with rapid transition from paralic sediments of the Clarence and Raukumara-aged Waihere Bay Group (see Fig. 4.2). It consists primarily of green to brown-grey massive to well-bedded volcaniclastic sandstone, grit, and scoriaceous conglomerate and lapilli tuff of basaltic composition. Some higher beds are calcareous and more regularly bedded. The unit is fossiliferous and at least in large part is marine (Campbell *et al.* 1993), and it appears to be restricted to Pitt Island. It attains a maximum thickness of approximately 225 m and passes into the overlying Southern Volcanics of Campanian (Haumurian) age without noticeable break. The unit is poorly dated, and may range from late Coniacian to late Santonian (early Haumurian). Foraminifera are characteristic of an inner-shelf, open marine environment of near normal salinity, with water depths in the range 5–50 m, which is consistent with the macrofaunal indications. The unit represents the beginning of full marine transgression across the older paralic sediments of the Waihere Bay Group.

Most of the Canterbury region was subaerial at the beginning of Santonian (Piripauan) times, although in the south of the region onlap by marine sediments was proceeding. Piripauan sediments form the basal strata of the Katiki Formation, which is represented in outcrop at Shag Point, north Otago, and in the Endeavour-1, Clipper-1 and Galleon-1 drillholes (Fig. 6.2). They typically comprise brownish-grey marine siltstone and mudstone, with less common sandstone that contains shallow marine microfossils. In the Clipper and Galleon drillholes the basal Piripauan sediments rest unconformably on older Cretaceous sediments or on basement. The Piripauan sediments in the Clipper-1 drillhole record a rapid relative rise in sea level (Field *et al.* 1989).

In the Great South Basin, marine transgression developed across the basin from the north and east close to the Coniacian–Santonian boundary (basal Piripauan). Paralic sands developed during these marine incursions, with coal swamps on the landward side. By the end of Piripauan times a restricted shallow-marine environment persisted in the Central Sub-basin, in which marine shales were deposited. A regional unconformity of late Santonian (late Piripauan) age over the southern, eastern and central parts of the basin separates the largely terrestrial Hoiho Sequence from late Santonian–Paleocene (Piripauan–Teurian) deposits, which are referred to as the 'Pakaha Group' (Cook *et al.* 1999). The unconformity is succeeded by a late Santonian (late Piripauan) transgressive shallow-marine sandstone (Kawau Sandstone), reaching at least 150 m in thickness, over most of the basin. This consists of grey, kaolinitic, quartzose, moderately well-sorted, mainly medium sand, and is a lateral equivalent of the shallow marine Katiki Formation (see above) at Shag Point (Fig. 6.4). Along the northwest margin of the basin, however, terrestrial deposition (including coal measures) persisted and recognition of the sequence boundary is difficult.

6.4. CAMPANIAN–PALEOCENE (HAUMURIAN–TEURIAN) SUCCESSION: SEDIMENTATION DURING SEA-FLOOR SPREADING (~84–~56 Ma)

6.4.1 Introduction

The first sedimentary deposits on the late Santonian (early Haumurian) erosion surface recorded a dramatic landward move of the shoreline as transgression accelerated as a result of increased subsidence due to separation from Australo-Antarctica. Marine sediments were dominated initially by a fine-grained terrigenous lithofacies derived from a low-lying landmass with little relief. Regional subsidence and transgression continued until the cessation of sea-floor spreading in the latest Paleocene (~56 Ma).

During this period of spreading two significant events occurred, each of which had dramatic effects on geological evolution of the region. The first was a tectonic event at ~70 Ma that caused reactivation of fault systems and sedimentary basins, mainly in the western and southern portions of New Zealand;

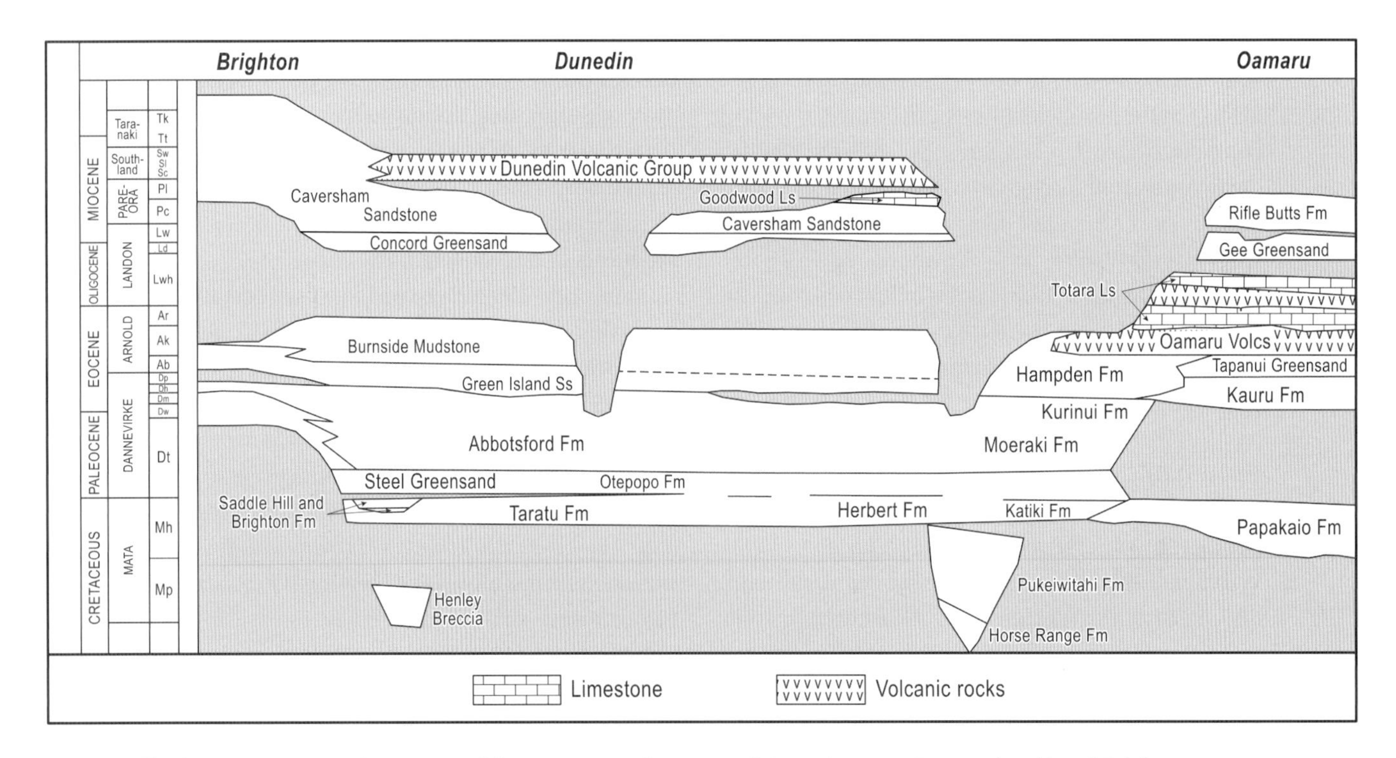

Fig. 6.4. Time-stratigraphic section of the Cenozoic rocks in coastal Otago between Oamaru (north) and Brighton (southwest of Dunedin). For lithologies of named formations, see section 6.3.2.3. Fm, Formation; Ls, Limestone; Ss, Sandstone; Volcs, Volcanics. (Adapted from Field *et al.* 1989.)

and the second was the mass extinction of species at the Cretaceous–Tertiary (K–T) boundary (see section 9.2.2).

6.4.2 Maastrichtian–Paleocene (Haumurian–late Teurian) sediments of eastern New Zealand

6.4.2.1 Whangai facies

The most widespread lithofacies occurring above the basal transgressive sands of mainly Santonian (Piripauan) age is of marine mud-rich, commonly siliceous lithologies of dominantly late Santonian to Paleocene (Haumurian to Teurian) age, which form a regional blanket stretching over most of eastern New Zealand from eastern Northland to the Great South Basin (Fig. 6.1a,b). The lithofacies is inferred to represent the sedimentary response to the initiation of sea-floor spreading and consequent rapid subsidence and transgression. The mud-rich lithologies are referred to as various facies of the Whangai Formation over much of the region, although separate formation names are used locally. Whangai Formation (Moore, 1988) is named from the Whangai Range, southern Hawke's Bay, where it consists of up to 500 m of medium-grey siliceous or calcareous mudstone of Haumurian to Teurian age. Although the mudstones have several local lithological variants, they will here be referred to collectively as the 'Whangai facies'. Over most of the region the facies extends from the Late Cretaceous into the Paleocene. However, in southwestern Wairarapa, it is present only below the K–T boundary (Laird *et al.* 2003), and in Marlborough it is restricted to a position beneath the cherty Mead Hill Formation of Maastrichtian to Early Paleocene (late Haumurian to Teurian) age (Field *et al.* 1997).

Northland

In eastern Northland, siliceous mudstone and calcareous mudstone and chert up to 300 m thick of the Whangai facies is widespread in the Mangakahia Complex (see Fig. 4.7) of the Northland Allochthon (Isaac *et al.* 1994). The mudstones have yielded Campanian–Paleocene (Haumurian–Teurian) foraminiferal faunas, suggesting deposition at bathyal or abyssal depths. In autochthonous Northland, the correlative Waiari Formation comprises 120–160 m of dark-grey to black centimetre to decimetre well-bedded chert, and massive to moderately fissile, dark-grey to black siliceous mudstone, locally bioturbated and of Campanian to Paleocene (Haumurian to Teurian) age. The outcrop is restricted to the Karikari Peninsula in northeastern Northland, where the facies rests unconformably on older Cretaceous strata (Isaac *et al.* 1994).

Eastern North Island

In the North Island sector of the East Coast region, the 'type' Whangai facies has several lithological variants: the regionally extensive Rakaura, Upper Calcareous and Porangahau members (Moore, 1988, 1989a) and the localised Kirks Breccia and Te Uri members (Fig. 6.1a). The first three members occur in both the Western and the Eastern sub-belts. The Whangai facies in many instances is in gradational contact with underlying Piripauan strata, with the exception of the Kirks Breccia Member (see below). However, in some parts of the East Coast Allochthon it is inferred to rest unconformably on Piripauan strata of the Tikihore or Tapuwaeroa formations (Rait, 1992), and in Marlborough there is commonly an unconformity or hiatus between its equivalent (Herring Formation) and the underlying Paton Sandstone (see section 6.3.2.2). The age of the Whangai facies here is middle Santonian–Late Paleocene (late Piripauan–late Teurian), based on foraminifera and dinoflagellates (Crampton *et al.* 2001).

Typical Whangai Formation consists of hard, generally poorly bedded, rusty-weathering, dark- to medium-grey, slightly siliceous non-calcareous mudstone. It is found throughout the North Island portion of the East Coast region, and is usually the lower Whangai unit. A laterally equivalent, poorly bedded, medium-grey, hard, slightly to moderately calcareous mudstone is also widespread throughout the East Coast of the North Island, and occurs both in the Western and Eastern sub-belts, but more commonly in the east (Moore, 1988). The dominant variant of Paleocene (Teurian) age consists of well-bedded, light-grey to white, hard, moderately calcareous mudstone, with common glauconitic sandstone beds. In some instances the glauconite content increases and the unit consists of interbedded glauconitic sandstone and slightly calcareous, glauconitic siltstone.

Locally, major channel systems occur. Near Motu, in the Western Sub-belt of the Raukumara Peninsula, a basal matrix-supported breccia of Santonian (Piripauan to early Haumurian) age, which includes blocks up to at least 10 m (Kirks Breccia Member; Moore, 1989a), fills a channel, up to 200 m deep and at least 2 km wide, cut into older Cretaceous sediments (Field *et al.* 1997). At Tora, southeastern Wairarapa, a channel, up to 18 m deep and perhaps 4 km wide, filled with alternating thin-bedded sandstone and mudstone with thin conglomerate lenses and limestone beds, also occurs within the uppermost part of the Whangai Formation, just below the K–T boundary (Laird *et al.* 2003). Below the channel system the late Haumurian portion of the ~200-m-thick Whangai Formation includes zones of slumps and olistostromes, suggesting local tectonic activity. Hummocky cross-stratification is also present, suggesting storm wave activity in a shelf environment (Laird *et al.* 2003).

Most of the Whangai Formation on the East Coast of the North Island appears, from foraminiferal assemblages, to have accumulated at bathyal depths, although shelf- to upper-bathyal species occur locally (Field *et al.* 1997). The facies is likely to be prominent in offshore East Coast deposits, as at ODP site 1124 on the eastern margin of the Hikurangi Plateau a similar facies of light-greenish cherty nannofossil diatom ooze occupies the latest Cretaceous portion of the column (Carter *et al.* 1999).

Eastern South Island

In Marlborough the basal Piripauan glauconitic sandstone unit (Paton Formation) is followed with sharp, commonly unconformable contact, by late Santonian–Maastrichtian (late Piripauan–Haumurian) dark-grey or black and locally siliceous or carbonaceous sandy mudstone, with rare interbeds and dikes of well-sorted very fine-grained sandstone (Herring, Woolshed and Conway formations; see Fig. 6.1b), here referred to as the 'Whangai facies'. Burrowing occurs at many horizons, but is not as ubiquitous as in the underlying glauconitic sandstone. Successions are in many instances characterised by jarositic efflorescence, and locally by the presence of large scattered dolomitic concretions up to 5 m in diameter (Browne, 1985). Lithofacies show a systematic change from sandy mudstone with interbeds of well-sorted sandstone in the southwest, to massive unlaminated siliceous dark mudstone in the northeast, closely similar to the Whangai Formation of the East Coast of the North Island. In northeastern Marlborough, in the coastal area lying east of the London Hill Fault (see Fig. 4.3), mudstones otherwise similar in appearance to the Whangai facies to the west include packets of turbidites and turbidite-filled channels, and horizons of slumps, thick mass-flow sandstone units and breccias (Price, 1974; Laird & Schiøler, 2005; Hollis *et al.* 2005b). Close similarities between the stratigraphy of the Ward coastal area and that of Tora in the Wairarapa, which forms part of the Eastern Sub-belt, suggest the likelihood that the area east of the London Hill Fault may form an extension of the sub-belt into Marlborough (Laird & Schiøler, 2005).

Isopachs on the mudstones of the Whangai Formation here, like those of the underlying Paton Formation, show a general thickening of the strata from ~10 m in the southwest to 700+ m in the northeast (Laird, 1992; Crampton *et al.* 2003). After palinspastic reconstruction taking into account the effects of Neogene deformation, the corrected data show thickening of strata towards the northwest (Crampton *et al.* 2003). The bulk of the carbonaceous mudstones were probably deposited in a shallow-marine, possibly mid-shelf, poorly oxygenated, reducing environment (Schiøler *et al.* 2002). The presence of slump horizons and thick mass-flow sandstone bodies in the Ward coastal area suggests that the sediments here were deposited in a slope environment. Deepening of the basin to the northeast (present configuration) is clearly indicated. Southwest of Kaikoura, in north Canterbury (Figs 6.5 and 6.6 and Plate 9), massive jarositic siltstone of the Whangai facies (Conway Formation) passes laterally into jarositic sandstone and loses its Whangai-like identity (Field *et al.* 1997).

In the southern Canterbury Basin the Whangai facies is represented by Campanian–Early Paleocene (Haumurian–earliest Teurian) marine mudstone (Katiki Formation), which locally includes thin beds of hard microcrystalline limestone, and reaches a thickness of 1000 m at Clipper-1 petroleum well (Field *et al.* 1989). The basal part of the Katiki Formation in Clipper-1 and Endeavour-1 wells is shallow marine, but higher in the succession it grades into a bathyal environment in Clipper-1 (Field *et al.* 1989). In petroleum wells Endeavour-1 and Galleon-1 the Katiki

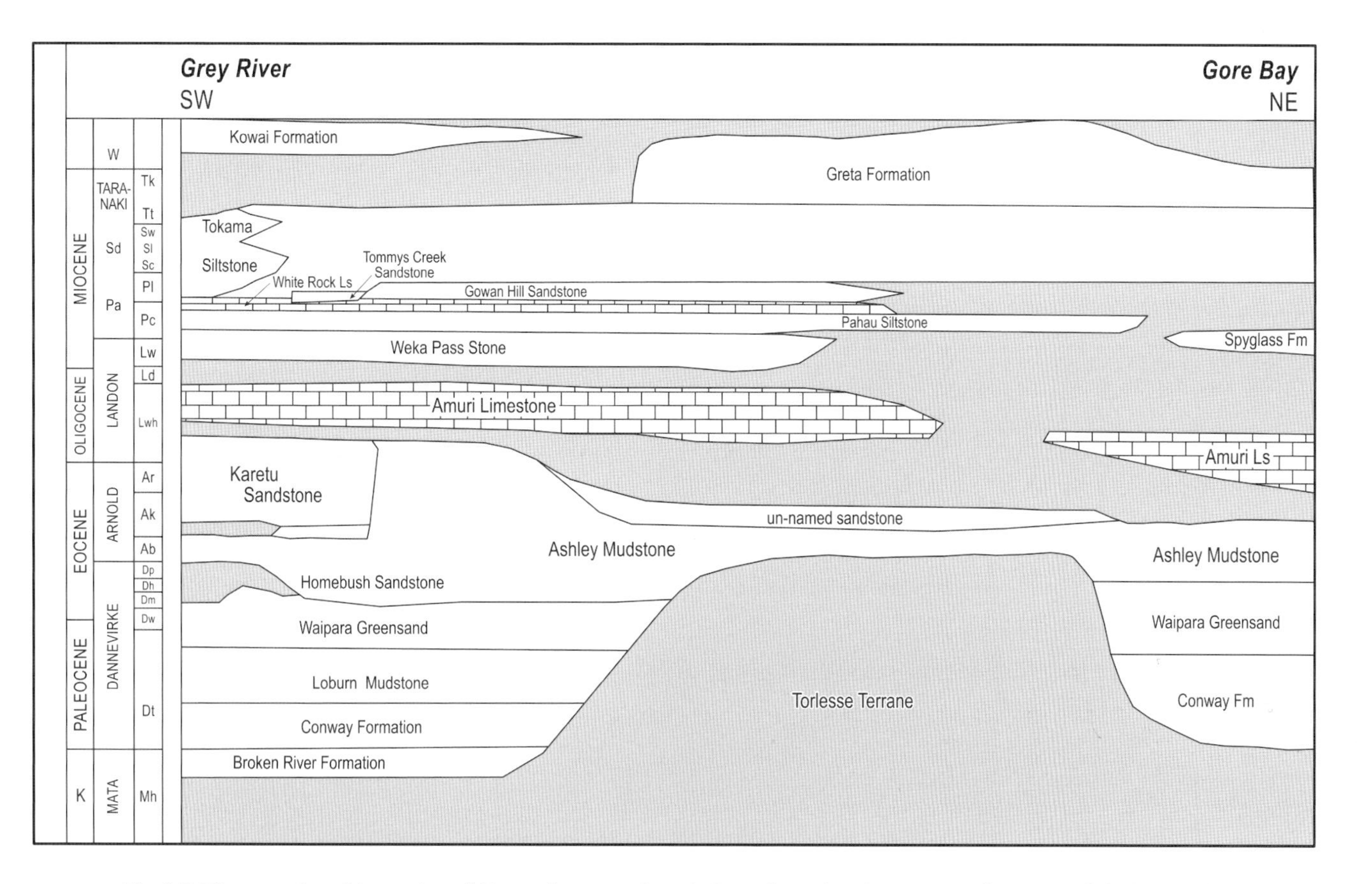

Fig. 6.5. Time-stratigraphic section of Cenozoic strata of north Canterbury (southwest to northeast). For lithologies of named formations, see section 6.4.3.2. Fm, Formation; Ls, Limestone. (Adapted from Field *et al.* 1989.)

Formation interfingers with and overlies non-marine beds. The Whangai facies is seen in outcrop only at Shag Point on the north Otago coast, where it consists of ~300 m of dark-grey to black mudstone and muddy sandstone containing numerous siliceous concretions (Figs 6.2, 6.5 and 6.6).

In the Great South Basin the Whangai facies is Santonian to Late Paleocene (late Piripauan to late Teurian) in age and represented by slightly glauconitic siliceous mudstones of the Wickliffe Formation (Cook *et al.* 1999). A nearshore to inner-shelf marine environment is indicated (Raine *et al.* 1993). The most widespread marine conditions appear to have been at the end of the Cretaceous, followed by shallowing to coastal environments during the Paleocene. On Campbell Island to the south, fluvial quartz-rich sands, resting unconformably on basement schist, are overlain by 30–100 m of massive dark-grey mudstones of the Whangai facies containing microfossils of Campanian to Paleocene (Haumurian to Teurian) age (Garden Cove Formation), suggesting an inner-shelf to brackish environment (Beggs, 1978; Hollis *et al.* 1997). DSDP site 275 of leg 21, lying on the southeastern margin of the Campbell Plateau, and nearby ODP site 1121 have very similar latest Cretaceous and Early Paleocene lithofacies respectively, consisting of slightly carbonaceous glauconitic silty clay (Andrews & Ovenshine, 1975). It is probable that much of the latest Cretaceous–Early Paleocene succession of the Campbell Plateau lying southeast of the Great South Basin is made up of the Whangai facies.

Western South Island, North Island and offshore

Whangai facies has not been specifically identified in western New Zealand, but has been recognised tentatively at DSDP drilling site 207 on the Lord Howe Rise. Here, transgressive silty clay and siliceous nannofossil ooze of Maastrichtian (late Haumurian) to Paleocene age that rest directly on continental crust and are associated with the initiation of sea-floor

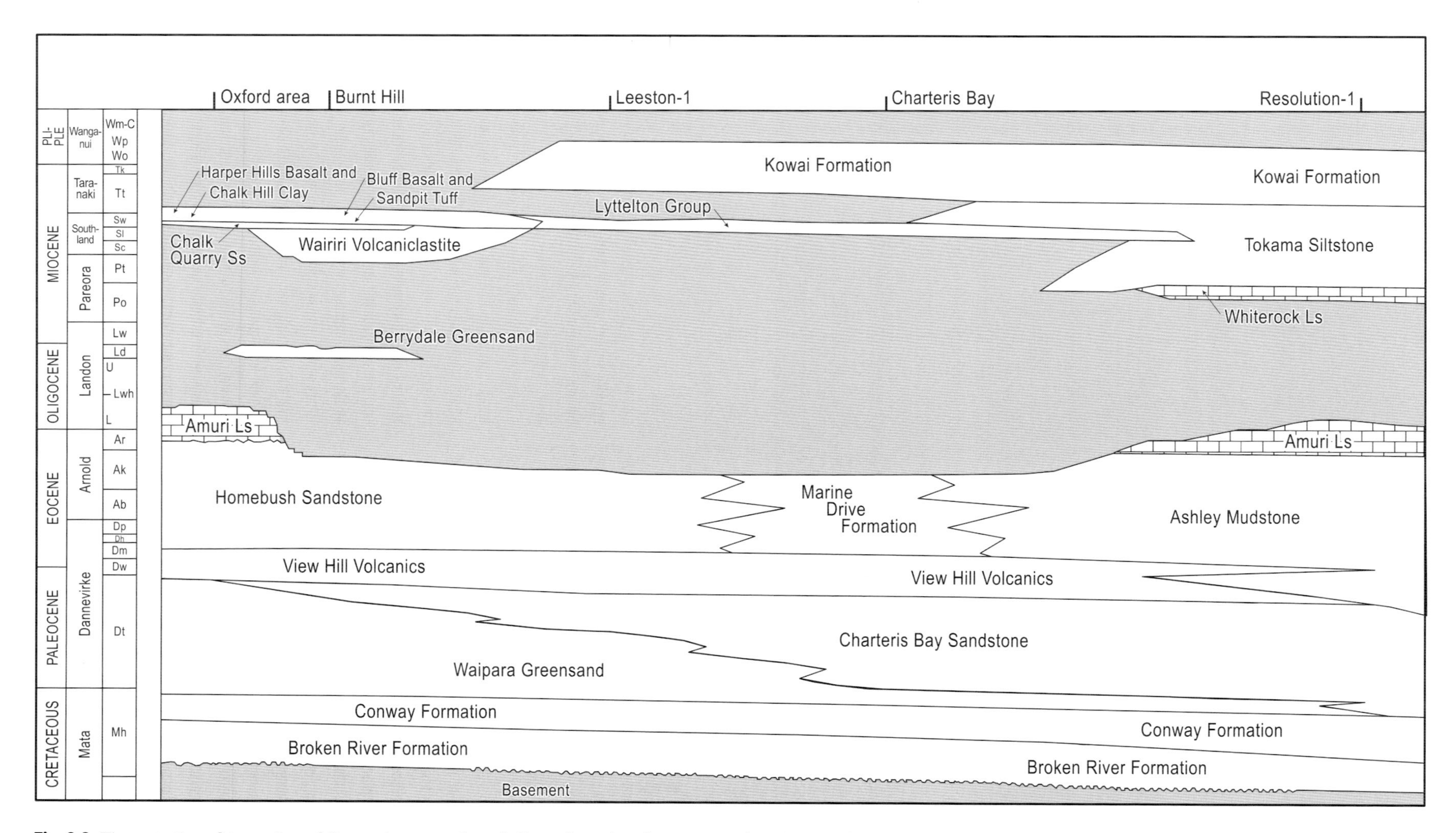

Fig. 6.6. Time-stratigraphic section of Cenozoic strata of north Canterbury (northwest to southeast). For lithologies of named formations, see section 6.4.3.2. Ls, Limestone; Ss, Sandstone. (Adapted from Field *et al.* 1989.)

spreading in the Tasman Sea have been correlated with the Whangai facies (Andrews & Ovenshine, 1975). Similar facies, which form the initial deposits at other drill sites in the Tasman Sea, are also inferred to be closely related to the history of Tasman seafloor spreading.

The widespread deposition of siliceous shale in the New Zealand region during the Late Cretaceous–Paleocene was attributed by Moore (1988) to transgression and trapping of coarser sediments near the strandline, coincident with development of a slowly subsiding basin or basins, and increase in productivity of siliceous organisms. The environment of deposition is inferred to be one with restricted circulation of ocean currents, which appeared to be a feature of the southwest Pacific region at this time (Andrews & Ovenshine, 1975; Moore, 1988).

6.4.3 Other lithofacies of eastern New Zealand

6.4.3.1 Northland

The Whangai facies is the main lithofacies of the Late Cretaceous–Paleocene sediments of the Northland Allochthon, although some sandstones of the Mangakahia Complex are age equivalents. In the autochthonous strata of onshore Northland, sediments of possible Paleocene age are known only from the bottom of the Waimamaku-2 petroleum exploration well, where Early Cretaceous mudstone is unconformably overlain by 48 m of glauconitic sandstone, with a thin basal conglomerate (Isaac *et al.* 1994). Sidewall cores contain reworked Late Cretaceous and in situ Paleocene–Eocene dinoflagellates.

In most of the North Island sector of the East Coast Basin, deposition of mudstone of the Whangai Formation continued on from Late Cretaceous until Paleocene times. The Late Paleocene up to the latest Paleocene 'Event' (see section 6.4.1) was dominated by the Waipawa 'Black Shale' facies (Fig. 6.1b), or, locally, by the laterally equivalent Whangai facies. The main exception is in the Eastern Sub-belt of southeastern Wairarapa, where rapid lateral and vertical facies changes occur in Paleocene sediments, which form lithological successions not recognised elsewhere on the East Coast of the North Island. In the Flat Point to Glenburn area, the Maastrichtian (late Haumurian) Whangai Formation is overlain gradationally by 90–100 m of Paleocene (Teurian) well-bedded glauconitic or quartzose sandstone and siltstone, calcareous siltstone and mudstone (Lee, 1995).

At Tora, 40 km to the southwest (see Fig. 4.3), similar Paleocene alternating sandstones and mudstones (Awhea Formation) are separated from the underlying Maastrichtian (Haumurian) Whangai Formation by a channel-fill complex (Manurewa Formation) consisting of limestone, alternating calcareous sandstone and siltstone, olistostromes and greensand, spanning the K–T boundary (Laird *et al.* 2003). Two major channelled units are represented. The oldest, which reaches a maximum of thickness of ~14 m and a width of possibly 4 km, erodes into the Whangai Formation, and is of late Maastrichtian (latest Haumurian) age. It consists of alternating thin sandstone and mudstone with thin limestone beds in the northeast, and calcareous sandstone and mudstone in the southwest. A quiet environment of deposition is inferred, probably deeper (bathyal?) than that of the underlying Whangai Formation.

The younger channel system, of Early Paleocene (early Teurian) age, consists dominantly of medium to coarse, thick-bedded, glauconitic sandstone, with local low-angle cross-stratification and microfloras suggesting low salinity. The unit includes olistostromes with megaclasts of limestone, and locally with clasts up to 1 m of dark-grey siltstone/very fine sandstone typical of the Whangai Formation. The younger channel system erodes into the older, and in the southwest it rests sharply on the Whangai Formation. The environment of deposition is inferred to be shallow-marine shelf, and implies a relative drop in sea level in the earliest Paleocene (Laird *et al.* 2003).

The channel-fill sediments pass upwards with rapid gradation into well-sorted bioturbated sandstones, locally with prominent low-angle cross-stratification, of the Paleocene Awhea Formation, which is interpreted as comprising very-shallow-marine probably nearshore deposits in the south (Laird *et al.* 2003). The thickness of the unit is variable, reaching a maximum of 290 m in the south of the area, and thinning and disappearing northwards (Field *et al.* 1997). Conformably overlying the unit here, and Whangai Formation 40 km to the northeast in the Glenburn area, is white, micritic limestone, up to 100 m thick, correlated with the Amuri Limestone of Marlborough (Field *et al.* 1997; and see section 6.7.2.2).

6.4.3.2 Northeastern South Island

In western and southern Marlborough the Whangai facies (Herring Formation) is sharply overlain by Maastrichtian (late Haumurian) well-sorted sandstone (Branch Sandstone; Fig. 6.1b), which commonly infills shallow channels or scours, locally with slight angular discordance (Laird, 1992; Reay & Strong, 1992). It is a massive, well-sorted, quartzo-feldspathic, mostly fine-grained sandstone, which is locally calcareous and slightly glauconitic. The base is in some places marked by scattered grit or small pebbles, and also locally by horizontal burrows. It shows rapid lateral thickness variations, ranges between 0.5 m and ~75 m, and generally thins and finally disappears to the northeast, but otherwise no pattern can be distinguished. The sandstone typically lacks body fossils, but it is highly bioturbated, with abundant trace fossils. The few faunas recovered indicate a shelf environment, with probably well-oxygenated bottom waters. Scattered phosphatic clasts and ubiquitous glauconite suggest slow rates of sediment supply. There is no evidence for subaerial erosion, and the unit is inferred to have been deposited as a migrating marine barrier bar consisting in part of sand reworked from the underlying Herring Formation (Reay & Strong, 1992). The Branch Sandstone is absent in the northeast of Marlborough.

Gradationally overlying the Branch Sandstone, or, locally, the Whangai facies, is the calcareous and siliceous Mead Hill Formation (Fig. 6.1b). The formation consists of decimetre-bedded, greenish-grey, strongly cemented, chert-rich micritic limestone with interbeds of medium to dark-grey, poorly indurated, calcareous smectitic mudstone. Sandstone dikes from the Branch Sandstone locally intrude the unit. Trace fossils are abundant. Hollis (1991) suggested a dominantly biogenic source for abundant silica within the formation, although Lawrence (1994) preferred an inorganic origin for most of the silica. The Mead Hill Formation, which is Maastrichtian to Late Paleocene (late Haumurian to late Teurian) in age and straddles the K–T boundary (see section 9.3.3), reaches its maximum thickness of 257+ m (of which 175+ m is Late Cretaceous) at its type section in Mead Stream. It is at its most extensive in the Clarence Valley, but is progressively truncated to the southwest beneath a latest Paleocene regional unconformity, and does not appear southwest of Bluff River. The Mead Hill Formation also occurs in the coastal areas between Cape Campbell and Kaikoura Peninsula, but does not extend further south (Morris, 1987).

Initial deposition of Mead Hill Formation over most of Marlborough appears to have been at about mid-shelf depths similar to those inferred for the underlying Branch Sandstone and in an environment with limited access to oceanic circulation. Common trace fossils indicate abundant bottom life, and normal oxygenation. High water-mass fertility (Hollis, 1991) supported a large siliceous biota, which, with diagenesis, contributed abundant silica to reprecipitate as the nodules and beds of chert that characterise the late Maastrichtian–Paleocene (late Haumurian–Teurian) rocks throughout much of Marlborough. Later in Haumurian times, depths increased relatively rapidly to outer-shelf to upper-bathyal, and connections to oceanic circulation rapidly became established. Where preserved, uppermost Haumurian siliceous limestones form a distinctive unit, with a fauna reflecting mid-bathyal depths, and slightly marginal to fully oceanic water mass conditions (Field *et al.* 1997).

Inland and southwest of Kaikoura the Haumurian succession is represented by fluvial conglomerate and paralic sandy siltstone (Stanton Conglomerate and Lagoon Stream formations), probably deposited in a half graben, which pass eastwards into massive grey shallow-marine siltstone of the Whangai facies, the Conway Formation of Field *et al.* (1989). Further south, in north Canterbury, the facies passes laterally into grey slightly calcareous, jarositic, bioturbated, slightly glauconitic silty fine-to-medium sandstone (Fig. 6.5). The Conway Formation is in turn overlain by ~20 m of moderately indurated, brown, sometimes concretionary slightly calcareous sandstone that may be a correlative of the Branch Sandstone further north. Mead Hill Formation is absent. Campanian–Maastrichtian (Haumurian) quartzose carbonaceous sands with thin conglomerate at the base (Broken River Formation) were deposited in terrestrial to neritic environments on either side of a 15-km-wide northwest–southeast-trending emergent high (Hurunui High). Local fault control of sedimentation is apparent, with deposition of up to 700 m of strata in a west-northwest–east-southeast-trending half graben (Nicol, 1993).

During the Paleocene a narrow seaway probably existed between the Hurunui High and the Chatham Rise, the western part of which, extending to just north of Banks Peninsula, was also probably emergent, as it is apparently free of Paleocene sediments (Field *et al.* 1989; and Fig. 6.6). Three depositional facies, laterally interfingering with each other, accumulated between the emergent Chatham Rise and the landmass to the west. In the northeast the sandy, slightly glauconitic Conway Formation (Whangai facies) passes upwards gradationally into similar mudstones (Loburn Mudstone) of Paleocene age, reaching a maximum thickness of 50–70 m over much of the southern part of its distribution (Browne & Field, 1985). The unit passes southwest laterally into, and is overlain by, richly glauconitic medium- to fine-grained quartzose sandstone (Waipara Greensand), which reaches a thickness of 80 m at its type locality at Waipara River, but thins considerably to the north and south, being only 1.5 m thick in the Kowai-1 petroleum well. To the west and south both the Waipara Greensand and the Loburn Formation pass laterally into a light-grey, commonly massive, well-sorted, medium to fine quartz sandstone (Charteris Bay Sandstone). This unit reaches its greatest thickness of 400 m in the west at Broken River, west of Christchurch, but also occurs on Banks Peninsula and in the offshore Resolution-1 petroleum well to the south (Browne & Field, 1985; Andrews *et al.* 1987). It was deposited in very-shallow-marine, possibly tidal, high-energy environments (Field *et al.* 1989). South of central Canterbury the Charteris Bay Sandstone merges to the west with fluvio-deltaic quartzose coal measures of the Broken River Formation, which south of Mount Somers is dominantly of Paleocene to Early Eocene (Teurian to Waipawan) age (Field & Browne, 1986). Offshore, in the Resolution-1 petroleum well, the Broken River Formation is paralic; it is at least 128 m thick, and is overlain by 120 m of marine sandstone (Conway Formation).

6.4.3.3 Chatham Islands and Chatham Rise

On the Chatham Islands the upper portion of the Kahuitara Tuff exposed on Pitt Island (see earlier) is dated as Campanian (early Haumurian) by Campbell *et al.* (1993). Deposits of this age consist of pale, yellow-brown, well-bedded, burrowed, fossiliferous and slightly calcareous very fine volcanic sandstone, fine pebbly sandstone, and grit. At northwestern Pitt Island a fissure cutting the tuff is filled with limestone containing a microfauna of late Maastrichtian (late Haumurian) age characteristic of an inner-shelf, open marine environment of near normal salinity, with water depths in the range 5–50 m (Strong, 1979). At eastern Pitt Island, deposition of Kahuitara Tuff was interrupted at about 80 Ma (early Campanian/early Haumurian) by the extrusion of a thick sequence of mildly alkaline olivine basalt (Southern Volcanics; see section 6.4.4). Fossiliferous limestone associated with pillow lavas on the southwest coast of Chatham Island is of a late Haumurian age similar to the fissure-fill limestone of Pitt Island (Campbell *et al.* 1993).

During latest Cretaceous and Paleocene times, parts of the Chatham Rise probably remained emergent, although slow subsidence was occurring. Much of the Paleocene sequence exposed on the Chatham Islands consists of thin discontinuous marine tuff and limestone (see Fig. 4.7), interrupted by unconformities. In northwest Chatham Island, Late Cretaceous subsidence was followed by latest Cretaceous–Paleocene deposition of thin phosphatic and glauconitic sandy silts, greensand and minor bioclastic limestone, united as the Tioriori Group (Campbell *et al.* 1988, 1993). This rests on a low-relief erosion surface cut into Chatham Schist, and is a maximum of 30 m thick. The age ranges from Campanian to earliest Eocene (early Haumurian to early Waipawan). Planktic foraminifera, siliceous microfossils, horizontal burrows, glauconite, phosphorite and lack of primary structures suggest mid- to outer-shelf, quiet water conditions for the basal part of the succession, while fossil evidence and sedimentary structures such as channels and cross-bedded sands suggest an upward shallowing sequence deposited in a mid- to inner-shelf environment for the upper part of the unit (Campbell *et al.* 1993). The lowest unit in the group, the Takatika Grit, rests on basement schist, and consists of up to 10 m of shallow-marine glauconitic quartzo-feldspathic grit, sandstone and siltstone, with horizons rich in phosphorite nodules. The unit hosts several bone horizons, mainly of marine reptiles, but also of theropods (Stilwell *et al.* 2006). The mixed age of dinoflagellate fossils (Campanian–Paleocene/early Haumurian–Teurian), and the presence of bones of terrestrial theropods, suggest reworking of uppermost Cretaceous fossils and sediments and redeposition in the Paleocene (Wilson *et al.* 2005; Stilwell *et al.* 2006).

6.4.3.4 Southeastern South Island and offshore

In the Clipper Basin (see Fig. 4.5) and on the site of the Galleon-1 petroleum well, Whangai facies (Katiki Formation) brownish-grey marine mudstone of Maastrichtian (late Haumurian) age passes upwards into lithologically similar sediments (Moeraki Formation) of Paleocene to Early Eocene (Teurian to Waipawan) age. This unit is characteristically a medium to dark-brown mudstone and is locally glauconitic, slightly carbonaceous and micaceous, with a maximum known thickness of Paleocene sediments of 485 m in the Clipper-1 petroleum well (Raine *et al.* 1993). Microfaunas and microfloras indicate neritic to upper-bathyal settings at Clipper-1, Galleon-1 and Endeavour-1 (Gibbons & Herridge, 1984). Onshore at Moeraki and Katiki beaches the Moeraki Formation consists of slightly glauconitic dark-grey to black mudstone, miospores indicating a probable inshore depositional site (Raine *et al.* 1993). Shallow marine conditions prevailed southwards to at least the site of Takapu-1A petroleum well in the Great South Basin (see Fig. 4.5).

During the Campanian and Maastrichtian (Haumurian), extensive floodplains extended south from Oamaru and eastwards offshore to the site of the Endeavour-1 drillhole, forming the western margin of the southern Canterbury Basin and the Great South Basin. These were the sites of quartz-rich, gritty coal measures. In coastal Otago, quartzose coal measures (Taratu Formation) deposited in a half graben, unconformably overlie weathered schist or Henley Breccia (late Cenomanian–Coniacian/Raukumara) to form a basal transgressive facies. The coal measures reach their greatest preserved thickness of about 600 m in the Kaitangata Coalfield, where deposition was strongly influenced by activity on the northeast-trending Castle Hill Fault. Elsewhere they are generally less than 20 m thick (Bishop & Turnbull, 1996). The coal measures are locally absent below marine sediments (e.g., in the offshore Takapu-1A petroleum well (HIPCO, 1978)), and in some areas are restricted to valleys cut into the schist basement (Landis, 1990). The Taratu Formation is almost entirely of Campanian to Maastrichtian (Haumurian) age and underlies latest Cretaceous to earliest Paleocene marine sediments; however, Early Paleocene ages for the coal measures have been determined in the northwestern portion of the Kaitangata Coalfield (Bishop & Turnbull, 1996).

Sand and conglomerate of the upper members of the Taratu Formation were inferred to have been deposited in a braided stream system on a swampy surface of low relief close to a shoreline (Duff, 1985; Raymond, 1985). Thick sets of inclined heterolithic stratification and other sedimentary structures were considered to be typical of meandering stream systems.

Conglomerate, consisting of massive to metre-bedded, fine pebbly to bouldery sandstone and semi-schist conglomerate with subordinate sandstone and carbonaceous siltstone (Blue Spur Conglomerate), is discontinuously preserved as a series of faulted inliers along the northwest-trending Tuapeka Fault Zone from Lawrence to the Tokomairiro area. The formation is latest Cretaceous in age, and has been correlated with the conglomeratic lower part of the Taratu Formation in the Kaitangata Coalfield (Bishop & Turnbull, 1996). The depositional environment is inferred to have been either a series of piedmont alluvial fans derived from the upthrown side of the adjacent then-active northwest-trending Tuapeka Fault Zone (Bishop & Turnbull, 1996), or fluvial deposits transported to the southwest across the fault scarp by a series of northeast–southwest-trending incised valleys (Els *et al.* 2003).

To the northeast and east in the southern Canterbury Basin the coal measures are inferred to pass into delta-front sands and silts, before grading eastwards into brownish-grey mudstone of Whangai facies (Katiki Formation). Locally, at the site of the Galleon-1 drillhole, volcanics, consisting of 9 m of calcareous and non-calcareous tuff, overlain by 23 m of white slightly dolomitic fine-grained limestone, occur.

In the Dunedin area, non-marine sediments of latest Cretaceous age, equivalent to the Moeraki Formation to the north, are overlain by, and interfinger with, well-sorted, commonly slightly glauconitic, micaceous quartzose marine sandstone and mudstone ranging up to 75 m in thickness (Wangaloa and Herbert formations), although locally greensand and siltstone (Brighton Formation and Saddle Hill Siltstone; Fig. 6.4) record the oldest portion of a regional marine transgression. In places, the Wangaloa Formation is highly bioturbated and exhibits swaley and hummocky cross-stratification, suggesting deposition in a shallow-marine well-agitated shoreface environment exposed to periodic storm waves (Lindqvist, 1996; Lindqvist & Douglas, 1987). Elsewhere in the area a shallow-marine tidally influenced environment, including estuarine, is indicated (Lindqvist, 1995).

In the Great South Basin the eastern offshore facies equivalent to the Haumurian western basin margin coal measures interfinger eastwards with siliceous and calcareous marine Whangai facies mudstone (Wickliffe Formation). This and other facies overlap the upper Cretaceous strata onto basement on all the basin's flanks. The thickest Paleocene accumulations are found in the Central Graben (800–1000 m). In the west of the basin the latest Cretaceous–Early Paleocene coal measure succession passes upwards into glauconitic sandstones inferred to represent a nearshore to paralic environment (Raine *et al.* 1993), similar to that of the Wangaloa Formation.

6.4.4 The Waipawa 'black shale' facies (Late Paleocene/late Teurian: ~58– ~56 Ma)

In many parts of New Zealand, particularly in eastern basins, the Late Paleocene contains an unusual and distinctive organic-rich unit, the 'Waipawa facies'. In the East Coast Basin, where the facies was first identified, the unit is formally defined as the Waipawa Formation (originally 'Waipawa Black Shale'; Moore, 1989b). Here it consists of mainly poorly bedded, dark brownish-black, non-calcareous micaceous siltstone. Locally, as in southern Hawke's Bay, there is a high percentage of glauconitic sandstone, and there are some intervals of Whangai-like calcareous mudstone. Though highly bioturbated, in places by abundant ?*Bathysiphon* tubes, there are few macrofossils other than small bivalves and rare gastropods. Dinoflagellates are relatively common but foraminiferal assemblages are of low diversity and are dominated by agglutinated forms.

6.4.4.1 Eastern North Island

The Waipawa Formation (Fig. 6.1a,b) ranges up to ~50 m in thickness. It is widespread over much of the East Coast Basin, but is absent in southeastern Wairarapa and from the area inland from Hawke Bay. In Marlborough it has been recognised only in the Mead Stream succession in the Clarence Valley (Hollis *et al.* 2005b). Although the unit is not so far intersected in petroleum exploration wells in the North Island, it is likely that it occurs in the subsurface offshore of the East Coast region (Field *et al.* 1997). Because of its high total organic carbon (TOC) and hydrogen index (HI), the Waipawa Formation has long been regarded as a prime potential source for hydrocarbons (Moore, 1989b; Field *et al.* 1997; Killops *et al.* 2000). Oil-impregnated greensands within the Waipawa Formation occur at several localities in the East Coast Basin of the North Island.

Biostratigraphic constraints at Mead Stream and at Tawanui in Hawke's Bay (Hollis *et al.* 2005b) indicate a depositional age of 58–57 Ma, although elsewhere the formation is poorly dated.

Very similar facies have been recorded in most other New Zealand basins, including the Northland, northern Taranaki, Canterbury and Great South basins and possibly southern Westland. The Waipawa facies is readily identified by its distinctive chocolate colour, high TOC content and other aspects of its geochemistry (Killops *et al.* 2000).

In the East Coast Basin of the North Island the basal contact of the Waipawa Formation with Whangai Formation is commonly gradational over a few centimetres to several metres. The upper contact with the overlying Wanstead Formation is sharp with local disconformity, but commonly conformable (references in Moore, 1989b).

6.4.4.2 Eastern South Island

In Marlborough the Waipawa facies is represented at Mead Stream in the Clarence Valley by a pair of thin dark-grey siliceous mudstone units, separated by 4.8 m of siliceous limestone, resting conformably on the siliceous Mead Hill Formation and overlain by the Amuri Limestone (Hollis *et al.* 2005b,c).

The Waipawa facies is well developed in the offshore south Canterbury petroleum exploration wells Clipper-1 and Endeavour-1, where it is represented by an organic-rich unit of mudstone within the Late Paleocene Moeraki Formation. In the Galleon-1 well it correlates with a carbonaceous mudstone unit near the top of the Katiki Formation (Field *et al.* 1997). The horizon is closely associated with a seismic reflector that can be traced over most of the offshore area (see also section 6.5). In the Great South Basin the facies is included in the Tartan Formation, which is represented in four drillholes by dark-brown Waipawa-like carbonaceous shale (Cook *et al.* 1999).

6.4.4.3 Western New Zealand

With the exception of a possible occurrence in south Westland (Nathan *et al.* 1986), the Waipawa facies appears to be absent from the West Coast of the South Island. Apart from a solitary occurrence in the Ariki-1 petroleum well in northern Taranaki it is also largely absent from the Taranaki Basin (King & Thrasher, 1996). It has, however, been recorded in the Waka Nui-1 petroleum well in offshore west Northland (Milne & Quick, 1999), and from one locality in the Northland Allochthon (Isaac *et al.* 1994).

Deposition of the Waipawa facies under dysaerobic conditions is indicated by very limited bioturbation in most exposures and pauperate benthic microfaunas (Killops *et al.* 2000). There are also various geochemical indicators, such as sulphate reduction, supporting the fossil evidence for oxygen depletion and the rapid development of anoxia. The facies occurs widely around the Zealandia continental margin at a water depth inferred by Killops *et al.* (2000) to be generally equivalent to the upper slope or shelf break. The high organic content and distribution suggest that regional upwelling, perhaps associated with changes in oceanic circulation, occurred in the Late Paleocene.

6.4.5 Haumurian–Teurian sediments of western New Zealand

6.4.5.1 The West Coast–Taranaki Rift

In the late Campanian–early Maastrichtian (late Haumurian), a tectonic event caused a rift zone, up to 150 km wide and approximately 800 km long, to develop on the west coast of the South Island and in the southern Taranaki Basin along a predominantly north-northeast–south-southwest trend (Laird, 1981, 1993, 1994; King & Thrasher, 1996; and Fig. 6.3 and section 9.3.3). The active faulting resulted in the formation of new sub-basins and half grabens that were relatively small (typically 10–50 km wide and 50–150 km long) and elongate parallel to the fault trend. The zone is termed the 'West Coast–Taranaki Rift', and trends sub-parallel to transform faults associated with the Tasman Sea spreading ridge, and is interpreted as a rejuvenated transtensional system that linked spreading in the Tasman Sea and rifting in the New Caledonia Basin (Laird, 1994; King & Thrasher, 1996). The new basins were filled initially with non-marine deposits of mainly late Campanian to Maastrichtian (late Haumurian) age, passing upwards into shallow marine sediments of latest Cretaceous to Early Paleocene age in the Taranaki Basin (see also Strogen *et al.* 2017 and Fig. 6.3).

Dates from basaltic lavas interbedded with near-basal sediments of the Paparoa Coal Measures have K/Ar ages of ~70 Ma, while fission track ages from rocks adjacent to the rift provide a mean of ~72 Ma, suggesting that there was an uplift and cooling event at that time (Seward, 1989; Kamp *et al.* 1996; and section 6.5.3). The initiation of the rift at ~70 Ma (early Maastrichtian/late Haumurian) is indicated and points to a separate period of tectonism later than the opening of the Tasman Sea at ~85 Ma. Paleontology does not provide good constraints on the infill of the West Coast–Taranaki Rift system, but pollen dates on initial infilling sediments in the western South Island and in most of the southern Taranaki Basin are restricted to the PM2 pollen zone (Raine, 1984). This correlates with the upper half of the Haumurian Stage (late Campanian–late Maastrichtian) and, in conjunction with the radiometric and fission track dates, supports a latest Campanian or earliest Maastrichtian age for initiation of the rift. However, some spore-pollen assemblages from petroleum wells in the central Taranaki Basin are from the underlying PM1b Zone, of early Haumurian (early Campanian) age, suggesting either that the base of the succession and onset of rifting may be older in more basinward localities, or that these sediments pre-date the formation of the West Coast–Taranaki Rift.

6.4.5.2 Sedimentary succession

Deposition was initially terrestrial with a marine transgression beginning in south Westland and the northern Taranaki Basin. A common sedimentary/tectonic theme runs throughout the late Haumurian. In all western basins for which information is available, mainly non-marine Maastrichtian (late Haumurian) strata rest unconformably on older rocks. In contrast to eastern New Zealand, deposition of marine strata associated with the Late Cretaceous transgression did not begin in western New Zealand until late Campanian to Maastrichtian (late Haumurian) times (Browne *et al.* 2008) The deposition is commonly fault controlled basins, lying within, or associated with, the West Coast–Taranaki Rift (previous section).

Southland and south Westland

The Campanian–Maastrichtian (Haumurian) Ohai Group of western Southland occupies a basin east of the Alpine Fault and forms the older rocks of the Ohai Coalfield. It comprises two fining-upwards sequences, probably separated by an unconformity (Warnes, 1990), and is overlain unconformably by Eocene non-marine sediments. The basal fining-upwards sequence (Wairio Formation), with a maximum thickness of ~50 m, consists of basal conglomerate passing up into trough-bedded fine-grained sandstone interbedded with carbonaceous mudstone, interpreted as an alluvial fan and proximal braided river system (Sykes, 1985). The overlying 400-m-thick succession consists of a basal thick conglomerate (New Brighton Conglomerate) fining upwards first into a 215-m-thick succession (Morley Formation) consisting of medium- to coarse-grained often cross-bedded sandstone, coal seams up to 18 m thick, and then cyclic, pebbly to fine-grained sandstone units with carbonaceous mudstone and thin coal seams. The environment was interpreted as proximal braided river passing up into meandering river floodplain, and a poorly drained backswamp/lacustrine complex (Sykes, 1988). The Ohai Group sediments occupy a half graben oriented approximately northwest–southeast, probably representing rejuvenation of a mid-Cretaceous structural trend. Sub-basins within the half graben are oriented both parallel and perpendicular to the margins of the half graben, this alignment probably relating to latest Cretaceous tectonics associated with the West Coast–Taranaki Rift (Shearer, 1995; and previous section).

Ohai Group sediments are inferred from seismic interpretation to extend eastwards beneath the Winton Basin (Fig. 6.7), and westwards into the Waiau Basin (Turnbull *et al.* 1993). However, the overall thickness and extent of the group beyond the coalfield is poorly known. To the west, in Fiordland, no Late Cretaceous sediments are preserved, but reworked Haumurian pollen in Late Eocene–Early Oligocene sandstones

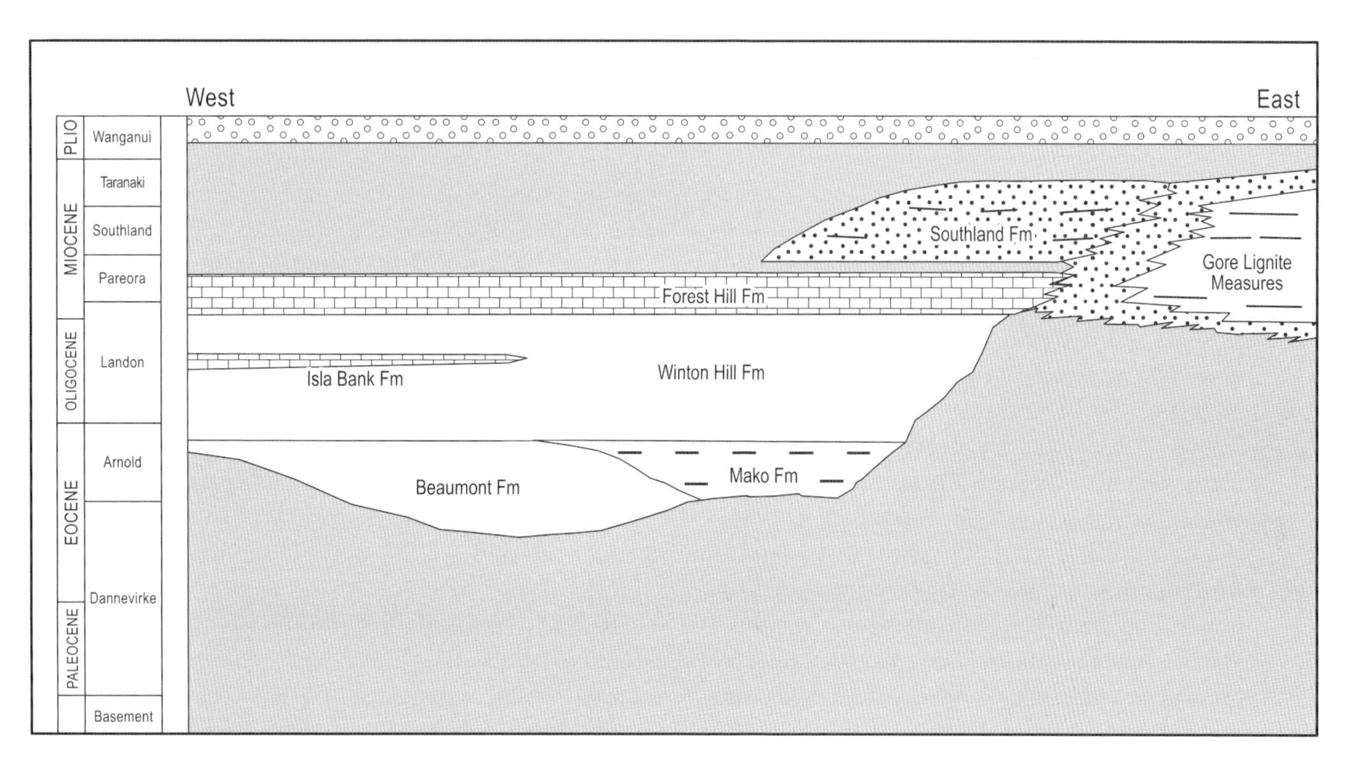

Fig. 6.7. Time-stratigraphic section of the Winton Basin, Southland. For lithologies of named formations, see sections 6.7.3.1, 6.10.3.1 and 8.2.4.2. Fm, Formation. (Adapted from Cahill, 1995, fig. 3.)

with probable Fiordland provenance suggests that Late Cretaceous sediments were originally deposited on the Fiordland block (Turnbull *et al.* 1993).

On the West Coast of the South Island, late Campanian–Paleocene (late Haumurian–Teurian) deposits appear to be restricted to four main depocentres: the south Westland embayment, the Paparoa and Grey Valley troughs, and the Pakawau and Greville basins. By the beginning of the Paleocene there was a general slowing down of tectonic activity on the West Coast of the South Island, indicated both by a marked increase in the quartz content of the youngest (Paleocene) sediments in the Paparoa Trough and the Pakawau Basin, and by the deposition of only a very thin Paleocene sequence in south Westland (Nathan *et al.* 1986).

Latest Cretaceous–Paleocene rocks in the south Westland embayment, representing the beginnings of marine transgression on the West Coast, have been identified only from a narrow coastal strip between Haast and Bruce Bay (Nathan *et al.* 1986). Seismic evidence, however, shows the presence of a thick offshore sedimentary succession, including an area of relatively thin (>500 m) sediments beneath a widespread Oligocene seismic reflector. Unlike onland successions to the northeast, the sub-Oligocene sediments apparently did not form in an enclosed basin, but appear to have an irregular eastern margin and to thicken towards the edge of the continental shelf. It is inferred that the oldest offshore sediments are latest Cretaceous and correlatives of the onshore succession.

The south Westland embayment includes the only rocks of Campanian (early Haumurian) age on the West Coast of the South Island (Adams, 1987; I. Raine, pers. comm. 2000). They are restricted to a small area near Paringa, and consist of 100 m of strata resting unconformably on non-marine Pororari Group equivalent strata of Clarence to Raukumara age. The early Haumurian unit consists of conglomerate, cross-bedded sandstone and carbonaceous mudstone with coaly layers, inferred to have been deposited in fluvial to lacustrine environments. It is disconformably overlain by a late Campanian–Maastrichtian (late Haumurian) succession (Fig. 6.8) consisting of basal non-marine pebble conglomerate, pebbly granular sandstone and carbonaceous massive silty sandstone (Tauperikaka Coal Measures), passing upwards into cross-bedded sandstones with mud drapes and flaser bedding, and then into interbedded burrowed and structureless glauconitic Whakapohai Sandstone (Adams, 1987). This transgressive succession is inferred to represent a transition from a coastal floodplain environment, through a tide-dominated coastline, to deposition in an open marine bay. Dinoflagellates, miospores and macrofossils from the upper portions of the succession confirm a late Haumurian age (Nathan, 1977). The thickness of the Cretaceous rocks is uncertain because of structural complexity, but a minimum is 70 m.

Conformably overlying the sedimentary succession is a 150–250-m-thick sequence of basaltic flows and tuffs (Arnott Basalt), which include pillow lavas and poorly preserved mollusca and shallow-water foraminifera of Maastrichtian (late Haumurian) age, suggesting dominantly submarine eruption (Aliprantis, 1987; and section 6.5.3). The Arnott Basalt is overlain conformably by shallow-marine volcanic conglomerate, calcareous sandstone, and glauconitic sandstone and siltstone (Tokakoriri Formation) of Early Paleocene to earliest Eocene (early Teurian to Waipawan) age that forms the lower part of an apparently continuous Paleogene marine succession. The basal sediments of the 42-m-thick unit include coarse granular to boulder conglomerate consisting mainly of volcanic clasts, which fines upwards into burrowed and shelly sandstone and siltstone. A U–Pb zircon date of 61.4±0.8 Ma on a rhyolite clast inferred to be approximately coeval with deposition supports the Paleocene age for the lower portion of the formation indicated by Teurian microfossils (Nathan, 1977; Phillips *et al.* 2005). Foraminifera from the upper portion of the formation give an earliest Eocene (Waipawan) age. The presence of tree trunks and carbonaceous sandstone in the basal portion of the unit suggests a nearshore environment, while foraminifera in the upper portion suggest deposition in deep water, probably outer neritic or upper bathyal (Nathan, 1977; Aliprantis, 1987; Phillips *et al.* 2005).

North Westland and Taranaki Basin

Elsewhere on the West Coast and in the southern Taranaki Basin, successions of late Haumurian age are almost entirely non-marine, and are mainly restricted to grabens or half grabens with a dominantly north-northeast trend. In north Westland, successions

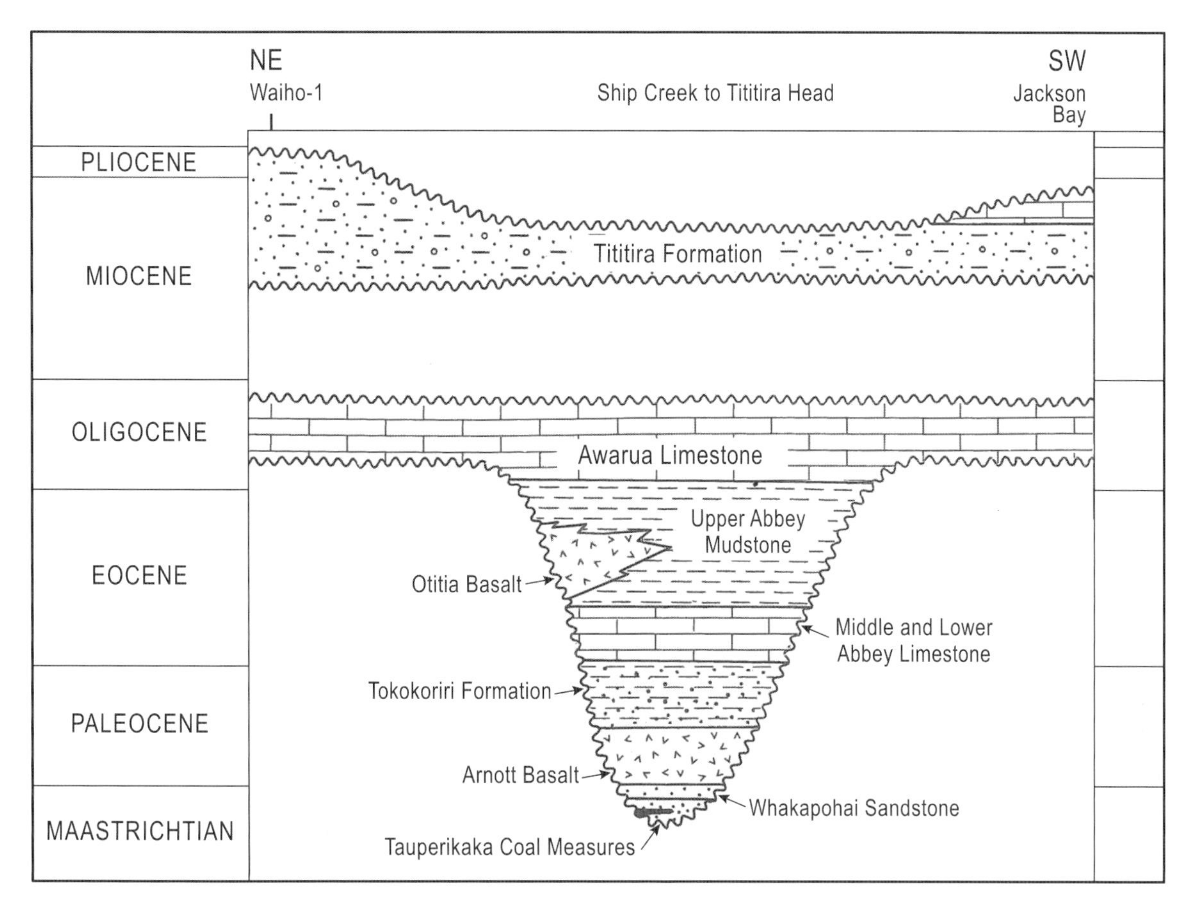

Fig. 6.8. Time-stratigraphic section of Cretaceous and Cenozoic rocks in south Westland between Jackson Bay and Waiho-1 drillhole. For lithologies of named formations, see sections 6.4.5.2, 6.7.3.1 and 6.10.3.1. (Adapted from Nathan *et al.* 1986.)

of late Campanian to Paleocene (late Haumurian to Teurian) age are preserved in two sub-parallel elongate north-northeast-trending basins. The eastern basin (Grey Valley Trough) is entirely subsurface and structural control is poorly known, but non-marine Paparoa Coal Measures have been penetrated in five exploration wells (Suggate & Waight, 1999). Drillhole data and seismic reflection profiles indicate that the preserved sequence occupies small basins to the north and south of the Neogene Kotuku Dome (Thrasher *et al.* 1996; Suggate & Waight, 1999). The Late Cretaceous succession rests unconformably on pre-Mesozoic basement, and is truncated by unconformably overlying Eocene strata. Paleocene deposits are absent.

The outcropping western succession was deposited in the major actively subsiding Paparoa Trough. The eastern margin of the trough was controlled by activity on the north-northeast-trending Paparoa Tectonic Zone (Laird, 1968), a part of the West Coast–Taranaki Rift. New fault-controlled basins within the Paparoa Trough were filled initially with non-marine Paparoa Coal Measures with minor basalts (see section 6.5.3.1) of Maastrichtian (late Haumurian) to Early Paleocene age, and rest unconformably on older rocks (Greenland Group, granite or mid-Cretaceous Pororari Group). Paparoa Coal Measures probably originally extended continuously along the Paparoa Trough for a distance of at least 100 km, shown by outliers at Koiterangi Hill, near Hokitika, in the south to Fox River in the north, and in several petroleum prospecting wells between Hokitika and Greymouth.

The best known succession of Paparoa Coal Measures is exposed in the Greymouth Coalfield, where seams of bituminous coal have been mined for over 120 years. In central parts of the coalfield the coal measures exhibit a well-defined alternation of fluvial and lacustrine members, attributed to periodic uplift along active faults and rejuvenation of source areas (Gage, 1952; Newman, 1985). Lithofacies include alluvial-fan pebble to boulder conglomerates with local debris-flow intervals, parallel-laminated and cross-bedded sandstones of braided stream origin, overbank siltstones and fine sandstones, and lacustrine sandstones, siltstones and mudstones (Sherwood *et al.* 1992). Coal seams occur in almost all lithofacies of fluvial origin. Maximum thickness is approximately 900 m.

Isopachs indicate that the oldest two members of the Paparoa Coal Measures (Jay and Ford members) accumulated in narrow northwest–southeast-trending basins (Gage, 1952). The geometry of the following Morgan and Waiomo members indicates a change to more widespread subsidence with a predominant north-northeast to northeast trend, accompanied by local volcanism. Following deposition of the Morgan and Waiomo members, thick fluvial deposits making up the Rewanui Coal Measure Member were deposited, mainly from the northwest (Ward, 1997). The western margin of the basin, now offshore, is fronted by a conglomerate facies that coarsens to the northwest and is interpreted as the alluvial apron of a repeatedly uplifted western block (Gage, 1952; Newman & Newman, 1992). These sediments are also mainly of Maastrichtian (late Haumurian) age, with the K–T boundary lying within a coal seam just beneath the top of the unit without notable stratigraphic break (Ward, 1997; Vajda *et al.* 2001).

A unit of lacustrine mudstone (Goldlight Member) follows the Rewanui Member, interfingering with and overlain by the fluvial Dunollie Coal Measure Member. This member is distinctly more quartzose than the underlying members, particularly towards the top, indicating increased chemical decomposition of unstable minerals as the tempo of uplift and erosion slowed in the Paleocene. In the axis of the trough the Dunollie Member passes conformably and gradationally up into a thin (15–20 m) Paleocene unit of quartzose sandstone similar in appearance to the Eocene Brunner Coal Measures (Nathan *et al.* 1986). Paparoa Coal Measures of Paleocene age have been identified in the Pike River Coalfield to the north, and in the Arahura-1 petroleum well to the south, the thickness varying up to a maximum of ~400 m in the axis of the Paparoa Trough. The top of the Paparoa Coal Measures is truncated by the Eocene Brunner Coal Measures over most of the region (Suggate & Waight, 1999).

Basin subsidence in the Paparoa Trough was substantial, particularly in the east, where more than 600 m of sediment accumulated along a north-north-east-trending axis little more than 1 km from the eastern basin margin during Paparoa Coal Measure deposition (Newman & Newman, 1992). Both margins are considered to have been controlled by faults bounding the basin, whose width was unlikely to have been more than 12 km (Newman & Newman, 1992; Sherwood *et al.* 1992). There has, however, been further debate on the eastern margin of the basin (see Kamp *et al.* response in Suggate *et al.* 2000) Although the dominant structural control on sedimentation within the Paparoa Trough following extrusion of the Morgan Volcanics at ~70 Ma was a system of north-northeast-trending faults parallel to the West Coast–Taranaki Rift (Newman & Newman, 1992; Ferguson, 1993), the older west-northwest to northwest pre-Morgan fault-set remained sporadically active throughout the remainder of the Cretaceous, and continued to exert control over sedimentation patterns in parts of at least the Greymouth Coalfield (Ward, 1997; Nathan *et al.* 2002; Moore *et al.* 2006).

The Pakawau Sub-basin in the northwest of the South Island, like the Paparoa Trough to the south, is bounded by major north-northeast-trending faults on its eastern and western sides, and represents a link between the South Island West Coast basins and the submerged Taranaki and western Northland segments of the West Coast–Taranaki Rift. It contains the northernmost outcropping Cretaceous rocks in western New Zealand, which continue offshore to the north and northeast into the southern Taranaki Basin (Fig. 6.3).

The Pakawau Group, originally defined to include all Cretaceous–Paleocene non-marine sedimentary rocks in the Pakawau Sub-basin, became restricted by oil company usage to include only latest Cretaceous sedimentary rocks in the Taranaki Basin (Thrasher, 1992; King & Thrasher, 1996). The group is up to 4200 m thick, and over most of its mapped distribution is at least 800 m thick. It is present in both isolated and interconnected depocentres and in sub-basins defined by major faults. The eastern known limit of latest Cretaceous deposition is the elongate Manaia Sub-basin, immediately west of the

Taranaki Fault. Except where latest Cretaceous sediments unconformably overlie the mid-Cretaceous Taniwha Formation, Pakawau Group strata directly overlie basement. The upper contact of the group is generally at or near the K–T boundary, although onshore in the Pakawau Basin the top of the Pakawau Group is marked by a minor unconformity within a section of Paleocene age (Bal, 1994; Bal & Lewis, 1994; Stark, 1996); thus, in this locality, the Pakawau Group extends into the earliest Paleocene. Offshore, the top Pakawau Group seismic reflector lies close to the paleontologically dated K–T boundary in petroleum exploration wells (King & Thrasher, 1996).

The subdivision of the Pakawau Group into Rakopi and North Cape formations is defined from seismic reflection mapping offshore, and from lithofacies character in exploration wells and in scattered outcrops in northwest Nelson (Thrasher, 1992). The contact between the formations marks a significant marine transgression, the oldest that can be defined regionally across the basin. Although the contact is generally conformable, a mild unconformity is evident locally on seismic reflection profiles (King & Thrasher, 1996).

The Rakopi Formation consists dominantly of non-marine fining-upwards successions, mainly sandstone, interbedded with carbonaceous mudstone, thin coal seams and rare conglomerate (Browne *et al.* 2008). The formation is locally up to 3000 m thick, and reaches thicknesses of up to 2500 m adjacent to the Taranaki, Wakamarama and other faults (Fig. 6.3). Elsewhere it is generally less than 1000 m thick. Thick conglomerate occurs locally at the base of the formation in the Pakawau Sub-basin. On seismic reflection profiles the interval that correlates with the Rakopi Formation is characterised by high-amplitude, hummocky and laterally discontinuous reflectors. The formation is interpreted as a fluvial coastal-plain succession, dominated by channels and their associated overbank and levee environments. Episodic marine influence during deposition is indicated by the presence of dinoflagellates, glauconite and elevated coal seam sulphur contents (Thrasher, 1992; Browne *et al.* 2008). Basal conglomerate was deposited as alluvial fans adjacent to the Wakamarama Fault scarp.

The overlying North Cape Formation is primarily distinguished from the Rakopi Formation by its marine and marginal-marine depositional setting and by a blander seismic signature (King & Thrasher, 1996). The unit is widespread in the Taranaki Basin (Fig. 6.3), as well as cropping out extensively in the Pakawau Sub-basin. Shallow marine siltstone, sandstone and silty sandstone are the predominant lithologies in most wells, but coal measures and conglomerate are intercalated locally, particularly in the east and onshore in the south. In the northern Taranaki Basin petroleum wells Ariki-1 and Wainui-1, uppermost North Cape strata are mud-rich and fully marine, passing to the south into shallow-marine transgressive sandstones. The North Cape Formation has a maximum thickness of about 1800 m in local sub-basins, and also appears to continue into the New Caledonia embayment, where it is 1400 m thick at the northwestern limit of mapping (King & Thrasher, 1996). The top reflector of the North Cape Formation can be traced as far as the site of DSDP drillhole 206 in the New Caledonia Trough (see Fig. 2.2), where it is stratigraphically just below the deepest level penetrated, occupied by a slumped unit containing Early and Middle Paleocene sediments. Elsewhere it is generally less than 1000 m thick. The thickest sections within individual sub-basins invariably occur adjacent to the main basin-controlling faults. In this and other respects, thickness patterns within the North Cape Formation are similar to those in the underlying Rakopi Formation (King & Thrasher, 1996). The North Cape Formation is thin and locally absent in the northeast corner of the Taranaki Basin where mid-Cretaceous Taniwha strata are erosionally truncated. However, a 50-m-thick interval of marine siltstones in the nearby Te Ranga-1 well indicates that at least part of the top Taniwha Formation unconformity surface was inundated by the end of the Cretaceous (Thrasher, 1992). Extension of the North Cape Formation to the north into the offshore western Northland Basin is uncertain, as seismic reflectors at the top and bottom of the sequence become indistinct (Stagpoole, 1997; see also Strogen *et al.* 2017 for more seismic reflection data).

The unit reflects a major south-directed transgression at the end of the Cretaceous, in which the sea flooded across coastal floodplains and into fault-controlled alluvial valleys to form an interconnected system of narrow tidal embayments (King & Thrasher, 1996). Conglomeratic facies within the outcropping North Cape Formation are inferred to have been deposited as part of braided fan complexes, while breccias derived from locally active fault scarps debouched directly into shallow marine environments. Paleocurrent directions in the Pakawau Sub-basin indicate sediment transport towards the northwest, away from the Wakamarama Fault (Wizevitch, 1994; Stark, 1996).

In the contiguous offshore western Northland Basin, early seismic surveys identified two seismic sequences (Isaac *et al.* 1994) tentatively tied to Late Cretaceous strata of the Taranaki Basin, with the lower one inferred to rest on basement (Gage & Kurata, 1996; Gage, 1998) and the upper correlated to the North Cape Formation. The first well drilled in the western Northland Basin (Waka Nui-1) intersected no Cretaceous strata, finding instead that Paleocene rocks rested unconformably on Murihiku Supergroup rocks of Jurassic age (Milne & Quick, 1999). However, the well was drilled on a structural 'high' and sedimentary sections imaged to the west and east of the high are likely to be Cretaceous rocks (Uruski *et al.* 2004). The presence of Pakawau Group sediments, forming the upper strata of a Cretaceous delta system, is also inferred in the southeastern portion of the Deepwater Taranaki Basin, at the head of the New Caledonia Basin (Uruski & Baillie, 2002).

Sedimentation patterns in the Taranaki Basin during the Paleocene, as in the West Coast to the south, reflected the structurally controlled basin setting of the Late Cretaceous, but were increasingly influenced by deepening of a marine basin to the northwest and by migration of the shoreline to the south and southwest. The Paleocene–Eocene succession is divided into two groups: the terrestrial to marginal marine Kapuni Group and the marine Moa Group. Isopach maps (King & Thrasher, 1996) of Paleocene–Eocene strata show that most of the accumulations more than 2000 m thick occur in fault-bounded depocentres associated with Paleocene rifting. In the d'Urville Sub-basin more than 2000 m of strata are inferred to be of Paleocene (mainly) and Eocene age, although Late Cretaceous strata may be present (Purcell, 1994).

The Farewell Formation is the basal formation of the Kapuni Group, cropping out in northwest Nelson in the Pakawau Sub-basin, where it unconformably overlies marginal marine strata of the uppermost Pakawau Group (Bal, 1994). It is almost entirely of Paleocene age, but in some southern parts of the Taranaki Basin the Farewell Formation extends into the earliest Eocene (Waipawan), and an intra-Waipawan unconformity marks the top of the formation (King & Thrasher, 1996).

In the mainly offshore Pakawau Sub-basin the Farewell Formation attains thicknesses of up to 750 m, and in parts of the central Taranaki Basin it exceeds 1600 m. Where the formation crops out onshore in northwest Nelson, it is lithologically similar to the coarsest-grained rocks of the underlying Pakawau Group, but is primarily regressive with respect to the underlying North Cape Formation. It is characterised by coalescing and stacked scours and channels infilled with pebbly quartzo-feldspathic sandstone and conglomerate, interpreted as high-energy deposition, either in a braided river system (Titheridge, 1977) or in a meandering river carrying a coarse sediment load (Bal, 1994). Conglomeratic lithofacies represent deposition in fault-controlled alluvial fans and braided rivers close to source. Offshore to the north the Farewell Formation is dominated by sandstone, or interbedded sandstone and mudstone with localised coal seams. Sand-rich coastal facies of the Farewell Formation are prevalent in the southern Taranaki Basin. West of the Taranaki Peninsula they thicken northwards and northwestwards to grade laterally into, and interfinger with, mudstones of the more offshore Turi Formation (King & Thrasher, 1996).

The Turi Formation (Fig. 6.3) is predominantly a non-calcareous, dark-coloured, micaceous, carbonaceous, marine mudstone that extends over much of the northern Taranaki Basin. It is generally uniform in lithology, except in the western part of the central Taranaki Basin, where it passes laterally into and locally encloses glauconitic siltstone with thin interbeds of limestone and poorly developed sandstone (Tane Member). The Tane Member was deposited in an inner- to mid-shelf environment, and is essentially a transgressive sequence. The top of the Tane Member is marked by a glauconitic horizon, probably representing a period of clastic sediment starvation, and a sharp break on well logs that marks an abrupt contact with overlying, slightly deeper-water mudstones of the Turi Formation. This horizon is inferred to be an important intra-Paleocene sequence boundary, the contact corresponding to a flooding surface (King & Thrasher, 1996).

The position of the shoreline in the northeast of the Taranaki Basin is uncertain. In this area the coastal sand belt is abruptly truncated by the Taranaki Fault, but it is likely to have extended further to the northeast in Paleocene times. In petroleum well Te Ranga-1, which lies southwest of Kawhia close to the present shoreline, a 170-m-thick Paleocene interval consists entirely of marine shelf siltstone, indicating that the paleo-shoreline was some distance further to the east.

In offshore west Northland, Paleogene strata above the basement or top Cretaceous reflector C1 (Isaac *et al.* 1994) are subdivided into two sequences by a regional marker reflector (P2). The lower sequence P2–C1 is present throughout the offshore Northland Basin, and appears to be continuous into the Taranaki Basin, where the equivalent beds constitute the Paleocene lower part of the Moa Group (Turi Formation) and the Kapuni Group (King & Thrasher, 1996; and see above). In offshore west Northland the sequence includes a ~530-m-thick Paleocene succession resting unconformably on basement Murihiku rocks in the Waka Nui-1 petroleum well. The Paleocene section comprises a basal non-marine ~70-m-thick conglomerate, overlain by ~460 m of Turi Formation claystone (Milne & Quick, 1999) deposited in a restricted bathyal marine environment (Strong *et al.* 1999). In a lobate area southwest of Kaipara Harbour, the sequence, inferred to be deltaic to shallow marine, is more than 1200 m thick. Onshore, the only correlative is 48 m of glauconitic sandstone with a thin basal conglomerate of Paleocene to Eocene (late Teurian to Waipawan) age, resting unconformably on Early Cretaceous probable Murihiku Supergroup rocks at the base of the Waimamaku-2 exploration well (Isaac *et al.* 1994).

The P2 reflector forming the upper boundary of the sequence is a consistent marker that is recognised on all seismic profiles throughout the region. It can be followed northwest for the length of the Reinga Basin, and for 1000 km along the New Caledonia Basin (Isaac *et al.* 1994).

The offshore west Northland basin appears to have been tectonically quiet, although subsidence appeared to continue in a fault basin northwest of Kaipara Harbour. Elsewhere, faulting was limited to less than 100 m of normal movement on some of the pre-existing large normal faults.

6.5. IGNEOUS ROCKS ASSOCIATED WITH BREAK-UP AND SEA-FLOOR SPREADING

6.5.1 North Island

In the East Coast region, excluding the Northland and East Cape allochthons, latest Cretaceous–Paleocene (Piripauan–Teurian) igneous rocks are known only from Wairarapa. Dike rocks at Cape Palliser have yielded $^{40}Ar/^{39}Ar$ data indicating a minimum age of ~80 Ma for emplacement and/or alteration events, but are likely to have been associated with an earlier, Cenomanian (late Ngaterian) tectono-magmatic event (see section 4.4.1.3). Further north the Ngahape Igneous Complex (Moore, 1980) includes basalt flows and dolerite sills inferred to be of Campanian–Maastrichtian (Haumurian) age. A dolerite sill intruding Glenburn Formation of probable Santonian (Piripauan) age in Kaiwhata Stream (Crampton, 1997) has an early Campanian (early Haumurian) age indicated by a $^{40}Ar/^{39}Ar$ defined plateau age of 81.6±0.8 Ma (W.C. McIntosh, pers. comm. 2002), which is likely to be close to the age of intrusion. Five kilometres to the northwest, the Red Hill Teschenite has a similar latest Campanian (late Haumurian) age indicated by a mean $^{40}Ar/^{39}Ar$ isotopic age of 70.9±2.2 Ma (W.C. McIntosh, pers. comm. 2002).

6.5.2 Eastern South Island and Chatham Islands

On eastern Pitt Island (Chatham Islands), deposition of Kahuitara Tuff was interrupted at about 80 Ma (middle Campanian/early Haumurian) by the extrusion of a thick pile of mildly alkaline olivine basalt (Southern Volcanics). In northwestern Pitt Island, deposition of tuff continued contemporaneously with the extrusive volcanism, both ceasing at about 70 Ma (Campbell *et al.* 1993). The Southern Volcanics crop out most extensively in the southern portion of Chatham Island (see Fig. 4.7). They show little evidence of deposition in marine conditions, and are considered to be largely terrestrial.

In the Canterbury Basin–Great South Basin region, acid volcanics of Campanian–Maastrichtian (Haumurian) age were reported from the offshore

Galleon-1 petroleum well, and intrusives and extrusives of similar age were mapped by Shell BP Todd (1984) at the offshore Barque prospect. Basaltic volcanism was widespread in central Canterbury during the Paleocene, locally lasting until the Eocene (View Hill Volcanics). Equivalent volcanics and volcaniclastics were found in the onshore Leeston-1 petroleum well, and in the offshore Resolution-1 petroleum well. Paleocene volcanics also have been recorded in the Endeavour-1 petroleum well, where up to 124 m of glassy basaltic tuff was intersected, and in the Clipper-1 and Galleon-1 petroleum wells (Field *et al.* 1989). No latest Cretaceous–Paleocene volcanism has been recognised onshore in Otago or Southland, or offshore in the Great South Basin.

6.5.3 Western South Island

In south Westland the Late Cretaceous sedimentary succession is overlain by a 150–250-m-thick sequence of basaltic flows and tuffs (Arnott Basalt; Fig. 6.8), which include pillow lavas and poorly preserved mollusca and shallow-water foraminifera, suggesting dominantly submarine eruption. No vents have been identified on land, but possible volcanic centres are indicated by a belt of offshore linear magnetic anomalies, parallel with the known outcrop area of the basalts and extending northwards for 50 km along or close to the mainly offshore Cape Foulwind Fault Zone (Nathan *et al.* 1986; Laird, 1993). Geochemical studies (Sewell & Nathan, 1987) show that the igneous suite is mildly alkaline and is classified as within-plate basalts, erupted during the period of Maastrichtian (late Haumurian) extension associated with the West Coast–Taranaki Rift (Laird, 1994; and section 6.4.3.1). The upper portion of the Arnott Basalt contains poorly preserved probable Campanian–Maastrichtian (Haumurian) foraminifera (Aliprantis, 1987). The base of the overlying Tokakoriri Formation, however, contains basaltic flows and clasts, probably derived from a similar source (Phillips *et al.* 2005), indicating eruption may have continued into the Early Paleocene. The basalts are inferred to correlate with basaltic lavas with similar geochemistry, in the Paparoa Coal Measures (section 6.5.2).

6.6 LATEST PALEOCENE EVENT (LATEST TEURIAN: ~58–55.5 Ma): THE CESSATION OF TASMAN SEA SPREADING

An apparently region-wide event in the Late Paleocene (late Teurian) led to a significant change in lithofacies. In places this change is marked by an unconformity, and in others there is apparent conformity but a rapid facies change. Some areas, such as some parts of the northern West Coast and southern Taranaki, the Wanganui, King Country and South Auckland basins, and interior Northland, showed little change, persisting as areas of emergent basement. Other areas, such as parts of western Southland and the central West Coast of the South Island, reverted to non-deposition. Locally, where strata of Paleocene age have been studied in detail (as in southeast Otago), several breaks in deposition of both local and regional extent have been recognised.

In the East Coast Basin of the North Island, dark, chocolate-coloured non-calcareous shales of the Late Paleocene (late Teurian) Waipawa Formation (Moore, 1988, 1989b; Field *et al.* 1997) are overlain sharply but in the main conformably by the calcareous Wanstead Formation (Fig. 6.1a,b). However, there is disconformity at the contact in the type section at Waipawa, and the presence locally of black mudstone clasts in overlying Paleocene–Eocene sediments indicates that there may be a significant erosion interval in some parts of the region (Moore, 1989b).

In Marlborough (Fig. 6.1b) an unconformity of Late Paleocene to Early Eocene age progressively truncates Teurian and then Haumurian strata southwestwards in the central Clarence Valley (Strong & Beggs, 1990), but has not been recognised in sections northeast of Muzzle Stream (Hancock *et al.* 2003). In Bluff and Muzzle streams the hiatus is accentuated by the occurrence of *Thallasinoides* burrows extending up to 50 cm from the overlying deposits down into the Paleocene Mead Hill Formation (Reay, 1993). A small, paleontologically unresolvable, stratigraphic break was inferred by Strong *et al.* (1995) to be present at the sharp Black Siltstone–Mead Hill Formation contact of Late Paleocene age in Mead Stream, but this is now discounted (Hancock *et al.* 2003; Hollis *et al.* 2005a,b).

A Late Paleocene unconformity is widespread in the north Canterbury Basin, where it is overlain by phosphatic sandstone. The onlap of sediments overlying the unconformity at the Hurunui River records a net relative rise in sea level (Field *et al.* 1989). The unconformity is inferred to be represented in offshore Canterbury by a widespread Late Paleocene seismic horizon, which records erosion on the western Chatham Rise, where the horizon pinches out on a base of Paleocene horizon (Wood *et al.* 1989). This horizon is probably the same as the widespread top of Paleocene seismic horizon recognised by Wood *et al.* (1989) as extending over much of the Chatham Rise. On the Chatham Islands it may be represented by the unconformity separating the base of the Late Paleocene–Early Eocene (late Teurian–late Waipawan) Red Bluff Tuff from older rocks.

In the offshore southern Canterbury Basin and in the Great South Basin the prominent Late Paleocene seismic reflector noted above appears tied to organic-rich mudstones of the Waipawa facies in petroleum wells, but no unconformity has been recorded. Onshore, south of Dunedin, a number of discontinuities have been recognised in successions of Paleocene age (Lindqvist, 1995; McMillan & Wilson, 1997), but only one of Late Paleocene (late Teurian) age can be shown to have more than local significance. In the northwest of the Great South Basin the Late Paleocene seismic horizon marks the lower boundary of southeastwards-prograding Eocene clinoforms, and to the southeast it marks an irregular erosion surface cut by channels. A significant unconformity has also been recorded on Campbell Island, where the latest Paleocene and Early Eocene (late Teurian–Waipawan) are absent (Hollis *et al.* 1997).

Over much of the West Coast of the South Island, Paleocene and Early Eocene deposits are absent, suggesting emergence at this time. No evidence of unconformity has been found in the Paleocene–Early Eocene marine succession in south Westland, but there is a stratigraphic break in the Greymouth and Pike River Coalfields of north Westland between quartzose sediments of Early Paleocene age and overlying Brunner Coal Measures of Middle to Late Eocene age, indicative of erosion or non-deposition (see Plate 10).

In the southern Taranaki Basin no significant depositional breaks have been noted in the bulk of the mainly Paleocene coarse-grained non-marine to paralic Farewell Formation (Fig. 6.3), which appears to be terminated by an Early Eocene intra-Waipawan unconformity. Similarly, the correlative fine-grained Turi Formation appears in the main homogeneous from Paleocene to Late Eocene times, although the top of the shallow marine Tane Member of the Farewell Formation is marked by a sharp break, which is inferred to be an important intra-Paleocene sequence boundary. This may correspond to the offshore west Northland Late Paleocene seismic reflector P2. In the northern Taranaki Basin petroleum well Ariki-1, reflector P2 coincides with a strong positive gamma ray, low-density event (probably a thin organic-rich shale), separating shales of different degrees of over-pressuring (King & Thrasher, 1996). In the offshore west Northland Basin the discontinuity represented by reflector P2 can be traced throughout this basin, and into the adjacent Reinga Basin and New Caledonia Trough (Isaac *et al.* 1994).

The unconformity and/or seismic reflector associated with the event appears to closely post-date the Waipawa Formation of the East Coast and lithological equivalents elsewhere. The facies is inferred to be broadly isochronous throughout its geographical range, and its age is inferred to lie between 58 and 57 Ma in the East Coast Basin (section 6.5.2). Where seen in outcrop or penetrated in petroleum wells the unconformity or equivalent seismic reflector also appears to lie close to but below the Paleocene–Eocene boundary, suggesting an age of between ~58 and 55.5 Ma. However, it is not well-constrained in all regions, and may be diachronous, occurring locally in the earliest Eocene.

These unconformities and strong reflectors lie close in time to the termination of sea-floor spreading in the Tasman Sea and its dramatic reduction on the Pacific–Antarctic Ridge shortly after chron 24o at 53.3 Ma (Gaina *et al.* 1998). It marked the beginning of a period of crustal cooling and slow passive continental-margin subsidence in the New Zealand region.

6.7 LATEST PALEOCENE-EOCENE SUCCESSIONS (~56- ~34 Ma)

6.7.1 Introduction

Over much of the present land area of New Zealand, Paleocene–Early Eocene strata are absent, suggesting that much of the axial portion of the landmass from Northland to central and western Southland was emergent and being eroded at this time. On both the east and west coasts of both islands, however, mainly shallow-marine sedimentation continued, and by Middle Eocene times transgression was occurring over previously emergent areas of both regions.

Although an unconformity is not always recognised, the Late Paleocene Event (section 6.6) was followed by a significant change in lithofacies recognisable throughout the eastern basins of New Zealand. The main Eocene units in the East Coast Basin show a change from the Late Cretaceous and Paleocene clastic deposits to dominantly calcareous sediments, as do the more offshore portions of the south Canterbury and Great South basins. The west coasts of both islands and the Northland, South Auckland and King Country basins saw marine transgression beginning in the Middle–Late Eocene, preceded by the formation of widespread coal measure deposits.

6.7.2 Latest Paleocene-earliest Oligocene sediments of eastern New Zealand

6.7.2.1 Eastern North Island

The main latest Paleocene–Eocene units of the East Coast region are dominated by calcareous sediments: the Wanstead Formation in the North Island and the Amuri Limestone in Marlborough (Fig. 6.1a,b). The Wanstead Formation, which rests sharply and locally with an erosion surface on the Late Paleocene Waipawa Formation, occurs over most of the North Island East Coast region. It consists mainly of poorly bedded to non-bedded, greenish-grey, soft to moderately hard, bioturbated calcareous mudstone, locally smectitic or siliceous, with intermittent beds of glauconitic sandstone. The unit varies considerably in thickness, reaching a maximum of 490 m in northern Wairarapa. Basal deposits are variable, and include glauconitic sandstone and mudstone. Locally, in the East Coast Allochthon, the formation shows intraformational slump folding and pebbly mudstone with large clasts of underlying Waipawa Formation dark shale (Mazengarb & Speden, 2000). Coarse, deep-water facies, including conglomerate beds, also occur in the early Middle Eocene (Heretaungan) in northern Wairarapa, and in the late Middle Eocene (?Bortonian) in southern Wairarapa (Field *et al.* 1997). The Wanstead Formation ranges in age from Late Paleocene to latest Eocene (late Teurian to early Whaingaroan), but the more usual range of age is Late Paleocene to late Middle Eocene. The formation is generally considered to represent the deepest facies of the latest Cretaceous–Paleogene of the East Coast with common mid-bathyal paleodepths. In southeastern Wairarapa, calcareous mudstone of the Wanstead Formation is inferred to pass laterally into Late Paleocene micritic limestone (Amuri Limestone; section 6.7.2.2) and the overlying sandy calcareous siltstone (Pukemuri Siltstone) and mudstone and marl of Early Eocene to late Middle Eocene (Waipawan to Bortonian) age.

In the upper part of the Eocene, sediments of the Wanstead Formation pass upwards locally into calcareous mudstones and limestone of the Weber Formation, which is the dominant unit of the Oligocene of the eastern North Island (Fig. 6.1a,b). The oldest Weber strata, late Middle to early Late Eocene (Bortonian to early Kaiatan) in age, are found in southern Hawke's Bay. In the northern Wairarapa, where the Eocene–Oligocene boundary is about 45 m above the base of the Weber Formation, the basal beds comprise Late Eocene calcareous, intensely bioturbated fine-grained sandstone and well-bedded glauconitic sandstone and sandy siltstone, grading up from the underlying Wanstead mudstone (Lee, 1995; Field *et al.* 1997). There are no Paleogene strata younger than Middle Eocene exposed in southeastern Wairarapa (Field *et al.* 1997).

6.7.2.2 Eastern South Island

Amuri Limestone

The most extensive lithostratigraphic unit in the latest Paleocene and Eocene of the eastern South Island is a white, bioturbated, micritic limestone, which, together with associated calcareous sediments, is known as the 'Amuri Limestone'. It attains its greatest thickness and variety in Marlborough, but extends north into southeastern Wairarapa, and south into the Canterbury and offshore Great South basins. The formation is inferred to have formed a single time-transgressive unit, younging to the south, its base varying in age from Late Paleocene (late Teurian) in southeast Wairarapa and Marlborough, to Eocene and earliest Oligocene (early Whaingaroan) in the Canterbury Basin.

In Marlborough, where it is best-developed and ranges in age from latest Paleocene to Middle Eocene (Fig. 6.1b), the formation is subdivisible vertically into five informal units traceable for several tens of kilometres (Morris, 1987; Reay, 1993; Strong *et al.* 1995), and three of more local extent. They thin and merge to the south, and, with the exception of the basal Teredo Limestone, are not recognised to the south of Kaikoura, where the Amuri Limestone is represented by largely uniform micritic limestone. From oldest to youngest, the major lithotypes are the Teredo Limestone, the Lower Limestone, the Lower Marl, the Upper Limestone and the Upper Marl. Subdivisions recognisable only locally include Grasseed Volcanics and Fells Greensand within the Upper Marl (see Plate 12a), and a locally occurring unit of graded sandstones and tuffs. The maximum recorded thickness of the Amuri Limestone is 400 m in Mead Stream, Clarence Valley (Rattenbury *et al.* 2006).

In Marlborough, the basal latest Paleocene Teredo Member consists of locally glauconitic, mainly calcareous sandstone, up to 24 m thick. To the southwest of Mead Stream in the Clarence Valley, an unconformity beneath the Teredo Member progressively truncates earlier Paleocene and then Late Cretaceous strata to the southwest (Strong & Beggs, 1990). Current flow to the northeast is suggested locally by low-angle cross-stratification in sandstone-filled shallow channels or scours (Reay, 1993). A shallow-marine, inner-shelf environment was inferred.

The overlying Lower Limestone, which consists of up to 80 m of predominantly light- to medium-grey strongly bioturbated limestone with common chert nodules, is of latest Paleocene to Early Eocene (late Teurian to early Waipawan) age. Early to Middle Eocene (late Waipawan to Heretaungan) times are represented by the Lower Marl, which consists of bioturbated tan to light greenish-grey decimetre-bedded cemented limestone beds regularly interbedded with less-indurated greenish-grey smectitic mudstone and marl. Coeval yellow-grey, poorly indurated, graded sandstones or coarse siltstones up to 150 m thick, interbedded with centimetre-thick mudstones and 1.5-m-thick massive, weathered tuffs and rare basaltic lavas, occur locally at Woodside Creek (Prebble, 1976; Field *et al.* 1997). The Upper Limestone, of Middle Eocene (late Heretaungan to early Bortonian) age, consists of cemented, centimetre- to decimetre-bedded, strongly bioturbated micritic limestone up to 150 m thick with very rare marl interbeds. Strata overlying the Upper Limestone consist mainly of poorly indurated, light greenish-grey to tan marls, with thin interbeds of limestone (Upper Marl), with a maximum recorded sedimentary thickness of 120 m. The Upper Marl, of late Middle to Late Eocene (Bortonian to Runangan) age, is overlain unconformably by Late Oligocene or Early Miocene deposits (see Plate 12a). Locally, fine- to medium-grained, glauconitic quartz sandstones and lenses of volcanics are interbedded within the Upper Limestone and Upper Marl.

In southern Marlborough the Amuri Limestone, at its type area at Haumuri Bluff, consists of about 100 m of cemented white micritic limestone with thin pale-grey marly interbeds. The rock consists largely of fine organic debris (principally coccoliths and foraminifera), with only a minor amount of terrigenous material. The upper limits are marked in most places by a deeply burrowed and usually phosphatised surface below an unconformity overlain by Neogene rocks (Warren, 1995). Locally, the Amuri Limestone has been intruded by bodies of well-sorted, soft, grey, unfossiliferous sandstone (Lewis *et al.* 1979; Warren, 1995).

The probable earliest Eocene (Waipawan) age for the Teredo Member at the base of the Amuri Limestone in this area indicates that limestone deposition began later here than further to the north, where it has a Late Paleocene (late Teurian) age. This is consistent with the regional pattern of a decreasing southward age for the onset of thick carbonate deposition in this part of the stratigraphic column. Most of the limestone appears to be of late Middle Eocene (Bortonian) age, but ages up to Late Eocene (Runangan) have been recorded within the highest beds (Warren, 1995).

The Amuri Limestone extends northwards to southeastern Wairarapa, where it is represented by the white, flinty micritic Late Paleocene Kaiwhata and Mungaroa limestones (Browne, 1987). The limestones range in thickness from 10 to 100 m and conformably overlie or locally interfinger with older Paleocene quartzose sandstone and calcareous siltstone, and are overlain by Eocene calcareous siltstone (Lee, 1995; Field *et al.* 1997).

In the northern part of the Canterbury Basin the Eocene, as in Marlborough, was dominated by the calcareous sediments of the Amuri Limestone. The Teredo Member forms the base in many places in north Canterbury, but has not been recognised south of the mouth of the Waiau River. The Eocene base of the Amuri Limestone rests in many places unconformably on Late Cretaceous (Haumurian) rocks (Warren, 1995). Although in Marlborough and the northern Canterbury Basin the Amuri Limestone is wholly pre-Oligocene in age, from the vicinity of Cheviot southwards the upper portion is of Early Oligocene (early Whaingaroan) age (Fig. 6.5). In central and southern Canterbury the lithofacies is wholly of Early Oligocene age. Originally, it was probably widespread in central Canterbury, although only erosional remnants now remain (Andrews *et al.* 1987).

Southeast South Island

Offshore, on the crest of the Chatham Rise, white, soft, nannofossil foraminiferal ooze and chalk, of mainly earliest Oligocene (early Whaingaroan) age, is correlated with the Amuri Limestone (Wood *et al.* 1989). Dredge samples indicate that it is a much deeper-water, more oceanic equivalent of the younger portion of the Late Eocene–earliest Oligocene (Kaiatan–early Whaingaroan) Te Whanga Limestone of Chatham Island, which was formed in a mid- to outer-shelf environment (Campbell *et al.* 1993).

In the southern Canterbury Basin, deposition of Amuri Limestone occurred locally, beginning in late Middle Eocene (latest Bortonian) times and continuing into the earliest Oligocene in the Clipper-1 petroleum well, where it is of neritic facies, while further south, at Resolution-1 well, Amuri Limestone was not deposited until the latest Eocene or earliest Oligocene (Field *et al.* 1989). In north Otago, near Oamaru, facies variants (Ototara Limestone) overlie and interfinger with tuff and pillow lava of the Oamaru Volcanics (Forsyth, 2001; Thompson *et al.* 2014). Southwest of Oamaru it is unknown in outcrop, probably because Miocene uplift removed most of the Oligocene sediments in this area. It does however occur in the offshore Galleon-1 petroleum well.

The Amuri Limestone in the Canterbury Basin and on the Chatham Rise is typically marl or wackestone. The thickness varies from zero to a maximum of 97 m in the Clipper-1 petroleum well and much of this variation is due to removal by erosion below a widespread late Whaingaroan unconformity (Marshall Unconformity; see section 9.3.5). In the Resolution-1 petroleum well the basal 3.5 m of the Amuri Limestone contains tuff, possibly related to Oamaru volcanism. Microfaunas, coupled with the fine texture of the unit, indicate an outer-shelf or bathyal paleoenvironment. The extent of the unit suggests the presence of an extensive Early Oligocene carbonate platform (Field *et al.* 1989).

In the Great South Basin, Eocene chalk cropping out on Campbell Island, and correlative nannofossil oozes in the eastern part of the basin (Rakiura Group), are also equated with the Amuri Limestone (Cook *et al.* 1999). At the site of DSDP drillhole 277, nannofossil ooze was deposited from Paleocene to Early Oligocene (Teurian to Whaingaroan) times (Kennett *et al.* 1975), suggesting that Eocene carbonates are widely developed on the Campbell Plateau.

Interpretation of the environment of deposition of the Amuri Limestone has long been a matter of debate. It seems likely that the unit results from the pelagic or hemipelagic deposition of calcareous ooze at bathyal or near bathyal depths (Morris, 1987; Warren, 1995). Siliceous productivity, which was high during the latest Cretaceous and Paleocene, and probably formed as a product of upwelling of cool, nutrient-rich waters, dwindled to insignificance during Early Eocene times, the highest silica-rich sediments occurring within the Lower Limestone lithology. The bulk of the Amuri Limestone was deposited in relatively warm low-fertility waters at bathyal depths (Strong *et al.* 1995).

6.7.2.3 Other sedimentary lithofacies of eastern South Island

Northeast South Island

At the beginning of the Eocene the Canterbury region was still largely divided into northern and southern sectors by the Chatham Rise, which was then a shallow submarine high or a very low-lying erosion surface (Field *et al.* 1989). In north Canterbury the older part of the Eocene Amuri Limestone passes laterally southwards into the clastic Ashley Mudstone. This glauconitic, sandy marine mudstone dominates the Early–Middle Eocene sediments in the central Canterbury region, and extends into the Late Eocene in the Resolution-1 petroleum well south of Banks Peninsula. The mudstone lies conformably, or disconformably (with a burrowed contact), on the Paleocene Waipara Greensand; the base of the mudstone is Paleocene–Early Eocene (Teurian–Waipawan), and the top is dated as Late Eocene (Browne & Field, 1985), although in the Waipara River the upper and lower contacts are obscured and only the Early–Middle Eocene (Waipawan–Bortonian) part of the Ashley Mudstone has been identified (Morgans *et al.* 2005). The paleoenvironment is inferred to be outer-shelf to upper-bathyal. The Ashley Mudstone passes gradationally westwards into Middle–Late Eocene sandy lithologies (Homebush and Karetu sandstones), which reflect the shallow marine conditions that prevailed around the western end of the Chatham Rise at that time. Much of the lower portion of the sandstone is composed of free-running, non-calcareous, massive, slightly glauconitic quartzose sand, inferred to represent a strandline or shoreface deposit (Andrews *et al.* 1987; Wood *et al.* 1989). The upper portion of the sandstone is dominated by bioturbated, medium- to fine-grained, muddy sandstone, probably deposited in a mid-shelf environment.

Subalkaline tholeiitic latest Paleocene volcanism (View Hill Volcanics) in the central Canterbury basin continued into Early Eocene (Mangaorapan) times accompanied by deposition of thin, shallow-marine limestone, and followed by Middle–Late Eocene shallow-marine quartzose and glauconitic sandstone units (Homebush Sandstone and Iron Creek Greensand). In the far west, marginal marine conglomerate of inferred Early to Middle Eocene age occurs (Field *et al.* 1989).

Chatham Islands

Late Paleocene deposition in the Chatham Islands region consisted mainly of basaltic volcanics (Red Bluff Tuff) and these deposits extended over much of Chatham Island and Pitt Island. In the northwest part of the island, Maastrichtian–earliest Eocene (Haumurian–Waipawan) glauconitic grits and sandstones (Tioriori Group) were being deposited (see section 6.4.2.3). The tuffs also occur on Pitt Island. Contemporaneous with, and succeeding, the emplacement of the volcanics, there was a long period of carbonate deposition that formed the Te Whanga Limestone of Late Paleocene to Early Oligocene age (Campbell *et al.* 1988, 1993; Wood *et al.* 1989). Two fundamentally different limestones exist: a packstone of Late Paleocene to late Middle Eocene (late Teurian to Bortonian) age, and a grainstone of Late Eocene to earliest Oligocene (Kaiatan to early Whaingaroan) age. All observed contacts between the two lithofacies are disconformable. The packstone formed contemporaneously with Red Bluff Tuff in many places, thin sheets of limestone commonly occurring between tuff intervals, locally wedging out on the flanks of tuff mounds. Where fully developed, it is at least 35 m thick, is moderately hard and porous, and is dominated by bryozoa, echinoids and benthic foraminifera. Deposition in a shallow, warm sea within or below the euphotic zone is inferred (Campbell *et al.* 1993).

The younger grainstone has not been recognised on Pitt Island, and appears to be confined to central and northern Chatham Island, where it disconformably overlies the packstone lithofacies or Red Bluff Tuff, and is in turn disconformably overlain by Neogene sediments. The limestone is a pale-yellow to white, locally very porous, massive, well-sorted bryozoan grainstone (Campbell *et al.* 1988, 1993). It reaches a maximum thickness of 25 m. At some localities glauconite and subangular to subrounded pebbles and cobbles of quartz, schist, basalt, phosphatised tuff and grainstone occur at the base, locally infilling karst-like cavities in the underlying packstone. Decimetre-scale cross-bedding in northern outcrops shows bimodal azimuths, with northwest-flowing currents dominant and easterly currents subsidiary, while in the south, azimuths are unimodal and indicate a west-flowing current system. The microfaunas suggest a mid- to outer-shelf environment, although the well-sorted grainstone nature of the limestone suggests that the depositional environment was probably a moderately shallow platform in an oceanic setting (Campbell *et al.* 1993).

Southeast South Island

In the western south Canterbury Basin between Mount Somers and Oamaru, most of the Early–Middle Eocene (Waipawan–Porangan) is represented locally by coal measures, and more commonly by paralic sediments (Kauru Formation) that typically consist of a condensed unit of pebbly shellbeds, glauconitic sandstone, and, in places, brown siltstone. This unit marks the transition between latest Cretaceous to earliest Eocene (Haumurian–Waipawan) terrestrial sediments and marine late Middle Eocene (Bortonian) glauconitic sandstones (e.g., Waihao Greensand). The Waihao Greensand locally grades westwards into a less glauconitic, yellowish-grey sandstone (Opawa Sandstone), and northwards into non-calcareous sandstone of nearshore facies (Homebush Sandstone). The Waihao Greensand is in most places overlain by the Ashley Mudstone, or locally by intervening siltstone (Kakahu Siltstone) of Late Eocene (upper Bortonian to Kaiatan) age, which ranges into the earliest Oligocene (early Whaingaroan).

In the Oamaru area (Fig. 6.4), mudstone was deposited during most of the Eocene period. Offshore to the east, at the Endeavour-1, Clipper-1 and Galleon-1 petroleum wells, Paleocene glauconitic micaceous mudstone (Moeraki Formation) is succeeded in the Early–early Middle Eocene (Waipawan–Heretaungan) by massive calcareous mudstone (Hampden Formation). Both units were deposited in outer neritic to bathyal settings. Deposition of calcareous mudstone continued at least until late Middle Eocene (Bortonian) times; south of Oamaru, this facies extends into the Late Eocene (Kaiatan). It is overlain by tuff and pillow lava (Deborah and Waiareka Volcanics) of Late Eocene (Kaiatan) age (see Plate 11), which has commonly associated diatomite. Deposition of Ototara Limestone occurred locally, associated with these volcanics (see section 6.7.2.2).

On the onshore portion of the Great South Basin in southeastern Otago the latest Paleocene–Middle Eocene sediments are fine-grained and shallow marine in character (Abbotsford Formation). South of Dunedin the sequence consists of glauconitic silty sandstone passing laterally into a mud-dominant succession with layers of glauconite-rich sandstone, which form highly bioturbated greensand–mudstone cycles inferred to result from a series of relative sea-level changes, probably in mid-shelf depths (Lindqvist, 1995; McMillan & Wilson, 1997). In the Dunedin area the Abbotsford Formation is overlain by sandstone (Green Island Sand) and calcareous and glauconitic ?mid-shelf mudstone (Burnside Mudstone) of Middle to Late Eocene age (Fig. 6.4), truncated by post-Eocene unconformities (Field *et al.* 1989).

In the offshore portion of the Great South Basin substantial subsidence took place during the Eocene such that fully marine and, locally, bathyal environments became permanently established. Deposits between latest Paleocene (latest Teurian) and earliest Oligocene are referred to as the Rakiura Group (Cook *et al.* 1999). Seismic facies mapping shows the progressive westward stepping of clastic shoreline-related facies along the western margin of the Great South Basin during deposition of the sequence; petroleum wells in this area (Toroa-1, Tara-1 and Rakiura-1) all contain sandy nearshore facies, passing upwards into fine-grained and increasingly calcareous facies from about the Middle Eocene. Further east, the transition to clastic-deficient shelf deposits occurs earlier, and the Rakiura Group is dominated by chalk (equivalent of the Amuri Limestone).

In the southern and southeastern parts of the basin, near the present shelf edge and on Campbell Island, the base of the Rakiura Group is defined by an unconformity, or the base of a condensed section, while in the central and western areas the correlative surface is a prominent seismic reflector. The top boundary is a strong regional reflector that commonly corresponds to a significant increase in carbonate content (Cook *et al.* 1999). In the southeastern part of the basin the oldest sediments of the Rakiura Group are shales, marls and minor dolomites (Laing Formation) overlain by marls (Tucker Cove Formation). To the northwest, interbedded sandstones become significant. In the central basin, part of the Middle Eocene (Porangan Stage) is missing, as it is in onshore Otago (McMillan & Wilson, 1997) and in the Canterbury Basin (Field *et al.* 1989).

The base of the Rakiura Group (see Fig. 4.6) corresponds to the transition from restricted marine circulation with organic-rich shale deposition in the Late Paleocene to open-ocean conditions in the Early–early Middle Eocene. Apart from a break in the northwest in Porangan times, sedimentation was continuous throughout the basin with a depocentre extending north–south in its central region. This was infilled from the Early–early Middle Eocene (Waipawan–Heretaungan) by a submarine fan sequence up to 1200 m thick, part of a broad prograding wedge formed at the shelf margin, from which a slope-basin floor fan complex developed, probably in outer-shelf to upper-bathyal depths (Raine *et al.* 1993). Pelagic carbonate sediments of the Tucker Cove Formation were first deposited in the southeast in the earliest Eocene, and extended progressively westwards with time.

In the Middle Eocene (Porangan) a marine transgression resulted in landward migration of the shoreline. The main depocentre remained in the central part of the basin, accumulating up to a further 800 m of submarine fan sediments. Most of the Great South Basin was now at bathyal depths, and the eastern side became starved of sediment, resulting in condensed successions and deposition of pelagic carbonates. During the Late Eocene (Kaiatan and Runangan), sedimentation rates dropped sharply and sedimentation was dominated by marine carbonates, particularly in the starved southern part of the basin, where only 200–300 m of carbonates accumulated.

Seismic data show extensive submarine canyon and valley systems developing within the Eocene succession, appearing at bathyal depths from at least the Early Eocene, and eroding into earlier Eocene and Paleocene sediment (Carter & Carter, 1987; Cook *et al.* 1999). This channelling eventually created a northeast-trending submarine channel complex about 200 km long and 10–50 km wide, cut 300–1000 m into the sea floor, which fed into the Bounty Channel System.

In inland east Otago, Cenozoic sedimentation did not begin until the late Middle Eocene, when coal-bearing non-marine sediments (Hogburn Formation) were deposited on deeply weathered Otago Schist (Bishop, 1974a). The formation comprises up to 300 m of quartzose granule and pebble conglomerate, sandstone, kaolin-rich silty clay, carbonaceous mudstone, and thin lignite of late Middle to Late Eocene age (Bishop, 1974a). Thin (up to 20 m), well-sorted sandy sediments overlying this unit contain features, such as *Ophiomorpha*-bearing hummocky-swaley cross-stratified coarse sandstone of late Middle Eocene (Porangan) age, which are typical of shallow-marine transgressive sands. Overlying this inferred shoreface, sandy facies are up to 40 m of calcareous siltstone and mudstone, including glauconite-rich horizons and concentrations of *Thalassinoides* burrows (Swinburn Formation) that record a progressive deepening to mid-shelf depths during Late Eocene (Kaiatan–Runangan) times (Lindqvist, 1995).

6.7.3 Latest Paleocene–earliest Oligocene sediments of western New Zealand

6.7.3.1 Southland onshore basins

There are several distinct basins and sub-basins recognised in this region (see Fig. 1.3), which show differing sedimentary patterns reflecting different active tectonic controls.

In central Southland, rocks of Eocene age are absent east of Gore, but, to the west, Late Eocene (Kaiatan–Runangan) coal measures (Mako Formation) were deposited in the Winton Basin in half grabens formed as a result of oblique extension (Cahill, 1995). The formation is correlated with the Beaumont Coal Measures of the Waiau Basin of western Southland (Fig. 6.7).

In western Southland a complex system of basins and sub-basins developed that strongly reflects differential subsidence and active faulting (Fig. 6.9). The earliest post-Cretaceous sediments are of Middle Eocene age and rest on Late Cretaceous deposits with low angular unconformity, suggesting Paleocene–Early Eocene uplift and erosion (Turnbull *et al.* 1993). Eocene sediments occur throughout the onshore Waiau, Te Anau and Balleny basins (Figs 6.10 and 6.11) and represent the onset of widespread Cenozoic subsidence and sedimentation. They are equally well represented offshore in the Solander Basin and in the offshore portion of the Balleny Basin, as inferred from seismic data, and confirmed in the Parara-1 petroleum well, which penetrated over 700 m of Eocene sediments (HIPCO, 1976). The Eocene succession commonly begins with a non-marine unit, often with a basal conglomerate, and in most areas is capped by a dark-brown mudstone facies, of latest Eocene (Kaiatan) age, which forms a useful marker horizon. In the Te Anau Basin, Eocene deposition was initially localised in fault-bounded marine basins that had high sedimentation rates, before being overlain by more-typical non-marine Eocene sediments (Turnbull *et al.* 1993).

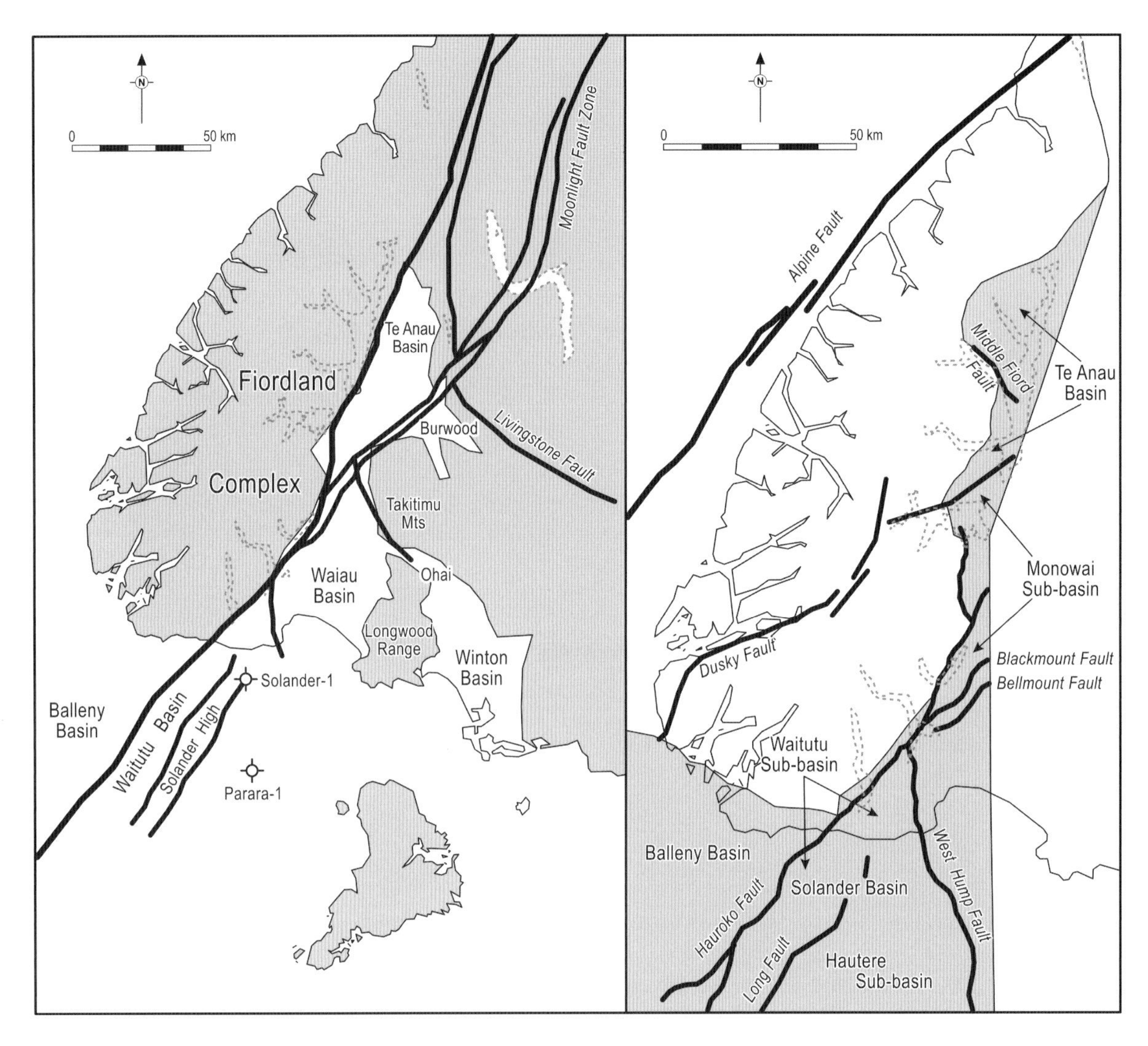

Fig. 6.9. General map of Southland basins (left) (based on Norris & Turnbull, 1993), and details of basins and sub-basins west of the Takitimu and Longwood mountains (right). (Adapted from Turnbull *et al.* 1993.)

In most places Eocene sediments are conformably overlain by Oligocene marine rocks, although in parts of the Balleny Basin and offshore Waitutu Sub-basin of the Solander Basin the contact is an angular unconformity on basement or on Cretaceous sediments. Eocene rocks are also absent from the eastern side of the Fiordland block adjacent to both the Waiau and southern Te Anau basins. These unconformable contacts indicate a period of Late Eocene tectonism. The offshore Solander Ridge and other basement highs remained emergent throughout the Eocene (Turnbull *et al.* 1993).

The oldest Eocene sediments, of late Middle Eocene (Bortonian) age, are found in the northern Te Anau Basin and in the Balleny Basin. Elsewhere, basal sediments are of Late Eocene (Kaiatan or Runangan) age. Much of the Eocene succession resembles the Brunner Coal Measures and Kaiata Mudstone facies association of the West Coast of the South Island (Nathan *et al.* 1986; and later in this section), and it is possible that some western Southland basins may have been continuous with those of the West Coast prior to Cenozoic movement on the Alpine Fault.

Eocene sediments are exposed throughout the eastern Waiau Basin and extend westwards to Hump Ridge and the Hauroko and Blackmount faults in the subsurface. West of the Hauroko Fault, Eocene sediments are absent beneath a thin Oligocene marine-shelf sequence. In the Waiau Basin, Eocene sediments are represented by the largely non-marine

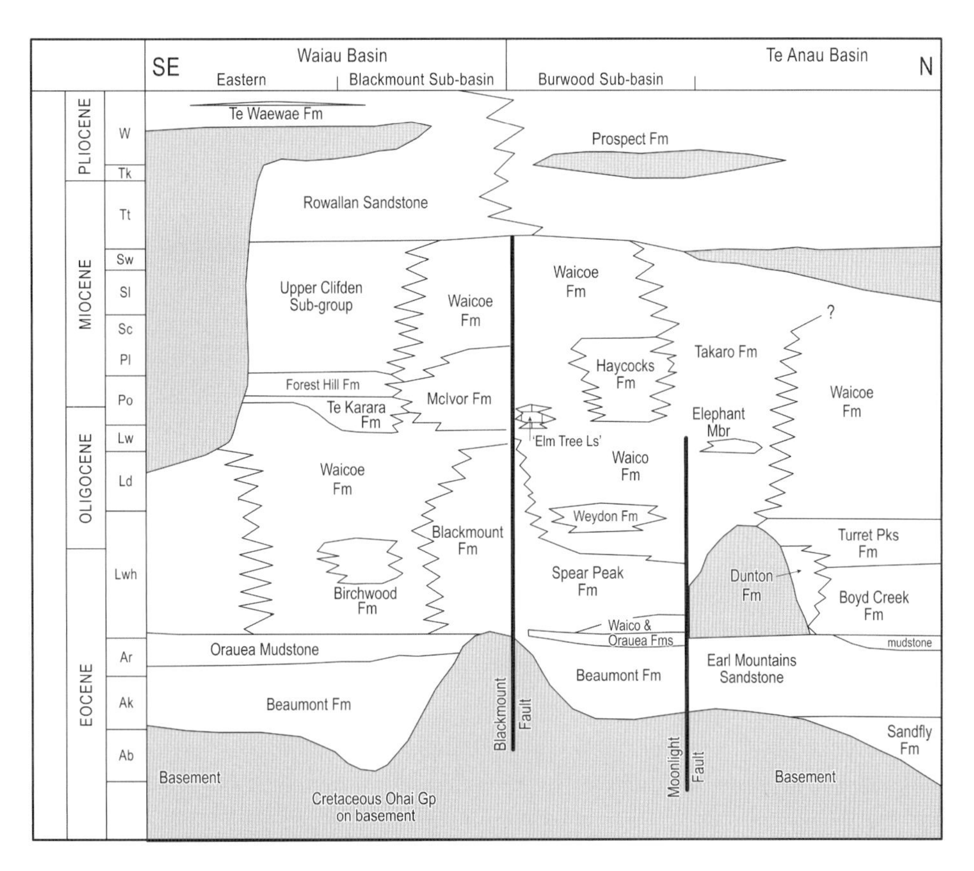

Fig. 6.10. Time-stratigraphic section through the Waiau and Te Anau basins of western Southland. For lithologies of named formations, see sections 6.10.3.1 and 8.2.4.3. Fm, Formation; Gp, Group; Ls, Limestone; Mbr, Member. (Adapted from Turnbull *et al*. 1993.)

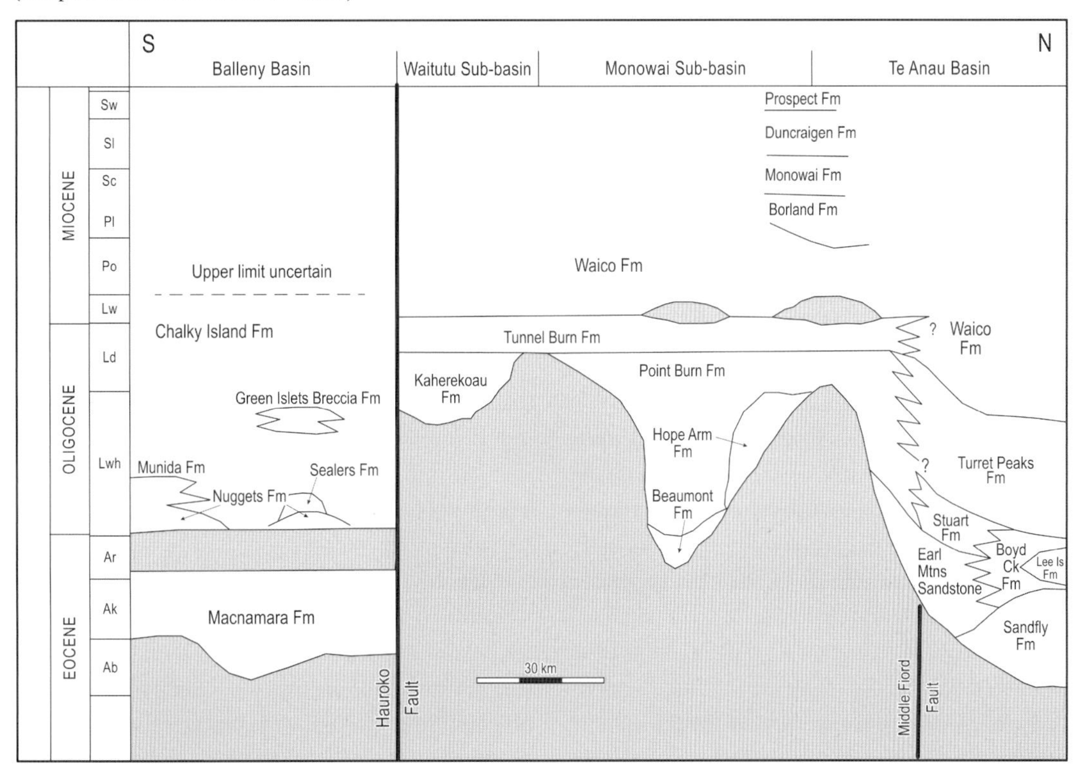

Fig. 6.11. Time-stratigraphic section through the Balleny and Te Anau basins, western Southland. For lithologies of named formations, see sections 6.7.3.1, 6.7.3.2 and 8.2.4.3. Fm, Formation. (Adapted from Turnbull *et al*. 1993.)

Nightcaps Group, consisting of Beaumont and Orauea formations in the east, and by the predominantly non-marine Hump Ridge and Sand Hill Point formations on the west side of the basin. The Eocene succession is only a few hundred metres thick where it is exposed in the east, but it thickens westwards in the subsurface; it may reach 1800 m on Hump Ridge (Turnbull *et al.* 1993).

Based on detailed work at Ohai (Sykes, 1985, 1988; Shearer, 1992) and near Blackmount (Griffith, 1983), the lithofacies of the two formations of the Nightcaps Group indicate fluvial and lacustrine environments. The older Beaumont Formation, which is up to 500 m thick, is characterised by coarse-grained, trough cross-bedded, slightly pebbly quartzo-feldspathic sandstone, with subordinate conglomerate and thin coal seams, and, locally, oil shale (Turnbull *et al.* 1993). At Ohai the succession is interpreted as forming part of a complex braided river system (Sykes, 1985, 1988), and in the Waiau River area a probable upper delta-plain or floodplain deposit is indicated. An overlying cyclic, coarsening-upwards, mudstone–sandstone succession is interpreted as a prograding delta facies (Pocknall & Turnbull, 1989). Marine indicators are lacking, and the deltas prograded into an extensive lake or lagoon represented by the Orauea Mudstone.

The widespread and distinctive Orauea Mudstone is massive, slightly carbonaceous and micaceous brown-grey in colour. Slumped horizons and graded sandstone beds are present locally. In most places it overlies the fluvial Beaumont Formation sharply but conformably, although in others (e.g., Waiau River) the two units interfinger. The unit appears to be thickest (500 m) in the Ohai area and may be extensive towards the southwest. Around the northern and eastern Takitimu Mountains the lithofacies is less than 100 m thick, and at Port Craig it may be less than 50 m. Sparse microfaunas, dinoflagellates and freshwater bivalves in the Orauea Mudstone indicate that it formed in a brackish lagoonal or lacustrine environment, showing a greater marine influence in its upper portion (Turnbull *et al.* 1993). Graded sandstone and slump packets are interpreted as delta front deposits.

The Late Eocene (Kaiatan–Runangan) southwestern Waiau Basin succession at Hump Ridge is dominated by slightly carbonaceous, quartzo-feldspathic, massive to planar and trough cross-stratified, coarse-grained sandstone, and subordinate thin carbonaceous siltstone beds and granite conglomerates (Hump Ridge Formation). The formation thins eastwards beneath the Waiau Basin where it apparently merges with the time-equivalent Beaumont Formation of the Nightcaps Group (Turnbull & Uruski, 1995). In the southern Hump Ridge area the unit is transitional into impure, bioturbated, bioclastic limestone, and there are rare, local occurrences of sub-bituminous coal seams up to 0.5 m thick. The Hump Ridge Formation is inferred to be a coastal-plain deposit, dominated by fluvial systems near the base, and grading through back-beach lagoons and sandy beaches into a shallow-marine sand-dominated shelf (Turnbull *et al.* 1993).

Paleocurrent data from the Beaumont Formation and from the Orauea Mudstone in the Waiau Basin indicate sediment transport towards the southeast and east; similar directions are reported from the Blackmount Sub-basin. Pebbles and sandstones of the Beaumont Formation and conglomerate clasts in the Hump Ridge Formation have a dominant Fiordland Complex provenance, implying a westerly derivation (Turnbull *et al.* 1993).

In the Te Anau Basin the Middle–Late Eocene succession is represented by the Annick Group. A connection between the upper Annick Group and the Nightcaps Group of the Waiau Basin is possible at the southern end of the basin, but the Te Anau Basin was a separate entity in the Middle Eocene (Turnbull *et al.* 1993). In contrast to the Waiau and Balleny basins, the oldest Eocene sediments in the Te Anau Basin are marine, and reflect considerable tectonic activity. These Eocene sediments probably extended over all of northern Fiordland to link with the West Coast of the South Island, as suggested by infaulted slivers of inferred Late Eocene sediments found near the Alpine Fault. The Annick Group rests unconformably on a variety of basement terranes across the basin, the lower part infilling paleovalleys eroded or downfaulted into basement. In the southeast of the basin the group is inferred to be overlain unconformably by Oligocene marine sediments.

The lower Annick Group (Sandfly Formation), which is over 3000 m thick, is characterised mainly by clast-supported cobble to boulder conglomerate in massive or metre-bedded successions that thicken into present-day topographic depressions, such as the Middle Fiord of Lake Te Anau and the Eglinton Valley. These rocks are interpreted variously as scree breccia originating from an adjacent scarp, and deposits of a high-energy fluvial environment (Turnbull *et al.* 1993). In the northern part of the Te Anau Basin the basal lithofacies is metre-bedded to massive, sandy, clast-supported cobble to boulder conglomerate up to 1000 m thick, with minor sandstone lenses. Scattered coalified logs occur near the base, and rare macrofossil fragments or shellbeds towards the top. This basal lithofacies grades laterally and vertically into metre-bedded massive and graded coarse sandstone, and then into dark grey-brown massive mudstone. This part of the succession is interpreted as a small-scale submarine fan complex deposited rapidly in a restricted marine basin, with paleocurrent data indicating southward-directed sediment transport. Thickness and distribution of many Sandfly Formation outcrops suggest deposition beside fault scarps or in major paleo-depressions parallel to present-day valleys (Turnbull, 1985; Turnbull *et al.* 1993).

The upper Annick Group (Earl Mountains Sandstone) consists of a variety of lithofacies, including sandy conglomerate, massive, thick-bedded, pebbly, coarse-grained sandstone beds, trough or planar cross-bedded medium- to coarse-grained sandstone, and laminated, rippled, carbonaceous siltstone and fine sandstone. The unit is up to 1500 m thick, and it fines and becomes thinner-bedded upwards. A carbonaceous mudstone lithofacies, which normally marks the top of the formation, closely resembles the Orauea Mudstone facies of the Waiau Basin. Rare coalified logs and thin sub-bituminous coal seams occur locally. The Earl Mountains Sandstone is inferred to be a major fluvial deposit, fining upwards from a relatively high-energy, channel-dominated braided system to a distal floodplain-dominated environment. Channelled conglomerates, including channels eroded in basement, are most common at the base of the unit. The upper carbonaceous mudstone is transitional from brackish to marginal marine, and may be lagoonal, as it contains rare marine macrofossils and is overlain by definite marine units. *Ophiomorpha* burrows occur locally.

In the Balleny Basin a late Middle Eocene–Early Oligocene (Bortonian–Whaingaroan) non-marine and shallow-marine succession, the Macnamara Formation (Bishop, 1986; Turnbull *et al.* 1989), is discontinuously preserved in fault blocks along the southwest Fiordland coast. The thickest onshore successions are more than 400 m thick, and they thicken offshore to the southwest. Outcropping Eocene sediments unconformably overlie either crystalline basement or Cretaceous strata, and similar relationships can be recognised on offshore seismic profiles. The Eocene sequence also clearly onlaps basement structural highs within the basin (Turnbull *et al.* 1993).

Three main lithofacies associations are recognised within the onshore Eocene succession of the Balleny Basin near Puysegur Point. The basal association consists of sandy cobble-to-boulder conglomerate, with blocks up to 4 m in diameter in beds 5–50 m thick, interspersed with sandy or carbonaceous mudstone and coal. It is inferred to represent an alluvial fan. The alluvial fan association is overlain by a 100–200-m-thick facies association composed of fining-upwards units from 2 to 15 m thick. Each unit consists of fine- to coarse-grained, slightly pebbly, channelled sandstone commonly consisting of sets of inclined, heterolithic, cross-stratified beds at the base, fining upwards into interbedded fine sandstone and mudstone, then carbonaceous mudstone and coal. This facies association is inferred to represent a meandering sand-laden river system with an extensive floodplain (Lindqvist, 1990). Overlying and locally interfingering with the alluvial plain association is a marginal marine association up to 150 m thick. It consists of dark-brown, fossiliferous, concretionary mudstone, interbedded with thin-bedded sandstone and mudstone, planar-laminated and cross-bedded sandstone, and minor conglomerate. This association was probably deposited in a wide range of environments, including gravelly to sandy beach, back-barrier lagoon, tidal flat, delta distributary channel and mouth bar, and shallow open-shelf environments (Lindqvist, 1986, 1990). Other lithofacies include basal conglomerates on Chalky Island that grade up into sandstones with shallow-marine trace fossils (*Ophiomorpha*, *Thalassinoides*). Near-basal deposits here also include hummocky cross-stratified sandstone (Turnbull *et al.* 1993).

6.7.3.2 Solander offshore basin

The mainly offshore Solander Basin consists of three sub-basins with Eocene strata. A large area in the Waitutu and Hautere sub-basins is floored by Eocene sediments that locally overlap onto basement. The distribution of the Eocene sediments closely coincides with the distribution of the unconformably underlying Cretaceous rocks. The easternmost Parara Sub-basin was initiated during the Late Eocene and was controlled by a set of north–south-trending en echelon faults. The Parara-1 petroleum well, sited on the western side of the sub-basin, penetrated a total of 748 m of non-marine sediments, including a basal granitic conglomerate and biotite-rich quartzo-feldspathic sandstone with coal seams near the top, of Late Eocene (Runangan) age. They could be correlatives of either the Nightcaps Group or the basal Balleny Group. In the remainder of the basin, Eocene sediments generally average less than 2 km in thickness. The adjacent Hautere Sub-basin is also controlled by north-northeast-trending en echelon faults, with Eocene sediments thickening northwards but rarely exceeding 2 km. In the Waitutu Sub-basin, seismic profiles show that the deposition of Eocene sediments was controlled by the Moonlight Fault System, downthrowing to the southeast along the offshore extension of the Hauroko Fault. The succession, which forms a single westwards-thickening wedge of reflectors against the fault, thins rapidly across it, from a maximum thickness of more than 3 km at the fault plane (Uruski, 1992; Turnbull *et al.* 1993).

6.7.3.3 Westland basins

Subaerial erosion dominated the West Coast of the South Island during the Early Eocene, the only record of essentially continuous sedimentation from Paleocene through to Late Eocene being in south Westland; elsewhere, sedimentation began in Middle Eocene times (Fig. 6.8).

In south Westland, Eocene sediments and interbedded volcanic rocks (uppermost Tokakoriri Formation and conformably overlying Abbey Formation) are up to 700 m thick (Fig. 6.8). They are exposed only in a small cove in south Westland in the vicinity of the Paringa River mouth and south of Jackson Bay, where they are unconformably overlain by Oligocene strata (Nathan, 1977; Adams, 1987; Aliprantis, 1987). The Abbey Formation is dominated by calcareous lithologies. The basal deposits consist of well-bedded to poorly bedded, fine-grained, slightly silty, foraminiferal calcilutite, locally intensely burrowed and slightly glauconitic. These pass upwards into massive, medium-brown, muddy limestone and calcareous mudstone interbedded with flows of alkaline basalt and breccia (Otitia Basalt) of mainly early Late Eocene (Kaiatan) age. The youngest beds consist of approximately 40 m of probably slumped calcareous mudstone containing scattered subangular to subrounded clasts, up to 1.5 m, of basalt and sedimentary rocks, including lower Abbey Limestone and Paleocene Tokakoriri Formation, indicating a short period of uplift and contemporary faulting (Nathan *et al.* 1986). The environment of deposition of the basal fine-grained lithofacies indicates a period of quiet carbonate sedimentation, the predominance of pelagic foraminifera indicating relatively deep-water sedimentation, probably in outer neritic to upper bathyal depths. Stratigraphically higher units represent gradually increasing terrigenous mud content, probably due to uplift and erosion in the source area. Isolated and poorly exposed outcrops of coal-bearing strata of early Late Eocene (Kaiatan) age also occur in this area (Adams, 1987), but their relationship to the correlative marine upper Abbey Formation is unknown.

In northern Westland the oldest Eocene lithology is almost always coal measures, formed largely as a consequence of rising base levels as the sea transgressed over the low-lying terrain. Being largely derived from the deeply weathered rocks underlying the erosion surface, the coal measures are generally highly quartzose. Various local names have been proposed for the coal measures, but as they essentially form a single unit at the base of a transgressive sequence the oldest and most widely accepted name is ‘Brunner Coal Measures’. The coal measures are overlain conformably (and in many sections gradationally) by shallow marine sediments.

Eocene deposition in the central and northern West Coast is strongly influenced by the development of north-northeast to northeast-trending fault-controlled basins, commonly formed by the rejuvenation of faults active during the latest Cretaceous. Thickness of Eocene sediments exceeds 2000 m in fault-controlled basins, while in interbasinal areas sediment thicknesses rarely exceed 200–300 m (Nathan *et al.* 1986).

The largest subsiding basin was the Paparoa Trough (Fig. 6.12), which was developed by the rejuvenated Paparoa Tectonic Zone, a major structural feature initiated in the latest Cretaceous as part of the West Coast–Taranaki Rift (Laird, 1968, 1994;

and section 6.4.4.3). Renewed movement along Late Cretaceous normal faults formed a narrow, elongate, north-northeast-trending, complexly faulted basin over 180 km long and 20 km wide extending from Ross in the south to the Mokihinui River mouth in the north. In the Greymouth Coalfield area there is a clearly marked change between the regionally extensive Brunner Coal Measures and a narrow basin containing the underlying Late Cretaceous–Early Paleocene Paparoa Coal Measures.

Typical Brunner Coal Measures consist predominantly of moderately to well-sorted micaceous quartz sandstone, commonly cross-bedded and locally containing scattered layers and lenses of quartz and greywacke conglomerate (see Plate 10b). Coal seams are sporadic throughout the unit, but in many places there is only a single major seam that occurs near the top. Gage (1952) showed that Brunner Coal Measures within the Greymouth Coalfield were thickest (up to 150 m) along the axis of the Paparoa Trough, thinning rapidly to only a few metres near the margins, where locally the formation is represented only by a single coal seam resting almost directly on pre-Eocene rocks (Nathan *et al.* 1986; Laird, 1988). A similar pattern occurs in the northern Paparoa Trough, north of the Buller River (see Plate 10a,b), where the thickness of Brunner Coal Measures close to the axis of the trough ranges from 150 to 270 m, thinning gradually eastwards over 15 km to less than 20 m (Nathan *et al.* 1986).

Tectonic, topographic and base-level controls have each been invoked as having had an influence on sedimentation in the Brunner Coal Measures. Titheridge (1993) showed there was a close relationship between isopachs on the main coal seam and Neogene faults with north-northeast and northwest orientations, suggesting that the Neogene faults were rejuvenated structures active during deposition of the Brunner Coal Measures. Syn-depositional tilting of small half graben blocks may also have influenced the migration of river channels and distribution of sandstone and locally thick mudstone units. Alternatively, Flores and Sykes (1996) suggest that the lower part of the coal measures in the Buller Coalfield was deposited

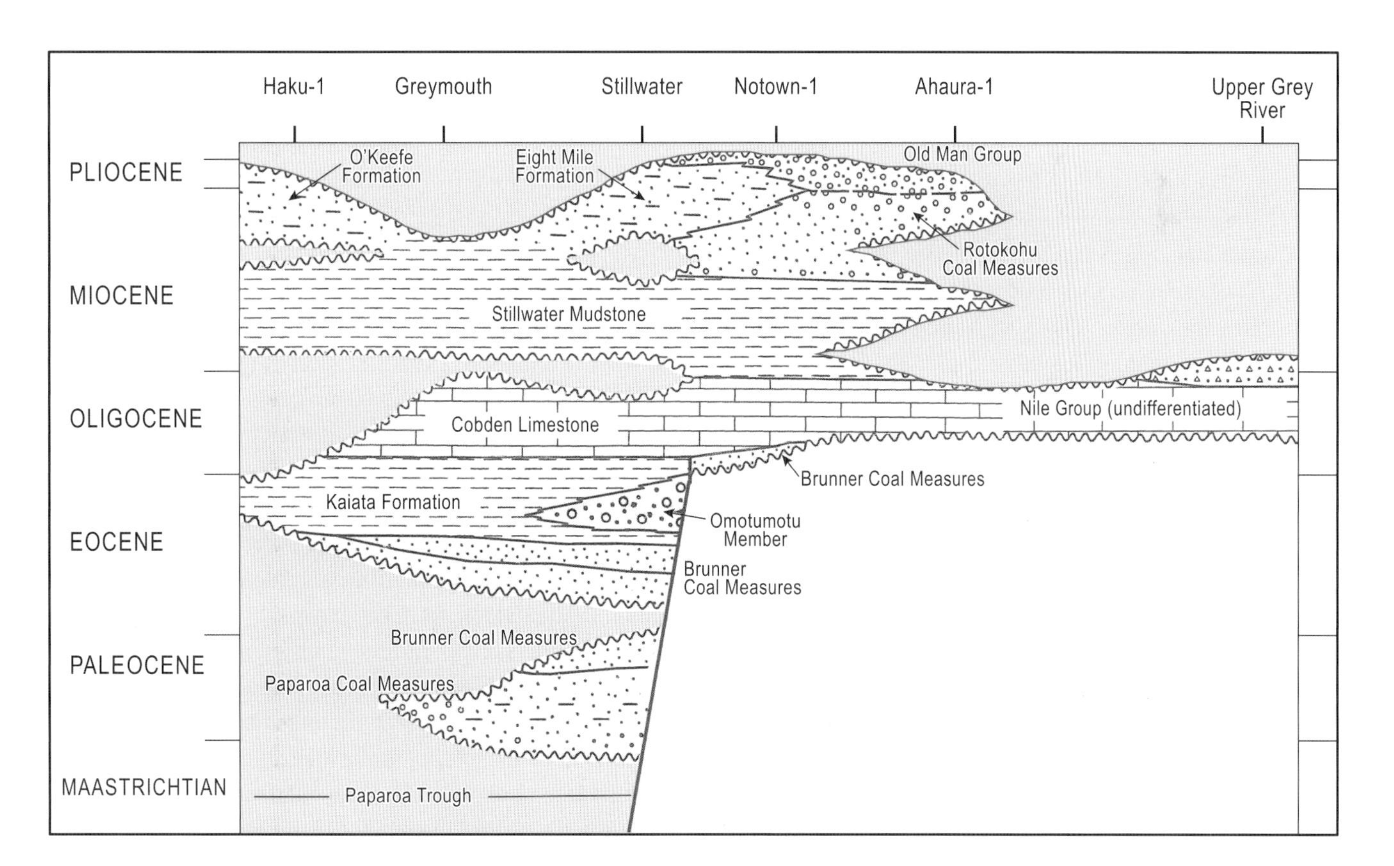

Fig. 6.12. Southwest to northeast time-stratigraphic section of Late Cretaceous and Cenozoic rocks in the Grey Valley region. (Adapted from Nathan *et al.* 1986.)

in paleovalleys incised into an undulating peneplain surface. They infer that thickening and areal distribution of the main coal seam was controlled by the nature of clastic distribution and peat-forming mire systems, marine transgression and local tidal incursion. The upper part of the coal measure is mainly of fluvial origin and comprises sandstone, bioturbated sandstone, mudstone and organic-rich lithofacies. It was inferred to represent the deposits of fluvial, paralic and shallow marine environments controlled by fluctuations of relative sea level within overall marine transgression (see Plate 10).

The age of Brunner Coal Measures of the Paparoa Trough shows a decrease to the northeast, as reflected by the age of the overlying marine beds. The oldest ages (Early to Middle Eocene) are found at the southern end of the Paparoa Trough, while the oldest coal measures in the north are late Middle Eocene to early Late Eocene in age. Continuing marine transgression also led to deposition of progressively younger coal measures outwards from the basin axis, being as young as Late Eocene on the margins of the trough in the Buller Coalfield, and late Late Eocene to Oligocene on the eastern margin of the Greymouth Coalfield, considerably younger than the Middle Eocene age of coal measures in the axis of the trough (Nathan *et al*. 1986).

In addition to the Paparoa Trough, smaller basins developed during the Eocene at Reefton, Murchison and Nelson, all including Brunner Coal Measures as the basal unit. Over 300 m has been recorded in the Reefton Basin, 400 m at Murchison and 500 m at Nelson, thinning rapidly towards the basin margins. The age is early Late Eocene (Kaiatan) in the Reefton and Nelson basins, and late Late Eocene (Runangan) in the Murchison Basin (Nathan *et al.* 1986).

The coal measures are overlain conformably (and in most sections gradationally) by shallow marine sediments, which are included in the Rapahoe Group (Nathan, 1974). The marine strata comprise a number of sometimes complexly interfingering lithofacies, the major ones including uncemented sands, calcareous sandstone, mudstone, and mass-flow deposits, which have formed in response to regional marine transgression and intermittent tectonic activity.

The most widespread lithofacies of the Rapahoe Group is a dark-brown, massive or faintly bedded, extensively burrowed, slightly to moderately calcareous, carbonaceous mudstone (Kaiata Formation). It has been recognised widely over a region between Greymouth in the south to Nelson in the north, a distance of over 200 km. The Kaiata Formation is also by far the thickest Eocene unit (up to 2000 m). It directly and conformably overlies the coal measures in many places with a sharp or gradational contact.

South of the Fox River mouth the lower part of the Kaiata Formation grades laterally into a grey-brown, massive, fine to very fine, calcareous marine sandstone, highly bioturbated, and commonly containing plant debris (Island Sandstone). The sandstone thickens southwards, reaching a maximum of 850 m in the axial region of the Paparoa Trough inland from Punakaiki (Laird, 1988), after which it thins to the south. The lithofacies, which extends at least as far south as Hokitika, is everywhere overlain conformably by mudstone of the Kaiata Formation. The environment of deposition of the sandstone is inferred to be shallow marine, probably inner- to mid-shelf. Occupying a similar stratigraphic position in the Grey Valley Trough to the east is another coarse lithofacies, consisting of basal conglomerate commonly resting on basement or Cretaceous rocks, or, locally Brunner Coal Measures, passing upwards into muddy to quartzose sandstone with dinoflagellates (Mawhera Formation). This formation passes up into Oligocene limestone, or, locally, into mudstone of the Kaiata Formation (Suggate & Waight, 1999). The age of the Island Sandstone in the Greymouth area is late Middle Eocene (Bortonian), while the age of the Mawhera Formation is late Late Eocene (Runangan), indicating continuing eastward transgression.

To the north, near Charleston, the Kaiata Formation mudstone lithofacies grades laterally into uncemented medium to fine, well-sorted, mica-feldspar-quartz sand with mainly wavy bedding and some large-scale cross-stratification and rare thin conglomerate layers (Little Totara Sand). At Charleston this unit occupies the entire Eocene succession above the Brunner Coal Measures, and reaches a maximum thickness of 210 m. It thins rapidly to the north and south where it grades laterally into, or forms discontinuous pods within, Kaiata Formation mudstone. Its northernmost exposure is at Cape Foulwind, where the sand discordantly overlies mudstone of the Kaiata Formation. The depositional environment is inferred to be nearshore marine or shoreline, perhaps tidal sandbar or beach, and locally subaerial (Nathan, 1975; Laird, 1988; Lever, 2001).

At the Cape Foulwind quarry west of Westport (Fig. 6.13) the entire Late Eocene is represented by a localised body up to 40 m thick of algal limestone and a few metres of calcareous mudstone, which interfingers with Kaiata mudstone, the lateral equivalent of a much thicker sequence of Kaiata Formation exposed on the coast only a few hundred metres further north (Lever, 2001). It is inferred that the algal limestone was deposited in a shoal on a basement high that was bypassed by silt and clay-sized material that was deposited further offshore. To the east and north-east of Westport, mudstone of the Kaiata Formation overlies and interfingers with Brunner Coal Measures in the Buller Coalfield, where it reaches a thickness of at least 700 m. The environment of deposition of the Kaiata Formation is inferred to be shallow-marine offshore, probably mid- to outer-shelf, but with restricted access to the open sea (Nathan, 1975; Laird, 1988).

In three areas of the Paparoa Trough, lenses or interbeds of coarse clastic material deposited by various mass transport mechanisms occur within the Late Eocene (Kaiatan–Runangan) portion of the Kaiata Formation. In the Greymouth area a thick clastic lens between 600 and 900 m thick (Omotumotu Member) consisting of conglomerate and breccia–conglomerate layers up to 3 m thick interbedded with typical Kaiata mudstone, occurs on the eastern side of the Paparoa Trough, thickening rapidly towards the eastern margin (Nathan, 1978b; Nathan *et al.* 1986). Similar coarse clastic deposits up to 140 m thick occur in fault-bounded strips in the central Paparoa Range east of Punakaiki. The lithofacies consists of pebbly mudstone, pebbly sandstone, fine to medium graded-bedded sandstone, and breccia, containing clasts up to 60 cm (Laird, 1988). At the northern end of the Paparoa Trough, coarse deposits consist of intercalations of breccia and other coarse clastic sediments in mudstone of the Kaiata Formation cropping out on the west-facing scarp of the Papahaua Range (Laird & Hope, 1968; Laird & Nathan, 1988). Locally, large olistoliths of granitoid rocks tens and perhaps hundreds of metres in diameter occur in Kaiata Mudstone (Nathan, 1996) adjacent to the present trace of the Kongahu Fault, where the wedge of coarse clastic deposits is also thickest. These occurrences of coarse clastic sediments interbedded with mudstone of the Kaiata Formation are inferred to represent the deposits of a variety of mass transport mechanisms,

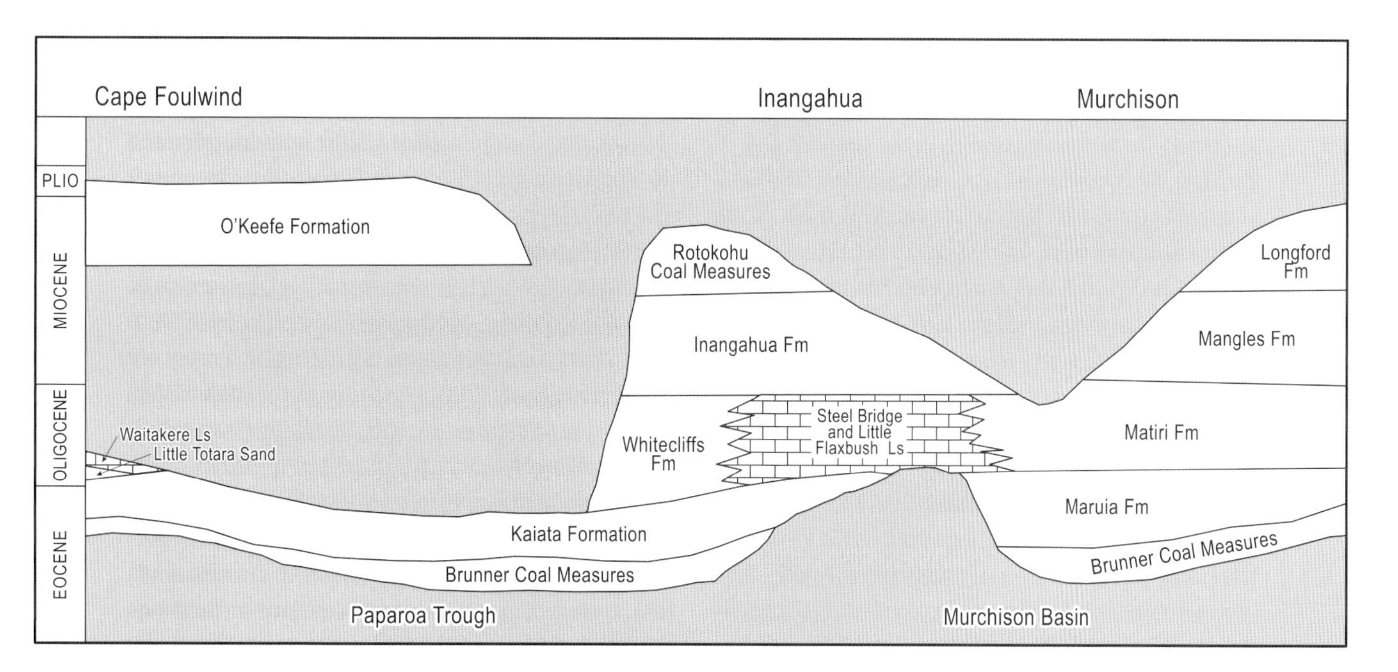

Fig. 6.13. West to east time-stratigraphic section through the Cenozoic deposits in the Buller Valley region. Fm, Formation; Ls, Limestone. (Adapted from Nathan *et al.* 1986.)

including debris flows, slumps and turbidites. They were generated by a phase of Late Eocene movement on separate faults within the Paparoa Tectonic Zone, leading to erosion and possible collapse of the footwall.

At Reefton the Eocene marine sediments overlying the coal measures consist mainly of massive dark-brown carbonaceous mudstone (Kaiata Formation), although sandstone is common as interbeds, and scattered conglomerate occurs locally. Erosion has removed the upper part of the formation in most areas, except on the eastern and western sides of the basin where thicknesses reach up to 800 m, but it seems likely that the maximum thickness in the centre of the basin exceeded 1000 m (Nathan *et al.* 1986).

Narrow elongate land areas almost completely separated the Murchison Basin from the Reefton Basin in the south and the Paparoa Trough to the west. In the northern part of the Murchison Basin, where subsidence was relatively slow and total Eocene thicknesses are generally less than 200 m, the coal measures are overlain by massive, dark-brown mudstone typical of the Kaiata Formation. To the south the Eocene succession gradually becomes thicker, and includes an increasing proportion of granite-derived sand. The 600+ m of Late Eocene sediment overlying the coal measures consists dominantly of bioturbated arkosic sandstone with interbedded carbonaceous mudstone. Thin-bedded graded sandstone–mudstone couplets are common, and the sand-dominated succession is inferred to represent a non-marine to marine fan, triggered by uplift of granitic rocks immediately to the east or northeast of the basin margin. The submarine fan extended almost to the west side of the Murchison Basin (Nathan *et al.* 1986). Immediately west of the White Creek Fault, along the western margin of the Murchison Basin, Late Eocene (Kaiatan–Runangan) sediments consist of several metres of poorly exposed granitic conglomerate, with clasts up to 0.3 m (Nathan, 1978d), deposited as a series of debris flows, suggesting that the fault may also have been active in the Late Eocene.

A small isolated area around Nelson City contains the northernmost known outcrops of marine Eocene sediments in the region. The coal measures here are conformably overlain by slightly calcareous massive dark-brown carbonaceous mudstone typical of the Kaiata Formation, which is overlain by up to 400 m of locally derived conglomerate and sandstone. Similar stratigraphy continues northwards into the southern Taranaki Basin, which is essentially a geological extension of the West Coast basins. There is an absence of Early and early Middle Eocene sediments over most of the region between Nelson and the Taranaki Peninsula. The earliest sediments are coal measures (Mangahewa Formation) of late Middle to Late Eocene (Bortonian to Runangan) age, correlative in age and lithology and merging southwards into the Brunner Coal Measures of the West Coast (King & Thrasher, 1996).

6.7.3.4 Taranaki Basin

To the northwest, towards Taranaki, the coal measures merge into marine strata, and there appears to have been northward tilting, causing marine flooding, over this part of the basin during at least the Early Eocene. The pattern in the Taranaki Basin reflects that in the southern West Coast region, including an unconformity developed across Paleocene strata in former rift sub-basins, and showing steady transgression throughout the Eocene. In contrast to the mainly northeast-directed transgression in the West Coast region, that in Taranaki was from the northwest towards the southeast (Fig. 6.3).

The Eocene succession continues the grouping used for Paleocene sediments (Fig. 6.3), and is divided into the upper parts of the terrestrial to marginal marine Kapuni Group and the marine Moa Group, which are, in the main, age equivalent. The Eocene portion of the Kapuni Group, unlike the coarse-grained clastic deposits of the Paleocene lower part (Farewell Formation), is characterised by a near-absence of high-energy, non-marine depositional environments. Dominant lithofacies reflect paralic, coastal-plain and alluvial-plain environments. The area of non-marine sedimentation gradually decreased throughout the Eocene, reflecting a long-lived but intermittent marine transgression and rise in relative base level. The northern marine part of the basin rapidly deepened as some former shelf areas subsided to bathyal depths.

Early Eocene strata are less than 100 m thick where they were deposited adjacent to the southern erosion surface and in upper bathyal areas to the far north. Elsewhere they generally range from 100 m to 400 m thick. The distribution represents a clastic wedge, thickening to the northwest from inner-shelf to mid-outer-shelf depocentres, and then thinning markedly beyond the contemporaneous shelf edge.

The oldest Eocene deposits of the Kapuni Group form the Early–early Middle Eocene (Waipawan–Heretaungan) Kaimiro Formation (Fig. 6.3), made up primarily of sandstone, siltstone, mudstone and, less commonly, coal, frequently in cyclic succession. The unit represents a range of lower alluvial plain, delta or coastal-plain and marginal marine lithofacies along the margins of a shelf embayment. The base of the Kaimiro Formation rests unconformably on the Paleocene–earliest Eocene Farewell Formation, and represents a basinward shift of lithofacies. This regression may be related to mild uplift and erosion in the southern part of the basin and hinterland. After this initial regression, regional transgression resumed and in the southern and eastern parts of the northern Taranaki Basin the mainly non-marine Kaimiro Formation is overstepped in the northwest by marine mudstone of the Moa Group, with which it also interfingers. Where the Moa Group is absent or not identified the Kaimiro Formation extends up to an unconformity in the early Middle Eocene (at the top of the Heretaungan Stage), which is overlain by the Middle–Late Eocene (latest Heretaungan–Kaiatan) Mangahewa Formation of the Kapuni Group. In the southern Taranaki Basin the Mangahewa Formation rests unconformably on basement rocks (see earlier).

The Mangahewa Formation is thickest (400–830 m) in the Taranaki Peninsula area, where the formation is dominantly terrestrial and typically consists of interbedded sandstone, coal, carbonaceous mudstone, and siltstone. Locally the formation consists of alternating cycles of stacked erosion-based sandstones, capped by coastal-plain coals and mudstones. These probably developed in response to changes in relative sea level, particularly as there is evidence of marine incursions (King & Thrasher, 1996).

Middle–Late Eocene facies with marked marginal marine affinities are present in petroleum wells close to or on the present land between the Maui field and the Te Ranga-1 petroleum well. In the northern wells, particularly well-developed sandstones of late Middle to early Late Eocene (Bortonian to Kaiatan) age are present at the top of a coastal facies interval. These sandstones may belong to a single transgressive shoreline system. In the Maui field the Middle–Late Eocene interval exhibits an increasingly marine influence up-sequence. Strata near the base are dominantly fluvial distributary-channel and tidal-channel sandstones and lagoonal mudstones, but near the top they are mainly shoreface sandstones (Bryant & Bartlett, 1991). The Mangahewa Formation passes laterally and upwards into marine mudstones (Turi Formation), except onshore on the present Taranaki Peninsula, where it passes upwards conformably into sandstones of the McKee Formation.

Shoreline and shelf depositional systems were very narrow in the northeast corner of the basin. This partly reflects post-depositional truncation by the Taranaki Fault, but probably also reflects control of the original basin topography by the proto-Taranaki Fault (King & Thrasher, 1996). That this fault was active during the Late Eocene is also suggested by thin sandy intervals of Kaiatan age in the succession west of the fault in north Taranaki containing abundant re-worked pollen of wide-ranging Mesozoic age. The pollen was probably eroded from uplifted Late Triassic–Late Jurassic Murihiku Terrane rocks to the east (King & Thrasher, 1996).

The youngest unit of the Kapuni Group in the Taranaki Basin (McKee Formation) comprises shallow-marine to shoreline sands of Late Eocene (Kaiatan to Runangan) age, which overlie the Mangahewa Formation conformably. It represents a continuation of marine transgression over the largely shoreline and lower coastal plain deposits of the older part of the Kapuni Group, and is associated with the last transgressive phase of the Eocene depositional system. The deposits are usually less than 100 m thick, with a maximum of 170 m, and consist dominantly of generally massive, pale-coloured, quartzose, fine- to coarse-grained sandstone. The McKee Formation has been identified with certainty only on the Taranaki Peninsula. However, at the southern extreme of the Taranaki Basin, strata of late Late Eocene (Runangan) age cropping out in Golden Bay (Leask, 1993) are predominantly marginal marine, indicating that the sea had encroached this far south by then. It is possible that these Runangan sediments belonged to the same shoreline system as the McKee Formation (King & Thrasher, 1996).

The Moa Group is entirely marine, and is the product of a major regional transgression across the Taranaki Basin during the Paleocene and Eocene. The unit is distributed mainly in the west and northwest of the Taranaki Basin. A paraconformity at the top of the Moa Group reflects waning subsidence and sediment supply, and marks the culmination of this marine transgressive phase in the Late Eocene–Early Oligocene (Fig. 6.14).

The main Eocene portion of the Moa Group (Turi Formation) is predominantly a uniform non-calcareous, dark-coloured, micaceous, carbonaceous, marine mudstone extending over much of the Taranaki Basin. Total thickness (including the Paleocene portion) reaches a maximum of 800 m+ in the northeast of the basin, while in most other parts the unit is generally of relatively uniform thickness and between 220 and 500 m thick. The paleoenvironmental setting was shelf and bathyal within the gradually deepening basin to the north and northwest of the coastal strand plain represented by the Kapuni Group sediments. With the onset of more regionally widespread subsidence and southward encroachment of the sea, the rift topography developed in the Paleocene was gradually onlapped and drowned.

Locally, in the northern part of the Taranaki Basin northwest of the Taranaki Peninsula (Fig. 6.14), a succession of deep marine sandstones (Tangaroa Formation) up to 200 m thick and of Late Eocene to Early Oligocene (late Kaiatan? to Runangan to early Whaingaroan) age is present within the upper part of the Turi Formation. Core samples suggest that the sands were deposited by sediment gravity flows, and seismic reflection mapping indicates that the formation has a fan lobe geometry. The fans appear to be detached from coeval shoreline sand depocentres, and their origin is uncertain (King & Thrasher, 1996). Strogen *et al.* (2017) provide a well-illustrated interpretation of the Paleogene rocks of the Taranaki region based on greatly increased seismic data.

6.7.3.5 Northland Basin

To the north of the Taranaki Basin the seismic sequence P1–P2 (Fig. 6.15) represents the Eocene succession in the offshore west Northland Basin (Isaac *et al.* 1994). The sequence is very widespread, extending without interruption as a thin, easily recognised unit of remarkably uniform thickness from the northern Taranaki Basin throughout the area west of the Northland Peninsula (Fig. 6.15). In the northern Taranaki Basin petroleum well Ariki-1, the basal reflector P2 coincides with a low-density unit with a strong positive gamma ray signature (probably a thin organic-rich shale), separating shales of differing degrees of over-pressuring. Reflector P1

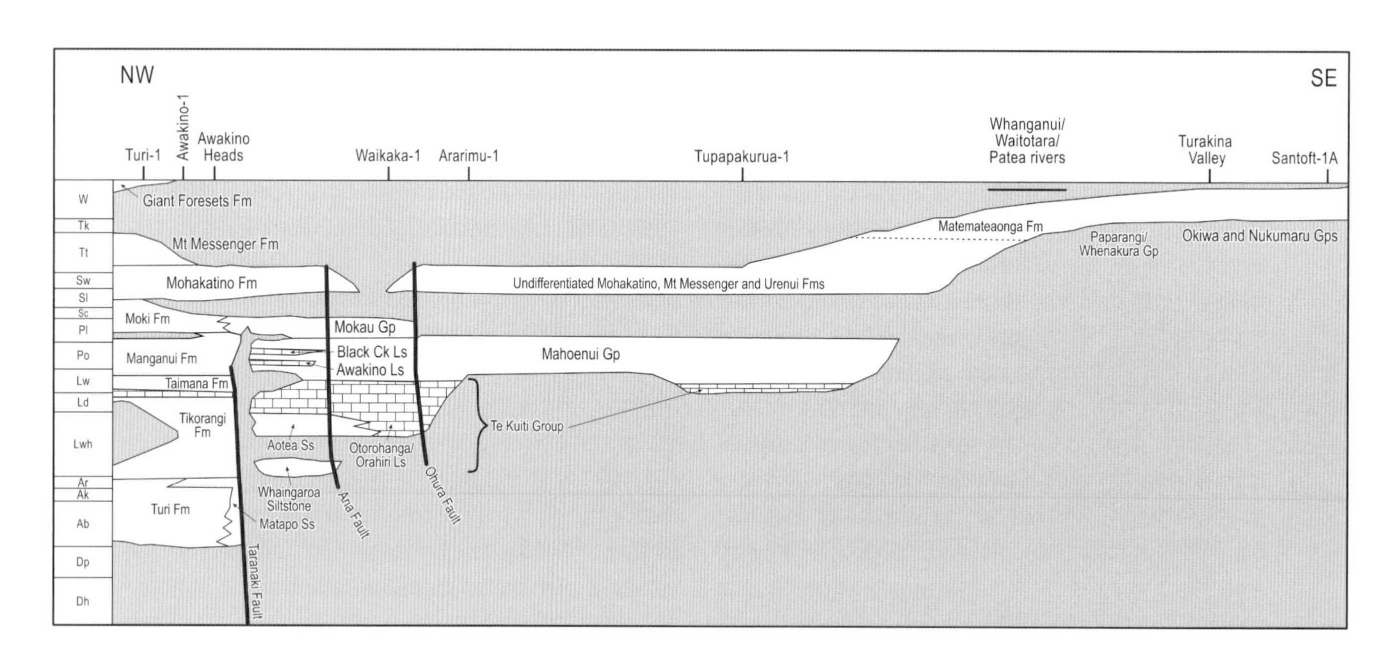

Fig. 6.14. Northwest to southeast time-stratigraphic section across northern Taranaki. For lithologies of named formations, see sections 6.7.3.4 and 8.2.6. Fm, Formation; Ft, Fault; Gp, Group; Ls, Limestone; Ss, Sandstone. (Adapted from Isaac *et al.* 1994.)

at the top of the sequence is caused by an abrupt downward increase in acoustic velocity at the base of the Oligocene–earliest Miocene Ngatoro Group limestone. The sequence P1–P2 is therefore mainly Eocene in age, and equivalent to the upper Moa and upper Kapuni Group strata of the Taranaki Basin (Fig. 6.3).

A lobate thickening of the sequence similar to that described for the underlying Paleocene interval, but offset northwards, is present west of the entrance to Kaipara Harbour. The configuration shown on seismic profiles suggests a very gentle nearshore-to-shelf-basin profile, interpreted as a delta (Isaac *et al.* 1994). In the Waka Nui-1 petroleum well, drilled 100 km west of the mouth of the Kaipara Harbour and lying just west of the inferred delta lobe, the ~200-m-thick Eocene succession consists of ~130 m of Early–Middle Eocene argillaceous limestone passing upwards into Late Eocene calcareous claystone, both indicating fully oceanic conditions in a lower bathyal environment (Milne & Quick, 1999; Strong *et al.* 1999).

The sequence P1–P2 can be traced across the Reinga Basin, and for at least 1000 km to DSDP drillhole 206 in the New Caledonia Basin. Here, and also in DSDP drillholes 207 and 208 on the Lord

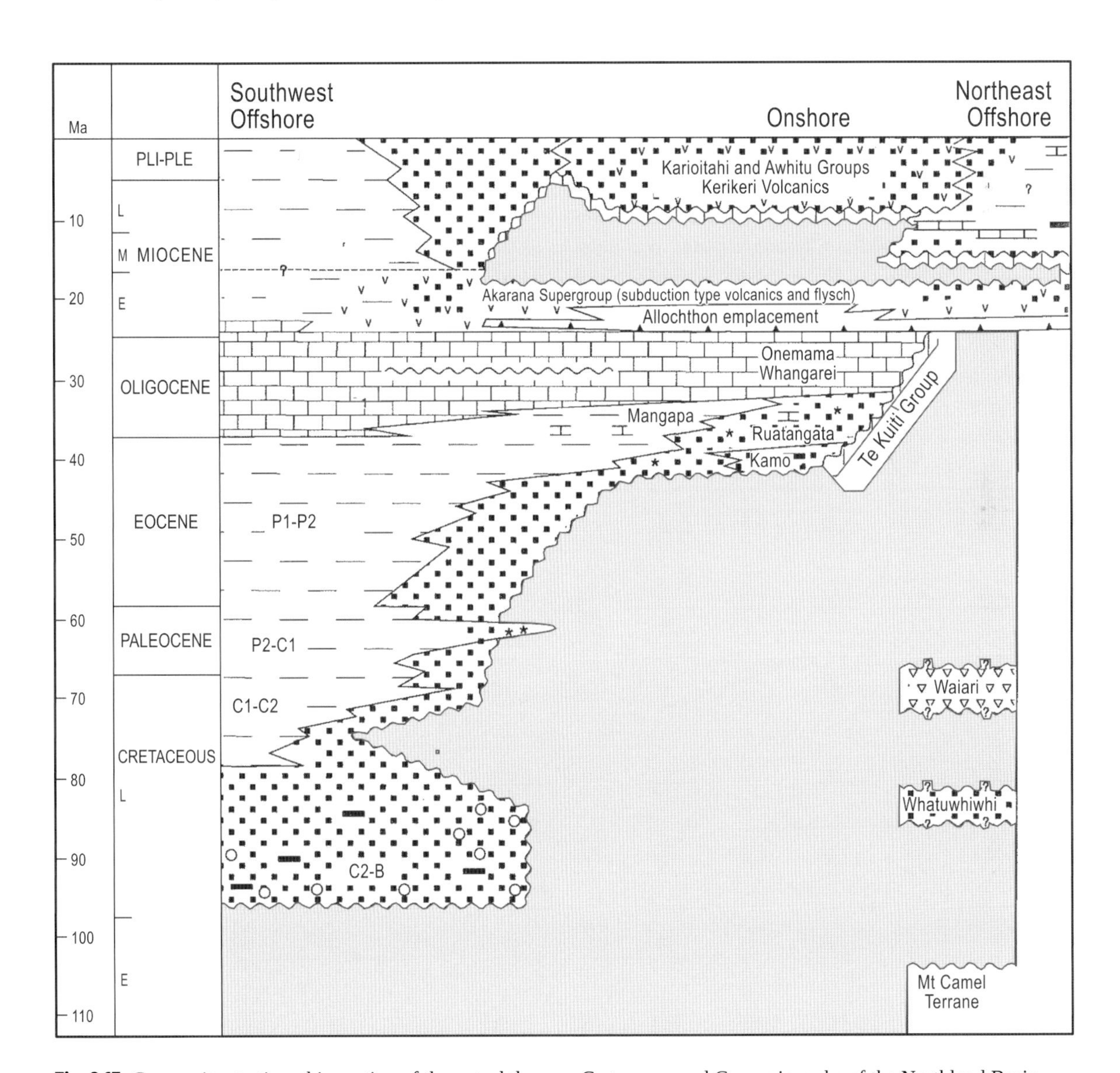

Fig. 6.15. Composite stratigraphic section of the autochthonous Cretaceous and Cenozoic rocks of the Northland Basin and Northland Peninsula. For lithologies of named formations, see sections 6.7.3.5 and 8.2.7. (Adapted from Isaac *et al.* 1994.)

Howe Rise, the Eocene sediments consist of calcareous biogenic ooze with chert layers (van der Lingen, 1973). The sequence pinches out eastwards 20–40 km west of the present Northland coast, and 20 km west of the Three Kings Islands, and has not been identified east of Northland. West of the Northland Peninsula the pinchout is an erosional truncation, rather than onlap. There seems to have been practically no faulting during the deposition of the Eocene sequence.

6.7.3.6 Onshore North Island basins

During the Eocene the onshore basins (Northland and South Auckland) lying east of the Taranaki and offshore west Northland basins followed a similar transgressive trend (Figs 6.14 and 6.15), beginning with deposition of non-marine coal measures and shallow-marine non-calcareous clastic strata in Late Eocene times (Fig. 6.16). The future sites of the Wanganui and King Country basins lying to the south, like the adjacent southern Taranaki Basin, remained emergent. The term 'Te Kuiti Group' is used to refer to the Eocene and Oligocene successions in both the Northland and south Auckland regions (Edbrooke *et al.* 1994; Isaac *et al.* 1994). Between the Brynderwyn Hills (south of Whangarei) and Auckland 110 km to the south, no in situ Te Kuiti Group rocks occur and Neogene rocks rest directly on basement (Isaac *et al.* 1994). It is uncertain whether the Te Kuiti Group was never deposited or was removed by the pre-Neogene. Although similar, the Northland and south Auckland regions are treated here as two separate basins in which Te Kuiti Group was deposited.

In eastern central Northland the mainly Late Eocene lower Te Kuiti Group comprises basal Kamo Coal Measures and Ruatangata Sandstone. North of the Hokianga Harbour there are no known coal measures, and Ruatangata Sandstone passes gradationally up into the calcareous Mangapa Mudstone (Fig. 6.16). The thickness patterns of the coal measures suggest that they may have accumulated mainly in half grabens that were oriented west-southwest–east-northeast. They reach a maximum thickness of 100 m, and rest unconformably on Mesozoic basement and are overlain either by Ruatangata Sandstone or by the basal décollement of the Northland Allochthon. Drilling and mapping demonstrate that the coal measures are discontinuous, and, in places, upper Te Kuiti Group limestone of Oligocene age directly overlies Waipapa Terrane basement. Palynofloras indicate an early Late Eocene (Kaiatan) age for the Kamo Coal Measures.

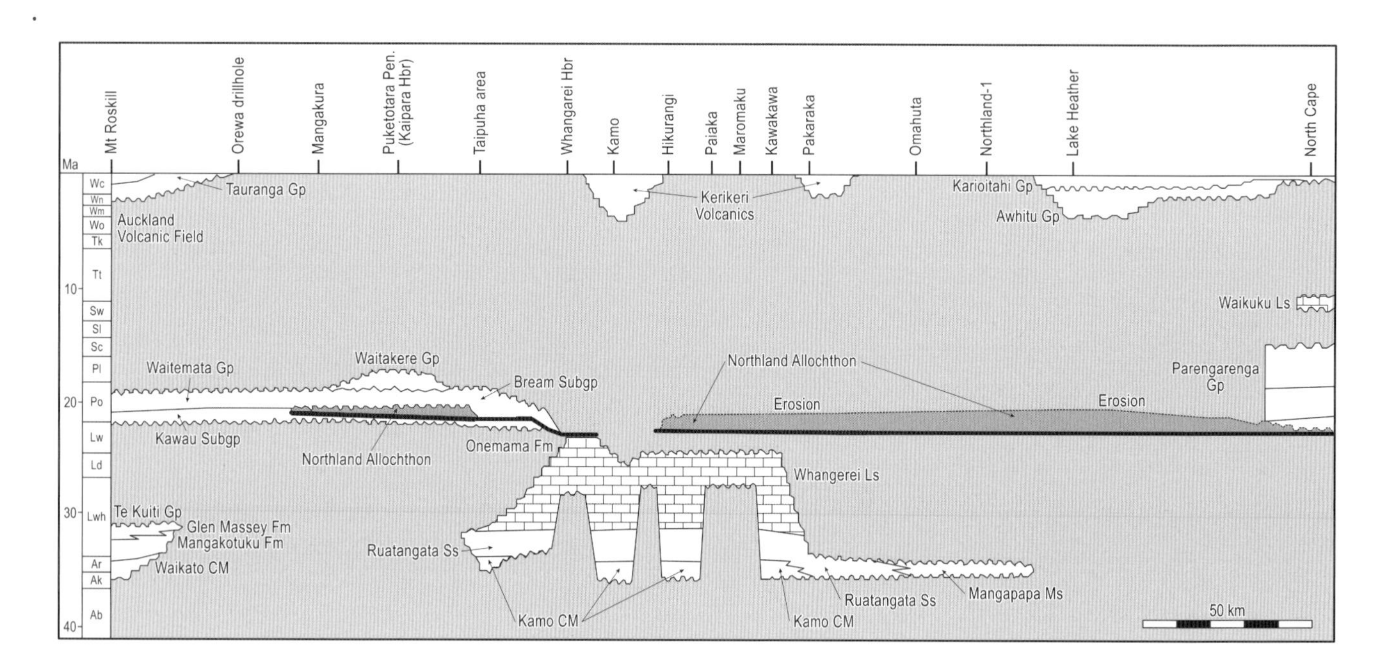

Fig. 6.16. Northwest to southeast time-stratigraphic section of the autochthonous Cenozoic sediments of the Northland Peninsula. For lithologies of named formations, see sections 6.7.3.6, 8.2.7 and 8.6.6. CM, Coal Measures; Fm, Formation; Gp, Group; Ls, Limestone; Ms, Mudstone; Ss, Sandstone; Subgp, Subgroup. (Adapted from Isaac *et al.* 1994.)

In the South Auckland Basin, coal measures of the Lower Te Kuiti Group extend southwards discontinuously from Auckland for some 170 km to Benneydale and the Awakino Gorge in the King Country (Edbrooke *et al.* 1994, 1998). The eastern limit of the depositional basin is marked by depositional thinning and disappearance of the coal measures against basement ranges (Waipapa Terrane) rocks, from the Hunua Ranges in the north to the Hauhungaroa Range in the south. In many areas coal measures have been eroded from the flanks of these ranges, which were uplifted by Neogene block-faulting. The basin is separated from the Taranaki Basin to the west by the eastern flanks of Murihiku Terrane basement hills and ranges, from Kaketu Range in the north to Herangi Range in the south. Most of the coal measures, which may have covered much of the basement rocks east of the highest ranges, have been removed by late Neogene erosion. North of the Waikato River mouth the western coal region boundary is concealed by late Neogene cover.

The basal Waikato Coal Measures, which unconformably overlie an undulating surface cut into weathered Mesozoic basement rocks, are typically less than 100 m thick, but may exceed 200 m locally. The palynomorph biozones show that the coal measures are Late Eocene (Kaiatan–Runangan) in the north (Huntly–Rotowaro area) and young southwards to mid-Oligocene at Benneydale in the south. They were deposited in a north to north-northwest-trending valley system, about 35 km wide and 200 km long, parallel to the regional structural grain. The earliest sedimentation was as a series of anastomosing and meandering fluvial drainage systems (Bryant & Bartlett, 1991) and the accumulation of thick peat beds on the northern alluvial plain in the Late Eocene (Kaiatan–Runangan). Synsedimentary faulting is inferred during earliest deposition in the north, but it did not persist. The major influences on deposition were regional subsidence and basement topography, the latter strongly influenced by regional structures in basement rocks (Edbrooke *et al.* 1994). Almost all the Waikato Coal Measures in the south Auckland region are younger than the Kamo Coal Measures of central and northern Northland, and are lateral equivalents of Northland marine strata, indicating marine transgression from north to south during Late Eocene and earliest Oligocene times (Edbrooke *et al.* 1994; Isaac *et al.* 1994).

In Northland the overlying marine strata (Ruatangata Sandstone) consist typically of slightly calcareous, glauconitic, muddy, fine-grained sandstone. Locally the formation unconformably overlies Waipapa Terrane basement, and elsewhere it conformably overlies Kamo Coal Measures; north of Whangarei, Kamo Coal Measures and Ruatangata Sandstone are intercalated. South of the Hokianga Harbour the facies include algal limestone, conglomerate and glauconitic sandstone, and are laterally variable over short distances. They are consistent with deposition in the range of inner- to outer-shelf environments suggested by foraminifera (Isaac *et al.* 1994). Glauconitic limestone, calcareous muddy sandstone, and conglomerate occuring between 2905 and 3040 m depth in the Waimamaku-2 petroleum well on the west coast of Northland are also included in the Ruatangata Sandstone. On the shores of Whangarei Harbour, hummocky cross-stratified, shelly, medium- to fine-grained sandstones are interbedded with metre-thick, *Thalassinoides*-burrowed fine-grained sandstone or siltstone, suggesting alternating deposition from storms and from background sedimentation respectively. The changing facies indicate an increase in the depth of deposition with time. No fauna is definitely older than Late Eocene (early Runangan), and the youngest known is Early Oligocene (early Whaingaroan). The unit varies widely in thickness, but reaches a maximum of at least 204 m east of Whangarei.

Northeast of the Hokianga Harbour, Ruatangata Sandstone is conformably and gradationally overlain by massive to poorly stratified calcareous, blue-grey mudstone (Mangapa Mudstone). The unit reaches a maximum observed thickness of 170 m, but the formation is truncated at the basal décollement of the Northland Allochthon. Foraminiferal faunas suggest deposition at mid-shelf to upper-bathyal depths, and indicate a Late Eocene (early Kaiatan to Runangan) age.

Portions of the Northland Allochthon contain Eocene–Early Oligocene marine strata that are correlative with the early Te Kuiti Group. North of the Hokianga Harbour, and locally in southern Northland, the Motatau Complex includes laminated to massive, slightly to moderately calcareous mudstone (Taipa

Mudstone) of Early Eocene to Early Oligocene (Waipawan to Whaingaroan) age (Isaac *et al.* 1994). Well-bedded, colour-banded mudstone is another facies variant, including intercalated centimetre- to decimetre-thick fine sandstones with sharp bases and normal grading, with Bouma ABCD intervals. Where massive, the Taipa Mudstone is identical in appearance to, and spans a similar age range to, the autochthonous Mangapa Mudstone. The Waipawan fauna is mid- to lower-bathyal, and almost all others are in the range ?outer-shelf to deep-bathyal, suggesting a somewhat further offshore environment of deposition than the Mangapa Mudstone.

The Northland Allochthon also includes a range of calcareous sandstone-dominated facies typically of early Middle Eocene to earliest Oligocene (Bortonian to early Whaingaroan) age, which have been grouped as Omahuta Sandstone (Isaac *et al.* 1994). This consists of decimetre- to metre-thick, well-bedded, green to grey, calcareous, glauconitic, medium- to fine-grained sandstones alternating with centimetre- to decimetre-thick, blue-grey calcareous mudstones. Sandstones have sharp bases and sharp to gradational top contacts. Micritic limestone interbeds occur locally. Foraminiferal faunas suggest an outer-shelf to upper-bathyal environment of deposition, and give a total age range of Middle Eocene to Late Oligocene.

6.8. LATE PALEOCENE–EOCENE VOLCANISM

At Woodside Creek (Marlborough) there is a section, up to 150 m thick, of yellow-grey, poorly indurated, graded sandstones or coarse siltstones interbedded with centimetre-thick mudstones and 1.5-m-thick massive, weathered tuffs and rare basaltic lavas (Prebble, 1976). It is in probable faulted contact with the underlying limestone. Foraminiferal and palynological samples from the unit have yielded late Early to early Middle Eocene (Mangaorapan to Heretaungan) ages, indicating that this unit is largely coeval with the Lower Marl of the Amuri Limestone (Field *et al.* 1997). A similar bathyal depth of accumulation is indicated. Incompatible trace element concentrations suggest an alkaline within-plate basic affinity for intercalated basalts (Warner, 1990).

Also in Marlborough, the Grasseed Volcanics (Reay, 1993) are intercalated as discrete lenses within the Upper Limestone and Upper Marl units at several localities in the middle Clarence Valley, and basaltic flows, including pillow lava, volcaniclastic deposits and associated dikes and sills occur in the Blue Mountain Stream area. Thickness of the unit reaches a maximum of 160 m at the type locality at Grasseed Stream in the Clarence Valley, but outcrops are lenticular and thin rapidly laterally. Marl intercalated with the volcanics contains foraminifera, which are diagnostic of bathyal oceanic depths, and give a late Middle Eocene (late Bortonian) age (Morris, 1987). K/Ar dating of phlogopite from a peridotite sill indicates an age of 43±0.8 Ma to 52.2±0.9 Ma, equivalent to a biostratigraphic range between Middle and Late Eocene (Reay, 1993). Incompatible trace element concentrations suggest that the volcanics are alkaline within-plate basalts (Warner, 1990; Levy, 1992; Reay, 1993).

Subalkaline tholeiitic latest Paleocene volcanism (View Hill Volcanics) in the central Canterbury Basin continued into Early Eocene (Mangaorapan) times accompanied by deposition of thin shallow-marine limestone. In the far west, marginal marine conglomerate, pillow lava and pillow breccia occur, and, although poorly dated, are inferred to be Early–Middle Eocene (Field *et al.* 1989).

Apart from the northwest part of Chatham Island, where Maastrichtian to earliest Eocene (Haumurian–Waipawan) glauconitic grits and sandstones of the Tioriori Group were being deposited, Cenozoic sedimentation in the Chatham Islands region began with a lengthy period of basaltic volcanism, in part coeval with the View Hill Volcanics of central Canterbury. The deposits (Red Bluff Tuff) extended over much of Chatham and Pitt islands. Red Bluff Tuff is a predominantly compact, coarse-grained, palagonitic tuff composed of fragments of highly vesicular glassy basalt altered to montmorillonite. It includes horizons of lapilli stone and agglomerate, and at a few horizons is very fossiliferous. At most localities it is water-sorted and well-bedded; in places cross-bedded. The formation is at least 100 m thick in southern Chatham Island and Pitt Island, but thins to about 12 m in northern Chatham Island. It unconformably overlies all older units. The age is Late Paleocene to Early Eocene (late Teurian to late Waipawan), and it is in part contemporaneous with the upper part of the Tioriori Group. The formation is fully marine, and a foraminiferal fauna dominated by benthics suggests mid- to inner-shelf depths (Campbell *et al.* 1988, 1993; Wood *et al.* 1989; see also Fig. 4.2).

A major period of Late Eocene (late Kaiatan–Runangan) volcanism in the southern Canterbury Basin is represented in the Oamaru area by basaltic tuff, agglomerate and pillow lavas, basaltic to doleritic dikes and sills (see Plate 11b), and layers of coeval sediment (Waiareka Volcanics of Gage, 1957). The greatest known thickness is ~200 m. Volcanic debris flows with fragments of pillows, dikes and bombs are present and pyroclastic surge deposits occur locally. Spectacular exposures of tholeiitic pillow lava, basanitic tuffs and impure calcareous beds occur at Boatmans Harbour, near Oamaru. Most of the basaltic volcanism was tholeiitic, and appears to be of normal intraplate composition (Coombs *et al.* 1986; Weaver & Smith, 1989). A K/Ar whole-rock date from the Waiareka Volcanics of 40 Ma is consistent with the biostratigraphic age (Forsyth, 2001).

The Middle–Late Eocene Otitia Basalt of south Westland was erupted during a period of crustal extension, starting about 45 Ma, and indicated onshore by the formation of small, fault-bounded basins as well as regional subsidence leading to marine transgression (Nathan *et al.* 1986). These lavas also have intraplate characteristics and are similar to the Arnott Basalt (Nathan *et al.* 1986; and section 6.5.3).

6.9 EOCENE–OLIGOCENE BOUNDARY EVENT (~33.7 MA)

The Eocene–Oligocene boundary in New Zealand generally corresponds with the beginning of the period of maximum Cenozoic submergence of the landmass, and is marked by the expansion of mainly carbonate deposition. In some areas the boundary is marked by an angular unconformity, and locally by coarse mass-emplaced deposits suggesting ongoing tectonic activity. Well-developed greensands are commonly present at this stratigraphic level, indicating a reduction in terrigenous supply caused by the rapid marine transgression.

By latest Eocene times the New Zealand landmass was mainly low-lying, and characterised by extensive coal swamps. Flooding close to the Eocene–Oligocene boundary left only small portions of the old landmass above sea level, mainly in west Otago and the central and southwestern North Island.

In parts of New Zealand, such as the onshore basins of Northland and south Auckland, the transition to marine conditions was usually gradational, and coal measures of Late Eocene age, or, locally, basement rocks, are overlain by earliest Oligocene (early Whaingaroan) calcareous deposits dominated by limestone. In the offshore Taranaki Basin a strong and pervasive seismic reflector, which is interpreted as being the base of the Oligocene, corresponds to a major unconformity that is locally angular. The unconformity became established across former Eocene coastal-plain and shelf areas, and effectively merges with a longer-lived unconformity to the south in the western South Island.

To the north and northwest of the Taranaki Basin the unconformity at the base of the Oligocene continues to be marked by a very strong and pervasive seismic reflector (P1), which can be traced continuously from north Taranaki to offshore west Northland (Isaac *et al.* 1994; Stagpoole, 1997). It represents a considerable hiatus in the seismic stratigraphic record, and is equated with the base of a regional limestone unit. It can be traced further northwards into the Reinga Basin (Herzer *et al.* 1997), and to the northwest of the Taranaki Basin for at least 1000 km to DSDP drillhole 206 in the New Caledonia Basin (Uruski & Wood, 1991), where the Oligocene is composed of biogenic ooze (van der Lingen, 1973). A hiatus is also present beneath the Oligocene in DSDP drillhole 208 on the northern Lord Howe Rise. In DSDP drillhole 207 on the southern portion of the rise the entire Oligocene is absent.

The Taranaki Basin passes southwards into the West Coast basins of the South Island, where, locally, in the Westport–Karamea area, earliest Oligocene (early Whaingaroan) deposits include debris flows linked to contemporaneous tectonic activity (see Plate 13a,b), in places overlying Late Eocene strata with slight angular unconformity (German, 1976; Carter *et al.* 1982; Riordan *et al.* 2014; Riordan, 2016). Contemporaneous local emergence and subaerial erosion in the coastal region between Punakaiki and Westport is probably associated with the same tectonic event (Lever, 2001). In the main subsiding basins of the West Coast the change to deposition of more calcareous sediments is characterised by a colour change in outcrops, and by a prominent seismic reflector of Early Oligocene (Whaingaroan) age (Bishop, 1992). Outside the main areas of subsidence Oligocene sediments are separated by an erosion surface from older rocks.

In the western Southland basins, Early Oligocene sedimentation patterns are mainly controlled by movement on the Moonlight Fault System, and that included rapid subsidence. In places, the base of the Oligocene is an unconformity, or is marked by an influx of coarse detritus, especially adjacent to active faults (Turnbull & Uruski, 1993).

In most of the East Coast Basin of the North Island the basal Oligocene appears to be in gradational contact with older strata, although in places it rests disconformably on Eocene rocks, and locally includes debris flow units inferred to derive from adjacent active faults (Moore, 1988). Oligocene strata generally comprise bioturbated, calcareous massive mudstones, commonly with a glauconitic lithofacies forming the basal portion. In southern Wairarapa and in Marlborough, however, the Oligocene is absent, and strata of Early Miocene (Waitakian) age rest unconformably on Eocene strata (Field *et al.* 1997).

In the central and western Canterbury Basin, basal Oligocene sediments rest directly on Torlesse basement rocks, or discordantly on earlier Cenozoic strata as a result of differential uplift and erosion (McLennan & Bradshaw, 1984); and in the west the change is marked by sediment becoming coarser or more glauconitic up-sequence. In the east of the basin, sedimentation is continuous across the Eocene–Oligocene boundary, but is marked by a change to calcareous mudstone and micrite. To the south, in the contiguous Great South Basin, the Eocene–Oligocene boundary is represented by a strong regional reflector, which commonly corresponds to a significant increase in carbonate content and a marked decrease in lithification of sediment across the horizon (Cook *et al.* 1999). The boundary is a regional unconformity near the present shelf edge, with Oligocene sediments progressively onlapping from the east.

The Ocean Drilling Program (ODP) leg 181, which drilled several holes south and east of the New Zealand continental area, penetrated the Eocene–Oligocene boundary in two drillholes (Carter, R.M. *et al.* 2004). At site 1123 on the northeastern slopes of the Chatham Rise the boundary appears to be transitional, with no detectable change in lithology. However, at site 1124, on the eastern margin of the submerged Hikurangi Plateau, Early Oligocene (early ?Whaingaroan) nannofossil chalk is separated by a hiatus from Middle Eocene brown mudstone.

6.9.1 Origin of the event

The passage from the Eocene to the Oligocene is recognised globally as marking one of the largest extinctions of marine invertebrates in the Cenozoic period, with a related abrupt increase in the isotope $\delta^{18}O$, inferred to be caused by the onset of ice-sheet formation in Antarctica (Ivany *et al.* 2000). The increase in ice volume caused a significant global fall in sea level close to the boundary (Haq *et al.* 1987; Abreu & Anderson, 1998).

Conversely, in New Zealand the latest Eocene–Early Oligocene is marked by rapid marine flooding, which has long been considered to represent a pre-modern plate boundary culmination of post-rift passive margin development. However, the presence of widespread basal unconformities, including local angular discordances, and evidence of widespread rapid subsidence beginning in the earliest Oligocene, indicate that, at least in western regions, tectonic activity started to overprint slow regional subsidence. This faulting may have had a regional cause, as it mirrors the earliest Oligocene rapid subsidence and collapse of the continental margin around the Tasmanian and southeastern Australian margins at ~33.5 Ma (Exon *et al.* 2002) and in the western Ross Sea in nearby Antarctica (Cande *et al.* 2000; Cape Roberts Science Team, 2000; Hamilton *et al.* 2001).

The abrupt change of sedimentation from largely terrigenous to largely biopelagic carbonate deposition at or just above the Eocene–Oligocene boundary is attributed to the severance at this time of the Tasmania–Antarctica landbridge and the opening of the Tasmanian deep ocean gateway (Lawver & Gahagan, 2003). This resulted in the initiation of the Antarctic Circumpolar Current (ACC), accompanied by global cooling and the beginning of Antarctic ice sheet formation (Cape Roberts Science Team, 2000; Exon *et al.* 2002). On the South Tasman Rise, on the northern flank of the Tasmanian gateway, the transition from the latest Eocene into earliest Oligocene was associated with increasing glauconite, reflecting an upward increase in bottom-current activity and winnowing. The final transition to pelagic carbonates in the earliest Oligocene (~33.5 Ma) represents a change from tranquil to moderately dynamic environments, and from relatively warm to cool climatic conditions (Exon *et al.* 2002). The shallowly submergent New Zealand plateau, centred then at latitude ~55°, lay in the path of the developing ACC (Carter & Landis, 1972), and it experienced a similar change from siliciclastic to carbonate deposition, as well as significant cooling (Hollis *et al.* 1997; Carter, R.M. *et al.* 2004). Erosional unconformities separating Early Oligocene from Late Eocene deposits may be evidence of increased erosional current activity due to the initiation of the ACC, as well as to local tectonic activity.

6.10 OLIGOCENE SUCCESSIONS

6.10.1 Introduction

By Late Eocene times much of the New Zealand landmass was still subaerially exposed or low-lying coal swamps. In the Oligocene, seas flooded across this subdued Eocene surface and it became the time of the most geographically widespread carbonate deposition in New Zealand's geological history (Lee *et al.* 2014; Strogen *et al.* 2014). The Oligocene also saw the earliest movements of the Kaikoura Orogeny, which ultimately caused regression and a change to dominantly terrigenous deposition.

6.10.2 Oligocene sediments of eastern New Zealand

The main Oligocene unit in the East Coast region is the Weber Formation (Fig. 6.1a,b), which occurs throughout the region except in southeastern Wairarapa and Marlborough, where erosion has removed Oligocene strata. The formation is generally a moderately hard, light-grey, bioturbated, calcareous (40–70%) massive mudstone, although well-bedded sandstones also occur. The base is glauconitic in places. In parts of the Raukumara Peninsula the formation is more calcareous, and locally it is represented by 400 m+ of hard grey-white limestone. Sandier lithofacies are present in the northwest of the peninsula (Joass, 1987). A change in the upper part of the Weber Formation from light-coloured fine-grained lithologies to medium- to dark-coloured coarser sediments marks the onset of renewed tectonism in the Early Miocene (Field *et al.* 1997).

The Formation ranges in thickness from 30 m to 900 m but commonly is about 200–300 m thick. The maximum thickness in the East Coast Allochthon is 900 m in central Raukumara Peninsula. The Weber strata range in age from Middle–early Late Eocene (Bortonian–early Kaiatan) to latest Oligocene (early Waitakian), but are usually of Oligocene age.

The paleoecology of the Weber Formation is generally quite uniform, with high percentages of planktic foraminifera indicating oceanic conditions and generally mid-bathyal paleodepths. While there is evidence for shallower deposition, most of the shelf faunas are considered to have been reworked downslope. The Weber Formation was possibly deposited at slightly shallower depths than some of the underlying Eocene strata, suggesting a slight regional regression.

In the northern and central Canterbury Basin the sedimentation pattern of the Late Eocene was disrupted in latest Eocene or earliest Oligocene times. The change is marked in the west by sediment becoming coarser or more glauconitic up-sequence (Field *et al.* 1989). Differential uplift and erosion occurred, at least locally, and in central Canterbury, basal Oligocene sediments rest directly on Torlesse basement rocks, or discordantly on earlier Cenozoic strata (McLennan & Bradshaw, 1984). The base of the Oligocene is marked locally by conglomerate, but more generally by an influx of quartz-rich sandstone or by phosphatic glauconitic sandstone. Further south, in the western part of the basin, Early Oligocene sediments are also glauconitic (Nessing Greensand) and locally contain pebbles of Torlesse basement.

In north Canterbury, shallow-marine glauconitic sandstone (Feary Greensand) of earliest Oligocene (early Whaingaroan) age unconformably overlies older units, the basal contact being marked by a concentration of phosphatised and bored pebbles of sandstone, and a few fish bones and teeth. The greensand passes southwestwards in central Canterbury into quartz-rich, fine-grained sandstone (Coleridge Sandstone), which reaches a maximum thickness of 240 m (Browne &

Field, 1985). The formation shows west to east sediment transport, and is inferred to also represent deposition in a marginal marine environment (Browne & Field, 1985; Field & Browne, 1986; Figs 6.5 and 6.6).

The shallow-marine to nearshore glauconitic and quartzose sandstones on the western margin of the northern and central Canterbury Basin pass gradationally upwards into Early Oligocene (early Whaingaroan) Ashley Mudstone or Amuri Limestone, or are overlain unconformably by younger units.

On the Chatham Rise and on the Chatham Islands, Late Oligocene sediments are absent, suggesting a major hiatus. A localised occurrence of loose carbonate sand (*Victoriella* Limestone) of Early Oligocene (early Whaingaroan) age in northern Chatham Island rests unconformably on lithified Eocene Te Whanga Limestone (Campbell *et al.* 1988). This unconformity correlates with local unconformities and changes of sedimentation patterns at the Eocene–Oligocene boundary in the western Canterbury Basin. In north Canterbury the hiatus is also present, but the Late Oligocene (Duntroonian–early Waitakian) is represented by greensand and limestone (Omihi Formation) resting unconformably on Early Oligocene Amuri Limestone. The greensand and limestone lithologies, which grade laterally into each other (Andrews, 1963), are separated by a zone of non-deposition or erosion in the Motunau–Hurunui River area from probably laterally equivalent calcareous deposits in northernmost Canterbury. Only in the western part of the central Canterbury Basin are Late Oligocene non-marine beds (White Rock Coal Measures) preserved (Field *et al.* 1989).

In the onland southern Canterbury Basin, as in north Canterbury and the Chatham Islands, the Early Oligocene (early Whaingaroan) unconformity is also present and overlain by the calcareous, very glauconitic Nessing Greensand (Field & Browne, 1986). The sandstone, which is locally phosphatised and was deposited in a nearshore environment, passes upwards into outer-neritic fine-grained calcareous sandstone, suggesting subsidence and transgression. In the east, sedimentation, dominated by calcareous mudstone and micrite (Amuri Limestone), continued across the Eocene–Oligocene boundary without apparent break. At offshore petroleum well Clipper-1 the setting was outer-shelf to upper-bathyal. Submarine volcanism continued in the Oamaru area.

In the southern part of the Canterbury Basin, Late Oligocene–Early Miocene sediments are absent along a northeast-trending zone extending offshore of the present coastline between the Oamaru area and Banks Peninsula, suggesting the presence of the Endeavour High in this area (Field *et al.* 1989). The occurrence of an angular unconformity between sediments of Early and Late Oligocene (Whaingaroan and Duntroonian) age, and the presence of paleokarst in the top surface of Whaingaroan limestone in the Oamaru area, suggests that tectonic deformation and at least local emergence took place along the Endeavour High during early Late Oligocene times (Lewis & Bellis, 1984). The high is inferred to have begun to develop in Late Oligocene (late Whaingaroan to early Duntroonian) times and remained a prominent feature well into the Early Miocene (Fig. 6.4).

In south Canterbury, Late Oligocene calcareous, glauconitic sandstone and overlying limestone form two major sheet-like units resting on a mid-Oligocene unconformity (Marshall Unconformity). The glauconitic sandstone (Kokoamu Greensand) is almost wholly Duntroonian, and the limestone (Otekaike Limestone) is Duntroonian to Waitakian in age. The Kokoamu Greensand is the most widespread glauconitic sandstone of the early Late Oligocene (Duntroonian), occurring between Oamaru and central Canterbury. Though it is generally only a few metres thick, locally it exceeds 260 m (Bishop, 1974a). In the southwest, at Naseby (north Otago), greensand interfingers with non-marine sediments (Bishop, 1974a), and fine-grained coal measure sedimentation of mainly Early Oligocene (Whaingaroan) age may have persisted into the Duntroonian in western central Canterbury. Elsewhere all preserved sediments of Duntroonian age are marine; most greensands contain inner-shelf faunas (Field *et al.* 1989).

The texture of the Otekaike Limestone ranges laterally from wackestone to grainstone and, in some localities, also vertically. The bioclastic assemblage is echinoderm–foraminifera–brachiopod–mollusc with echinoderm fragments being the dominant component. The inferred paleogeographic setting in the central Canterbury area is an intra-shelf basin, bounded to the east by the Endeavour High. Locally the limestone is anomalously thick (>150 m), due to local subsidence. In south Canterbury the limestone is extensively channelled and cross-bedded (see Plate 12b) with evidence of intermittent strong storm-generated longshore drift from the south in a barrier or shoal environment (Ward & Lewis, 1975). Similar Late Oligocene (Duntroonian) sediments are recorded east of the Endeavour High in the offshore Clipper-1 petroleum well. Here 9 m of quartzose, slightly glauconitic sandstone is overlain by 80 m of outer-neritic to bathyal chalky limestone similar to the underlying Early Oligocene Amuri Limestone (Field *et al.* 1989).

The thin veneer of mainly carbonate sediments of Oligocene and younger age in the Great South Basin was termed the 'Penrod Group' by Cook *et al.* (1999). Because the Penrod Group is restricted mainly to the shallowest part of the section, where well records are incomplete, its composition is poorly known. The lower boundary of the group is the regional Eocene–Oligocene boundary, represented by a prominent seismic reflector and typically characterised by an increase in carbonate content above it. The Oligocene portion of the Penrod Group consists mainly of foraminiferal ooze and chalk that contains chert and rare pyrite. Strata of both Early and early Late Oligocene (Whaingaroan and Duntroonian) age are present, although, in the Hoiho-1 petroleum well in the south of the basin, upper Whaingaroan rocks are absent and Duntroonian sediments rest unconformably on the lower Whaingaroan rocks. Oligocene strata are absent in the Tara-1 petroleum well in the northwest of the basin, where strata of Middle Miocene age rest directly on the Late Eocene. In the three other wells where Oligocene rocks were sampled, apparently complete successions of rocks of this age are present. Oligocene limestone is also present on Campbell Island, where Whaingaroan–Duntroonian microfossils have been collected (Hollis *et al.* 1997). An unconformity of mid-Oligocene age is recognised as being a missing or condensed interval in petroleum well Kawau-1A, but the horizon does not mark any significant change in lithology or sedimentation style (Cook *et al.* 1999).

The sediments thicken to the east of the Great South Basin, and appear to consist entirely of carbonates deposited as deep-water turbidites (see Fig. 4.6). Mid-shelf to upper-bathyal depths prevailed over much of the Oligocene, although, in the Hoiho-1 well, foraminifera suggest bathyal environments, with water depths in excess of 1200 m (Raine *et al.* 1993). Seismic mapping shows that Oligocene strata lap onto the western margin of the basin. Onshore, in southeastern Otago, the Early Oligocene is absent, and greensand and limestone of latest Oligocene to earliest Miocene age belonging to the Kekenodon Group rest unconformably on Eocene rocks (Raine *et al.* 1993). The group is generally less than 30 m thick, consisting of coarse glauconitic sandstone and fine-grained, crystalline, sandy or glauconitic limestone deposited in shallow to inner-shelf carbonate platform environments.

6.10.3 Oligocene rocks of southern and western New Zealand

6.10.3.1 Eastern Southland

Over most of eastern Southland the Permian–Jurassic basement is unconformably overlain by a succession of Late Oligocene–?Middle Miocene deposits belonging to the East Southland Group (Isaac & Lindqvist, 1990). The group consists of the basal and geographically and stratigraphically restricted Pomahaka Formation, and the aerially and vertically more extensive marine Chatton Formation and the non-marine Gore Lignite Measures (Fig. 6.7) that overlie it. The three units have complex relationships with each other, and probably interfinger. All are of Late Oligocene to earliest Miocene (Duntroonian to Waitakian) age in their basal portions. The bulk of the Chatton Formation and the Gore Lignite Measures are Miocene in age, and are discussed in Chapter 7.

The basal deposits in the northeastern portion of the region comprise intercalated mudstone, sandstone and thin lignite beds (Pomahaka Formation) interpreted to represent estuarine, lagoonal, intertidal, supratidal and swamp environments of probable local extent. The sediments show facies repetition in a series of transgressive–regressive facies cycles, which are interpreted to have been part of a delta interdistributary bay complex. The facies association and the generally very fine grain size indicate protected, low-energy, shallow-marine to freshwater swamp environments of deposition (Isaac & Lindqvist, 1990). The unit reaches a thickness in excess of 95 m, but its full extent is unknown. It may pass shorewards laterally into the Gore Lignite Measures. No age-specific fauna or flora has been obtained from the Pomahaka Formation, but a Late Oligocene to Early

Miocene (Duntroonian to Waitakian) age is inferred from the stratigraphic relationships.

A buried topographic high (Isaac & Lindqvist, 1990; Cahill, 1995) separates the Late Oligocene and younger deposits of eastern Southland from the basins of western Southland, which contain a much more complete stratigraphic succession, ranging from a maximum age of Late Eocene in the immediately adjacent Winton Basin to the west, to a maximum Cretaceous age in the more western basins.

6.10.3.2 Western Southland

Oligocene sediments in all the western Southland basins (Balleny, Solander, Waiau and Te Anau; Figs 6.9–6.11, see also Fig. 1.3) reflect the beginning of major movement on the Moonlight Fault System, and the onset of marine conditions over most of the region. The very varied successions with many formations are represented in Figs 6.10 and 6.11. Early Oligocene sedimentation patterns are varied and usually represented by asymmetric infilling of fault-controlled basins. Initially, marine sedimentation was concentrated in narrow basins adjacent to major faults of the Moonlight Fault System, but by the Late Oligocene the marine transgression had spread over the entire area. The basins were connected by a seaway, which may have extended into the West Coast region, at that time adjacent to western Southland across the proto-Alpine Fault.

The change from Eocene to Oligocene sedimentation varied from basin to basin throughout the region. In onshore parts of the Balleny Basin, Early Oligocene deep-marine sediments overlie non-marine Eocene sediments unconformably. In the Waiau Basin the boundary is marked in the west by at least 1000 m of latest Eocene–Early Oligocene conglomerate that cumulated adjacent to the Moonlight Fault System. The base of the Oligocene succession in the main Te Anau Basin is within the Orauea Mudstone or its equivalents, and is gradational. Beyond the main basin, Oligocene sediments are unconformable on unweathered basement (Turnbull *et al.* 1993).

As noted earlier (section 6.7.3), from the Late Eocene to the Early Oligocene, the eastern Waiau Basin, including the Ohai region, was a restricted, slowly subsiding lacustrine area with minimal terrigenous input. In earliest Oligocene times marine influence increased, but there were no major lithologic changes, and the marine Waicoe Formation gradually succeeded the lacustrine Orauea Formation. The Oligocene Waicoe Formation mudstone in the Waiau Basin is a blue-grey, slightly calcareous to non-calcareous, slightly carbonaceous mudstone deposited in quiet-water conditions. Microfaunas indicate a steadily deepening environment, from inner- to outer-neritic or upper-bathyal conditions, sheltered from full open-ocean influence (Turnbull *et al.* 1993). The thickness of the Oligocene part of the formation is unknown, but may be up to 1000 m. The unit extends east of Ohai into the Winton Basin, where its correlative (Winton Hill Formation) is an Oligocene–earliest Miocene (Whaingaroan–Waitakian) marine succession up to 1200 m thick, characterised by calcareous to silty mudstone. Paleodepths here were mid-outer-shelf and bathyal (Cahill, 1995).

On both eastern and western margins of the Waiau Basin a number of coarser units form packets up to 2000 m thick within the lower part of the Waicoe Formation mudstone. The successions typically are sand- or conglomerate-rich fining-upwards units with abundant sedimentary structures in the coarser beds, indicating deposition by a variety of sediment gravity-flow mechanisms. They are inferred to represent Early Oligocene submarine fan complexes (Turnbull *et al.* 1989).

By contrast, the southeastern margin of the Waiau Basin, near the Longwood Range, was a local high in the Oligocene. It was overlain by a 100–300-m-thick shallow marine sequence (Waihoaka Formation), of Early Oligocene to earliest Miocene (Whaingaroan to Waitakian) age, consisting of basal conglomerate, macrofossiliferous sandstone and mudstone, and limestone lenses lying on or close to basement. The bioclastic limestone facies probably extended north over the Longwood Range to Ohai. Micro- and macrofaunas from the unit indicate very shallow, inner-shelf conditions at the base, deepening through mid- to outer-shelf, then probably to bathyal at the top, in the Early Miocene (Turnbull *et al.* 1993).

Along the western margin of the Waiau Basin (Fig. 6.10) the upper part of the mainly Eocene Hump Ridge Formation contains scattered macrofossils and sparse microfaunas of Early Oligocene (Whaingaroan) age. This part of the formation consists of coarse, quartzo-feldspathic, slightly carbonaceous sandstone with minor carbonaceous mudstone and conglomerate, and sandy to pebbly impure bioclastic limestone. To the east the top of the formation passes rapidly into a graded sandstone–mudstone sequence of earliest Oligocene age, which progressively fines and thins upwards into background basinal Waicoe Formation mudstone (Turnbull *et al.* 1993). The sedimentary setting is inferred to be a shallow marine shelf. Locally these shallow marine deposits pass upwards into submarine fan complexes.

In the northwest of the Waiau Basin, two sub-basins, the Monowai Sub-basin and the Blackmount Sub-basin (Figs 6.9–6.11) developed within and adjacent to the Moonlight Fault System in the Oligocene. Two contrasting sedimentary facies, resting unconformably on Fiordland basement rocks, were developed in the Monowai Sub-basin: a western shelf and shallow-marine sequence, and an eastern coarse clastic sequence. The western shallow-marine facies, of Late Oligocene (late Whaingaroan to Waitakian) age, is up to 200 m thick, lies west of the Hauroko Fault on the margin of the Fiordland block, and continues northwards into the Te Anau Basin, extending up the western side of the basin as far as the Middle Fiord. The basal portion comprises a boulder layer overlain by loose, well-sorted, locally derived coarse-grained sandstone. The formation becomes increasingly calcareous upwards, grading first into bioclastic grainstone, and then into interbedded sandy limestone and Waicoe Formation mudstone of Late Oligocene (Duntroonian) age.

The eastern sedimentary facies of the Monowai Sub-basin, which lies between the Hauroko and Blackmount Faults, consists of more than 1000 m of conglomerate and sandstone. The conglomerate is probably alluvial in the north, but contains scattered marine macrofossils in the south (Turnbull *et al.* 1989). Slump sheets of sandstone and conglomerate 50–100 m thick occur in the southern part of the sub-basin, suggesting that relatively steep slopes were present. The oldest microfauna from the overlying sediments is Early Oligocene (Whaingaroan) and indicates bathyal conditions, so subsidence rates were extremely high. In the south the conglomerate fines upwards into pebbly sandstone, graded sandstone and, finally, Waicoe Formation mudstone up to 1000 m thick, representing a submarine fan complex. In the north it passes up into a succession of cross-bedded sandstone and limestone representing a shallow marine environment (Turnbull *et al.* 1993).

The contiguous Blackmount Sub-basin is bounded by the Blackmount Fault to the northwest and the high-standing Takitimu Mountains block to the northeast. Oligocene strata are absent on the eastern side of the sub-basin, but on the western side, against the Blackmount Fault, is a 2500 m succession of Early–mid-Oligocene Blackmount Formation comprising breccia, conglomerate, sandstone and mudstone (Carter & Norris, 1977), resting on Fiordland basement, and grading up into Late Oligocene Waicoe Formation mudstone. The coarse units form fining-upwards successions, including graded sandstones with flute and groove casts, and Bouma sequences. Slump folds occur throughout, and rare macrofossils are present. The formation, like the adjacent eastern facies of the Monowai Sub-basin, is interpreted as a submarine fan complex, with a progressively more distal environment towards the middle of the sub-basin (Turnbull *et al.* 1993). Similar and correlative deposits in the southern Monowai Sub-basin and at Hump Ridge suggest that the fan complex was areally widespread. In the northern part of the basin, west of the Bellmount Fault, seismic records show that a maximum thickness of more than 3.5 km of Oligocene Blackmount Formation is preserved, forming a westwards-thickening wedge of clastic submarine fan sediments, shed into the Blackmount Sub-basin from the Moonlight Fault System. Oligocene sedimentation along the western margin of the Waiau Basin appears to have been controlled by waning normal movement along these faults.

The Moonlight Fault System, accompanied by marine transgression, propagated northeast along the trend of the Blackmount Fault into the Te Anau Basin in Early Oligocene times, forming a local depocentre – the Burwood Sub-basin – which lasted until the Late Miocene. The Burwood Sub-basin records the same Early Oligocene marine transgression seen in the Waiau Basin. Eocene non-marine carbonaceous mudstone is conformably but sharply overlain by an Early Oligocene (Whaingaroan) marine clastic unit, consisting of up to 400 m of unsorted pebble-and-boulder conglomerate and breccia infilling channels passing laterally into regularly bedded, graded sandstone and mudstone. The unit, which becomes thinner-bedded and finer upwards, grading into the Waicoe

Formation mudstone, is interpreted as a series of small coalescing submarine fans, building northwest from the Takitimu Mountains. By the mid-Oligocene (mid-Whaingaroan), the central Burwood Sub-basin had subsided to outer-shelf or upper-bathyal depths and a thick sequence of Waicoe Formation mudstone accumulated. Submarine fan sedimentation recommenced in the mid- to Late Oligocene.

A separate sub-basin (Dunton Sub-basin) in the northeastern Te Anau Basin was controlled by renewed movements along pre-existing faults of the north-trending Hollyford zone. The Dunton Sub-basin filled with more than 700 m of Early–Late Oligocene marine sediments. These sharply overlie Eocene Orauea Formation mudstone, and consist mainly of 10–100-m-thick, fining-upwards, graded sandstone and mudstone cycles. The sediments in places reek of hydrocarbons (Turnbull *et al.* 1993). Sedimentary structures are compatible with a submarine fan setting for most of the unit, above a shallow marine base.

In the bulk of the Te Anau Basin, Oligocene sedimentation largely followed pre-existing Eocene structures in the northern and central portions, and a Late Oligocene marine transgression affected a much wider area beyond these. Lithofacies patterns are similar to those of the Waiau Basin, with submarine fan sediments within background mudstone in the basins, and shallow-marine carbonate-rich shelf facies on the basin margins. The base of the Oligocene succession in the Te Anau Basin is within the Orauea Mudstone and equivalents, and is gradational on the Eocene wherever seen. Beyond the main basin, Oligocene sediments are sharply unconformable on unweathered basement. The Oligocene succession passes upwards conformably into Miocene sediments in the northern part of the basin, but in the southwest of the basin, and over much of the southeast side in the subsurface, the upper contact is an unconformity.

Two depositional settings can be recognised: a central basin, and shallow-marine western and eastern shelves that developed in the later Oligocene. Early Oligocene (Whaingaroan) central basin sediments in the northern Te Anau Basin form a complex marine succession of mudstone, graded sandstone and massive to cross-bedded sandstone, with a total thickness of up to 500 m, and containing sparse shallow-marine to outer-shelf microfaunas of Eocene to Oligocene age. The depositional environment is inferred to be a restricted, poorly oxygenated lagoon or partly enclosed marine basin at the base, becoming deeper and more fully marine upwards. Overlying and interbedded with this lithofacies are fining-upwards graded sandstone and mudstone cycles interrupted by lenticular trough cross-bedded sandstone bodies up to 2 km wide and 1–300 m thick. The graded sandstones have abundant sedimentary structures, such as slump folds, sole markings and Bouma sequences. The microfauna resembles that of the Kaiata Formation of the West Coast, and indicates outer-shelf to slope depths in an embayment with restricted circulation (Turnbull *et al.* 1993).

Overlying this complex succession is an Early–early Late Oligocene (Whaingaroan) sandstone unit with scattered macrofossils (Turret Peaks Formation), which forms a simple fining-upwards sequence reaching 1700 m in thickness and filling the whole of the northern Te Anau Basin. Within it are large channels up to 100 m deep and 1–2 km wide, filled with massive or graded metre-bedded sandstone or trough cross-bedded sandstone. Towards the top, sedimentary structures, including sole marks, grading and partial Bouma sequences, become more common, and beds become thinner. This sandstone unit is inferred to represent a very large and active submarine fan, or series of fans, which gradually retreated northwards, or became starved of sand. Fan sedimentation was replaced in the Late Oligocene (Duntroonian) by Waicoe Formation mudstone, which covers the whole central and northern Te Anau Basin. The mudstone contains macro- and microfaunas indicating gradually deepening conditions, down to bathyal in the Early Miocene.

Shallow marine shelves developed on the eastern and western flanks of the basin during the Late Oligocene (Duntroonian), the western shelf facies being a continuation of that described from the Monowai Sub-basin to the south. The upper limestone unit becomes sandier northwards, and it is assumed to 'feather out' over the Middle Fiord depocentre. In the north, east of the Eglinton Valley, the eastern shelf facies consists of metre-bedded, trough cross-bedded, medium- to very coarse-graded sandstones infilling channels, overlain by sandstone with planar and hummocky cross-stratification and flaser-bedded sandstone that may be storm deposits. This passes laterally into a 500-m-thick unit of convolute to parallel-laminated, decimetre-bedded coarse sandstone with scattered macrofossils, occasional slump horizons of fossiliferous pebbly mudstone, and some thick-bedded massive structureless sandstones. The depositional environment of these sediments is inferred to be a subsiding shallow-marine shelf. These sediments are of Early and Late Oligocene (Whaingaroan to Duntroonian) age, and apparently grade laterally westwards into the sandstones of the Turret Peaks Formation.

6.10.3.3 Offshore Southland basins

The mostly offshore Solander Basin, which lies to the south of the Waiau Basin, and whose western margin was also controlled by the Moonlight Fault System (Hauroko Fault), formed Oligocene depocentres in three sub-basins: the Waitutu, Hautere and Parara sub-basins. The last two were separated by basement highs in the Eocene, but merged in the Oligocene to create one larger feature.

Only the Waitutu Sub-basin extended onshore, and it contains both basinal and shelf facies. The oldest sediments onshore are poorly exposed Early Oligocene (Whaingaroan) mudstone and graded calcareous sandstone in the east. On the western margin of the sub-basin, Late Oligocene (Duntroonian) limestone occurs at several localities, typically consisting of graded, pebbly bioclastic limestone with cobbly breccia beds in some places, grading up into Early Miocene mudstone. These outcrops are probably equivalents of the shelf limestone inferred to have originally lain to the west of the Hauroko Fault (Turnbull *et al.* 1993). West of the fault at the north end of the Waitutu Sub-basin a 600 m succession of locally derived, metre-bedded cobble to boulder conglomerate and sandstone is overlain by 5 m of pebbly bioclastic limestone, of inferred Late Oligocene (Duntroonian) age. It is correlated with similar facies in the Monowai Sub-basin of the Waiau Basin. Further north, shallow-water sandstone and limestone are inferred to represent an extension of the Monowai Sub-basin western shelf.

In the north of the offshore portion of the Waitutu Sub-basin the Oligocene succession forms a westwards-thickening gently folded wedge, controlled by continued normal movement on the Hauroko Fault. The main sediment supply was from the Fiordland basement block to the northwest, and the Oligocene sequence progrades from the fault scarp into the basin, downlapping on the top Eocene reflector. The sub-basin contains more than 1 km of Oligocene sediments adjacent to the Hauroko Fault, thinning to zero near the Solander Ridge. The influence of the Moonlight Fault System on Oligocene sedimentation died out to the southwest, at about the latitude of Solander Island. The Waitutu Sub-basin and the Balleny Basin merge in this region, and the sedimentological patterns in the two basins are similar, with thin sediments draped over earlier Eocene structures.

The now-combined Parara and Hautere sub-basins were separated from the Waitutu Sub-basin by the Solander Ridge basement high. The two petroleum wells drilled to date in these sub-basins both penetrated an Oligocene succession. Parara-1 penetrated approximately 100 m of marine Oligocene sediments overlying Eocene coal measures, while in Solander-1 approximately 130 m of Oligocene sediments were encountered. The map of the top Oligocene surface shows that the sub-basins within the eastern Solander Basin gradually migrated during the Late Eocene transgression, until they were fully connected to the north by the end of the Oligocene. Oligocene sediments are relatively thin compared with onshore basins and are mainly less than 1 km thick. In the latest Oligocene in Parara-1 petroleum well the environment was inferred to be a shallow marine shelf, based on foraminifera, with glauconite suggesting low rates of sedimentation.

The Balleny Basin is the westernmost of the western Southland basins, lying between the present-day Alpine Fault and the Moonlight Fault System. It is mainly offshore, but has an onshore component in southwest Fiordland. This is Early Oligocene (earliest to middle Whaingaroan) in age, and unconformably overlies basement rocks of Eocene–earliest Oligocene Macnamara Formation. Offshore, the Oligocene succession in many places lies unconformably on the Eocene sequence, and Oligocene sediments are draped over Eocene fault blocks.

The exposed onshore Oligocene sediments form a progressively deepening succession, with a basal granitic breccia-conglomerate overlain by thick-bedded sandstone grading into a regularly bedded fining-upwards graded sandstone unit, locally channelled, and capped by chalk and calcareous mudstone.

The total thickness is ~1000 m. Lithofacies of the Oligocene sediments represent environments changing from nearshore submarine fans and channels, through more distal submarine fans to deep-marine sea floor. Sedimentation was probably controlled by active faults (Turnbull *et al.* 1993).

In the offshore Balleny Basin, seismic data show extensive accumulation of Oligocene sediments. By the end of the Oligocene the Balleny Basin had coalesced with the Waitutu Sub-basin of the Solander Basin, forming one large basin west of the Solander Ridge. The Oligocene sequence thickens westwards and appears to onlap Eocene and older highs. Sediment thickness locally exceeds 2 km, but, in general, it averages between 0.5 and 1 km.

6.10.3.4 West Coast South Island

On the west coast of the South Island, west of the present Alpine Fault, sedimentation was essentially continuous from latest Eocene to Oligocene times within the main basins established during the Eocene, with only local unconformities (Laird, 1988). Elsewhere, outside the main areas of subsidence, Oligocene sediments rest unconformably on older rocks. The change from Eocene to Oligocene sedimentation was characterised not only by an increase in carbonate, but also by a marked microfaunal change, from the relatively restricted and distinctive faunas of the carbonaceous mudstone typical of the Eocene Kaiata facies to normal open-water assemblages containing a mixture of planktic and benthic foraminifera. This indicates a major paleo-oceanographic change from sedimentation in small relatively isolated basins, common in the Late Eocene, to fully oceanic circulation (Nathan *et al.* 1986).

The change to deposition of more calcareous sediments is characterised by a prominent seismic reflector recognised throughout most of the region (Bishop, 1992). Dating of the horizon from outcrop or drillhole data shows that it is of dominantly Early Oligocene (Whaingaroan) age in Westland and Buller, and dominantly of Late Oligocene (Duntroonian) age in the Karamea Peninsula. This could be explained by subsidence propagating from south to north.

Most of the region consisted of tectonically relatively stable platform areas with thin successions (generally less than 100 m), which contrast with the still rapidly subsiding basins – Murchison and Paparoa – where maximum thicknesses of Oligocene sediment are in excess of 1000 m (Nathan *et al.* 1986; Laird, 1988).

The Oligocene extent of the Paparoa Trough is poorly known, as sediments in the inferred axial area (based on Brunner coal rank studies) have been entirely eroded. However, rapid eastward thickening of strata towards the inferred axis in the Punakaiki area (Laird, 1988) indicates that subsidence along the Paparoa Tectonic Zone continued into the Oligocene, though probably at a slower pace than in the Eocene. Deposition of limestone continued throughout most of the trough, with combined macro- and microfossil evidence indicating deposition at depths equivalent to the present-day outer shelf (Nathan *et al.* 1986). At the northern end of the trough the dominant lithology is grey-brown calcareous mudstone containing deep-water microfaunas, locally passing into light-brown muddy sandstone.

Close to the present coastline north of Westport, between the mouths of the Mokihinui and Little Wanganui rivers, mass-flow breccia and breccia-conglomerate of Early Oligocene (Whaingaroan) age containing abraded transported shallow-water fossils are locally interbedded within the carbonate mudstone succession (see Plate 13a,b). Paleocurrent evidence indicates that the coarse facies is derived from the western (upthrown) side of the faulted basin margin, an extension of the Paparoa Tectonic Zone, immediately west of the present coastline (German, 1976; Riordan, 2016). Near the eastern margin of the Paparoa Trough at Whitecliffs, Lower Buller Gorge, the Oligocene deposits rest unconformably on Late Eocene Kaiata Formation, and locally include at the base up to 30 m of angular and polymict breccia. Clasts up to 30 cm in diameter are common, more rarely attaining a size of 1.5 m or more (Carter *et al.* 1982). This breccia lithofacies is overlain by a sequence of sandy, glauconitic, algal biosparites, passing up into finer-grained and more massive limestone. The succession ranges in age from Early to Late Oligocene (Whaingaroan to Duntroonian). As at Little Wanganui River, the unconformity and local mass-flow breccia indicate that active faulting continued into the Early Oligocene.

The sediments filling the other major depocentre during the Oligocene, the Murchison Basin, consist dominantly of blue-grey to grey-brown calcareous mudstone or muddy limestone (Matiri Formation). Although the succession appears massive in most outcrops, weathering or minor bedding features, such as lenses of fine sand or silt, outline abundant slump folds and other indications of hydroplastic deformation resulting from very rapid subsidence. The unit is dominated by an assemblage of deep-water foraminifera, and shows a rapid increase in water depth from shelf to bathyal during the Early Oligocene (Whaingaroan), with bathyal depths being maintained through the Late Oligocene to earliest Miocene (Waitakian).

Scattered graded beds of redeposited bioclastic limestone occur locally in the upper part of the succession, and increase in abundance upwards until the formation changes into a calc-flysch sequence, interpreted as a submarine fan assemblage (Nathan *et al.* 1986). Limestone beds contain abundant redeposited shallow-water bioclastic detritus, while the mudstone interbeds contain deep-water foraminifera. Rapid thickness changes (Fig. 6.13) from thin shelf sequences to thick muddy basinal successions near the basin margins suggest that they were fault-controlled. The presence of a wedge of mass-flow beds, mainly conglomerate, grit and sandstone and including redeposited shallow-water faunas, within the Matiri Formation on the western margin of the Murchison Basin, lends support to this hypothesis.

In the areas outside the rapidly subsiding Paparoa and Murchison basins, a platform facies, consisting predominantly of shallow-water bioclastic limestone, and averaging about 100 m in thickness, was dominant (Nathan *et al.* 1986). Despite the apparent variability, a small number of distinctive lithofacies recur throughout the region. These include bioclastic grainstone, algal packstone or grainstone, sandy limestone or calcareous sandstone, and muddy limestone or calcareous mudstone. The bioclastic grainstone is generally thinly bedded and of characteristically flaggy appearance, and frequently contains large-scale cross-bedding. The well-known 'pancake' limestone at Punakaiki (Laird, 1988) is typical of this lithofacies. The sorting, fragmentation and abrasion of the sand-sized material, together with the lack of any finer matrix, suggest extensive winnowing and reworking on the sea floor, possibly by tidal currents or wave action. The fauna indicates shallow, clear water above wave base (i.e., less than 50 m). Algae are minor but common components of this lithofacies, but at some localities algal rhodoliths are the most conspicuous bioclastic component, for example in the Waitakere Limestone of the Charleston area (Nathan, 1975; Riordan *et al.* 2014). By analogy with modern environments, the algal-rich lithofacies was probably deposited in very shallow clear water, with intermittent current action. The sandy limestone or calcareous sandstone lithofacies is characterised by the presence of abundant terrigenous sand, for example in the Takaka Limestone of northwest Nelson (Leask, 1980). Where it occurs, this lithofacies is invariably at the base of the platform facies sequences, and is interpreted as the product of either shallow-water reworking of underlying units or as sand deposited on an erosion surface during the marine transgression. The muddy limestone or calcareous mudstone lithofacies appears to be a lateral variant of the muddy basinal facies, containing microfaunas ranging from mainly benthic to mainly planktic, for example the Tiropahi Limestone of the Charleston area (Nathan, 1975; Riordan *et al.* 2014).

6.10.3.5 Taranaki Basin

In the Taranaki Basin to the north, the Oligocene–earliest Miocene deposits (Ngatoro Group) are also characterised by a high calcium carbonate content (King & Thrasher, 1996). In general, the base of the Ngatoro Group corresponds to a major unconformity, which has different origins in different parts of the basin. In some southern areas the unconformity spans much or all of the Eocene–Early Oligocene: it is angular in the Manaia Sub-basin, where it is evidently related to Late Eocene tilting and erosion; and in central and northern parts of the basin the absence of latest Eocene–Early Oligocene sediments reflects non-deposition. In distal northwestern regions, calcareous oozes and muddy limestones of the Ngatoro Group (Tikorangi and Otaraoa formations; Fig. 6.3) overlie the Eocene Moa Group with apparent conformity.

Deposition above the basal unconformity of greensands of mid-Oligocene (late Whaingaroan) age marked the beginning of renewed subsidence and a period of widespread marine inundation in the basin. Rapid platform or foreland subsidence in the mid- to Late Oligocene resulted in greater water depths than during the preceding Eocene transgression, causing renewed and accelerated transgression around the basin's margins. Subsidence was particularly dramatic in the east adjacent to the Taranaki Fault, locally resulting in foundering from subaerial or shallow marine environments to become a bathyal foredeep trough (King & Thrasher, 1996).

The Ngatoro Group carbonates are diachronous, ranging in age from Early Oligocene to late Early Miocene in the west, and Late Oligocene to earliest Miocene in the east. In the west, thin, highly condensed carbonate sequences were deposited in a starved basin setting. In most petroleum wells offshore, the Ngatoro Group interval is less than 200 m and commonly only about 20–30 m thick. Elsewhere, the group consists of shelf and slope carbonate assemblages and highly calcareous clastic rocks. In the region of the eastern foredeep trough, by contrast, the Ngatoro Group succession has a stratigraphic thickness of about 1200 m.

Here, along the eastern flank of the Taranaki Basin, the bulk of the Ngatoro Group consists of a sequence of calcareous siltstones and sandstones, the Otaraoa Formation, everywhere exceeding 100 m in thickness and deposited mainly in outer-shelf to upper-bathyal water depths (Palmer & Andrews, 1993; King & Thrasher, 1996). The base of the succession consists of a mid-Oligocene (late Whaingaroan) fine-grained, glauconitic sandstone, locally up to 50 m thick. It is overlain along the eastern flank of the Taranaki Basin by a distinct sandstone unit of similar age (Tariki Sandstone Member) that consists of thick interbedded quartzo-feldspathic sandstone (sometimes graded), calcareous mudstone, and limestone up to 300 m thick. The sandstones are interpreted as turbidites forming a submarine fan system restricted to the eastern portion of the Taranaki Peninsula adjacent to the Taranaki Fault (de Bock *et al.* 1990). The remainder of the Otaraoa Formation consists of grey, very calcareous mudstone.

The turbidites of the Tariki Sandstone Member were deposited in a rapidly deepening trough oriented north–south along the basin's eastern margin, at bathyal depths (King & Thrasher, 1996). Sediment supply was almost sufficient to keep pace with subsidence. The character and distribution of the Tariki Sandstones suggest that they were supplied from elevated areas to the east, and then transported southwards. Contemporaneous rejuvenation and erosion of the adjacent hinterland, driven by uplift along the Taranaki Fault, is inferred. Uplift on the Taranaki Fault and subsidence within the adjacent trough were linked, so that the ultimate control on Tariki Member deposition was local tectonism. Evidence of tectonism of similar age was reported by Nelson *et al.* (1994) from the King Country Basin immediately to the east (see below, this section).

Elsewhere in the southern Taranaki Basin and in onshore northwest Nelson, age- and stratigraphically equivalent shelf deposits consist mainly of calcareous siltstones and sandstones, with minor limestones (Bishop, 1971). Potential correlatives of the Tariki Sandstone Member occur at Nelson, where age-equivalent sandstones and conglomerates (Magazine Point Formation) were deposited in a tectonically controlled, rapidly subsiding basin, similar to the setting of the Tariki Sandstone depocentre west of the Taranaki Fault. The 140-m-thick succession of coarse deposits was inferred to be emplaced by storm-driven currents and/or turbidity currents (Lewis, 1980). Just beyond the southeastern margin of the Taranaki Basin, at Picton, deposition of an Oligocene mainly coal-measure succession took place within fluvial to inner-shelf environments, and was also controlled by synsedimentary faulting (Nicol & Campbell, 1990).

The terrigenous lithologies of the Otaraoa Formation grade laterally to the northwest into, and are gradationally overlain to the east by, highly calcareous outer-shelf to upper-slope deposits of restricted late Late Oligocene (early Waitakian) age assigned to the platform facies of the Tikorangi Formation. The thickest (about 100–200 m) limestones (mainly grainstones) in this region coincide with the Tariki Sandstone Member depocentre, suggesting that subsidence was still occurring adjacent to the Taranaki Fault. In central and southwestern areas of the Taranaki Peninsula the Tikorangi Formation is absent. Age-equivalent strata are represented by outer-shelf to upper-bathyal marine mudstones, similar to those of the Otaraoa Formation.

Limestones of equivalent age are also well developed in offshore regions along the northeastern margin of the Taranaki Basin, and also to the northwest, reaching thicknesses of 200 m. In more distal western and northwestern regions the Oligocene succession consists of thin, condensed, deep-water carbonate sequences belonging to the basinal facies of the Tikorangi Formation. Predominant lithologies include foraminiferal limestones, marl, and calcareous mudstones. In the Maui field area, interbedded chalk and marl are well developed, and represent a transition between the platform and basinal carbonate facies. Similar limestones, which crop out in northwest Nelson (Leask, 1980), also have an early Waitakian age, and are generally less than 50 m thick. Benthic foraminifera indicate inner-shelf depositional environments, in some instances deepening up-sequence to outer-shelf.

In western and northwestern regions, thin deep-water carbonate sequences of the Tikorangi Formation represent long periods of clastic sediment starvation in a distal basin setting. The interval is highly condensed and, in places, spans the entire Oligocene: Early Oligocene (early Whaingaroan) strata are absent over much of the western basin and some areas further east, resulting in a marked disconformity between the Tikorangi Formation and underlying marine clastic strata of the Eocene Moa Group.

A highly calcareous interval dominated by marls (Taimana Formation; Fig. 6.3) occurs between the Tikorangi Formation and overlying Miocene strata. It is generally less than 250 m thick. In onshore wells the Taimana Formation is predominantly silty with varying proportions of fine sandstone, mudstone, and thin limestone: calcium carbonate content usually increases with depth. The top of the formation in onshore wells is always within an indeterminate Early Miocene (Waitakian to Otaian) age range, and the base is usually just within the top of the Late Oligocene. In offshore wells the formation consists mainly of marl, closely comparable in lithology to the Otaraoa Formation. Taimana Formation strata are youngest in the west of the basin, where they are of Early to Middle Miocene (Altonian to Lillburnian) age.

In proximal eastern parts of the basin the Taimana Formation represents the first minor influx of clastic sediments into the carbonate-dominated foredeep, effectively ending deposition of the highly calcareous Tikorangi Formation. An abrupt top to the Taimana Formation in the east denotes a sudden influx of terrigenous detritus associated with overthrusting on the Taranaki Fault and uplift in the basin's hinterland.

In distal western parts of the basin the Taimana Formation is a diachronous and very condensed sequence that reflects a general absence of clastic sediment input for long periods. Over a wide area offshore, a depositional hiatus within the formation spans much of the Early Miocene (late Waitakian–early Altonian) interval. This sediment starvation was partly due to distance from clastic source areas, and partly due to the preferential infilling of the subsiding foredeep in the eastern basin. In the west the transition from condensed carbonate-dominated sequences of the Ngatoro Group to deep-water terrigenous sequences of the Miocene Wai-iti Group was more gradational and longer lived than in the eastern basin.

6.10.3.6 Northland Basin

In the offshore west Northland Basin, to the north of the Taranaki Basin, at least 185 m of Ngatoro Group limestone and glauconitic mudstone of Early to Late Oligocene (Whaingaroan to early Waitakian) age was penetrated, although some micropaleontological ages are in conflict and the thickness may be as much as 360 m (Milne & Quick, 1999; Strong *et al.* 1999). The paleoenvironment was inferred to be lower bathyal, suggesting a paleodepth of at least 1000 m.

In the remainder of the basin the age is closely constrained below by the age of the P1 seismic reflector in Taranaki Basin wells, and above by Miocene volcanic rocks. Reflector P1 is the strongest and most pervasive regional marker horizon in the Taranaki and Northland basins. In the latter it represents an unconformity and/or a condensed sequence equivalent to the carbonate-dominated Ngatoro Group. Lying directly on the P1 reflector in places is a volcanic-dominated sequence, associated with the Waitakere Volcanic Arc (Herzer, 1995). The earliest dated eruptive activity of the volcanic arc is 25 Ma (latest Oligocene), although the main phase of eruptions did not start until 22 Ma (Early Miocene). The Oligocene–earliest Miocene strata are so thin that they are represented by only one or two reflection wavelets, and are inferred to be less than 200 m thick (Isaac *et al.* 1994; Herzer, 1995). The equivalent autochthonous facies onshore are the bioclastic limestone and calcareous sandstone of the Whangarei Limestone and Onemama Formation (Te Kuiti Group) (Fig. 6.15).

The Oligocene succession in the offshore Reinga Basin, lying to the northwest of the Northland Peninsula, is a continuation of the offshore west Northland Basin, with reflector P1 defining a thin paraconformable condensed Oligocene sequence, probably less than 100 m thick. Rare dredge samples of Oligocene rocks are confined to latest Oligocene to Early Miocene (Waitakian) age, and consist of bathyal foraminiferal limestone (Herzer *et al.* 1997).

6.10.3.7 Onshore North Island basins

In onshore Northland, Oligocene calcareous deposits of the upper Te Kuiti Group (Whangarei Limestone, Onemama Formation; Fig. 6.16) are widespread, overlying the Eocene Ruatangata Sandstone with sharp facies change, but otherwise commonly with apparent conformity, or onlapping Waipapa basement rocks directly. Locally, Whangarei Limestone is succeeded by calcareous flaggy sandstone and pebbly limestone of the Onemama Formation, and is commonly overthrust by a mudstone-dominated mélange of the Northland Allochthon (Isaac *et al.* 1994).

Where Whangarei Limestone overlies the Ruatangata Sandstone the unit is commonly conglomeratic at the base, including many Waipapa basement-derived pebbles and granules in a muddy micritic matrix. Stratigraphically higher the sediments become finer grained and contain planktic foraminifera, interpreted as a deeper-water facies (Isaac *et al.* 1994). The limestone varies locally from a bryozoan calcarenite grainstone to a packstone biomicrite and biosparite, inferred to have accumulated mainly at inner- to mid-shelf depths in a deepening basin. Maximum thickness is 114 m. Locally the limestone upper surface is a paleokarst with up to 3 m of relief, overlain by Northland Allochthon mélange and broken formation. The age ranges from Early Oligocene to possibly earliest Miocene (Whaingaroan to early Waitakian).

South of Whangarei Harbour, Whangarei Limestone is overlain by calcareous glauconitic fine-grained sandstone (Onemama Formation) with prominent *Scolicia* burrows. Stratigraphically higher beds are flaggy, calcareous fine-grained sandstone, with rare planar cross-beds up to 30 cm thick. A thick (up to 5.3 m) tuff unit overlies the succession, and is locally overlain by a bed of clast-supported pebble breccia, derived from both Waipapa Terrane basement and Northland Allochthon sources. In the east the flaggy calcareous sandstone oversteps on Waipapa Formation basement rather than on Whangarei Limestone. The latest Oligocene to earliest Miocene (early Waitakian) Onemama Formation is considered to be the uppermost unit of the Te Kuiti Group in Northland. Faunas are mainly bathyal, but some suggest deposition at shelf depths (Isaac *et al.* 1994).

That the younger Te Kuiti Group strata were formerly more extensive is indicated by small areas of Te Kuiti beds outcropping in the northern Coromandel Peninsula. Near Coromandel, Mesozoic basement is unconformably overlain by approximately 100 m of coal measures, conglomerate, carbonaceous sandstone, and siltstone; calcareous siltstone and shelly sandstone; and glauconitic, bioclastic and sandy limestone (Kear, 1955). Other erosional remnants of Te Kuiti Group are present to the north and to the southeast. An Early Oligocene (early Whaingaroan) palynoflora has been obtained from the coal measures, equivalent in age to those from south Waikato and Kawhia, the southernmost and youngest Waikato Coal Measure now preserved (Edbrooke *et al.* 1994); and see below). Calcareous sandstone underlying the limestone has yielded a Late Oligocene inner-shelf foraminiferal fauna (Isaac *et al.* 1994).

Te Kuiti Group is unknown in Northland south of Waipu, its next known occurrence to the south being in a drillhole in Auckland city (Edbrooke *et al.* 1994). This represents the northern extent of the Te Kuiti Group in the South Auckland Basin. It was inferred by Nelson (1978) that the Te Kuiti Group in this basin was deposited in a 50–100-km-wide, north-trending seaway that was bounded on the east and west by paleobathymetric highs. Most prominent amongst these were the Herangi High to the west and the Rangitoto High to the east. This seaway deepened to the north and was probably open to the Tasman Sea to the southwest (Anastas *et al.* 1998; see also Anastas *et al.* 2006).

In the South Auckland Basin the oldest marine sediments of the Te Kuiti Group consist of non-calcareous siltstone and claystone passing gradationally upwards from the basal non-marine Waikato Coal Measures. In the northern part of the Te Kuiti Basin the coal measures are Eocene in age (section 6.7.3), but young to the south, and from Kawhia southwards they are of Early Oligocene (early Whaingaroan) age. The immediately overlying marine sediments (Mangakotuku Formation) are of Early Oligocene (early Whaingaroan) age in the northern part of the basin, but, in a similar fashion to the coal measures, young to the south, and are late

Early Oligocene (mid-Whaingaroan) in the southern portion of the basin. The youngest units of the Te Kuiti Group are of latest Oligocene to earliest Miocene (early Waitakian) age. The marine part of the group is dominated in the north by siltstone, but south of Raglan Harbour sandstone predominates (White & Waterhouse, 1993). Limestone occurs throughout the south Auckland region, the facies thickening to the south, until limestones dominate south of Waitomo (Nelson, 1978). The group thins to zero immediately south of Te Kuiti, but narrow zones of thin Te Kuiti Group sediments extend to just east of Awakino, and to the east of Taumarunui, in the northern part of the King Country Basin. The southernmost known occurrence of Te Kuiti Group is in the Tupapakurua-1 petroleum well, west of National Park, which bottomed in 208 m of carbonaceous, shelly mudstone and sandstone of Oligocene age (Gerrard, 1971).

The basal marine unit of the Te Kuiti Group (Mangakotuku Formation) consists of unbedded, typically non-calcareous siltstone to claystone, with glauconitic muddy sandstone beds in the lower part. From a thickness of up to 200 m a few kilometres south of Auckland it thins southwards to the Kawhia area and is absent further south. The sediments accumulated in sheltered shallow-marine environments, ranging from coastal lagoon, tidal flat and estuarine settings, through to an inner-shelf, sheltered gulf environment (Edbrooke *et al.* 1994).

In the northern part of the South Auckland Basin, from Auckland to Kawhia, the fine-grained sediments of the Mangakotuku Formation are overlain conformably or disconformably by a succession of calcareous facies (Glen Massey Sandstone), which extends over most of the northern and central parts of the basin. The unit, which is thickest in the west and thins to the east and south, reaches a maximum thickness of 100 m. Thin glauconitic limestone usually makes up the base of the formation, followed by calcareous siltstone, and then by massive calcareous sandstone, which dominates the unit. It was deposited during the Early–mid-Oligocene (Whaingaroan) and represents a more open marine environment than the underlying Mangakotuku Formation. It records progressive deepening from a shoreface to inner-shelf environment, with the deposition of inner- to mid-shelf calcareous muds and offshore sands.

This dominantly sandy unit was replaced gradationally over most of the South Auckland Basin by massive calcareous siltstone (Whaingaroa Siltstone) with a thin, basal, glauconitic fine sandstone, or, in the south, limestone. South of the Kawhia area, where underlying sandstone and siltstone are absent, Whaingaroa Siltstone conformably or disconformably overlies Waikato Coal Measures or lies unconformably on Mesozoic basement. Lateral facies variations are complex, particularly in the south of the South Auckland Basin, leading to considerable local thickness variation in the Whaingaroa Siltstone. It thickens to the north, where it reaches 250 m. This mid-Oligocene (late Whaingaroan) unit was deposited in a mid- to outer-shelf environment, except for the inner-shelf basal limestone and glauconitic fine sandstone.

To the south of Hamilton the Whaingaroa Siltstone grades up into and interfingers laterally to the south with the Aotea Sandstone. In the extreme south of the basin the sandstone directly overlies Waikato Coal Measures, with probable disconformity, and lies unconformably on basement further south. It is composed of calcareous, fine to medium sandstone and siltstone, commonly with a very glauconitic, calcareous sandstone or crystalline limestone at its base. It was deposited in a range of inner- to mid-shelf environments during mid- to Late Oligocene (Whaingaroan to Duntroonian) times.

The youngest units of the Te Kuiti Group are of Late Oligocene to earliest Miocene (Duntroonian to Waitakian) age, and show lateral variations in lithology. North of Raglan Harbour the lithofacies consists dominantly of calcareous siltstone or very fine sandstone (Te Akatea Siltstone), and overlies Whaingaroa Siltstone or Aotea Sandstone conformably. Over most of this area the Te Akatea Siltstone has been removed by erosion, but is locally up to 50 m thick. Its depositional top is nowhere preserved. The southern equivalent of the siltstone is represented by a persistent limestone unit (Orahiri Limestone), which overlies or grades laterally into Aotea Sandstone. It is up to 50 m thick and consists of a variety of limestone types, which range from massive sandy to glauconitic and pebbly lithologies in the lower part to pure flaggy limestone and thick oyster beds higher up (Nelson, 1978). It accumulated in an inner-shelf environment during Late Oligocene (Duntroonian to early Waitakian) times.

The youngest lithological unit making up the Te Kuiti Group is a flaggy limestone (Otorohanga Limestone) of latest Oligocene to earliest Miocene (Waitakian) age. It is found almost entirely south of Raglan, where it directly overlies the Orahiri Limestone, a locally intervening calcareous glauconitic sandstone (Waitomo Sandstone), or Mesozoic basement rocks (Nelson, 1978). The Otorohanga Limestone was emplaced in a marine shelf environment, with continued shallowing in the southwest near Awakino in response to the emerging Herangi High (Waterhouse & White, 1994). In its southernmost known occurrence, at the base of the Tupapakurua-1 petroleum well, the Te Kuiti Group contains no limestone, but is represented dominantly by brown carbonaceous shale and siltstone containing thin beds of sandstone and coal. Shell fragments are common in the lower half. Microfaunas from the upper part of the succession suggest a mid- to outer-shelf environment of deposition (Gerrard, 1971).

The main limestone units contain unidirectional cross-stratification, inferred to have been produced by migrating subaqueous dunes within a generally northeast-oriented Oligocene seaway in water depths between 40 and 60 m (Anastas *et al.* 1997). In general, there was a relatively nearshore, high-energy marine environment to the south, and more offshore, low-energy conditions to the north.

At the southwestern extremity of the Te Kuiti Group outcrop at Awakino road tunnel, near Awakino, the group is generally thicker (>300 m) than elsewhere, has strong dips (25°–45° E), exhibits an up-section decrease in the amount of dip, and the capping Orahiri Limestone includes several thick (up to 3 m) mass-emplaced units containing a variety of 1–10-cm-sized calcareous lithoclasts of older Te Kuiti Group rocks. The lithoclasts produced were rounded by abrasion in shoal water before being periodically mass-emplaced from west to east onto a shelf accumulating coeval Orahiri Limestone lithofacies (Nelson *et al.* 1994). The lithoclasts were derived from the southern part of the present basement Herangi High, which otherwise separates the King Country Basin from the Taranaki Basin, but must have been submarine and accumulating Te Kuiti Group calcareous facies during the Early Oligocene (early Whaingaroan). Uplift of this depocentre was accompanied by synsedimentary eastward tilting of the Te Kuiti Group strata already deposited immediately east of the Herangi High. Inversion and tilting of the high began in the mid-Oligocene (late Whaingaroan), concomitant with the onset of rapid subsidence along the eastern Taranaki Basin margin directly west of the Herangi High. Uplift continued throughout the Late Oligocene (Duntroonian–Waitakian), when erosion possibly expanded into parts of the shelf, stripping out sections of the overlying Otorohanga Limestone, the topmost formation of the group. The uplift and partial emergence of the Herangi High is viewed as a topographic response to the initiation of basement overthrusting from the east along the Taranaki Fault Zone. This Late Oligocene phase of deformation developed in a mildly compressive regime, which corresponds to a time of proto-plate boundary development throughout New Zealand that preceded propagation through the country of the continuous and more localised present plate boundary at about 22–23 Ma (Nelson *et al.* 1994).

6.10.4 Oligocene volcanism

6.10.4.1 Northland Ophiolite and Northland Allochthon

The Tangihua Complex of Brook *et al.* (1988) is a component of the Northland Allochthon and comprises the Northland Ophiolite (Isaac *et al.* 1994), a suite of submarine basaltic extrusives, with basalt, dolerite, gabbro, and dioritic subvolcanic intrusives, tuff and tuff breccia of Oligocene age (Whattam *et al.* 2004, 2005) and an older igneous suite of Cretaceous–Paleocene age (Nicholson *et al.* 2000; Cluzel *et al.* 2010). Pelagic mudstone and micritic limestone are minor constituents together with older Cretaceous and Paleocene sediments (see section 4.3.2.1). The two episodes of magmatism are geochemically distinct, with the older suite showing a supra-subduction signature, while the young suite has oceanic affinities. The complex crops out as numerous discrete masses, which range in area from a few square metres to over 450 km^2. Most are in northern and central Northland. Tangihua massifs have also been identified in the offshore west Northland Basin from geophysical surveys and dredge samples (Isaac *et al.* 1994).

Rock types within the complex are dominated by basaltic pillow lavas, flows, breccias and hyaloclastites with lesser sheeted dikes. The basalts are dominantly tholeiitic (although calc-alkaline basalts are common) and relatively homogeneous. The North Cape mass includes the only thick and laterally extensive plutonic sequence, which consists of serpentinised peridotite overlain by gabbro hosting microgabbro and microdiorite sills. The lower part of the gabbro shows well-developed mineralogical layering and banding, with microgabbro and microdiorite sills becoming increasingly common upwards to form a sheeted intrusion complex at the top of the present exposure of the plutonics. Sill intrusion took place preferentially within sequences of intercalated lava and mudstone, and multiple intrusion in parts of the succession has given rise to sheeted sill complexes. Whattam *et al.* (2004) prefer the term 'Northland Ophiolite' for the Oligocene magmatic component of the allochthon, and suggest the older component may be autochthonous and belong with the Mount Camel Terrane. This question is discussed by Toy & Spörli (2008) and a parautochthonous status is also possible.

6.10.4.2 Other Oligocene igneous rocks

Marlborough contains the only known igneous rocks of Oligocene age in the East Coast Basin. In the middle Clarence Valley between Bluff River and Gore Stream the Eocene Amuri Limestone is unconformably overlain by the Cookson Volcanics (Reay, 1993). This unit consists of up to 46 m of vesicular basalt and fossiliferous tuffaceous sediment, with intercalated limestone and calcareous mudstone beds up to 8 m thick. No pillow lavas were observed. Foraminifera within the volcarenite indicate an Early Oligocene (Whaingaroan) age.

Late Oligocene (Duntroonian) volcanic rocks are voluminous and widely distributed in the central and northern Canterbury Basin. The Brothers Basalt of the Mount Somers area, consisting of tholeiitic olivine basalt and volcaniclastics interbedded with Oligocene limestone and greensand, was erupted in at least three episodes from two or more centres (Field & Browne, 1986). Olivine tholeiites of this age occur also in the Oxford and Castle Hill areas of inland central Canterbury (see Plate 11). These consist mainly of palagonitised tuffs and associated feeder dikes interbedded with Late Oligocene limestone. In the north of the region the Cookson Volcanics, here of Early and Late Oligocene (Whaingaroan to Duntroonian) age, form a thick succession up to 1200 m thick. They comprise dominantly alkalic, pillowed basalt flows, volcaniclastic breccia, tuff, conglomerate and minor crystalline limestone, deposited in shallow marine conditions (Browne & Field, 1985; Coote, 1987).

Early–mid-Oligocene volcanism in the southern Canterbury Basin was concentrated in the Oamaru area, where eruption of volcanics of the Alma Group (Gage, 1957) continued from the Late Eocene. The younger volcanics (Deborah Volcanics) comprise pyroclastic deposits, and several intrusive dolerite or basalt masses. Tuffs containing bombs and lapilli of basic volcanics, blocks of schist, xenocrysts and xenoliths were deposited in shallow water by mass-flow processes (Reay & Sipiera, 1987). Microfossils give an Early Oligocene (early Whaingaroan) age for the Deborah Volcanics. K/Ar whole-rock dates from Deborah tuffs of 32.4 Ma and 31.6 Ma (Coombs *et al.* 1986) are consistent with the biostratigraphic dates.

The volcanoes near Oamaru are inferred to have erupted on a relatively shallow continental shelf (Cas *et al.* 1989). Coombs *et al.* (1986) interpreted the tuffs as the products of Surtseyan eruptions from shallow submarine vents that ejected pyroclastic material into the atmosphere. Some materials were deposited directly around vents by fallout or as surges, but many deposits were reworked by wave and current action.

Similar basaltic volcanism straddling the Eocene–Oligocene boundary was recorded in DSDP drillhole 593 on the Challenger Plateau in the eastern Tasman Sea (Nelson *et al.* 1986a).

CHAPTER 7

Mid- to Late Cenozoic Tectonism: The Kaikoura Orogeny

7.1 INTRODUCTION

The concept of a late Cenozoic 'Kaikoura Orogeny' can be traced back to Cotton (1916), who used the expression in relation to the uplift of the Kaikoura ranges of the northern South Island. Since that time the term has been expanded to embrace all late Cenozoic earth movements in New Zealand. It is in this sense that Suggate *et al.* (1978) ascribe to it all periods of mainly compressive deformation starting in the Early Miocene. Within the context of plate tectonics the term 'orogeny' has tended to lose some of its significance. Tectonic activity is seen to be concentrated at or near plate boundaries, and plates themselves are in continuous relative motion. Styles of deformation may vary from compressional through strike-slip to extensional along a single plate boundary at the same time, and the present-day plate boundary through the New Zealand region is an excellent example of contemporaneous but different response along a relatively short segment of plate boundary. There are at least six different zones between the Tonga Trench and Macquarie Island, each with a distinctive style of deformation. Traditional 'orogenic episodes' correspond to local maxima in deformation or abrupt changes in style of deformation within an ongoing pattern of relative plate motion.

The current Kaikoura deformation is clearly tied to the present boundary between the Pacific and Australian plates that cuts through the New Zealand region. While linking the idea of a Kaikoura Orogeny to movement on the 'new' plate boundary is attractive, data from the southeast Tasman Sea (e.g., Wood *et al.* 1996) show that from the Late Paleocene a boundary between the Pacific and Australian plates must have existed within the New Zealand continental crust. Therefore the early history of the 'new' plate boundary zone presents a difficulty because some tectonic events that relate to it have not been considered part of the 'Kaikoura Orogeny' and indeed are not 'orogenic' in the conventional sense because they are mainly extensional. There is a marked change in relative plate motion shown by the magnetic anomalies in the southeast Tasman oceanic crust between anomalies 11 and 10 (Wood *et al.* 1996) at about 29 Ma (mid-Oligocene) that must have resulted in changes in the nature of movement on the plate boundary. The succeeding period (Late Oligocene), however, is the period of maximum inundation and carbonate sedimentation in New Zealand, suggesting that the effect of the change was probably simply from extension to transtension. Indications of deformation within southern New Zealand at this time are few, but there is evidence of uplift on the 'Herangi High' north of the rotation pole in Taranaki (see Chapter 8). Lewis (1992) figures small angular unconformities and infers mild folding at ~30 Ma in north Canterbury, although an alternative explanation is that these well-documented tectonic highs reflect footwall rebound associated with obliquely extensional faulting during a period of falling sea level. The Moonlight Fault (western Otago) was active in the Oligocene as a normal fault with strong westerly downthrow (Turnbull *et al.* 1993). Normal faulting was common on basin margins in Fiordland, Westland and southwest Nelson (see section 6.7.3).

Wood *et al.* (1996) show a second change in magnetic lineation after anomaly 9 (probably between 26 and 23 Ma) and continuing through the Miocene. The changes involve the development of more nearly east–west ridge segments and longer transforms, resulting in a southeastward shift of the rotation pole (Lamarche *et al.* 1997). LeBrun *et al.* (2000) equate the early intra-continental plate boundary with the Moonlight Fault and suggest the Alpine Fault was initiated at 23 Ma, close to the Oligocene–Miocene boundary. As noted above, however, normal faulting was widespread through the Oligocene and a broad plate boundary zone rather than a single fault is possible. Fault strands within this zone could lie below the overthrust Pacific plate at the present time.

The change in dynamics at the Oligocene–Miocene boundary is closer to the general perception of the beginning of the Kaikoura Orogeny as a period of generally transpressive tectonics extending to the Present. This second period of change has the advantage of conforming closely to the concept expressed by Cotton (1916), Suggate *et al.* (1978) and common usage. The redefined 'Kaikoura Orogeny' corresponds to a period of changing relative movement between the Australia and Pacific plates that is reflected in the ongoing movement to the southeast of the pole of relative movement between the two plates (King, 2000, fig. 1). The migration of the pole away from a position in central New Zealand near to the plate boundary explains an overall change to transpression over the last 23 m.y., and the sharp movement to the west in the last 5 m.y. is consistent with accelerated shortening in the Pliocene to Recent.

7.2 TECTONIC ARCHITECTURE OF THE KAIKOURA OROGENY

The tectonic changes described above are derived from the study of the southeast Tasman Sea. The earlier stages record the opening of a wedge-shaped piece of ocean floor between the Resolution Ridge and the western margin of the Campbell Plateau (Sutherland, 1995; Wood *et al.* 1996). Over time the original north-northeast-trending spreading ridge progressively changed through northeast to an east–west orientation with a marked sigmoidal twist in the transforms. By the late Middle Miocene the transforms had become very long and the ridge had become broken into very short segments. By the Late Miocene the northern portion of the plate boundary became parallel to the plate motion direction and thus strike-slip in character. Links with the Alpine Fault developed, possibly with a strike slip and a western loop with east-northeast-directed subduction on the southwest Fiordland margin (Sutherland & Melhuish, 2000). Spreading may have continued for a time in the south, but, through the Pliocene–Recent, change continued, with the strike-slip northern sector evolving into a slow subduction zone and the southern sector changing to a strike-slip boundary. These changes led to the development of the Puyseger Trench and the Macquarie Ridge Complex (Lamarche *et al.* 1997). This area is discussed further by Cande and Stock (2004).

The changes outlined above and the consequent changes in rotation pole can also be seen to the north of New Zealand. Changes begin before the 'Kaikoura Orogeny', starting with the more rapid northward movement of Australia at ~40 Ma and spreading in the Adare Trough (Cande *et al.* 2000; Eagles *et al.* 2004) on the Antarctic margin northeast of northern Victoria Land. These data point to convergence between the Australian and Pacific plates north of New Zealand in the Eocene–Oligocene and are consistent with an Australian–Pacific rotation pole in central New Zealand (Sutherland, 1995; Lamarche *et al.* 1997). Interpretation of tectonic events north of New Zealand is covered in a wide-ranging review by Schellart *et al.* (2006), who propose large-scale roll-back of the Australian–Pacific margin during the Cenozoic, a position supported by evidence from dredging in the South Fiji Basin (Mortimer *et al.* 2007). Immediately north of New Zealand subduction is indicated by arc-type rocks of the Three Kings Rise dated at 32–26 Ma (Oligocene). At scattered locations further east, arc-type rocks with ages between 22 and 18 Ma are interpreted as rifted arc lavas separated during a period of rapid roll-back of the Australian–Pacific margin (Mortimer *et al.* 2007).

At the beginning of the Miocene a remnant of Cretaceous ocean crust must have existed to the east of the Three Kings Rise as the parent of the Cretaceous Tangihua volcanic massifs and the basement for at least some of the Cretaceous–Paleocene sediments that now form part of the Northland Allochthon (Fig. 7.1). All of this Cretaceous crust was subducted at a trench east of the Three Kings Rise and its great age may have been a significant factor in facilitating subduction. Some authors invoke a southward extension from New Caledonia of an east- or northeast-dipping subduction zone (Cluzel *et al.* 2001; Schellart, 2007) and obduction of the allochthons (Fig. 7.2), while others believe that it can be achieved within the context of a westward-directed subduction of the Pacific plate. The development of the Van der Linden and Vening Meinesz fracture zones (Mortimer *et al.* 1998) from the Late Oligocene may also be a factor. Late Oligocene subduction towards the west was favoured by Bradshaw (2004), to the southwest by Mortimer *et al.* (2007) and to the north by Schellart (2007). This problem is not resolved.

Despite these uncertainties it is clear that the emplacement of Late Cretaceous and early Cenozoic rocks in the allochthons of Northland and the East Coast regions in the Early Miocene was followed by subduction towards the southwest along the northeastern side of the Northland Peninsula, as evidenced by the development of the Northland volcanic arc rocks soon after the emplacement of the allochthon. There was less convergence and no magmatism to the southeast in the East Cape area, although the emplacement of a large volume of allochthonous sediment and seafloor volcanics indicates significant convergence in the Early Miocene. The direction of emplacement is from the north-northeast, and the pile of slices has a substantial thickness of Cretaceous rocks in the northern (upper) slices, including 'exotic' ocean floor volcanics. The central and southern zones contain increasing proportions of Cenozoic, and the southern

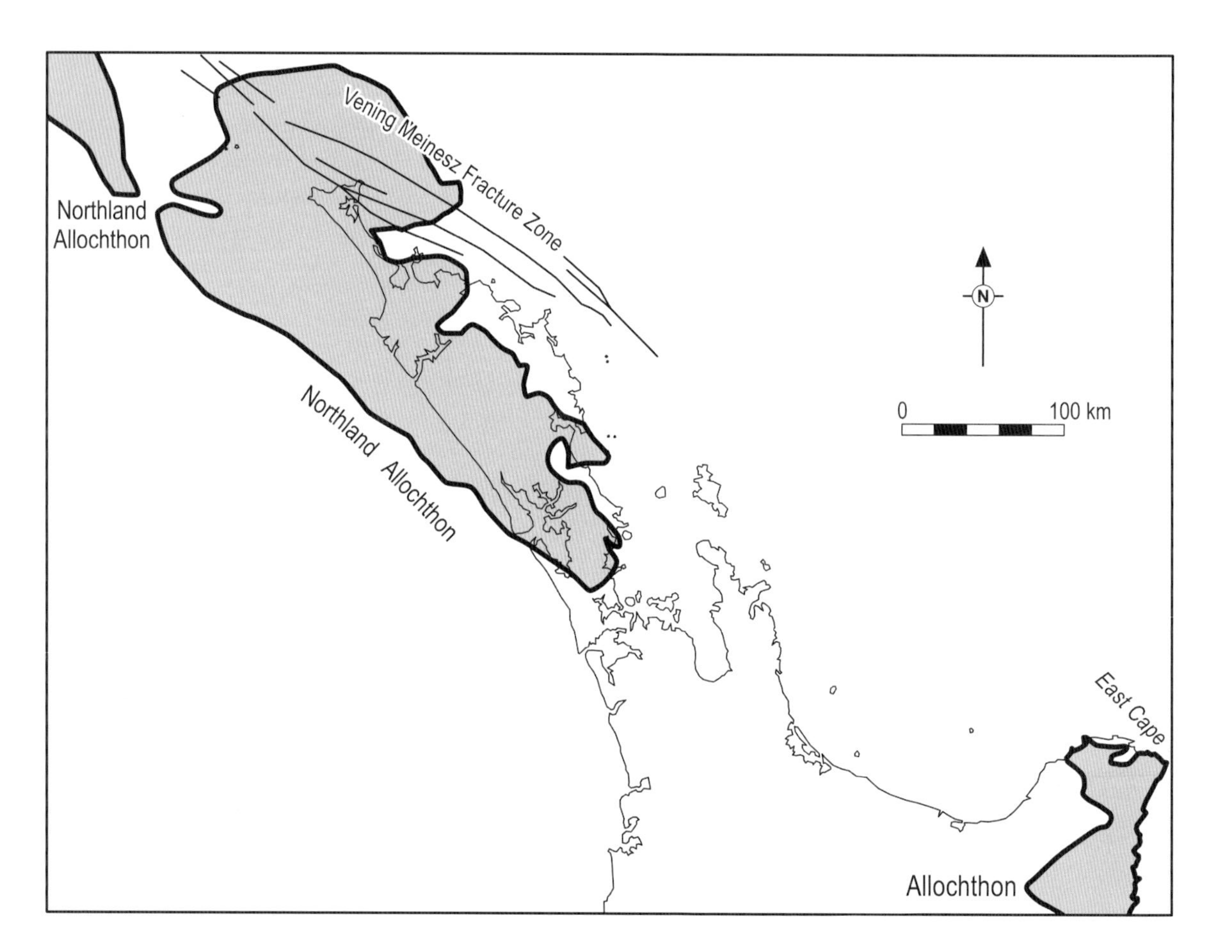

Fig. 7.1. Map of the distribution of the Northland Allochthon and related allochthon near East Cape. (Adapted from Brook & Hayward, 1989.)

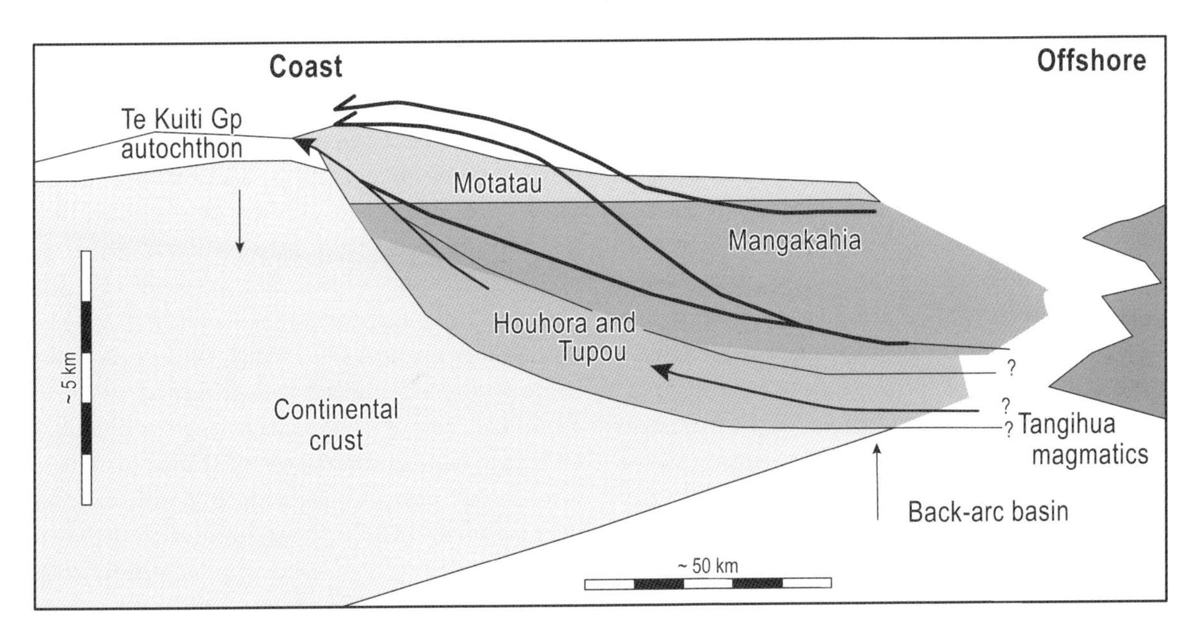

Fig. 7.2. Schematic diagram to show the emplacement of the sedimentary and volcanic rocks of the Northland Allochthon onto the Northland Peninsula. For details of component complexes, see section 8.2.1.

slices include rocks comparable to the autochthonous formations of the East Coast Basin (Field *et al.* 1997) and could be considered parautochthonous. This change is consistent with an Australia–Pacific pole position a moderate distance to the south.

The cessation of activity in the Northland arc at ~15 Ma indicates that the Three Kings Rise and the South Fiji Basin had become part of the Australian plate by that time. Movement between the Pacific and Australian plates was accommodated by the subduction that generated the main Colville Ridge. Enhanced subduction at this trench was followed by back-arc extension between the active Tonga–Kermadec arc and the now inactive Colville Ridge to form the Havre Trough.

7.3 THE HIKURANGI MARGIN

The sector of the Australian–Pacific boundary between East Cape and the Chatham Rise has a distinctive history. It is generally accepted that the Hikurangi margin started to develop in the Miocene with the emplacement of the East Coast Allochthon and other indications of crustal shortening in the Early Miocene. Ideas about the dynamics of the development of the margin have been strongly influenced by the suggestion of Walcott (1978) that 'the Hikurangi subduction developed by southward propagation from the Kermadec subduction zone'. Within this conceptual framework, the southward extension of the subducted slab is seen to have progressed so that the leading edge now lies under Marlborough at a line corresponding to the northern margin of the Chatham Rise. The Chatham Rise has rotated counterclockwise with the Pacific plate, leading to dextral strike-slip faults that exposed the eastern North Island to subduction and creating the northeast-trending leading edge of the continental crust (the Hikurangi margin). This perception can be graphically illustrated by maps in which the Australian plate is held stationary relative to a counterclockwise rotating Pacific plate (e.g., Walcott, 1984b; King, 2000; and many others). In these maps the trench and plate boundary appear to have rotated clockwise.

Of all the elements in this tectonic system, the subduction zone with a slab of ocean crust extending more than 200 km down into the mantle is the one least likely to rotate. The trench, however, simply marks intersection of the upper portion of the subducted slab with the surface and can relocate by a roll-back. If the point of inflexion of roll-back moves faster in one location than another the trench will appear to rotate. Roll-back is consistent with evolution of the complex ocean crust between the Three Kings Rise and the Tonga Trench to the northeast, with elements of the subduction system being progressively abandoned from west to east, and the development of back-arc extension. The weaker continental crust of the Australian plate accommodates to the change by compression, extension, relocation and rotation, and is probably detached from lithospheric mantle. There has been considerable debate about how these changes have occurred.

An alternative way of viewing the development of the Hikurangi margin is to consider the Pacific plate to be stationary (Vickery & Lamb, 1995, fig. 8). This involves a general motion of the Australian plate towards the north and northeastwards in such a way as to progressively override the plate boundary at depth. This alternative view involves no change in dynamics and is simply a change of view, but it does explain many interesting features, not least why the Benioff zone below Marlborough extends to a depth of more than 200 km, a little less than elsewhere under the North Island; yet within the Pacific-centred view this would be the youngest part of the subduction zone at the very southern limit of its propagation. Using the alternative framework it is possible to reconcile some of the difficulties that have been identified. The main latitudinal difference is that subduction at the south end has been consistently slower. A further variation suggests that subduction continues much further to the south and that a large part of the Hikurangi Plateau has been subducted (Reyners *et al.* 2011). Though controversial, a volume of lithosphere must be removed to accommodate shortening of continental crust across the South Island.

The structural evidence for deformation mechanisms at crustal level that resulted in the northeast-trending margin and the northeast structural grain of the Hikurangi margin is equivocal. Field *et al.* (1997) propose two end member scenarios, suggesting that the truth lies somewhere between. Model 'A' was developed by Walcott (1978) and Rait *et al.* (1991) and invokes a clockwise rotation of at least 60°. Rotation is supported by some paleomagnetic data (Lamb & Bibby, 1989; Vickery & Lamb, 1995; Walcott, 1998) that indicate a large number of paleomagnetic domains that have rotated at different times and by different amounts so that the rotation may be both significantly less or much greater than the re-orientation of the margin as a whole. Most data come from Miocene–Pliocene sections because many older rocks have unstable magnetisation. Field *et al.* (1997) suggest that this model implies a major shortening and thickening across central New Zealand, a view for which there is limited evidence. In Model 'B' the result is achieved by accumulated movement of a series of strike-slip faults. There is evidence for major strike slip on faults in the eastern North Island; however, the slices are wedge-shaped and would also produce thickening in the southwest. A factor that seems to have been overlooked by Field *et al.* (1997) is that in many well-constrained reconstructions (Walcott, 1978; Rait *et al.* 1991), the overall length of continental margin between Northland and the Chatham Rise is much the same after the rotation. In a framework with the Pacific plate stationary the overall deformation resembles ductile deformation on a grand scale (Fig. 7.3) and it is reasonable to assume that the shape of the Hikurangi margin is as determined by the volume of rocks involved. A 'fluid' combination of rotation and faulting on a wide variety of scales appears sufficient to explain the re-orientation (Fig. 7.3).

A thorough and comprehensive analysis of the geodynamics of Neogene convergence and its impact on the overriding plate is presented by Nicol *et al.* (2007).

7.4 THE WESTERN MARGIN OF THE KAIKOURA OROGEN

It is possible to define a western boundary to the region affected by the Kaikoura Orogeny and the boundary is most clearly seen in the Taranaki Basin where it is marked by the line between the Eastern Mobile Belt and the Western Stable Platform (King & Thrasher, 1996). The boundary trends south-southwest and separates the western South Island from the stable Challenger Plateau in the south. To the north, the Northland Peninsula appears to have become part of the stable zone after the Miocene.

Tectonism in the west also starts in the Early Miocene with westward thrusting on the Taranaki Fault. Evidence comes from seismic sections that show the fault truncates the Oligocene Tikorangi Formation (see section 6.7.3.5) and that movement is mainly confined to the Early Miocene. Displacement is substantial: the vertical component is up to 6 km and the leading edge of the thrust mass advanced approximately 10 km to the west. The thrust has its

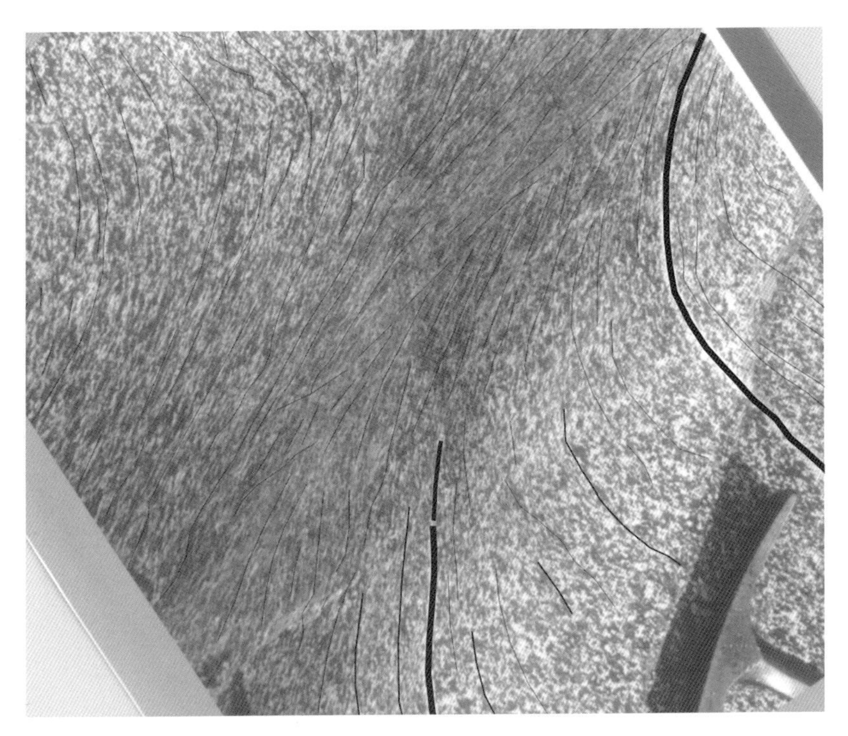

Fig. 7.3. Photograph of fabric in a deformed granite with a combination of brittle and ductile structures showing a resemblance to the oroclinal bending in Southland and the Alpine Fault zone. (Photo: J.D. Bradshaw)

lowest dip in the central portion and in this region the Tarata Thrust Zone is developed to the west. King and Thrasher (1996) attributed these thrusts to the squeezing out of material beneath the advancing front of the Taranaki Fault wedge. The northeastern margin of the basin continued to be active through the Middle Miocene (until about 12 Ma) and then began to subside rapidly (King & Thrasher, 1996), but the reason for this is unclear.

The southern part of the Taranaki Basin evolved in a different manner and became a zone of shortening through the Middle Miocene–Pliocene. During this period a number of early Cenozoic half grabens were inverted, a process that can be seen on land and continues in the northern South Island. Many of the faults involved have a northerly trend and converge on the Alpine Fault to the south. The south Taranaki–Westland region is much closer to the plate boundary than north Taranaki, which probably explains the ongoing shortening. The strong development of sedimentary basins (Murchison and Maruia) during the Late Miocene–Pliocene close to the plate boundary may be of the 'foreland basin' type and relates a westerly thrust slab of Pacific plate bounded by the Alpine–Wairau Fault section of the plate boundary. The south Westland Neogene basin has also been interpreted as a foreland basin (Kamp *et al.* 1992a).

7.5 TECTONIC HISTORY

7.5.1 Early–Middle Miocene (23–15 Ma)

The first clearly compressive episode is expressed in the earliest Miocene by a series of events in a zone from the Northland Peninsula to East Cape as outlined in the previous sections. The allochthons rest on autochthonous Mesozoic and Cenozoic rocks, including Oligocene sediments (Fig. 7.2). The provenance of the sediments of the Northland Allochthon is mainly the Mesozoic rocks of Northland and deposition was probably partly on the then edge of the quiescent Mesozoic accretionary complex and partly on ocean crust or back-arc basin crust (Tangihua component). Currently the assemblage is at least 3 km thick and may be up to 5 km thick (Isaac, 1996). It is made up of approximately six slices, with the younger and more proximal rocks towards the base and the more distal and older rocks above (Fig. 7.2). The generally inverse stacking is consistent with progressive uplift of the source and gravitational gliding. The original stacking has, however, been complicated by later reactivation and imbrication along certain thrusts (Evans, 1992; Rait, 2000). Emplacement is thought to be by gravitational gliding directed towards the southwest (Rait, 2000) into a basin that, at least locally, had reached bathyal depths. The transport direction appears to be almost perpendicular to the continental margin. If gravitational gliding is involved, then the key problem is how to elevate the older oceanic rocks to a point where they can slide onto the continental crust of Northland (Bradshaw, 2004). How this elevation was achieved remains unclear in either the east-dipping or west-dipping subduction models (Schellart *et al.* 2006; Mortimer *et al.* 2007).

The East Coast Allochthon has much in common with the Northland Allochthon (Ballance & Spörli, 1979) and was emplaced at much the same time. The separation from the Northland Allochthon may simply reflect back-arc extension and burial by the volcanic arc and back-arc deposits of the Taupo Volcanic Zone.

Rait *et al.* (1991) show regional scale thrusting and shortening during the latest Oligocene–Early Miocene, which they relate to convergence between the Pacific and Australian plates. Subduction is greatest in the northwest where there is subduction-related magmatism (Northland). There are no equivalent Miocene arc rocks to the southeast associated with the East Cape Allochthon where subduction was less. In Marlborough the thrusting is younger (late Early Miocene to early Middle Miocene) and may relate to a different episode. It is interesting that the directions of thrusting/sliding in Northland (Rait, 2000), the Raukumara Peninsula and Marlborough (Rait *et al.* 1991) are very similar. If thrusting was driven by subduction, it suggests that during the Early Miocene all three areas were on the same plate and that the Australian plate extended south of the Wairau Fault.

The suggestion of the initial position of the plate boundary being immediately east of Marlborough is consistent with other studies. The vertical axis rotations proposed by Little and Roberts (1997) also imply that Marlborough was close to the subducting margin in the Early Miocene, and was part of the Australian plate. Little and Roberts (1997) place the southern limit of the subduction zone and the northern edge of the Chatham Rise at the 'incipient Clarence Fault' at 18 Ma and only a few kilometres north of its present position. In the context of the fixed Australian plate model, this implies a very rapid southward extension of the subduction and motion of the Chatham Rise before 18 Ma and very slow motion thereafter (King, 2000, fig. 1). This is the opposite of that suggested by the migration of the rotation pole but is consistent with an overriding by the Australian plate and subsequent annexation of Marlborough by the Pacific plate by westward migration of the crustal plate boundary. Stratigraphic contrast and structural studies (Lamb & Bibby, 1989; Little & Roberts, 1997) all point to a major difference across a line approximating to the present Hope Fault, with the Marlborough region as part of the Australian plate in the Early Miocene. During the Middle Miocene the active boundary shifted progressively northwestwards to the present northern section of the Alpine and Wairau faults, possibly with oblique movement. In this way Marlborough was transferred to the Pacific plate and remained close to the Chatham Rise. Post-Pliocene changes have tended to shift the main movement back towards the Hope Fault. These changes are consistent with westward and eastward excursions from the mean shift of the rotation pole (King, 2000, fig. 1). Taken together they suggest that the Miocene plate boundary within the upper crust was marked by a fold-style rotation close to the present western limit of the Chatham Rise and the development of three or four sub-parallel faults (Hope, Clarence, Awatere and Wairau) and a large number of lesser fractures, including thrusts. These faults facilitate translation and rotation of crustal blocks. It is notable that the main basement structural blocks of the Hikurangi margin are oblique to the margin as a whole, consistent with the fault-fold analogy and with the suggestion that the shape of the Hikurangi margin is exactly what is required by the volume of rocks involved and that the problem identified by Field *et al.* (1997) may be an illusion.

Chanier *et al.* (1999) have drawn attention to evidence for crustal extension and basin formation between 15 and 5 Ma, during a period of oblique convergence of the plates. They attribute this to tectonic erosion. The recorded extension directions are diverse and may reflect the erratic advance of the overriding Australian crust perturbed by large volcanoes on the subducted slab. Alternatively, the development of Neogene basins could also represent the interaction of slow plate convergence and contemporaneous trench roll-back with change in the position of the initial inflexion point.

The estimates of principal horizontal stress at the present (e.g., Walcott, 1978; Pettinga & Wise, 1994), as west-northwest–east-southeast in the eastern North Island and northern South Island, are also consistent with the 'flow' of rocks on the eastern edge of the Australian plate onto the Pacific plate. It appears that a major feature of the plate boundary at crustal levels is dominantly westward thrusting along the Alpine Fault with subsidiary eastward thrusting on the Main Divide and related thrust in the south, but predominant eastward thrusting in the north. Both are equally developed through an intervening region in north Canterbury (Pettinga *et al.* 2001). Overall, a two-sided configuration is characteristic of the orogen from Otago in the south to the latitude of the Wanganui Basin in the north.

7.5.2 'Far field' Kaikoura events in New Zealand

The events of the plate boundaries outlined above are felt to differing degrees through much of New Zealand but their stratigraphic consequences are less easy to summarise. The ongoing nature of the changes means that initial structures become progressively less appropriate for the stress regime over time and are abruptly replaced or overprinted by new structures. In any one region this may give the impression of a succession of orogenic pulses that individually can be dated only with difficulty. We recognise major events and sedimentary cycles in New Zealand that we consider, on balance, to be largely the result of tectonic control (see Chapter 3). The Neogene Kaikoura Orogeny corresponds to the regressive portion of the first order cycle (Chapter 3). Within the Neogene we recognise three main phases, with boundaries at ~23 Ma, ~13 Ma and ~5 Ma respectively. There is, however, a distinct possibility that deformation events and their sedimentary consequences may vary in age from place to place and that in general sedimentary response may lag behind tectonic change.

Outside the Kaikoura orogen as outlined above, change at the beginning of the Miocene is muted on both plates. The change is marked by limited uplift as shown by widespread unconformities and a general change from carbonate to detrital sediments. In many sections, however, the erosion of Oligocene sediments could be the result of either this or later tectonic episodes.

In the western South Island, new sedimentary basins were developed during the Early Miocene in Grey Valley. The Stillwater Mudstone (mid-Early Miocene) and related units rest discordantly across folded and faulted Eocene–Oligocene rocks. Folding with significant erosion at the beginning of the Miocene is indicated by coaly detritus in the Welsh Formation near Punakaiki. The main phase of subsidence and sedimentation was later, at the beginning of the Middle Miocene.

Strictly speaking the Marlborough deformation discussed above in context of the general development of the plate boundary belongs to an episode close to the Early Miocene–Middle Miocene boundary (approximately the base of the Clifdenian Stage: ~16 Ma). To the south of the Flags Creek Thrust the Late Cretaceous–Oligocene rocks form a northeastwards-thickening succession, consistent with progressive subsidence of the margin of the Mesozoic accretionary complex (Field et al. 1997). This succession is stacked by several southwestwards-verging thrusts (Lamb & Bibby, 1989; Rait *et al.* 1991). In this area the Lower Miocene is abruptly overlain by extremely coarse conglomerate sourced from both Cretaceous–Cenozoic rocks and Torlesse basement (Lewis *et al.* 1980). Immediately north of the Flags Creek Thrust, as far as the Wairau (Alpine) Fault, Late Miocene or younger rocks typically rest directly on Torlesse basement except in the coastal strip where Middle Miocene rocks are unconformable on Cretaceous–Paleocene strata, a relationship consistent with regional scale uplift of Australian plate rocks. The period of deformation appears quite protracted, with Miocene conglomerates overridden by the later thrust sheets (Lamb & Bibby, 1989). The extent of this deformation beyond Marlborough is difficult to determine, but at some localities in the eastern Wairarapa the Whangai Formation is deeply eroded before the Middle Miocene. It is tempting to see this uplift as related to that in Marlborough, and if this interpretation is correct it would suggest that Marlborough and the Wairarapa were on the same (Australian) side of the plate boundary.

Elsewhere in the South island the initiation of the new Maruia Basin (Cutten, 1979), and the increase in uplift and erosion indicated by the conglomeratic sediments that infill the Maruia and Murchison basins (West Coast) began at this time. In south Westland the

Tititira Formation also marks an influx of detritus at this time, although the younger part could equally represent coarse sediment related to cycle ~15 Ma deformation. The conglomerates are made up of Western Province material and there is no schist or material from the Pacific plate at this time. The occurrence of conglomerate in southern Fiordland also reflects uplift of Western Province material. Deposition of the Penrod Group in the Great South Basin commences at about this time.

This 15 Ma pulse coincides with cessation of activity in the Northland arc and the shift of the locus of subduction to the Colville Trench.

7.5.3 Middle–Late Miocene (~13–11 Ma)

During this period the southward shift of the plate boundary rotation pole suggests an increased convergence in the north. In the southeast Tasman Sea, this period is marked by the development of long transforms in the ocean crust that eventually develop into a strike-slip plate boundary (Lamarche *et al.* 1997).

In the South Island a westward spread of detrital sedimentation is best marked by the continuing influx of thick conglomerate in the Murchison and Maruia basins (Cutten, 1979; Nathan *et al.* 1986). These conglomerates may reflect a westward jump of the functional plate boundary to the Wairau branch of the Alpine Fault System. Regression, with the local deposition of coal measures and the formation of the Grey–Inangahua Depression probably also relates to this episode.

Further to the south, conglomeratic sedimentation continued, and tectonic activity culminated in the Late Miocene with the emplacement onto the Tititira Formation of the allochthonous Eocene–Oligocene rocks of the 'Jackson Formation' (Sutherland *et al.* 1996). This event took place between 7 and 5 Ma when the emplacement site lay to the northwest of the Fiordland block. The 'Jackson Formation' allochthon is of deeper-water facies than Eocene–Oligocene rocks below the Tititira Formation, and may represent inversion of a deeper fault-bounded basin that lay to the southeast, and possibly now lies beneath the overthrust edge of the Pacific plate. Inversion of Eocene–Oligocene basins is also typical of north Westland. Details of the timing of inversion are to some extent controversial (see discussion in Suggate *et al.* 2000).

7.5.4 Latest Miocene–Pliocene

At the beginning of the Pliocene there was a marked eastward shift of the pole of rotation. The result was to increase the rate of convergence in the north and spread the zone of convergence southwards. The change is marked by accelerated uplift along the Southern Alps and greatly enhanced sediment input into the basins around New Zealand.

Deformation is evident in the Australian plate with clear evidence of folding and faulting in the Westland basins and southern Taranaki basins. In the former basin, inversion of sedimentary basins occurred, with the development of major uplifts commonly located along Eocene–Oligocene normal faults. Thrusts that placed early Cenozoic or basement rocks over Late Miocene sediments were widespread, and rocks as young as Pleistocene were involved in folding. Active faulting continued on mainly north-northeast-trending reverse faults. In the North Island, uplift and deformation with shortening were strongly developed in the southern sector of the Hikurangi margin and activity continues to the Present. The central North Island was dominated by the Pleistocene–Holocene development of a back-arc extensional basin related to the Taupo Volcanic Zone. South of the limit of active volcanism the actively subsiding Wanganui Basin has been interpreted as a foreland basin (Stern & Davey, 1989). The basin extended into the northern South Island, where it was responsible for the drowned coastlines of northern Nelson and Marlborough and the broad shallow South Taranaki Bight.

Active folding and faulting through the Pleistocene–Holocene was characteristic of the Pacific plate in a corridor parallel to the plate boundary and extending approximately parallel to the present coastline, except in the region between Banks Peninsula and north Otago. The Chatham Rise was not affected by compressive deformation (Barnes, 1994), and the eastern offshore area was not part of the orogen.

CHAPTER 8

The Final Convergent Margin Phase: The Neogene Assemblage

8.1 INTRODUCTION

The Neogene sedimentary rocks provide a record of the development of a plate boundary system in the New Zealand region at a time when the Pacific plate was being subducted southwestwards beneath the Australian plate. Although there were precursor indications of developing convergence during the Oligocene (see Chapter 7), the main body of evidence indicates that a convergent plate movement developed rapidly through the New Zealand continental region at around the Oligocene–Miocene boundary. Recently there has been a revival of the theory of a pre-Oligocene plate boundary through New Zealand with a large sinistral displacement in Cretaceous times (Lamb *et al.* 2016) and the related question of the extent of oroclinal bending that is of Miocene in age. There are data that do not support large Miocene oroclinal bending (Turner *et al.* 2012; Mortimer, 2014; Lamb *et al.* 2016), although as yet there are no detailed paleogeographic maps on this basis.

The beginning of the Miocene was also when obduction of the older (Late Cretaceous–Oligocene) passive margin sequence together with parts of its igneous basement occurred, with the emplacement of allochthonous nappes in the north and east of the North Island. In the north this was followed by voluminous calc-alkaline arc-related volcanism.

The new stress regime reconfigured the pattern of sedimentary basins with a profound change in sedimentation at the beginning of the Miocene. The renewed tectonic activity led to a New Zealand-wide change from mainly carbonate-rich sediments to the terrigenous clastic sediments that are characteristic of the Neogene. Subduction of the Pacific plate in the north with transpressional deformation to the south caused widespread uplift and emergence, and a period of regression was ushered in.

On the eastern seaboard of the North Island imbrication of the older passive margin sequence and the underlying wedge of Mesozoic accretionary complex rocks led to the development of small, thrust-controlled piggyback basins in a supra-subduction trench-slope setting. The slope basins were separated by narrow elongate structural ridges and infilled by flysch comprising sediment gravity flows, hemipelagic muds and arc-derived ash (van der Lingen, 1982). Though sometimes referred to as foreland basins, they do not conform to the foreland basin model (see DeCelles & Giles, 1996) and may be better viewed as fore-arc basins above an obliquely convergent subduction zone. The complex pattern of local basin formation and growth, contemporaneous faultings, and current active deformation makes coherent description difficult.

In western New Zealand the Taranaki and West Coast regions were the sites of retro-arc foreland basins, where west-directed thrusting of Cretaceous and Paleogene rocks over younger sediments created a foredeep. Continuing convergence on the main thrusts led to inversion and uplift along the whole belt.

Well to the southeast of the plate boundary, in the Great South Basin and southeast Canterbury Basin, passive margin subsidence continued from the Paleogene through the Neogene. Northwest, towards the plate boundary, the developing compressive transform system caused uplift and erosion of much of the sedimentary record.

Three distinct periods of tectonic activity with related sedimentary packages can be recognised within the Neogene and they are commonly separated by unconformities (see Chapter 7). The first package marks the beginning of uplift and regression at the start of the Miocene, and continues until Middle Miocene times. The second package marks a further significant tectonic pulse, lasting until the Miocene–Pliocene boundary; and the third phase, beginning in the Early Pliocene, and ongoing, was marked by accelerated uplift along the Southern Alps and a greatly increased outpouring of clastic sediments into most New Zealand basins.

8.2 EARLY TO MID-MIDDLE MIOCENE (LATE WAITAKIAN–LILLBURNIAN)

8.2.1 Northland and East Coast allochthons

Changes to the tectonic setting in the latest Oligocene, the precursor to the present west-dipping Hikurangi subduction system, resulted in the Cretaceous–Paleogene margin sedimentary wedge being deformed, uplifted and then displaced southwestwards onto northern and northeastern New Zealand (see Figs 7.1 and 7.2). The obducted material forms an allochthon covering much of Northland and parts of the East Coast (see section 7.5.1). The tectonics of this episode are complex and have been much debated (Schellart, 2007; Cluzel *et al.* 2010 and references therein) but appear to be a response to strong uplift northeast of the Northland Peninsula in the Early Miocene, an event recorded by the development of the Cavalli core complex (Mortimer *et al.* 2008). Along with the sediments, igneous rocks were also obducted to form the Tangihua and Matakaoa complexes and the Northland Ophiolite that forms a dismembered sheet at the top of the allochthon. Late Cenozoic tectonic events have split the allochthon into two discrete portions: the Northland Allochthon and the East Coast Allochthon (Ballance, 1993).

The Northland Allochthon, comprising between 40,000 and 60,000 km³ of Cretaceous–Paleogene rocks, was emplaced in the Northland Basin during the latest Oligocene–earliest Miocene (Waitakian–Otaian) as a series of nappes (Ballance & Spörli, 1979; Hayward *et al.* 1989). The allochthonous rocks crop out throughout western and central Northland between North Cape and Silverdale, and in places the stack of sheets exceeds 3 km in thickness. They extend 10–15 km beyond the present west coast, where seismic sections show the nappes wedging out southwestwards into the normal marine basin succession (Isaac *et al.* 1994). The Northland Allochthon presumably covered all of eastern Northland and much of the eastern continental shelf as well, but most of it has been removed by erosion (Hayward, 1993; Isaac *et al.* 1994).

Nappes within the Northland Allochthon are mostly composed of rocks from four assemblages:

1. Tupou Complex, which consists of highly deformed and indurated Early Cretaceous sandy and pebbly turbidites
2. Mangakahia Complex, which includes Late Cretaceous–Early Eocene sandy turbidites, siliceous mudstone, thin-bedded fine-grained limestone, and red, green and grey clay-rich, non-calcareous mudstone, all of which accumulated in deep bathyal or abyssal depths (see sections 4.2.2.1 and 4.3.2.1)
3. Motatau Complex, which consists of Middle Eocene–Oligocene calcareous mudstone and redeposited glauconitic sandstone and micritic limestone that mainly accumulated at bathyal depths
4. Tangihua Complex, which is a magmatic suite of pillow basalts with subordinate breccias, and multicoloured mudstone, and less common peridotite, gabbro and dolerite intrusions. These comprise two compositionally distinct suites: a Cretaceous–Paleocene subduction-related assemblage, and an Oligocene oceanic ophiolite assemblage, Northland Ophiolite (Whattam *et al.* 2004).

The first three complexes are inferred to have accumulated in a deep marine basin northeast of northern Northland, while the Tangihua Complex may have developed further to the east (Hayward, 1993; Cluzel *et al.* 2010).

The base of the allochthon is a décollement generally showing overall concordance with the little-deformed underlying autochthonous strata. Nappes were emplaced as separate thrust slices into a passive basinal setting. Almost everywhere, the Northland Basin sediments indicate rapid deepening (to at least mid-bathyal depths) immediately prior to emplacement of the earliest allochthon nappes. A combination of gravity sliding and thin-skinned tectonic thrusting was the probable mechanism for emplacement (Hayward *et al.* 1989). Basinal sequences that accumulated close to the advancing allochthon contain coarser sediments derived from the erosion of the advancing nappe front before they were overridden (Evans, 1989).

The orientation of major thrust faults and folds within the allochthon in northern Northland indicates that the main emplacement was to the southwest (Brook *et al.* 1988). In southern Northland the allochthonous sequences were emplaced towards the south and southeast, and the presence of serpentine pods within the allochthon in this area, east of their inferred source, also supports emplacement towards the south or southeast (Hayward, 1993). Dating of associated and included sedimentary sequences shows that southwestward emplacement began in the north in the latest Oligocene–earliest Miocene (early–mid-Waitakian), and south and southeastward advance began in the earliest Miocene (late Waitakian) and continued into the middle Early Miocene (Otaian) (Hayward, 1993). The minimum distance over which rocks of the Tangihua Complex were transported was estimated at ~300 km (Rait, 2000).

The East Coast Allochthon is recognised in the Raukumara Peninsula where it also consists of nappes of Cretaceous–Oligocene rocks emplaced from the north-northeast (Field *et al.* 1997). It is currently exposed from the tip of the Raukumara Peninsula to just east of Matawai in the west, and is inferred to extend southwards under a cover of younger Neogene strata as far as Mahia Peninsula. Elements have been recognised 300 km further south in eastern Wairarapa, but location here is more likely to be the result of strike-slip faulting from the vicinity of the Raukumara Peninsula (Delteil *et al.* 1996).

The basal detachment lies for most of its length at or near the top of the Whangai and Waipawa formations, facilitated by an overlying Eocene smectite mudstone that formed a glide horizon. Depth to the detachment is unknown, but a stratigraphic package originally 5–6 km thick is involved in the thrust sheets (Field *et al.* 1997).

Rait (2000) divided the East Coast Allochthon into three zones – northern, central and southern – on the basis of gross structural differences. The northern nappes are formed of the Mokoiwi Formation, the Matakaoa Volcanics and the upper Tikihore sheet. These rocks form southwest-displaced sheets that overlie and truncate southwest-dipping reverse faults. The oldest rocks of the allochthon are at the top and back of the thrust pile. The volcanic rocks consist largely of subalkaline basaltic lavas, but have intercalated fossiliferous sedimentary lenses (see Chapter 4) yielding late Early Cretaceous (Albian), Late Cretaceous and Paleocene–Eocene ages (Strong, 1976, 1980; Spörli & Aita, 1992). In age range and lithology they are very similar to the Tangihua Complex of the Northland Allochthon.

The central zone contains two allochthonous units: the Waitahaia sheet, a southwestern continuation of the Mokoiwi sheet; and the Te Rata unit, which consists of the Tikihore Formation (see section 4.3.2.2), mélange, and Eocene glauconitic sandstone and mudstone. The Waitahaia sheet was emplaced first and gently folded within the allochthon before the later Te Rata unit was emplaced across it.

The southern zone is structurally simpler, comprising Eocene, Oligocene and Early Miocene rocks tectonically stripped from the underlying Whangai Formation and telescoped into an imbricate fold and thrust belt fan (Kenny, 1984; Field *et al.* 1997). Stoneley's (1968) map shows 15 thrust slices in this area.

Estimates of horizontal transport and shortening for various units in the East Coast Allochthon were given by Rait (2000). The minimum distance over which the Matakaoa Volcanics were transported was estimated at ~280 km. Rait further suggested that tectonic transport decreased to the southwest, and was only 1–2 km at its southwestern limit. In the middle parts of the allochthon a minimum transport distance of 42 km can be inferred for the East Coast Allochthon.

In the East Coast region, deposition of limestone, marl and greensand, indicative of tectonic quiescence, persisted into the latest Oligocene (early Waitakian). The earliest indication of uplift and erosion occurs at the Oligocene–Miocene boundary (mid-Waitakian) and led to the allochthon being emplaced during a series of events lasting from early to late Early Miocene (mid-Waitakian to early Altonian) times. Throughout the central zone, early Early Miocene (late Waitakian–early Otaian) breccias, folds and thrusts are overlain by undeformed late Early Miocene (lower Altonian) strata (Field *et al.* 1997). Debris from the advancing allochthon was deposited in adjacent slope basins that were subsequently incorporated into the advancing thrust complex.

8.2.2 East Coast Basin

The Early–Middle Miocene succession of the North Island portion of the basin consists mainly of bathyal mudstone and flysch with local developments of conglomerate, neritic sandstone and limestone (see Fig. 6.1a,b). Early Miocene sediments are not present in the western Wairarapa, where Middle Miocene and younger sediments commonly rest on Mesozoic basement rocks. The equivalent succession in Marlborough is dominated by bathyal mudstone and mass-flow conglomerate. The interpretation of the rocks in individual piggyback basins, probably arranged en echelon, is difficult and further complicated by later strike-slip faulting.

The Miocene strata of the Raukumara Peninsula are included in the Tolaga Group (Fig. 8.1a). The earliest autochthonous strata of the group are early Early Miocene (late Waitakian–Otaian) in northeastern and central Raukumara Peninsula. They commonly comprise a coarse basal breccia and conglomerate up to 150 m thick, or shallow-marine bioclastic, sandy or muddy limestone with intercalations of breccia resting unconformably on allochthonous Cretaceous strata. These breccia-bearing units are considered to be synorogenic, deposited in piggyback basins during and immediately following emplacement of the East Coast Allochthon (Mazengarb & Speden, 2000). Basal greensands occur locally, extending into southern Raukumara Peninsula, where they appear to lie disconformably on latest Oligocene (early Waitakian) Weber Formation (Field *et al.* 1997). In central Raukumara Peninsula the basal beds of the Tolaga Group pass upwards into ~170 m of early Early Miocene (lower Otaian) mudstone. Locally, the mudstones are cut by 30-m-deep channels filled with thick sandstones that are inferred to record sedimentation in front of an advancing nappe. In the south, early Early Miocene (Waitakian–Otaian) sediments of the Mahia Peninsula area are dominated by bathyal mudstone, locally containing packets of alternating sandstone and mudstone (Field *et al.* 1997).

The late Early Miocene (Altonian) portion of the Tolaga Group in the Raukumara Peninsula consists mainly of bathyal mudstone and alternating mudstone and sandstone. In central Raukumara Peninsula it is relatively thin (<150 m; Mazengarb *et al.* 1991), but it thickens rapidly to the east where over 650 m of mainly bathyal mudstone cut by coarse channel-fill deposits passes laterally into over 2000 m of mainly shelf sediments (Stoneley, 1968). The rapid subsidence (over 900 m/m.y.) suggests active tectonism. Locally limestones occur that are up to 150 m thick and of shallow marine origin. To the north and west of Gisborne, late Early Miocene (Altonian) cross-bedded and locally glauconitic sandstone and sandy siltstone rest unconformably on Oligocene bathyal siltstone, or locally on Eocene rocks (Strong *et al.* 1993; Field *et al.* 1997). Inland to the west, near Lake Waikaremoana, Altonian sediments appear to be absent. In southern Raukumara Peninsula, bathyal glauconitic boulder–pebble conglomerate of Altonian age locally overlies mudstone in an inferred channel-fill system (Field *et al.* 1997).

In the East Coast Allochthon, by contrast, the earliest Miocene rocks in northern Raukumara Peninsula consist mainly of bathyal mudstone deposits of early Early Miocene (late Waitakian to Otaian) age. Thickness of the deposits, which locally rest on Oligocene limestone, reaches a maximum of ~270 m (Pick, 1962). The Early Miocene succession is overlain by up to 400 m of late Early Miocene (Altonian) mudstone, deposited in outer-shelf to bathyal depths (Mazengarb *et al.* 1991). Further south, however, in eastern central Raukumara Peninsula, the Early Miocene mudstone passes into 500 m of mudstone and flysch.

In the Raukumara Peninsula, coarse deposits are most common in the early Middle Miocene (Clifdenian) portion (Fig. 8.1a) of the Tolaga Group that may correspond to a relative sea-level lowstand (Field *et al.* 1997). No sediments of this age have been recorded in the far north of the peninsula, but southwest of Ruatoria, the early Middle Miocene is 700–760 m thick and includes at least two bathyal channel-fill deposits (Field & Scott, 1993). These consist generally of conglomeratic and gritty sandstone at the base, passing up-section through turbidites into mudstone. Sparse paleocurrent records indicate flow to the southeast or northwest. About 600 m or more of equivalent-aged mudstone occurs northwest of Gisborne, but the succession appears to thin towards the east. At Lake Waikaremoana where the Early Miocene is absent, the earliest Middle Miocene consists of about 400 m of neritic sandstone and limestone with bathyal mudstone and flysch that thickens towards the west. Here 600–800 m of Clifdenian sandstone was deposited in mainly shelf conditions. This succession points to rapid subsidence with high rates of sedimentation, and the proximity of this depocentre to major faults of the present axial ranges suggests contemporaneous activity on these faults.

Although Middle Miocene (Lillburnian) strata of the Tolaga Group are not extensive in the far north of the Raukumara Peninsula, they reach over 1000 m in parts of the central and southern peninsula. These are some of the thickest Lillburnian successions known from the region, and comprise mainly bathyal mudstone or alternating sandstone and mudstone, the lower portion consisting in part of thin-bedded turbidites. Locally, by contrast, 100–200 m of bathyal mudstone of upper Lillburnian age rests unconformably on possible Cretaceous rocks. Further south the early Middle Miocene includes limestone and sandstone, and, locally, glauconitic siltstone (Moore, 1957).

In parts of eastern central Raukumara Peninsula, up to 10 m of Lillburnian micritic, sulphurous and macrofossiliferous limestone forms lenticular bodies up to 400 m in extent. It appears to represent in situ bathyal biocommunities resembling bioherms, and may have been associated with sea-floor vents and/or deep-sea condensed sections (Mazengarb *et al.* 1991; Field *et al.* 1997; Campbell *et al.* 2008). Further southwest towards Gisborne, at least 400–500 m, and possibly over 1500 m, of dark blue-grey probably bathyal Lillburnian mudstone occurs, although, in the vicinity of Gisborne itself, alternating sandstone and mudstone is exposed (Ridd, 1964, 1967; Hoskins, 1978). Paleocurrent data suggest transport towards the north to northeast. Locally, an unconformity separates the Lillburnian mudstone from overlying late Middle Miocene (Waiauan) sediments.

The southwest of the peninsula, the region of Lake Waikaremoana, was an area of mainly turbidite deposition during the mid-Middle Miocene (late Lillburnian). Up to 1200 m of these sediments were deposited in bathyal settings, with paleoflow to the northeast in the north, and to the southeast in the south. The great thickness of strata and apparently continuous deep-water deposition suggest high rates of both sedimentation and subsidence.

Little of the Early–Middle Miocene is preserved in western Hawke's Bay and Wairarapa. The oldest Neogene unit exposed in western Hawke's Bay, near Mohaka, is a 350–750-m-thick Whakamarino Formation of late Early Miocene (mid-Otaian to Altonian) age consisting of bathyal mudstone and siltstone with sparse thin interbeds of fine sandstone (Cutten, 1994). The base of the unit is not exposed in outcrop, but it rests unconformably on Cretaceous rocks at petroleum well Te Hoe-1 (Morgans *et al.* 1990), suggesting that lower Otaian strata are not preserved in this area. In the northeast near Wairoa, bathyal late Early Miocene (Altonian) faunas occur in mudstone and alternating sandstone and mudstone. The upper Altonian appears to have been partly or wholly removed at an unconformity beneath the Middle Miocene (Cutten, 1994). The preserved sedimentary thickness varies, but it reaches 300 m in places. Further to the southwest, Altonian rocks have not been preserved.

Earliest Miocene (late Waitakian) strata are, however, preserved in the coastal ranges from near Hastings in southern Hawke's Bay to Castlepoint. Near Makara, mélange zones and debris flows were developed, with condensed successions of mudstones on basin highs, and Early Miocene flysch (Puhokio Formation) in an adjacent basin that was fed longitudinally from the north (Pettinga, 1982, 1990).

The middle Early Miocene (Otaian) record is discontinuous south of Napier, and over most of the Napier–Dannevirke area Otaian strata have not been positively identified. An exception is a local

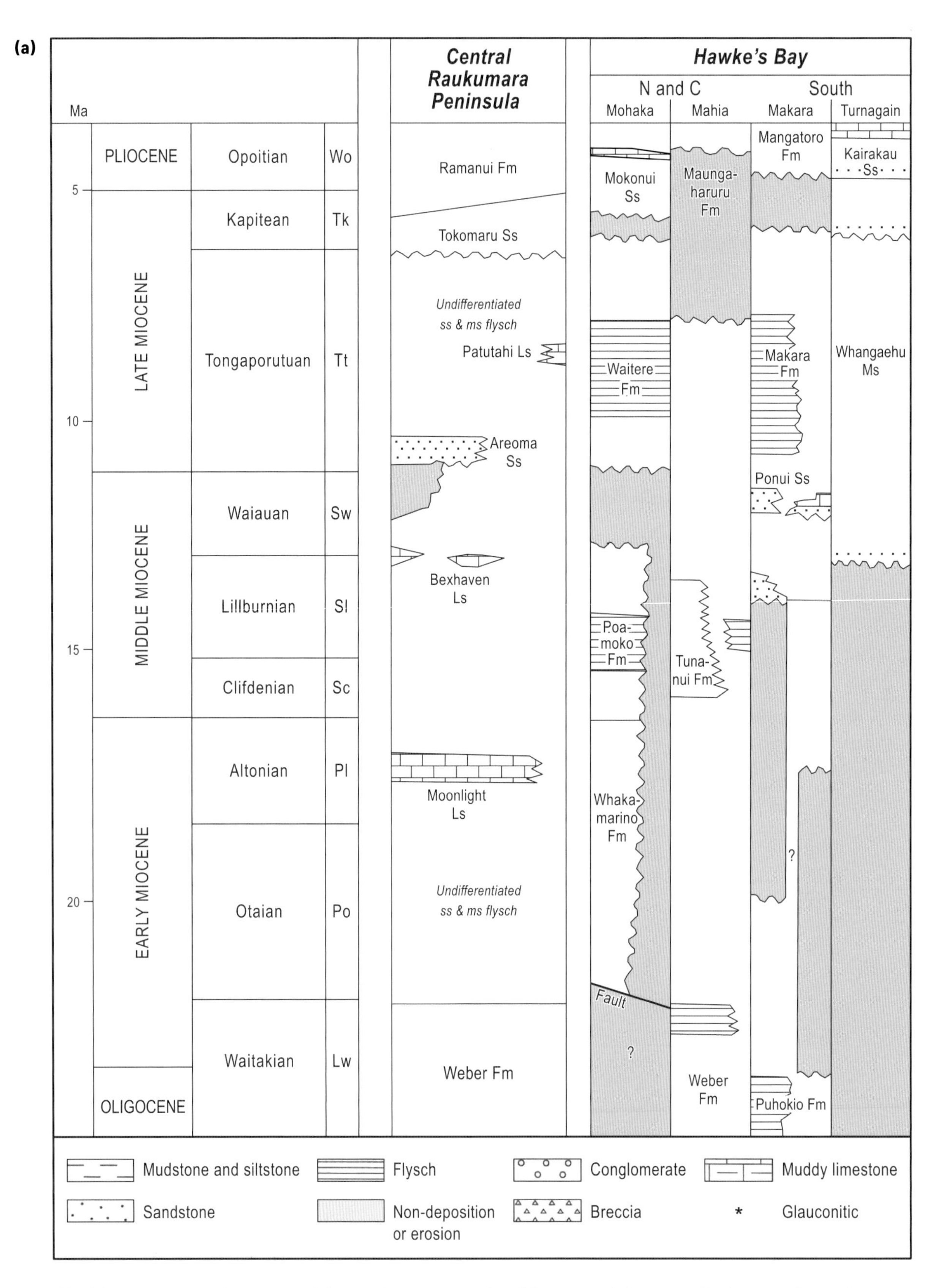

Fig. 8.1. Time-stratigraphic section of the mainly Miocene rocks of the supra-subduction wedge between the Raukumara Peninsula and Hawke's Bay (a), and Wairarapa and Marlborough (b). Cng, Conglomerate; Fm, Formation; Gnsd, Greensand; Ls, Limestone; Mbr, Member; Ms, Mudstone; Ss, Sandstone. (Adapted from Field *et al.* 1997.)

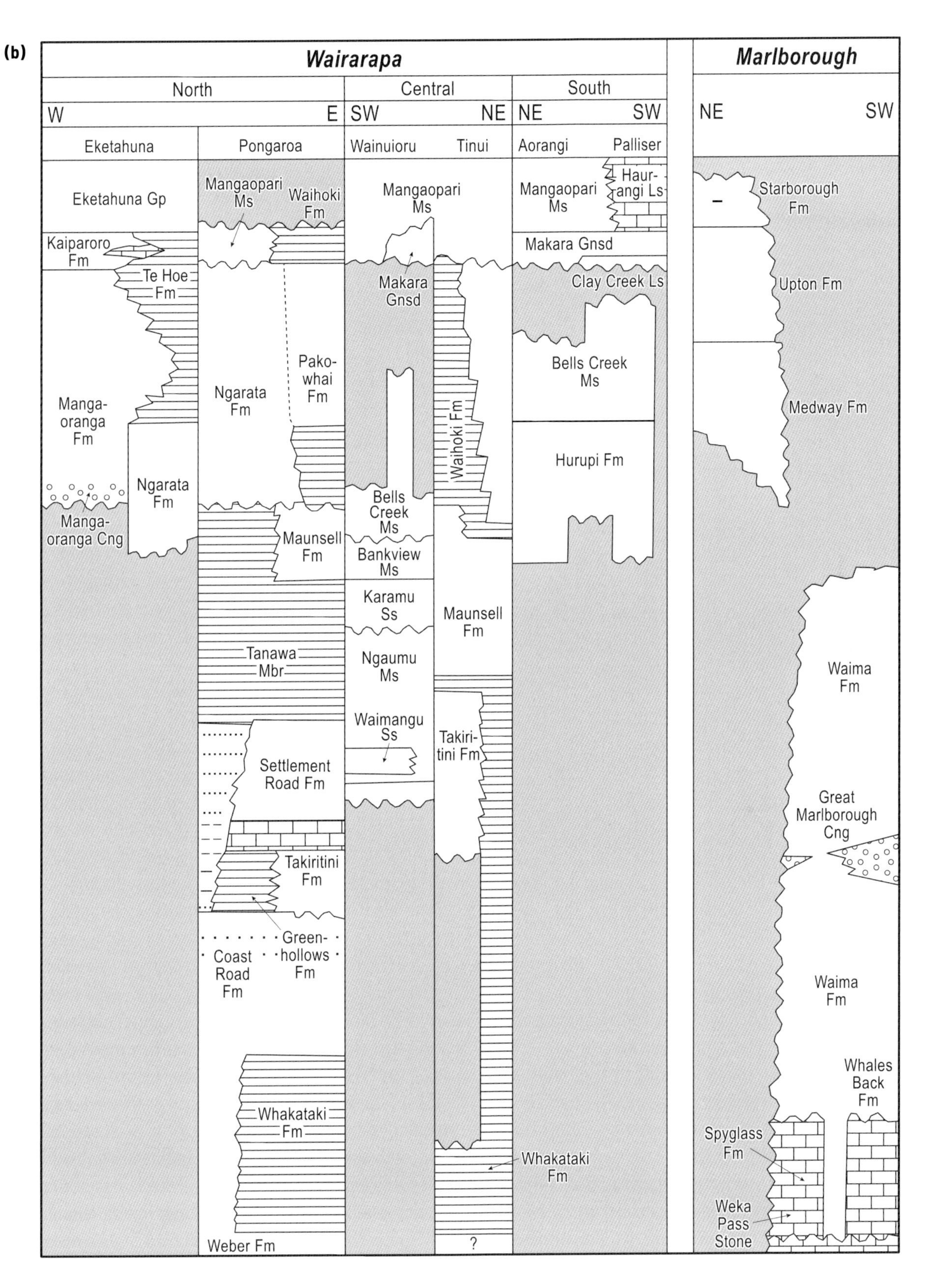
(b)
Wairarapa
North
Central
South
W
E
SW
NE
NE
SW
Eketahuna
Pongaroa
Wainuioru
Tinui
Aorangi
Palliser
Eketahuna Gp
Mangaopari Ms
Waihoki Fm
Mangaopari Ms
Mangaopari Ms
Haur-angi Ls
Kaiparoro Fm
Te Hoe Fm
Makara Gnsd
Makara Gnsd
Clay Creek Ls
Pako-whai Fm
Ngarata Fm
Manga-oranga Fm
Waihoki Fm
Bells Creek Ms
Hurupi Fm
Ngarata Fm
Manga-oranga Cng
Bells Creek Ms
Maunsell Fm
Bankview Ms
Karamu Ss
Maunsell Fm
Tanawa Mbr
Ngaumu Ms
Waimangu Ss
Takiri-tini Fm
Settlement Road Fm
Takiritini Fm
Green-hollows Fm
Coast Road Fm
Whakataki Fm
Whakataki Fm
Weber Fm
?
Marlborough
NE
SW
Starborough Fm
Upton Fm
Medway Fm
Waima Fm
Great Marlborough Cng
Waima Fm
Whales Back Fm
Spyglass Fm
Weka Pass Stone

Fig. 8.1. (b)

occurrence, south of Hastings, where ~50 m of mid- to lower-bathyal calcareous mudstone and alternating sandstone and mudstone is present (Lillie, 1953). East of Dannevirke there are earliest Miocene (Waitakian) calcareous mudstones, with limestones in places containing a shallow marine fauna that indicates shelf conditions. Locally, these Waitakian and overlying Otaian rocks are absent, and late Early Miocene (Altonian) mudstone rests on Mesozoic accretionary basement.

Early–Middle Miocene strata are not found west of the Wairarapa Fault, in western Wairarapa, but are widely distributed in central and eastern Wairarapa, where they are divided by a major structural feature – the Adams-Tinui Fault, which was active during the Early Miocene (Field *et al.* 1997). Alternating sandstone and mudstone of the Early–earliest Middle Miocene (late Waitakian–Clifdenian) Whakataki Formation are distributed along the eastern Wairarapa coast in a discontinuous >100-km-long belt from Cape Turnagain to the Pahaoa River mouth (Lee & Begg, 2002). Scattered outcrops of the formation are also present inland in central Wairarapa, on the western side of the Adams-Tinui Fault, as far west as Pongaroa. The Whakatiki Formation is also represented in the offshore Titihaoa-1 petroleum well, where high gas levels were recorded (Biros *et al.* 1995). The succession may be up to 1500 m thick in coastal areas, but is probably only several hundred metres thick in central Wairarapa. It is locally conformable on Late Oligocene strata, but in most places is unconformable on older rocks (Field *et al.* 1997; Lee & Begg, 2002).

Whakataki Formation (Fig. 8.1b) is characterised by centimetre- to metre-bedded alternating sandstone and mudstone, the sandstone beds displaying Bouma sequences typical of turbidite deposition. Minor lithologies include bioclastic (algal) limestone, and matrix-supported pebble conglomerate, particularly near the base of the formation (Johnston, 1980). Channel-fill facies, consisting of metre-thick, normally graded packets of amalgamated beds of sandstone, with minor conglomerate, are locally present (Field *et al.* 1997). Matrix-supported polymict breccias up to several hundred metres thick are intercalated with basal or near-basal alternating sandstone in coastal areas and in places inland. Clasts are typically centimetres to metres in diameter, but blocks and megaclasts up to tens of metres, and occasionally up to several hundred metres, in size occur. Clasts and blocks are of mainly Cretaceous and Paleogene strata. The coarse deposits are interpreted as olistostromes and debris flows (Neef, 1995; Delteil *et al.* 1996). Sedimentology, trace fossils, shell fragments and microfauna suggest deposition at upper bathyal depths (Crundwell, 1997). Paleocurrents indicate a northeast to east-northeast-directed flow.

To the west, at Pongaroa, the alternating sandstone and mudstone of the Whakataki Formation passes into successions dominated by bathyal mudstone. In the central Wairarapa area the Whakataki Formation thins, and is probably only several hundred metres thick (Delteil, 1992). A unit consisting of medium- to coarse-grained calcareous sandstone, algal limestone and fossiliferous mudstone (Takiritini Formation) is recognised around Tinui and Pongaroa in central Wairarapa, where it unconformably overlies Whakatiki Formation or older Cenozoic or basement rocks. The thickness of the formation is at least 640 m, and may approach 1000 m locally. Sparse fossils indicate a late Early Miocene (mid-Altonian) age.

Most of the remaining Early Miocene sedimentary rocks are restricted to central Wairarapa. Typical lithologies are alternating mudstone and graded-bedded sandstone, massive concretionary mudstone and massive very fine-grained sandstone. Mapping indicates thicknesses between 300 and 800 m, and foraminifera and macrofauna suggest deposition at shelf to mid-bathyal depths (Lee & Begg, 2002).

The early Middle Miocene succession appears to be thinner in Hawke's Bay and Wairarapa than at the Raukumara Peninsula. In northern Hawke's Bay, in the Mahia Peninsula area, up to 800 m of early Middle Miocene rocks, consisting of probably bathyal mudstone and sandstone, was recorded (Field *et al.* 1997). The succession passes westwards into bathyal alternating sandstone and mudstone near Mohaka, and then further west into 150 m of fossiliferous sandstone with interbeds of siltstone representing outer-shelf conditions.

Immediately south of Napier the basal Middle Miocene (Clifdenian) is thin or absent, comprising calcareous, glauconitic, shelly medium sandstones unconformably overlying Paleocene (Teurian) rocks (Kingma, 1971). The sediments thicken to the south into the Makara Basin, which started to form at this time (Pettinga, 1982). The sediments are generally only about 20 m thick and of restricted distribution in the north, but they reach about 2000 m thick in the middle of the depocentre (Kingma, 1971). Lithofacies include glauconitic, sandy siltstone, calcareous mudstone and alternating sandstones and mudstones. Locally, early Middle Miocene shelf or upper bathyal sandstone rests unconformably on bathyal middle Early Miocene mudstone.

In the Pongaroa–Tinui area, while basal Middle Miocene (Clifdenian) deposits are either thin or have not been recorded, the overlying Lillburnian succession is possibly more than 500 m thick, and comprises mainly bathyal alternating sandstone and mudstone (Johnston, 1980; Neef, 1992, 1997). Locally, water depths were shallow enough to allow shell banks to form, but turbidites, deposited from currents flowing to the north-northeast and west-northwest, are dominant (Johnston, 1980).

Further south, in the central Wairarapa southeast of Masterton, Early Miocene and older Tertiary strata are absent, and early Middle Miocene sediments rest directly on Cretaceous rocks. A basal transgressive inner-shelf deposit (Fig. 8.1b) consists of up to 300 m of bedded fine to medium sandstone fining up to sandy siltstone, locally with *Ostrea* near the base, disseminated carbonaceous material, and scour structures, and grading up into 700 m of overlying middle- to outer-shelf mudstone (Ngaumu Mudstone), which extends in age into the Lillburnian. This basal unit thins to the west and to the northeast where it rests unconformably, with a basal conglomerate 60–70 m thick, on Cretaceous rocks (Crundwell, 1997). To the southeast, west of Flat Point, it passes into 330 m of alternating sandstone and mudstone (Van den Heuvel, 1960) and locally into trough cross-bedded inner-shelf sandstone (Crundwell, 1997). Contemporary uplift along the adjacent Adams-Tinui Fault is inferred to have occurred at this time (Moore, 1980). To the west of the Adams-Tinui Fault near Flat Point, early Middle Miocene (Lillburnian) sediments rest directly on Cretaceous–Paleogene rocks (Field *et al.* 1997).

Early–early Middle Miocene rocks are not present in southern Wairarapa south of Flat Point. They are also absent in northern Marlborough, the result of movements on the Marlborough Fault System and regional uplift in northern Marlborough in the Early–Middle Miocene (Browne, 1995b).

In southern Marlborough, and extending into the northern Canterbury Basin (see Figs 6.2, 6.5 and 6.6), the early Neogene succession ranges in age from possibly latest Oligocene to mid-Middle Miocene (early Waitakian to Lillburnian). Latest Oligocene–Early Miocene rocks consist of predominantly fine-grained limestone, calcareous mudstone and calcareous sandstone that rest disconformably, or with slight angular unconformity, on the micritic Late Eocene Amuri Limestone, or conformably on Oligocene basalt of the Cookson Volcanics (Browne, 1995b). Two depocentres are recognised during the Early–late Middle Miocene of southern Marlborough. The western depocentre contains slightly coarser-grained rocks than the eastern depocentre, which may reflect an eastward trend towards increasing water depths. These depocentres may have been separated by a northeast-trending paleogeographic high (Browne, 1995b).

The western depocentre comprises calcareous lithofacies deposited in shelf to bathyal depths. The basal unit (Weka Pass Stone) is a creamy grey, very calcareous, well-sorted, very fine- to fine-grained, bioturbated, glauconitic sandstone up to 90 m in thickness, and lithologically similar to, and contemporaneous with, calcareous sandstones of the same name deposited further to the southwest in the Canterbury Basin (see section 8.2.3). The formation rests conformably on basaltic and volcaniclastic sediments of the Cookson Volcanics (Reay, 1993), or, more commonly, on a highly glauconitic, burrowed omission surface (with abundant *Thalassinoides*), correlated with the Marshall Unconformity (see section 9.3.5). In Marlborough, strata above and below the omission surface are concordant, or show mild angular discordance. Locally, karsting is apparent below the surface, indicating emergence followed by submergence after karstification (Browne, 1995b). The overlying Whales Back Limestone (Fig. 8.1b) is an alternating creamy grey, decimetre-bedded, glauconitic, bioturbated wackestone, and calcareous mudstone. Thickness of the unit is up to 90 m, and both it and the Weka Pass Stone are of latest Oligocene or earliest Miocene (early Waitakian) age (Browne, 1995b). The equivalent beds in the eastern depocentre consist dominantly of stylobedded wackestone Spyglass Formation, (see Plate 12) reaching up to ~100 m in thickness, deposited in a mid-bathyal environment (Warren, 1995).

Interbedded with the wackestone are well-sorted very fine- to medium-grained glauconitic sandstones with comminuted carbonaceous debris. They thicken towards the west. The Spyglass Formation rests on the Marshall Unconformity, the contact being marked by high concentrations of glauconite and by phosphatised pebbles of Amuri Limestone.

The bulk of the succession overlying the basal calcareous units in both depocentres comprises pale blue-grey, calcareous, micaceous, fine- to medium-grained sandy bioturbated siltstone (Waima Formation), ranging in age from Early–Middle Miocene (late Waitakian–Lillburnian). The formation is up to 360 m thick (see Plate 14b), foraminifera indicating outer-shelf to bathyal settings throughout the formation (Reay, 1993). The Waima Formation forms the background sediment into which thick conglomerate units (Great Marlborough Conglomerate), locally associated with sandstones, were deposited as debris flows, mainly during the late Early Miocene (Altonian) (Lewis *et al.* 1980). The conglomerate lithofacies consists of angular to well-rounded Torlesse Terrane-derived greywacke clasts, with lesser, variable proportions of Amuri Limestone and alkaline basalts, at least some of which have been derived from the Cretaceous Lookout Volcanics, now exposed in the Awatere Valley (Warner, 1990). Clast size varies from granule-to-pebble (common) to rare megaclasts up to 30 m by 100 m (Prebble, 1980). Conglomerate units are lenticular over large strike distances, and can be seen in some instances to infill broad channels. Paleocurrent indicators show a transport direction towards the southeast. The influx of Great Marlborough Conglomerate marks the first regionally significant uplift and erosion of fault blocks in the Marlborough region (Lewis *et al.* 1980; see Plate 15).

8.2.3 Canterbury, Chatham Rise and Great South basins

The Early–Middle Miocene sedimentary basin of southern Marlborough extended south into the northern portion of the northern Canterbury Basin (see Figs 4.2 and 4.6). The Spyglass Formation thins and finally disappears to the south, its base here being of Late Oligocene (Duntroonian to Waitakian) age (Browne & Field, 1985). As in Marlborough, a burrowed omission surface (Marshall Unconformity) separates the formation from the underlying Amuri Limestone (Lewis, 1992). To the south, Weka Pass Stone interfingers with and is finally replaced by a greensand member of the Omihi Formation (see Figs 6.2, 6.5 and 6.6; Plate 9).

A shallow submarine 'high' extending from Chatham Rise probably existed near Cheviot during most of the Late Oligocene (Duntroonian–Waitakian), effectively separating areas of siltstone and limestone to the north from greensand packstone and siltstone (Omihi Formation) to the south. The high was onlapped by sandy siltstones in the earliest Miocene (late Waitakian) (Andrews, 1968).

South of the 'high', in southeastern north Canterbury, the oldest Miocene (Otaian) sediments are glauconitic and rest on an unconformity of earliest Miocene (late Waitakian) age. The sediments were dominated by a succession up to 520 m thick of tidal and inner neritic calcareous siltstone and fine sandstone (Waikari Formation, Pahau and Scargill siltstone members; see Fig. 6.5), which passed basinwards into glauconitic sandy siltstone (Andrews, 1968). They are the equivalent of the Spyglass and Waima formations to the north of the 'high'. The Waikari Formation is overlain by a sandstone and limestone succession (Mount Brown Formation), up to 830 m thick, which locally includes a probably tectonically induced calcareous cobbly mass-flow deposit, and over 100 m of sandy siltstone, sandstone and bioclastic limestone. This in turn is overlain by late Early Miocene (Otaian–Altonian) siltstone in the south (Wilson, 1963; McCulloch, 1981). In the late Early Miocene (Altonian), mainly middle to outer neritic sandstone, conglomerate, grainstone and packstone, and calcareous fine sandy siltstone, were deposited to the east, with inner-neritic coquina shellbeds and medium-grained sandstone dominating in the west.

Earliest to middle Early Miocene (Otaian) sediments may not have been deposited in central Canterbury (see Fig. 6.6). The Oligocene, where preserved, is overlain unconformably by late Early Miocene (Altonian) coastal sand and phosphatic greensand. The unconformity is locally angular and is in places a paleokarst surface (van der Lingen *et al.* 1978; Field, 1985). It is associated with pebbles of basalt that record erosion and reworking of

Oligocene volcanics or volcaniclastites. Erosion, probably associated with faulting, at Banks Peninsula left significant topographic relief and locally exposed Torlesse basement. Onlap onto this erosion surface in late Early Miocene (Altonian) times continued into the early Middle Miocene in the northwest as paralic-estuarine deposits (Brechin Formation). In the far west of central Canterbury the earliest Miocene (Waitakian) is marked by a return to siliciclastic sedimentation, with the accumulation of a prograding wedge of sandstone (Swin Sandstone).

On the Chatham Islands, other than an unnamed limestone and a tuff of late Early Miocene (Altonian) age, there is no evidence for post-Paleogene deposition until the beginning of the Pliocene (Wood *et al.* 1989; and see Fig. 4.2). The fossiliferous Altonian limestone occurs as loose blocks of glauconitic, pale-yellow, bryozoan–foraminfera–echinoderm packstone of variable grain size, indicating deposition in an outer-shelf or deeper environment. The unit is apparently less than 5 m thick (Campbell *et al.* 1993). Phosphoritic limestone and marl of similar age have also been dredged from the Chatham Rise (Wood *et al.* 1989). A thin fossiliferous tuff of similar age, indicating deposition at shelf depths, has also been recorded from a single borehole locality in northern Chatham Island (Campbell *et al.* 1993).

Over most of the southern Canterbury shelf basin, deposition of grainstone and packstone (Otekaike Limestone) continued into the Miocene, and sandstone with lignite (Wedderburn Formation) records terrestrial settings in western north Otago. Lewis and Belliss (1984) documented a paleokarst surface close to the Oligocene–Miocene boundary (intra-Waitakian) within the Otekaike Limestone in this region, and deduced that it was at least partly due to deformation. In eastern north Otago and at Endeavour-1 and Galleon-1 petroleum wells the basal Miocene sediments are apparently absent (see Fig. 6.2), presumably because of the presence of a 'high' (Endeavour High) in the coastal area. To the east of the 'high', at Clipper-1 well, deposition of limestone appears to have continued from the Oligocene into the Miocene (see Fig. 6.2).

In western south Canterbury the Early Miocene (Otaian) was marked by cross-bedded shallow shelf sand (Southburn Sand), which prograded eastwards over silty fine sandstone and siltstone (Tokama Siltstone), extending to eastern onshore Canterbury by late Early Miocene (Altonian) times. East of the present coastline, at Clipper-1 petroleum well, 100 m of Middle–Early Miocene (Otaian) calcareous sandstone and siltstone occurs, but no late Early Miocene or early Middle Miocene (Altonian–Clifdenian) sediment was recorded, suggesting the presence of an unconformity (Eaton in Crux *et al.*, 1984). Similar Early Miocene sediments occur at Galleon-1 petroleum well, with a similar truncating unconformity overlain by 4 m of late Middle Miocene (Waiauan) sandstone. The sedimentary record appears complete, however, at petroleum wells Endeavour-1 and Resolution-1. At Endeavour-1, Altonian siltstone overlying limestone corresponds to the basal part of the prograding wedge of shelf sediments that characterises the Miocene above an Early–Middle Miocene seismic horizon. At DSDP site 594, nannofossil ooze (now chalk) was deposited apparently without break from the Early to Middle Miocene (Nelson *et al.* 1986b).

In the Oamaru to Dunedin area, where earliest Miocene erosion was sufficient to remove any Oligocene sediment, the Waitakian unconformity is overlain by phosphatic, highly glauconitic greensands (Gee and Concord greensands), which suggests that the area was starved of sediment and that the Endeavour High might have extended as far south as Dunedin in the Early Miocene (see Fig. 6.4). These Early Miocene deposits form an onshore extension of the Great South and southern Canterbury basins, and are almost entirely restricted to a thin coastal strip about 50 km long from Shag Point to south of Dunedin (see Fig. 6.4). The Waitakian greensands are overlain by a younger Early Miocene (Otaian–Altonian) assemblage of marine terrigenous–carbonate sandstone and shelly mudstone, and impure limestone facies interpreted as a regressive succession. The most extensive unit (Caversham Formation), which reaches a thickness of ~250 m, is composed mainly of fine to medium quartzo-feldspathic calcite-cemented sandstone. Bedding is poorly defined by differentially cemented intervals that contain vestiges of *Thalassinoides*-like trace fossils, and by rare scour fills (Lindqvist, 1995). To the north of Dunedin the sandstone is sharply overlain by silty limestone alternating with shelly siltstone (Goodwood Limestone), considered to be a deeper-water, offshore facies of late Early Miocene (Altonian) age. The unit is commonly cross-bedded and slumped, and includes large-scale channel features (Lindqvist, 1995). A similar and probably correlative limestone (Dowling Bay Limestone) crops out near the entrance to Otago Harbour.

South of Dunedin near Milton, a small area of earliest Miocene (Waitakian) Milburn Limestone, up to 35 m thick, lies disconformably on late Middle Eocene strata. It is glauconitic at the base, passing up into pure, white, flaggy crystalline limestone, and then into increasingly terrigenous, quartz-rich material. A karstic surface caps the formation, with solution pits up to 25 m deep, infilled by overlying phosphatic and quartz-rich sandstone (Clarendon Sandstone) of similar Waitakian age. The sandstone is inferred to have been formed as a residual deposit by dissolution of the upper portion of the Milburn Limestone (Bishop, 1994).

Uplift in southeastern Otago caused extensive erosion in Altonian times, and was followed locally by the eruption of the late Early Miocene Dunedin Volcanic Group (Sikumbang, 1978; see section 8.7.1). In offshore petroleum well Takapu-lA, 30 km south of Dunedin, the only Miocene lithology sampled was Altonian ?mid-shelf siltstone (Hunt Petroleum, 1978). This unit, which could include unrecognised sediments of Otaian age, is inferred to rest unconformably on Eocene strata.

By the late Early Miocene, non-marine to marginal marine coal measures (Wedderburn Formation; White Rock Coal Measures) were being deposited in north Otago and in western south Canterbury (Bishop, 1974a; Field & Pocknall, 1984). In eastern Central Otago the Early–early Middle Miocene succession comprises quartzose fluvial sediments and lake deposits of the Manuherikia Group, reaching a thickness of 1000 m. The fluvial sediments (Dunstan Formation) were divided by Douglas (1986) into five complexly interrelated lithofacies representing braided river, alluvial plain, anastomosing river, delta plain and marginal lacustrine environments. Prominent coal seams between 33 and 90 m thick accumulated as peats in long-lived backswamps of the upper delta plain–lower alluvial plain environments. The fluvial deposits interfinger with, and are overlain by, largely fine-grained sediments (Bannockburn Formation) comprising massive or cross-stratified fine to medium sand interbedded with massive mudstone that contains a variegated freshwater fauna (Douglas, 1986), and, locally, stromatolites (Lindqvist, 1994). The Bannockburn Formation is inferred to represent deposits of a lake, which progressively onlapped older lake-margin and alluvial plain strata, depositing a thick (~700 m) pile of sandy and muddy sediments, blanketing an area in excess of 5600 km^2. The Manuherikia Group lies below a basalt flow with a K/Ar age of 13.4±0.3 Ma, i.e., mid-Middle Miocene (Youngson *et al.* 1998), and the palynology is compatible with a Middle Miocene age (Pocknall & Turnbull, 1989).

In the Great South Basin, Miocene sediments were included in the Penrod Group by Cook *et al.* (1999; see also Fig. 4.6). The unit was poorly sampled in exploration wells, and information on the group's stratigraphy and age is sparse. Miocene sediments commonly show different seismic characteristics from older strata, and in the northern part of the basin indicate a greater clastic input. Along the northwestern margin of the basin, increased sediment supply during Miocene to Recent times formed a thick (up to 800 m) shelf wedge characterised by southeastwards-prograding clinoform strata. Foraminiferal limestone accumulated in southeastern parts of the basin, reworked by a strong northeasterly current that caused a complex pattern of bedforms.

8.2.4 Southland basins

Early Miocene sediments form a substantial portion of the Eastern Southland, Winton and western Southland sedimentary basins (see Figs 6.9–6.11). In the Eastern Southland Basin, east of the Mataura River, the Miocene beds onlap Triassic–Jurassic Murihiku basement, and in the west they are separated by a low basement sill from deposits in the adjacent Winton Basin. Deposits are regressive, shallowing upwards from shallow marine to non-marine.

The Winton Basin, which started to develop in the Eocene, contains calcareous marine Early Miocene sediments that are inferred to interfinger with those in the Eastern Southland Basin. The basin is separated from the Waiau and Te Anau basins in the west by Mesozoic and Permian metavolcanics and intrusives of the Longwood and Takitimu ranges, although the

Ohai Depression provides a narrow connecting corridor with the Waiau Basin.

Miocene sediments are exposed in the Waiau and Te Anau basins and the onshore sector of the Waitutu Sub-basin, but not in the onshore Balleny Basin. Offshore they extend throughout the Solander and Balleny basins. Miocene outcrops are now restricted onshore compared with underlying Oligocene sediments as a result of Pliocene uplift and erosion, but their original extent was probably as great as that of the Oligocene sequence.

Miocene sedimentation occurred in two phases: an Early–early Middle Miocene continuation of Oligocene regional subsidence, with predominantly mudstone deposition; and a late Middle–Late Miocene phase marked by increasing tectonic activity and uplift that led to erosion in several parts of the region and an end to marine sedimentation in the Te Anau Basin. This activity produced a wider range of types of sediments and is inferred to be a result of changing motion on the Alpine Fault, from strike-slip to oblique compressive slip (Norris & Carter, 1980, 1982). Throughout Miocene times the Moonlight Fault System strongly influenced sedimentation, producing very thick successions in the western Waiau Basin and locally in the Burwood and Waitutu sub-basins.

The lithostratigraphy of the Early–early Middle Miocene Waiau Group in the Waiau and Te Anau basins resembles that of the Oligocene. Deposition of the basin-wide Waicoe Formation mudstone (see Chapter 6), with minor units of sandstone or conglomerate, continued through the Miocene. Lateral thickness and facies changes are common. Shelf conditions persisted in the east of both basins, but the western shelf sediments from the Waitutu Sub-basin to the Te Anau area were buried by deep marine Waicoe Formation mudstone and turbidites. The Monowai Sub-basin contains a distinct succession of units not recognised further south. The central Waitutu Sub-basin, although dominated by Waicoe Formation mudstone, also includes several distinctive units.

8.2.4.1 Eastern Southland Basin

This basin has a sedimentary fill, often less than 300 m thick, consisting of shallow marine and non-marine deposits of Late Oligocene–?Middle Miocene age resting unconformably on Permian–Jurassic basement. The Early Miocene succession consists of the upper part of the Chatton Formation, and the bulk of the Gore Lignite Measures (see Fig. 6.7).

As noted earlier (Chapter 6) the Chatton Formation consists of shallow-marine, often calcareous and glauconitic fossiliferous sandstone and conglomerate near the base of the sedimentary succession. Locally it consists of sandy limestone or calcareous sandstone, and it reaches a maximum thickness of 145 m. Only the upper part of the unit is of Early Miocene age, the youngest foraminifera giving a late Early Miocene (early to mid-Altonian) age (Isaac & Lindqvist, 1990).

The Gore Lignite Measures interfinger with and overlie the Chatton Formation. They consist of carbonaceous and non-carbonaceous mudstone, sandstone and conglomerate with lignite seams up to 18 m in thickness. They are overlain unconformably by Quaternary gravels. Upward-fining units are common, as is channelling. Thickness is uncertain, but composite stratigraphic columns indicate a thickness of at least 500 m over much of the basin (Isaac & Lindqvist, 1990). The known age range is Late Oligocene–Early Miocene (Duntroonian–Altonian), but the unit may extend into the Middle Miocene.

The Chatton Formation represents a transgression in response either to subsidence or sea-level rise. It is time-transgressive, ranging in age from Late Oligocene (Duntroonian) in the north to Early Miocene (Altonian) in the south. The Gore Lignite Measures were mainly deposited in a range of fluvial channel, overbank splay, floodplain and swamp environments, indicated by the presence of in situ terrestrial plant remains, lateral persistence of the multiple coal seams, common root-penetrated seat earths, large-scale upwards-fining clastic sequences typical of fluvial cyclothems, and the paucity of marine fauna. The depositional setting was a prograding deltaic plain, which advanced across a shallow marine Chatton shelf during Late Oligocene to Early Miocene times (Isaac & Lindqvist, 1990).

8.2.4.2 Winton Basin

In the Early Miocene, shallowing of the Winton Basin and tectonic quiescence led to deposition of calcareous deposits (see Fig. 6.7). The Early Miocene (Otaian) Forest Hill Formation, which rests conformably on older sediments, is a sandy, bryozoan-rich shelf limestone (Hyden, 1979). It ranges in thickness up to 240 m, and foraminifera indicate an inner- to outer-shelf environment (Cahill, 1995). The unit also overlaps into the adjacent eastern Waiau Basin to the west. To the east it probably interfingers with the upper part of the calcareous sandstone of the Chatton Formation in the Eastern Southland Basin.

8.2.4.3 Waiau Basin

Sedimentation continued uninterrupted from the Late Oligocene into the Early Miocene (Waitakian to Altonian stages) in most of the Waiau Basin (see Fig. 6.10), and the Oligocene–Miocene boundary is seldom marked by major lithologic breaks except in the south-central part of the basin, where Miocene sediments rest directly on Eocene rocks in the subsurface.

A seismic reflector was mapped at the top of the Early Miocene, and represents the top of the carbonate-rich part of the Clifden Subgroup, the Forest Hill Formation on the eastern side of the basin and the top of the McIvor Formation on the western side of the basin (Turnbull *et al.* 1993). In the southern part of the Waiau Basin the seismic reflector represents an erosion surface that has cut a large channel into the underlying shelf sediments represented by the Clifden Subgroup. The channel, which was eroded in late Early Miocene times, trends north-northeast from Te Waewae Bay, and is approximately 3 km wide.

Throughout the Early–early Middle Miocene the eastern Waiau Basin was a shallow marine shelf, blanketed by terrigenous and bioclastic sediments (Clifden Subgroup). West of the shelf, Miocene sediments change to outer-shelf or bathyal mudstone. The shelf edge migrated westwards throughout the Early Miocene, extending west as far as Helmet Hill (see below), with widespread development of shelf limestone facies.

The Clifden Subgroup varies from 100 to 400 m in thickness. The lowest unit (Forest Hill Formation) is a bioclastic grainstone of fairly uniform thickness (100 m). A thin basal conglomerate, inferred to represent channelised mass-flow deposits (Hyden, 1980), passes up into a limestone with bioclasts including a varied macrofauna, and minor glauconite and terrigenous clasts. Shallow-water foraminifera are abundant locally. Beds are extensively bioturbated, thick and planar, and locally cross-bedded. Lateral equivalents include a 60-m-thick, late Early Miocene pebbly bioclastic limestone within Waicoe Formation mudstones at Helmet Hill, 20 km southwest of the Clifden area (Wood, 1969). Lithofacies in the upper 175 m of the Clifden Subgroup include carbonaceous siltstones and concretionary mudstones, thinly laminated carbonaceous mudstones and siltstones with plane-parallel and ripple cross-laminae and lignite, and sparsely fossiliferous sandstone (Fleming *et al.* 1969; Hyden, 1980). The early–late Middle Miocene boundary (Lillburnian–Waiauan boundary) at Clifden, on the Waiau River, is represented by a sharp contact. The overlying late Middle Miocene sediments consist of probably mass-flow transported pebbles and broken shells, and show a marked microfaunal change from those below the boundary, suggesting the presence of a hiatus (Fleming *et al.* 1969).

Faunas from the Clifden Subgroup indicate a shallow inner-shelf marine setting (Fleming *et al.* 1969; Arafin, 1982), while age-equivalent strata at Helmet Hill and Goldie Hill to the southwest show dramatic shallowing in the Early Miocene, from bathyal through outer-shelf, to inner-shelf depths in the early Middle Miocene (Turnbull *et al.* 1993). Paleoenvironments of sediments overlying the limestone include offshore bars, beaches and tidal lagoons (Turnbull *et al.* 1993), with possibly some local mass-flow deposits (Fleming *et al.* 1969). Although no Miocene rocks remain in the Ohai Depression, which connects the Waiau and Winton basins, sediments further east in the Winton Basin are direct correlatives (Hyden, 1979; Turnbull *et al.* 1989). Presumably, the Clifden sequence was continuous through the Ohai area and probably covered the Longwood Range (Wood, 1966).

The western Waiau Basin was the main depocentre in the Early Miocene, and contains a thick sequence, locally exceeding 2000 m, dominated by deep-marine outer-shelf to bathyal mudstone (see Figs 6.9 and 6.10), the McIvor Formation of Turnbull *et al.* (1993). It is massive, soft, calcareous, slightly micaceous and carbonaceous, and rich in planktic foraminifera. Interbedded graded sandstone and mudstone units up to 200 m thick occur, often with significant amounts of macrofossil debris. In the Blackmount area the McIvor Formation is a distinctive Early Miocene facies, consisting of up to 350 m of thickly bedded bioclastic limestone with interbedded background basin mudstone, overlain by up to 450 m of terrigenous alternating mudstone and graded sandstone. Fining-upwards cycles and spectacular slump deposits also occur (Hicks, 1974; Carter & Norris, 1977; Turnbull *et al.* 1989). The formation fines and becomes thinner-bedded and less calcareous southwestwards, where only undifferentiated graded calcareous sandstone and mudstone is mapped. Micro- and macrofauna imply a bathyal environment for these rocks, with abundant planktic species (Turnbull *et al.* 1993). This deep-water environment persisted throughout the Early–Middle Miocene, and only in the Late Miocene did any significant shallowing occur in the western Waiau Basin. The unit has been interpreted as a submarine fan deposit, with a paleoslope towards the west (Carter & Norris, 1977; Turnbull *et al.* 1993).

The Monowai Sub-basin had merged with the northern end of the Waiau Basin in the Oligocene, and by the Middle Miocene the northern part of the sub-basin had merged with the Te Anau Basin (Turnbull, 1985). During the Late Oligocene the sub-basin was an area of shallow-marine limestone sedimentation, a situation that changed abruptly in the Early Miocene, when the sub-basin subsided rapidly and a mudstone and sandstone succession, the Waicoe and Borland formations, was deposited unconformably on the limestone. The oldest Miocene sediments in the sub-basin are massive glauconitic sandy mudstones or muddy sandstones up to 30 m thick, which grade up into typical micaceous, soft, blue-grey Waicoe Formation mudstone (Carter *et al.* 1982). Mudstone was replaced in the middle Early Miocene by a terrigenous, graded sandstone and mudstone succession (Borland Formation) up to 1500 m thick (Carter & Norris, 1977; Turnbull *et al.* 1989). Sandstone beds range from very coarse to very fine-grained, are relatively well-sorted, and include a variety of sedimentary structures typical of turbidites. The formation coarsens northwards into the western Te Anau Basin (Jamieson, 1979; Turnbull, 1985).

Microfaunas from the Waicoe and Borland formations indicate a bathyal environment of deposition of those formations, probably consisting of several turbidite fans emanating from the west side of the basin and coalescing into an axially directed fill to the south (Turnbull *et al.* 1993). Paleocurrents in the Borland Formation are generally directed to the south, southeast and south-southwest in the Monowai Sub-basin, although there is wide variation (Carter & Norris, 1977; Jamieson, 1979).

The early Middle Miocene (Clifdenian–Lillburnian) in the Monowai Sub-basin is represented by the Monowai Formation, which is dominated by up to 500 m of massive, pebbly, clast-supported conglomerate with a medium to granular sandstone matrix (Carter & Norris, 1977). Other lithofacies include pebbly mudstone, bioturbated muddy sandstone, dark carbonaceous mudstone, and ripple-, flaser- and lenticular-bedded sandstone–mudstone units. Carter and Norris (1977) interpreted the Monowai Formation as a delta-front to delta-top deposit, with slide sheets and mass-flow conglomerates at the base and a tidally influenced shallow-marine facies at the top. Redeposited macrofaunas in the formation include shallow-marine bivalves and gastropods, usually abraded. The microfaunas are relatively poor in benthic forms.

Paleoslope indicators in the Monowai Formation are towards the southeast, suggesting that the alluvial system that must have fed the Monowai Formation delta in Middle Miocene times followed the eastern side of Fiordland. The provenance of the Monowai Formation is dominantly semischist and Caples-type metagreywacke, with a minor Fiordland-derived component.

8.2.4.4 Te Anau Basin

At the end of the Oligocene the axial region of the Te Anau Basin (see Figs 6.9 and 6.10) was a deep marine depocentre with shallow eastern and western margins. The western half of the basin (by then part of the Monowai Sub-basin) subsided rapidly in earliest Miocene times, and the shelf limestone was buried beneath deep marine mudstone that extended over eastern Fiordland. The enlarged western side of the basin was then filled by the Early–Middle Miocene formations of the Monowai Sub-basin (Jamieson, 1979; Turnbull, 1985).

No Middle Miocene sediments are preserved in the central Te Anau Basin. The eastern shallow-marine shelf persisted from the Oligocene throughout the Early–Middle Miocene. The Dunton Sub-basin to the northeast was still in existence in the earliest Miocene (Waitakian), but younger sediments are no longer preserved. South of the shelf the Burwood Sub-basin subsided rapidly in the Early Miocene along the Moonlight Fault System, and was filled by a thick, northerly derived sandstone and mudstone unit (Haycocks Formation). Further to the northeast, sediments deposited in the Late Oligocene of the Wakatipu area were already being uplifted and eroded in the Early Miocene (Turnbull *et al.* 1985).

The Upukerora-1 petroleum well in the central Te Anau Basin (Carter & Rainey, 1988) intersected latest Early Miocene mudstone (Takaro Formation), immediately underlying Late Miocene Prospect Formation conglomerate. A pause in sedimentation, or erosion, is indicated. Carbonaceous material and limestone were commonly present within the background mudstone.

Throughout the central and western Te Anau Basin several hundred metres of Miocene mudstone (Waicoe Formation) rests conformably on the sandstones of the Oligocene Turret Peaks and Dunton formations. The mudstone was deposited in a deep, sheltered, marine basin with little coarse-grained clastic input. Planktic microfaunas indicate bathyal depths in the west, and shallower outer-shelf conditions in the north. On the western basin margin the basal contact is unconformable over the Oligocene Tunnel Burn Formation limestone. Here, the lower Waicoe Formation is sandier and slightly glauconitic, and contains limestone intraclasts. The unit grades into the Borland Formation in the southwest, but in the centre of the basin it is eroded and abruptly overlain by Late Miocene–Pliocene Prospect Formation conglomerates (Turnbull, 1985, 1986). The formation coarsens eastwards into the Early–Middle Miocene Takaro Formation, where macrofauna and microfauna indicate that conditions shallowed from bathyal in the earliest Miocene, to inner-shelf by late Early Miocene (Morgans & Raine, 1988). The Early–Middle Miocene Takoro Formation shelf sediments near the eastern side of the basin include up to 1000 m of muddy macrofossiliferous siltstone and fine-grained sandstone, as well as shellbeds and slightly pebbly bioclastic limestone units up to 30 m thick (Turnbull, 1986). Bioturbation has destroyed most sedimentary structures, although rare cross-lamination is still preserved. The formation is increasingly sandy and pebbly upwards into the Middle Miocene, with thin pebbly sandstone lenses and pebble trains infilling shallow channels.

The western side of the basin, the extension of the Monowai Sub-basin, remained relatively deep during deposition of the Borland Formation, then shallowed in the Middle Miocene when the Monowai Formation was deposited.

Outcrops in the Burwood Sub-basin, adjacent to the Moonlight Fault, comprise an Early Miocene sandstone and mudstone succession (Haycocks Formation) in a vertically dipping, fault-bounded unit up to 3 km thick (Carter & Norris, 1977; Harrington, 1982; Turnbull *et al.* 1993). Several smaller-scale, fining-upwards cycles occur within an overall fining-upwards trend. Sandstone beds range from coarse- to fine-grained, reaching 2 m in thickness, and include abundant bioclastic material, mainly foraminifera. Bioturbation and sedimentary structures typical of turbidites are common. Microfaunas from mudstone interbeds indicate outer-shelf to bathyal conditions. Sedimentary structures and cycles strongly suggest the formation is a submarine fan sequence, with south-directed turbidity currents (Harrington, 1982).

8.2.4.5 Solander and Balleny basins

Miocene evolution of the Solander and Balleny basins (see Figs 6.9 and 6.11) was a continuation of earlier events. The Eocene–Oligocene Hautere and Parara sub-basins had merged into a single eastern Solander Basin in the Late Oligocene, with the Waitutu Sub-basin still separate and to the west of the Solander Ridge. The Solander and Balleny basins became fully connected during the Early Miocene, although several basement highs remained devoid of Early Miocene sediments and probably contributed debris to adjacent basins. Miocene sediments do not overlap the Hump Ridge–Stewart Island Shelf, and this area is believed to have stayed emergent for much of the Miocene, shedding clastic sediments into the Solander Basin.

Some control on the offshore succession is afforded by Miocene sediments exposed in the onshore part of the Waitutu Sub-basin (see Figs 6.9 and 6.11). These are dominated by mudstone, but include several graded sandstone and mudstone successions, and a distinctive unit containing many slump horizons. The Miocene Waitutu Sub-basin was controlled by the Hauroko Fault and the uplifted Hump Ridge, although Late Miocene–Pliocene overthrusting of the ridge has obscured much of the eastern part of the onshore basin.

A continuous succession at least 3000 m thick, from earliest Miocene (Waitakian) to Late Miocene (Kapitean) in age, is exposed between Hump Ridge and the Hauroko Fault. The Miocene succession rests conformably on Late Oligocene sediments. Waicoe Formation mudstone forms the bulk of the succession, with a slumped, pebbly, mudstone facies in the west (Turnbull *et al.* 1993). Within these units are thinner sandstone-rich units and olistostromes, and pebbly bioclastic limestone up to 20 m thick occurs near the base of the section. Microfaunas in this sequence indicate bathyal conditions for the Early–Middle Miocene.

To the east, graded sandstone and bioclastic limestone units become more prominent. Olistostromes occur in the same area, with isolated masses up to 1 km long and 100–200 m thick within late Early Miocene thin-bedded graded sandstone and mudstone. To the west, a 3.5-km-wide major channel within Waicoe Formation mudstone is infilled with a sandy, greywacke–pebble conglomerate of undated but probable Middle Miocene age. It is approximately 500 m thick and a similar Middle Miocene conglomeratic unit was intersected in the Solander-1 petroleum well (Turnbull, 1993).

Slump horizons or debris flows, deposited in outer-shelf to bathyal environments (Bishop, 1986), characterise the Early–Middle Miocene succession (Knife and Steel Formation) adjacent to the Hauroko Fault. They consist mainly of deformed clasts of laminated to massive mudstone, from cobble to large boulder size, and less common angular Fiordland plutonic boulders, in a mudstone matrix. The slump facies is interbedded with packets – some slump-folded – up to 100 m thick of graded fine- to coarse-grained sandstone and mudstone. The abundant slump facies suggest both tectonic activity and a steep slope. Paleocurrent and slump-fold data confirm a westerly source and a paleoslope towards the southeast. A reflector marking the top Early Miocene was mapped over the whole offshore region, as it was onshore.

In the offshore Solander Basin, Miocene sediments occurring in the Solander-1 petroleum well consist mainly of calcareous mudstone with occasional coal fragments (see Figs 6.9 and 6.11). This mudstone overlies 20 m of limestone of early Early Miocene (latest Waitakian) age, which in turn overlies some 500 m of earliest Miocene sandstone and conglomerate. The whole of the Early Miocene succession is shallow to marginal marine in the well (Clarke, 1986). The middle Early Miocene (Otaian) section is very thin, and possibly unconformable on the Waitakian limestone, such as is seen on the western margins of the Te Anau and Monowai basins. In the Parara-1 petroleum well the 260-m-thick Early Miocene succession is dominated by mudstone with thin limestone beds. The microfaunas indicate deposition in a mid- to outer-shelf environment (Hornibrook *et al.* 1976).

A maximum thickness of more than 1400 m of Early Miocene sediments in the Solander Basin is preserved in a small half graben occupying the re-entrant between the Stewart Island and Hump Ridge portions of the shelf. The fault controlling this half graben was reactivated by Miocene compression. South-southwest of Solander Island, a large depocentre near the head of the Solander Trough contains over 1200 m of Early Miocene sediments. Thicknesses vary markedly across faults, suggesting active syn-depositional faulting. In general, Early Miocene sediments are thicker near the coast, thin southwards across relatively shallow basement, and then thicken again to the south. The main Early Miocene Solander Basin developed northeast from Solander Island.

The offshore Waitutu Sub-basin had a similar geography in the Early Miocene to that of the Oligocene sub-basin, and subsidence and sedimentation continued. The basin trends southwest for at least 100 km and is approximately 18 km wide; it is bounded by the Hauroko Fault along its northwest margin. In the southern part of the basin the Waitutu Sub-basin merges with the Balleny Basin.

Seismic data show that the maximum thickness of the Early Miocene succession exceeds 1600 m in a large lobe that thins to the south and west, off the present coastline north of the Solander Islands. A second depocentre southwest of the Solander Islands contains more than 800 m of sediment. The thicker sediment lobe is interpreted as a seawards-prograding fan, possibly a distal equivalent of submarine fan sediments seen onshore (Turnbull *et al.* 1993).

The en echelon nature of the Hauroko Fault makes it possible to tie the top Early Miocene reflector of the Waitutu Sub-basin to a similar reflector in the Balleny Basin, although the continuity and amplitude of the Early Miocene reflector package vary considerably.

Early Miocene sediment infills lows and drapes over highs, tending to smooth the physiography. The top Early Miocene reflector itself is often undulating, suggesting post-depositional faulting or reworking by strong currents.

8.2.5 West Coast basins

Early Miocene sediments on the West Coast of the South Island are wholly marine and fairly uniform in lithology, in contrast to the overlying Middle Miocene–Pliocene succession, which is more varied. The southern West Coast area was emergent during the Early Miocene, and progressive submergence, beginning in the earliest Middle Miocene, continued into the Middle Miocene (Lillburnian). The basal Middle Miocene sediments, the Tititira Formation (Nathan, 1978a), of undifferentiated Middle Miocene (late Lillburnian to Waiauan) age, rest unconformably on the Oligocene Awarua Limestone and its equivalents (see Fig. 6.8). They consist dominantly of bathyal mudstones, with sandstones and conglomerates of mass-flow origin in the upper portion. Further to the northeast, similar bathyal mudstones of Middle Miocene (Clifdenian to Lillburnian) age occur in the onshore petroleum wells Waiho-1 and Harihari-1 and the offshore well Mikonui-1.

In the northern West Coast and in Nelson, sedimentation was continuous from the Late Oligocene to Early Miocene only in a relatively small area, including the rapidly subsiding Murchison Basin (see Fig. 6.13) and a separate area, mainly offshore, to the north of Nelson. By the end of Early Miocene (Altonian) times the subsiding area had expanded so that the whole of northwest Nelson was submerged as well as much of the area around and immediately south of Greymouth. The Paparoa Trough that had persisted from the Late Cretaceous through to the Oligocene ceased to exist at the end of the Oligocene, but a new basin (the Grey–Inangahua Depression) started to subside immediately to the east (Gage, 1949; Dibble & Suggate, 1956).

The main Early Miocene lithologies fall into two distinct groups. Early Miocene sediments in the Murchison Basin (Mangles Formation) consist predominantly of a deep-water flysch-type sequence passing up into shallow-water sandstone (Fyfe, 1968; Suggate, 1984). Elsewhere, the Early Miocene consists dominantly of grey-brown calcareous mudstone, containing varying amounts of sand.

The Grey–Inangahua Depression (see Figs 6.12 and 6.13) was infilled mainly by late Cenozoic sediments, and bounded by uplifted basement rocks to the west (Paparoa Range) and the east the (Victoria and Hohonu ranges). The southern, rapidly subsiding, part of the depression is named the 'Grey Valley Trough', while the section to the north is known as the 'Inangahua Depression'.

In the Inangahau Depression, Nathan (1978b) included all Early Miocene sediments in the Inangahua Formation. The basal part of the sequence (latest Waitakian), resting conformably on the underlying Oligocene bioclastic limestone, consists of highly calcareous grey-brown mudstone containing interbeds of bioclastic limestone. Towards the top, microfaunas indicate rapid shallowing, and interbeds of granite-derived arkosic sand appear and rapidly increase in abundance to become the dominant lithology, until the formation grades conformably up into non-marine sandstone, conglomerate and carbonaceous mudstone, with thick lensoid coal seams (Rotokohu Coal Measures). The top of the Inangahua Formation is dated as late Early Miocene (Altonian), so the base of the Rotokohu Coal Measures is either late Early Miocene or earliest Middle Miocene; its upper portion ranges into the Pliocene (Nathan *et al.* 1986).

To the south in the Grey Valley Trough, the Inangahua Formation, which here also rests conformably on underlying Oligocene strata, has a similar lithology of brown-grey calcareous mudstone, supplemented by local thin sandstone and limestone layers (Suggate & Waight, 1999). Thickness variations are dramatic, ranging from a minimum of 30 m to a maximum of 722 m close to the inferred axis of the Grey Valley Trough. Locally, granite-derived basal conglomerate (Hohonu Conglomerate Member), associated with a fault zone, reaches a thickness of 1662 m (Suggate & Waite, 1999). It indicates rapid erosion of the rising Hohonu Range.

Variations within the microfauna of the Inangahua Formation indicate different depths of deposition that show a pattern of differential subsidence within the Grey Valley Trough, with subsidence exceeding sedimentation in places. Overall the formation shows a rapid increase in depth when compared to the underlying Nile Group.

At the end of the Early Miocene, in latest Altonian to earliest Clifdenian times, the Grey Valley Trough expanded to the south and west, the expansion accompanied by a gradational lithological change from the grey-brown calcareous mudstone of the Inangahua Formation upwards into nearly ubiquitous blue-grey calcareous mudstone (Stillwater Mudstone), reflecting a slight reduction in organic carbon, possibly because of increased circulation as the trough expanded. Stillwater Mudstone is generally massive and featureless in outcrop, although a few thin bands of *Amphistegina* limestone are known at or near the base where it rests unconformably on older rocks. Locally, the earliest Middle Miocene (Clifdenian) part of the formation contains thick interbeds of pebbly mudstone that include clasts of Cobden Limestone up to 0.3 m, and broken shallow-water macrofossils. These were interpreted (Nathan, 1978b) as mass-flow beds from a shallow-water source, possibly close to the then-emergent western margin of the basin. The Stillwater Mudstone may reach a maximum thickness of ~3500 m and ranges in age from late Early to mid-Late Miocene (late Altonian to late Tongaporutuan). The mudstone has the large foraminiferal microfauna indicating deposition in neritic to bathyal depths (Nathan *et al.* 1986; Suggate & Waight, 1999).

In the southern part of the Grey Valley Trough, several petroleum wells (Niagara-1 and 2, and Hohonu-1) penetrated arkosic and micaceous turbidite sandstones and a channel-slump conglomerate. The unit, Niagara Sandstone Member (Matthews, 1990; Suggate & Waight, 1999), has not been traced beyond the wells and is inferred to be of local extent.

In the Murchison Basin a dramatic increase in the rate of subsidence at the beginning of the Miocene initiated a new phase of deposition. The calcareous Oligocene Matiri Formation grades rapidly upwards into the overlying marine Early Miocene Mangles Formation, marked by the incoming of feldspathic sandstone (Suggate, 1984). Initially, a flysch-type succession of alternating deep-water mudstone and graded sandstones up to 1600 m thick (Tutaki Member) was deposited over the whole basin except in the northwest. There is a gradual upward increase in the proportion of sandstone until the flysch-type succession grades into the predominantly massive sandstone of the overlying members. Combined sedimentological and paleontological evidence suggests sedimentation in a prograding bathyal fan system. The few paleocurrent measurements indicate current flow from east to west in the lower part of the Tutaki Member, with a change in flow from south to north in the upper part. The base of the Tutaki Member is probably earliest Miocene (latest Waitakian) in most places. The top contains late Early Miocene (late Otaian) microfaunas (Suggate, 1984).

The Tutaki Member was succeeded in the east by a coarser wedge of feldspathic sandstone and conglomerate derived from the Separation Point Granite. This wedge thins southwards and interdigitates with uniform, dark, muddy, fine sandstone that later extended over the whole basin and forms the dominant Mangles lithology (see Fig. 6.13). The sandstone contains inner-shelf to middle-shelf microfaunas that are markedly shallower than in the Tutaki Member, indicating rapid infilling of the basin. The younger sandstones are litharenites, containing 30% or more altered volcaniclastic lithic fragments derived from the Maitai and Caples terranes to the east or northeast (Cutten, 1979; Smale, 1980; Watters, 1982). Conglomerate horizons are minor, but become important in the upper part of the formation where some beds contain rounded cobbles up to 230 mm across. Thicknesses reach 1900 m in the Owen Valley. The abrupt incoming of volcaniclastic lithic sand in the upper part of the Mangles Formation is inferred to mark the beginning of rapid uplift of Permian rocks on the east side of the Waimea and Alpine faults. The rapid shallowing is therefore attributed to infilling of the basin by a huge influx of material from the rising eastern source area.

Continued subsidence of the Murchison Basin caused it to extend southwards for at least a further 30 km. A rapid shallowing and increase in volcanogenic material was seen in the upper part of the Mangles Formation and continued into the early Middle Miocene (Clifdenian). At this time, an influx of fluvial deposits dominated by sandstone and conglomerate, with lensoid coal seams near the base (Longford Formation and Rappahannock Group), resting conformably on Mangles Formation or unconformably on older rocks, led to rapid infilling of the basin (Cutten, 1979; Nathan *et al.* 2002). The conglomerate clasts, some up to 50 mm in diameter, are dominated by Permian sediments derived from across the Alpine Fault to the east, with a minor component of locally derived granite.

Northeast of the Murchison Basin a 500-m-thick sequence of flysch-like alternating sandstone and mudstone (Burmeister Formation), of similar lithology and age to the Tutaki Member, is interbedded within massive mudstone containing shelf microfaunas (Johnston, 1971). Sole markings suggest sediment transport to the west-southwest, and the succession may represent turbidites derived from a granitic source to the east, and transported west-southwest along a shelf channel into the Murchison Basin (Nathan *et al.* 1986). West and north of the Murchison Basin, Early Miocene sediments consist predominantly of massive, grey-brown to greenish-grey calcareous mudstone, locally interbedded with sandstone and sandy limestone. Microfaunas indicate deposition in open water, at depths equivalent to outer-shelf or deeper (Nathan *et al.* 1986).

In the coastal area between the Punakaiki River and Westport, Early Miocene sediments (Welsh Formation) overlie Oligocene limestone mainly conformably but with local hiatus (see Fig. 6.13). The strata are notably calcareous, consisting mainly of sandstone and mudstone, but including a prominent limestone lens south of Punakaiki. A single offshore drillhole, Haku-1, which penetrated the succession, shows that the formation there ranges up into the early Middle Miocene (Clifdenian–Lillburnian), and consists entirely of massive blue-grey calcareous mudstone containing a deep-water microfauna dominated by planktic foraminifera (Hematite Petroleum, 1970). The sediments are truncated by a sub-Waiauan unconformity, and the upper part of the Lillburnian Stage is missing. Onshore, this unconformity has resulted in removal of all sediments of early Middle Miocene (Clifdenian to Lillburnian) age (Laird, 1988).

In northwest Nelson, sedimentation appears to have been continuous from the Early to the Middle Miocene, but with limited exposure on land. In the Takaka area the lower (Early Miocene) part of the Tarakohe Mudstone, which consists predominantly of massive grey-brown mudstone, forms a regressive sequence with an upwards-increasing proportion of sandstone and microfaunal evidence of progressive shallowing (Leask, 1980). In the upper Takaka Valley the Tarakohe Mudstone passes upwards into massive shallow-water sandstone (Waitui Sandstone) of early Middle Miocene (Clifdenian to Lillburnian) age, which contains pebbles of locally derived schist and granite.

8.2.6 Taranaki and western offshore Northland basins

The base of the Miocene succession in the Taranaki and western offshore Northland basins, like that of the West Coast basins to the south, is marked by a regional change from dominantly carbonate to dominantly clastic deposition (see Fig. 6.3). Sedimentation in the Taranaki Basin was initially accompanied by rapid foreland subsidence, and initial infilling of a foredeep trough developed along the basin's eastern margin, as a result of basement being thrust westwards along the rejuvenated Taranaki Fault (King & Thrasher, 1996). Basin bathymetry and transgression reached a maximum in the Early Miocene (see Figs 6.3, 6.14 and 6.15).

The Miocene succession is represented by the Wai-iti Group (King, 1988), and the Lower Miocene portion of the Taranaki sequence is included in the lower Manganui Formation (see Fig. 6.3). The foredeep trough trapped most of the available sediment supplied from the east, with at least 900 m of lower Manganui Formation accumulating in the Taranaki Peninsula region. A combination of rapid foredeep subsidence rates and low sediment supply rates in the Early Miocene, however, led to a net increase in water depths to mid-bathyal, and precluded the complete infilling of the foredeep trough. As a result, only minimal clastic detritus reached more distal central and western parts of the basin. The presence of the calcareous Taimana Formation rather than lower Manganui Formation in the west indicates that there was a considerable time lag before the influx of terrigenous sediments from the south and east reached distal western parts of the basin. Bathyal depths across most of the Taranaki Basin in the Early Miocene persisted in the Middle–Late Miocene, and sedimentation was dominated by fine-grained deposits of the Manganui Formation.

The most widespread lithology of the Manganui Formation is non-calcareous mudstone. In central parts of the basin the unit grades laterally into, or interfingers with, the Taimana Formation, which is predominant in the west (King & Thrasher, 1996). In the subsurface Taranaki Peninsula the contact between the Taimana Formation and the overlying Manganui Formation corresponds to a gradational but distinct up-sequence increase in terrigenous clastic sediments within the early Early Miocene (late Waitakian–Otaian). Onshore, this contact corresponds to a pronounced and widespread GR log break (decrease downhole), which marks the onset of compression along the Taranaki Fault. The contact offshore is more diffuse and diachronous, and the basal age of the Manganui Formation offshore varies from late Early Miocene to early Middle Miocene (Altonian to Lillburnian). Correlatives of the lower Manganui Formation exposed in northwest Nelson include the Early Miocene Kaipuke Siltstone and Tarakohe Mudstone. Nearshore lateral equivalents include mid- to late-Early Miocene (Otaian–Altonian) sandstones exposed near Awakino. Coal measures in the Surville-1 petroleum well are also age equivalent (see below).

In the latest Early Miocene (late Altonian), sand was intermittently reworked off the southern shelf into deeper-water areas immediately to the north. These pulses of coarse-grained sediment roughly coincided with those in the eastern foredeep, and were precursors to widespread sand turbidite deposition in the Middle–Late Miocene.

Regressive coastal-plain coal measures of late Early Miocene (Altonian) age, unconformably overlain by Plio–Pleistocene strata, are recorded in the Surville-l petroleum well, and are inferred to have been present in the proto-Tasman Bay region at the beginning of the Middle Miocene. These are the only known Miocene coal measures in the Taranaki Basin, and their position immediately beneath the unconformity may indicate that uplift and regression began in the southern basin by the start of the Middle Miocene, at the same time that coarse-grained clastic detritus began entering the basin.

Along the eastern margin of the basin, seismic reflection profiles reveal that these deep-water strata progressively lap eastwards onto the inclined and planed upper surface of the basement rock. The basement onlap succession crops out inland from Awakino where inner-shelf sandstones of late Early Miocene (Otaian to Altonian) age rest unconformably on steeply dipping basement (Happy, 1971; King *et al.* 1993; Wilson, 1994). The sandstones in this area are thick-bedded and appear massive, but they contain fine parallel lamination and some indication of hummocky cross-stratification (King & Thrasher, 1996). These sandstones are the coastal equivalents of paralic Mokau Formation sandstones exposed in the King Country Basin (see section 8.2.7). Manganui Formation exposed near Awakino consists of outer-shelf to upper-bathyal massive mudstones with concretion horizons and thin stringers of sandstone (Wilson, 1994). The predominantly fine-grained nature of the Manganui Formation both west and east of the Taranaki Fault implies that there was minimal fault-induced elevation on the basin margin.

An abrupt increase in the supply of coarse clastic sediment into the Taranaki Basin, and the infilling and initial over-spilling of the basin in the eastern foredeep, began in the latest Early–earliest Middle Miocene (late Altonian–early Clifdenian). Submarine fan sandstones (Moki Formation) were deposited by turbidity currents and as fluidised mass flows at the base of the prograding slope, the position of which dictated their position through time (King & Thrasher, 1996).

The older Moki Formation ranges in age from latest Early to mid-Middle Miocene (latest Altonian to mid-Lillburnian), and is thickest and best developed in the southwest of the Taranaki Basin. It comprises interbedded sandstones, siltstones and mudstones, with common limestone stringers and concretion horizons. The sandstones are usually very fine- to fine-grained, and are commonly argillaceous. The Moki Formation is usually readily identifiable as a distinctive sand-rich interval with abrupt upper and lower boundaries. Wireline log facies are indicative of well-sorted, massive, thick-bedded, or amalgamated channel and sheet sands, deposited by turbidity currents and fluidised flows. Both coarsening- and fining-upwards sand bodies, representing a series of prograding submarine fan lobes and lobe fringe or mid-fan channel deposits, have been identified on GR logs in the wells (de Bock, 1994; King & Thrasher, 1996). Some tabular sandstone packages are up to 100 m thick.

In the Maui field the Moki Formation is about 300 m thick, but an overall thinning of sandstones to the north or northwest is evident, and three-dimensional seismic data reveal the presence of highly sinuous, meandering, suprafan channels (Bussell, 1994). A source to the south is inferred (de Bock, 1994). Moki Formation sandstones extend northwards as far as the central Taranaki Peninsula and offshore areas north of the peninsula. They are generally not as well developed as the southern offshore type area, although the gross Moki interval is several hundred metres thick in many wells. Sand-dominated successions reach 350 m in thickness.

Sandstone intervals of similar age also occur in the north and northwest of the Taranaki Basin. These sandstones are separated from the main Moki Formation submarine fans by central basinal areas of dominantly mudstone deposition (Manganui Formation), and as such are probably derived from shelf areas located to the northeast. Poorly developed Moki sandstones are also encountered in intervening deep-water areas in the northeast. The Moki Formation represents the first main pulse of coarse-grained clastics deposited in the Taranaki Basin. The various individual deep-water basin marginal troughs down which Moki Formation sands were funnelled apparently converged onto a basin floor that sloped towards the abyssal depths of the New Caledonia Basin.

Early Miocene events affecting the offshore western Northland Basin at the newly active convergent margin resulted in significant differences in geology from the offshore Taranaki and West Coast basins to the south. The earliest Miocene (Waitakian) obduction event, which led to emplacement of the Northland Allochthon (see section 8.2.1), coincided with the beginnings of subduction-related volcanism in the Northland region (see Fig. 6.15). The allochthon continued to move during the early part of the Early Miocene, shedding sediment locally into flanking basins. Volcanism was widespread, and it continued throughout the Early Miocene (see section 8.7.2).

The Early Miocene (late Waitakian–Altonian) rocks of western offshore Northland are almost entirely of volcanic and volcaniclastic origin. To the east and north they abut, interfinger with, and onlap the Northland Allochthon. To the west, volcaniclastics grade into probable volcanopelagic sediments. In the southeast the volcaniclastics interfinger with the Waitemata Group (see section 8.2.7) of the Auckland region (Ballance *et al.* 1977).

The main volcanic massifs dominate the Early Miocene stratigraphy, and are dealt with separately in section 8.7.2. Each of the volcanoes is surrounded by an apron, which is gently sloping, and a ring-plain, which is the flat region beyond the apron. Within the apron facies, high-amplitude sub-parallel seismic reflectors on the higher slopes are probably caused by proximal pyroclastic flows, thin lavas and base surge deposits. They pass laterally downslope into lenticular, discontinuous, subhorizontal reflectors, interpreted as medial debris flows. Where each gently sloping volcanic apron passes laterally into the almost flat ring-plain, the apron facies reflectors give way to smooth, even, continuous, parallel, high-amplitude reflectors typical of turbidites. With increasing distance from the volcano these reflectors become weaker, and, finally, transparent. They are interpreted to be distal turbidites, passing laterally into hemipelagic deposits (Isaac *et al.* 1994; Herzer, 1995).

There are three intra-arc basins (Herzer, 1995). In the northern basin, the Hokianga, volcaniclastic sediments are very thick (up to 18,000 m). The basin is effectively surrounded by volcanoes and the allochthon, and the basin floor is locally depressed by approximately 2000 m. On the central Waipoua Plain, by contrast, there was no ring of volcanoes and no concave volcanic geometry to trap sediment. The volcaniclastic section is comparatively thin, averaging only 500–600 m over most of its area. In the basin off Kaipara Harbour to the south (the Kaipara Basin) the volcaniclastic section is ~900 m thick (Herzer, 1995).

In the offshore Waka Nui-1 petroleum well the Early Miocene is dominated by volcaniclastic sediments, volcanic tuffs, and sandstones and siltstones of the Wai-iti Group. A late Early Miocene (early–mid-Altonian) unit, 70 m thick, is separated by unconformities from underlying Late Oligocene Ngatoro Group limestone, and overlying similar volcaniclastic sediments of early Middle to mid-Middle Miocene (Lillburnian to ?Clifdenian) age (Milne & Quick, 1999; Strong *et al.* 1999).

8.2.7 Onshore Northland, South Auckland and King Country basins

Parts of the onshore Northland Basin contain thin transgressive sedimentary sequences that record rapid subsidence in the latest Oligocene and earliest Miocene, immediately prior to and accompanying emplacement of the Northland Allochthon (see Figs 6.16 and 7.1). At several localities shallow-marine calcareous sandstone, 0–20 m thick in places, containing sandy, coarse bioclastic limestone, passes rapidly upwards into deep bathyal turbidite deposits up to 200 m thick. Locally the shallow marine deposits are sandwiched between Oligocene Te Kuiti Group and a basal Northland Allochthon décollement. Several 5–30-m-thick olistostromes of Northland Allochthon lithologies occur within the shelf sandstone sequence on the south shore of Whangarei Harbour. These sequences document rapid earliest Miocene (mid-Waitakian) foundering, followed soon after by progressive advance of one or more Northland Allochthon nappes into a newly formed bathyal basin (Hayward, 1993).

In the onshore southern parts of the Northland Basin, in the south Auckland region and on the northern Coromandel Peninsula, strata of the Te Kuiti Group show that this area had subsided to bathyal depths during the Oligocene but was uplifted and severely eroded subaerially in the earliest Miocene (Hayward & Brook, 1984; Hayward *et al.* 1989).

Thin transgressive sequences of the basal Waitemata Group (Ballance *et al.* 1977) record subsequent rapid subsidence of the southern half of the Northland Basin in the middle Early Miocene (early Otaian), forming the Waitemata Basin (see Fig. 6.15). This basin was divided by Ballance (1974) into a central flysch basin, flanked by northern and southern shelves. It is bounded to the east by a Coromandel–Whangarei basement sedimentary and volcanic high and to the west by the Early Miocene western volcanic arc complex.

The transgressive successions lap onto and partly bury an irregular paleo-shoreline of Waipapa Terrane basement (Hayward & Brook, 1984; Ricketts *et al.* 1989), and in the south unconformably overlie the eroded remnants of the local Oligocene Te Kuiti Group. They consist of up to 60 m of locally derived breccia and conglomerate, shelly calcareous sandstone, and coarse bioclastic limestone that accumulated between intertidal and uppermost bathyal depths. Rock fragments and heavy minerals in these sediments indicate almost exclusive derivation from the underlying Waipapa Terrane (Hayward & Smale, 1992).

These mainly shallow-water sediments are either directly overlain by mid-bathyal turbidites, or separated from them by at most a few metres of mudstone, indicating that much of the basin was starved of sediment during the middle Early Miocene (early Otaian) period of rapid subsidence of 1000 m or more (Ricketts *et al.* 1989). The turbidites that provide the bulk of the Waitemata Group accumulated in a deep bathyal basin formed around the southern toe of the advancing Northland Allochthon (Hayward, 1993). The succession has few marker horizons and a complex structure, and correlation of the basin sediments is uncertain. Overall there is a general regional tilt and younging towards the west, with the oldest units outcropping in the east. The central part of the Waitemata Basin is filled with an unknown thickness (probably 1–2 km) of interbedded turbidite sandstone and pelagic siltstone, which accumulated in bathyal submarine fan and basin floor settings.

Several geographically or stratigraphically separate lithofacies have been distinguished, based in part on the proportion of volcanic detritus (Hayward, 1993; Isaac *et al.* 1994). Turbidites poor in volcanic detritus largely occupy the east and south of the Waitemata Basin, and they extend as far south as the Hunua Ranges in the south Auckland region. In the east they conformably overlie the transgressive sequences. Sandstones beds are typically 0.3–1.0 m thick, with trace fossils of the deep-water *Nereites* ichnofacies (Seilacher, 1967). Sporadic volcaniclastic submarine mass-flow deposits are present within the southern and eastern (older) parts, and up to 1 km of canyon, channel and submarine fan conglomerates (Albany Conglomerate) is present in the northwest (Ballance, 1974; Sprott, 1997). The facies contains disrupted thrust wedges or blocks of Northland Allochthon sedimentary and (less commonly) igneous rocks.

A thick-bedded volcanic-rich turbidite facies crops out in the northern part of the central basin, and almost across to the Kaipara Harbour in the west. The facies is more proximal in character than the coeval volcanic-poor facies, and consists mainly of 10–30-m-thick packets of graded, medium- to coarse-grained turbidite sandstones 1–4 m thick, with thin intervals of laminated siltstone. Some volcaniclastic submarine mass-flow deposits (Parnell Grit) are present in the south, and they are common in the west. The grit beds are generally graded, poorly sorted, sandy breccia to pebbly sandstone, with angular to subrounded clasts, rarely up to 30 m across. Transport direction suggests a western source (Gregory, 1969; Ballance, 1974). Most fossil material is derived from shallow (less than 100 m) depths, and, rarely, it is a mixture of shelf and bathyal forms. The grit beds were inferred to have been deposited by subaqueous mass flows that originated on unstable slopes of volcanoes lying to the west of the central Waitemata Basin. They commonly flowed 40–60 km, with final deposition on the basin floor at bathyal depths.

A somewhat different facies, consisting dominantly of centimetre-bedded mudstone, sandy siltstone, and rippled, muddy, fine-grained sandstone occurs around the eastern shores of the central Kaipara Harbour. The facies was deposited over the toe of the Northland Allochthon on the northwestern slopes of the Waitemata Basin. These thin-bedded units, which accumulated in bathyal depths, are inferred to represent a slope environment. They enclose up to 300 m of clast-supported channel- and canyon-fill conglomerate, which is locally associated with lenses of slumped Northland Allochthon. The thin-bedded facies and the conglomerate are at least 800 m thick, and are inferred to have been deposited on the northwestern slopes of the central Waitemata Basin. Lithofacies distribution, thickness variation, clast size and imbrication all indicate conglomerate derivation from the northwest (Hayward, 1993; Sprott, 1997). This is consistent with paleocurrent data from the rest of the Waitemata Group sediments, which indicate current flow towards the east-southeast and southeast (Ballance, 1974). The stratigraphically lower parts of the unit are inferred to have accumulated in northeast-trending fault-angle depressions, developed on top of the Northland Allochthon. Channel-fill deposits and the upper parts of the unit are inferred to have accumulated in and on the flanks of a south-southeast-trending depression, which funnelled sediment southwards into the Waitemata Basin (Brook, 1983).

Small piggyback basins, none larger than ~60 km by 30 km, also occurred locally on the surface of the Northland Allochthon, particularly where there were significant marine depressions. The Parengarenga Basin (see Fig. 6.15) near North Cape has also been identified offshore (Brook & Thrasher, 1991). It consists of a number of elongate fault-bounded depressions that were tectonically depressed at varying rates throughout most of the Early Miocene (early Waitakian–mid-Altonian). The basin sediments (Parengarenga Group), which consist locally of over 2 km of calcareous mudstone and muddy fine sandstone, with channelised redeposited conglomerate, were derived by erosion of nearby uplifted portions of the Northland Allochthon, and from a contemporaneous andesitic volcanic centre to the southeast. Most sediments were deposited at upper bathyal depths. Rocks of the Otaua Basin crop out to the north and south of the Hokianga Harbour, and are identified in seismic surveys offshore to the northwest. These basin sediments were also derived almost exclusively from erosion of the surrounding uplifted Northland Allochthon, and consist of over 1 km of massive to poorly bedded mudstone and muddy fine sandstone, and conglomerate, of Early Miocene (late Waitakian to late Otaian) age.

On land, sedimentary rocks of early Middle–mid-Middle Miocene (Lillburnian to Waiauan) age are restricted to the North Cape area, where several metres of subhorizontal sandy limestone (Waikuku Limestone) is inferred to unconformably overlie folded Early Miocene strata (Brook, 1989). The limestone consists mainly of bryozoan fragments, with abundant benthic foraminifera. Calcareous volcanic sandstone and conglomeratic limestone of similar age have also been dredged from offshore north of Cape Reinga (Isaac *et al.* 1994).

Waitemata Group extends southwards into the South Auckland Basin as far as the northern flank of Pirongia Mountain (Kear & Schofield, 1965). Further south, Early Miocene rocks are included in the age-equivalent Mahoenui Group (Fig. 8.2). The relationship between these two units is one of lateral transition within similar environments, distinguished mainly by the lack of volcanism in the south (Ballance, 1976). A gap in outcrop at Pirongia Mountain provides a convenient boundary between the two.

The Waitemata Group in the South Auckland Basin is subdivided into two subgroups. The older Kawau Subgroup (Hayward & Brook, 1984) is a heterogeneous assemblage of basal Waitemata Group rocks, including shallow-water shelf limestone and sandstone sediments. The subgroup is present only in the north and northwest of the area. These sediments are overlain by the Meremere Subgroup, which oversteps onto the Oligocene Te Kuiti Group in the south. The base of the Meremere Subgroup consists of massive, calcareous, glauconitic, fine to medium sandstone, which grades up into massive, calcareous siltstone with interbedded fine sandstone, which is in turn conformably overlain by massive, non-calcareous, fine to

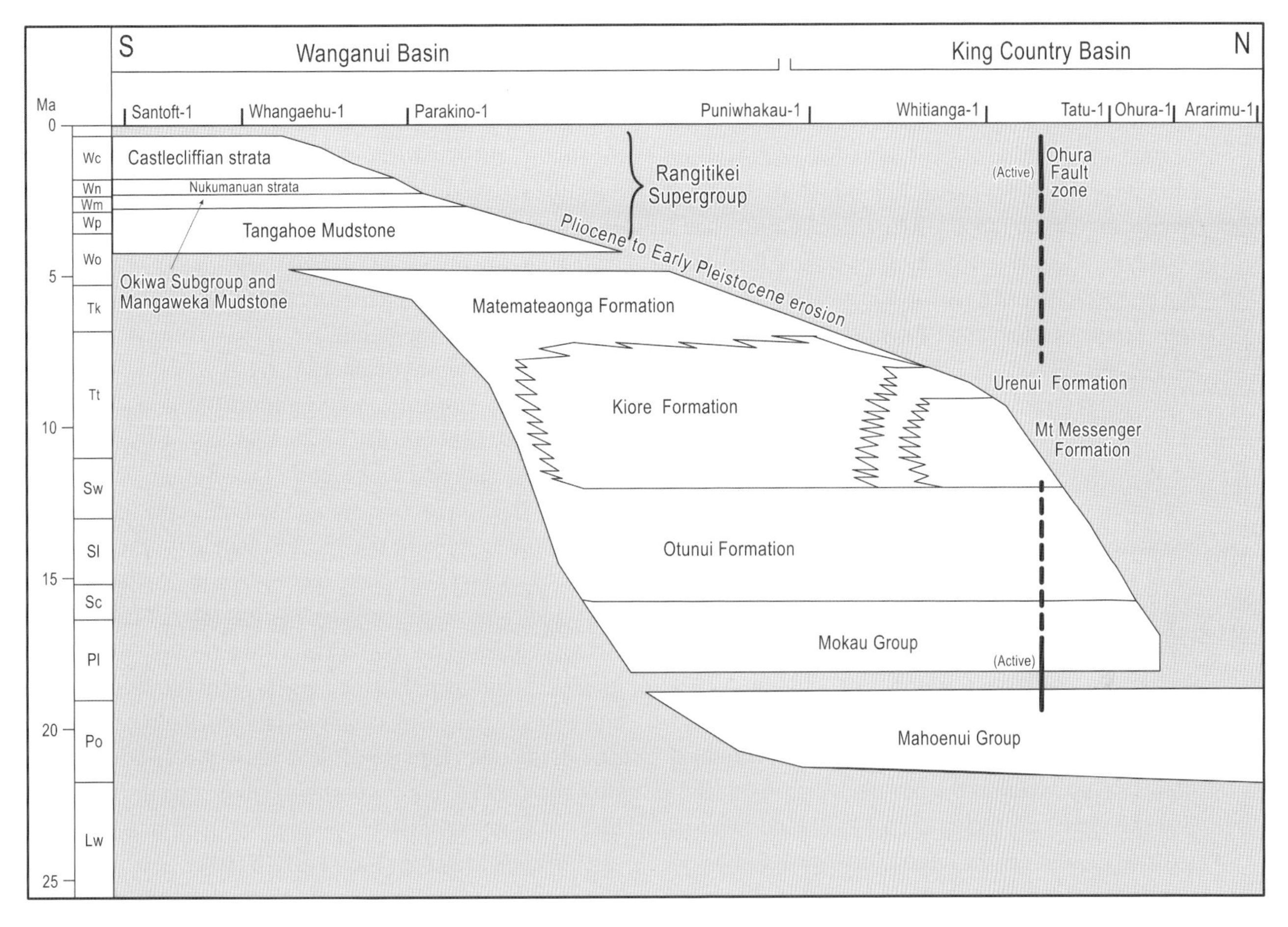

Fig. 8.2. Time-stratigraphic section of the sediments of the Wanganui and King Country basins. For lithologies of named formations, see section 8.4.6. (Adapted from Kamp *et al.* 2004, fig. 6.)

medium sandstone with thin interbedded sandy mudstone. The uppermost beds of the subgroup consist of alternating mudstones and graded turbidite sandstones. The whole subgroup accumulated in a bathyal environment (Hayward & Brook, 1984). Exposure of the Waitemata Group is generally poor, and it has been extensively eroded over much of the region. In the vicinity of Hamilton, ages range throughout the Early Miocene (from Otaian to Altonian) but elsewhere the group is restricted to the Otaian.

South of Pirongia Mountain the equivalent Mahoenui Group unconformably overlies the Oligocene Te Kuiti Group, except in the southern part of the region, where the contact is gradational. The group is characterised by a thick (up to 1200 m) succession of mud-dominated siliciclastic rocks that lack the volcanic content of Waitemata Group sediments to the north. The rocks are early Early Miocene (Waitakian to Otaian) in age (Simms & Nelson, 1998).

The principal depocentre extends from Waitomo in the north to Taumarunui in the south, bounded by paleohighs that today form the Herangi Range in the west and the Rangitoto and Hauhungaroa ranges in the east (Simms & Nelson, 1998).

The Mahoenui Group (Fig. 8.2) comprises two main lithofacies: a massive mudstone facies (Taumatamaire Formation), and a laterally equivalent, alternating sandstone and mudstone turbidite facies near Taumarunui (Taumarunui Formation) (Nelson & Hume, 1977).

The Taumatamaire Formation consists primarily of up to 800 m of bioturbated, massive to cryptically bedded mudstone deposited in outer-shelf to bathyal environments. Along the western side of the basin in particular, marginal to the Herangi High, flaggy to massive skeletal limestones are locally interbedded with the mudstone. Some are of shallow shelf origin, while others are interpreted to have been mass-emplaced into deeper offshore sites (Simms & Nelson, 1998). At the Awakino Gorge a foraminiferal, pebbly limestone (Lower Awakino Limestone), which occurs near the base of the Taumatamaire Formation, rests directly on basement. The limestone is widely distributed throughout the King Country Basin and is found at Piopio and at Te Kuiti. Younger limestones are restricted to the Awakino Gorge where they form one or two conspicuous foraminiferal sandy limestone sheets (Topping, 1978; Simms & Nelson, 1998). The Taumarunui Formation is restricted to the eastern half of the basin and consists of up to 870 m of interbedded graded sandstones and mudstones (turbidites) within an envelope of Taumatamaire mudstone. The turbidites were mainly sourced from the north and were deposited with fan-like geometry in a narrow submarine trough created by the basin east of the Ohura Fault subsiding to lower bathyal depths in the earliest Miocene (Glennie, 1959; Topping, 1978; Simms & Nelson, 1998).

Petroleum exploration wells (e.g., Ararimu-1, Tatu-1, Tupapakurua-1 and Puniwhakau-1) show that the Mahoenui Group occurs in the subsurface at least as far south as the latitude of Cape Egmont (see Fig. 6.14), beyond which the unit is inferred to pinch out. In the Puniwhakau-1 petroleum well, close to the southern margin of the King Country Basin, the Mahoenui Group is 315 m thick, while only 17 km to the northwest, on the flanks of the subsurface Patea–Tongaporutu High, the unit is absent. It thickens eastwards, and has its greatest recorded thickness of 880 m in the Tupapakurua-1 well close to a fault marking the eastern limit of the basin. The Mahoenui Group in the southern (subsurface) part of the basin is mainly a mudstone with minor limestone beds.

The Mahoenui Group is overlain unconformably by transgressive sandstones of the Mokau Group (Fig. 8.2), except to the east of the Ohura Fault, where the block remained above sea level throughout the remainder of the Early Miocene, providing a source for sediments of the Mokau Group (Kamp *et al.* 2004).

The Mokau Group comprises shallow marine sandstones and mudstones, with associated shoreface and tidally influenced back-barrier deposits that include coal measures (Vonk & Nelson, 1998). Outcrop thickness of the Mokau Group varies from 60 m at Waitomo in the north to a maximum of 280 m at its southernmost extent. It has been informally divided into two units, both dominated by sandstone, referred to as the Upper and Lower Mokau sandstones (Nelson & Hume, 1977; Gerritsen, 1994). The lower unit consists mainly of massive sandstones, some calcite-cemented, with some interbedded sandstone and mudstone containing shelly lenses, and coal seams or carbonaceous lenses (Gerritsen, 1994). The upper unit consists of a basal conglomerate, followed by massive sandstone with lenses of shelly material. Locally, thin lenses of limestone up to 60 cm thick occur. Lying conformably between the Lower and Upper Mokau sandstones, and best developed in the eastern King Country Basin, are coal measures (Maryville Coal Measures), up to 50 m thick and including sandstone intervals up to 12 m thick, coal seams up to 1.3 m thick and carbonaceous mudstone. The upper Mokau Group comprises up to 1300 m of fine- to very fine-grained, light-grey, marine, argillaceous sandstone and grey siltstone, with conglomerate horizons occurring in the east. The age of the Mokau Group is not well constrained, but is considered to be mainly late Early Miocene (Altonian), possibly ranging into the early Middle Miocene (Kamp *et al.* 2004).

Away from the basin edge the contact between Lower Mokau Sandstone and mudstone of the Mahoenui Formation appears concordant. The basal portion, however, oversteps westwards against the Patea–Tongaporutu High onto progressively older units in the Mahoenui Group and then rests unconformably on, successively, siltstone of the Oligocene Te Kuiti Group and Mesozoic basement (Cochrane, 1988).

The Mokau Group sandstones were deposited on a continental shelf system confined to the east of the Patea–Tongaporutu High, which was active and uplifted as the group was deposited. Associated with this uplift was a regional down-to-the-east rotation to form a fault-angle depression. As a result, the Lower Mokau Sandstone was eventually exposed and eroded in the west as the Maryville Coal Measures were being deposited in the east of the depression as an easterly thickening wedge. The upper sandstone was deposited unconformably on the lower unit, but the hiatus decreases, and eventually disappears, eastwards. The Mokau Group laps onto the western high, being absent from the Kiore-1 well on the western flank of the basin, but is present in outcrop to the north and in wells on the axis of the basin. It is absent from the eastern wells (e.g., Tupapakurua-1) and outcrop, in an area of a Middle Miocene high.

The Mokau Group is separated from the overlying transgressive but conformable Otunui Formation by a marine flooding surface (Kamp *et al.* 2004; and Fig. 8.2). The basal facies of the early Middle Miocene (Clifdenian–early Waiauan) Otunui Formation are commonly characterised by an onlap shellbed, which passes into a gritty glauconitic calcareous sandstone at the Awakino River mouth (King *et al.* 1993). The Otunui formation is 100–200 m thick, and comprises crudely bedded silty fine sandstone and sandy siltstone, locally with conglomeratic channels.

The controls on sedimentation of the Early–early Middle Miocene strata are tectonics and eustatic sea-level changes. The main tectonic effects are associated with movements of the Ohura Fault, which, apart from the Taranaki Fault, is the dominant fault in the region. It is a high-angle fault, trending north-northeast, which at present has downthrown to the east; the area is currently under extension (King & Thrasher, 1992). The fault had a major effect on Miocene sedimentation in the King Country region. It has changed the nature of its throw several times, the first major movement recording a downthrow of up to 3000 m to the east. This resulted in deposition of Mahoenui Group sediments in bathyal to abyssal depths on the eastern side of the fault in a trough-like feature (McQuillan, 1977; Topping, 1978). How far to the east this trough extended is not certain. Movement on the fault occurred after the end of Te Kuiti Group deposition and before the start of Mahoenui Group deposition, an interval generally placed in the latest Oligocene (early Waitakian).

At the end of Mahoenui Group deposition (late Otaian), reverse movement on the Ohura Fault caused the eastern side to be uplifted and subaerially exposed (Kamp *et al.* 2002). This is suggested by the lack of Mokau Group east of the Ohura Fault. There is a slight angular unconformity between the Early Miocene Mahoenui Group here, and the overlying Middle–Late Miocene Mohakatino Group, implying that the Mahoenui sediments have been tilted and eroded before deposition of the younger sediments. Stainton and Gibson (1964) estimated that upwards of 300 m of Mahoenui Group sediments had been removed from the area east of the Ohura Fault. This reversal of throw uplifted the eastern side about 3000 m, bringing the Mahoenui Group sediments up from abyssal depths to above sea level.

It appears that the eastern side of the Ohura Fault was uplifted and that there was downward displacement of the western side, as shown by the shallow-water assemblages found in the Mokau Group sediments. The western side of the Ohura Fault shows a general trend of shallowing from middle to late Early Miocene (mid-Otaian to mid-Altonian). Uplift on the eastern side of the fault reflects compression and crustal shortening.

Sedimentation in the adjacent Taranaki and King Country basins was linked, particularly from the Miocene on. During this time, gaps in the Patea–Tongaporutu–Herangi High allowed water and sediments to move between the two basins (Topping, 1978; King & Robinson, 1988).

Immediately west of the Patea–Tongaporutu High, in the easternmost Taranaki Basin, the oldest post-basement sediments in the study area are sandstones and mudstones of mid- to late Early Miocene (?Otaian to Altonian) age, laterally equivalent to the Mokau Group (King *et al.* 1993). About 5 km east of the Awakino River mouth Mokau Group sandstones unconformably overlie Triassic Murihiku Supergroup basement rocks of the Herangi Range. Altonian-aged sediments are intermittently exposed west of this locality as far as the coast. They comprise up to 80 m of massive ferruginous sandstone, gradationally overlain by up to 150 m of alternating ferruginous sandstone and mudstone, capped by blue-grey calcareous mudstone and minor calcareous fine sandstone (Happy, 1971). King *et al.* (1993) included the lower sandstone-dominated interval within the Mokau Formation, and the upper mudstone-dominated interval with the Manganui Formation. These Early Miocene units are inferred to represent more distal (seaward) equivalents of the Mokau Group in the King Country region. Within the mudstones, benthic foraminiferal assemblages indicate that depositional sites deepened from outer-shelf to mid-bathyal water depths; these outcrops are inferred to be the onshore expression of the very widespread bathyal mudstones of the Manganui Formation documented by King and Thrasher (1996) in the subsurface Taranaki Basin (see section 8.2.5).

8.3 MIDDLE MIOCENE EVENT

A second major change in tectono-sedimentary patterns occurred during the Middle Miocene throughout most of the New Zealand region. This mid-Middle Miocene event is thought to relate to a change in plate boundary configuration at ~16 Ma, resulting in a significant change in volcanic style and orientation in the north and west of the North Island, and the initiation of subduction along the Puysegur Trench. Volcanism in the Early Miocene northwest-trending Northland arc had ceased by ~15 Ma and was almost immediately followed or overlapped by volcanism in the late Middle–Late Miocene in a new arc inferred to have a north-northeast orientation (Herzer, 1995). Major increases in sedimentation rate that occurred in south Westland at ~15–13 Ma (Sircombe & Kamp, 1998) record the beginning of reverse movement on local faults, considered by Sutherland (1996) to be linked to changes in the Australian–Pacific plate vector.

In western Southland and in western New Zealand basins (see Figs 6.10 and 6.11) this event appears to have occurred close to the Lillburnian–Waiauan boundary (~13 Ma) where there was a change to non-marine sedimentation in the Te Anau Basin, which was flooded by up to 3000 m of non-marine Prospect Formation gravels, with the oldest sediments dated as Waiauan–Tongaporutuan. The basal contact is mainly conformable, but locally it is erosional (Turnbull, 1986). Elsewhere, it appears to coincide with the less well-dated Middle Miocene change in volcanic style and orientation in the northern North Island and the initiation of submarine andesitic volcanism and associated sedimentation in the Taranaki Basin. Local inversion and erosion on the West Coast of the South Island started at this time, as did extension and subsidence in the East Coast Basin.

In south Westland the first appearance of clastic sediments in the Middle Miocene (~12 Ma) in the Waiho-1 petroleum well coincides with uplift southeast of the Alpine Fault and reverse faulting on adjacent faults, events linked to changes in the Australian–Pacific plate vector (Sutherland, 1996; Wood *et al.* 2000). In north Westland and Buller (see Fig. 6.13), a sub-Waiauan unconformity is strongly developed, and locally sediments of Clifdenian and Lillburnian age are missing (Laird, 1988). There is a slight angular discordance at the base of the Waiauan, the depth of erosion increasing towards the western edge of the Paparoa and Papahaua ranges, implying major pre-Waiauan uplift of the area east of the Paparoa Tectonic Zone.

In the northern Taranaki Basin, Middle Miocene andesitic volcanic activity began in the mid-Lillburnian (~14.5 Ma) and reached a peak of activity in the Waiauan. It is balanced by the cessation of subduction-related Early Miocene volcanic activity in the Northland region, which became extinct at ~16 Ma (Herzer, 1995). The Early Miocene arc volcanoes of Northland are distinguished not only by age, but also by difference in trend from the Middle–Late Miocene Taranaki volcanoes. The older Northland volcanic belts trend northwest–southeast, while the younger Taranaki volcanic belts trend north-northeast–south-southwest, almost at right angles to each other. Evidently the Middle Miocene marked a change from a Lower Miocene Northland arc facing the southwestwards-subducting South Fiji Basin, to a separate west-northwest-facing arc – the Lau–Colville Ridge – whose effects penetrated as far southwestwards as the northern Taranaki Basin (Herzer, 1995).

Middle–Late Miocene sedimentary rocks are mainly absent onshore in Northland, and are entirely absent from the south Auckland region, as a result of erosion due to post-Early Miocene uplift and westward tilting of the Northland Peninsula. In a small area near North Cape, limestone of Middle Miocene (Lillburnian to Waiauan) age rests unconformably on folded Early Miocene strata. Only south of Te Kuiti, in the King Country Basin, do sediments of Middle Miocene age occur. Although the mid-Middle Miocene sedimentary succession shows no stratigraphic breaks, rapid late Middle Miocene subsidence is evident (Kamp *et al.* 2002).

In the East Coast region of the North Island, excluding the northern part of the East Coast Basin, the Middle Miocene event appears to be represented by a major extensional episode, which is inferred to have begun at ~15 Ma (Lillburnian) and which continued until the Miocene–Pliocene boundary (5 Ma) (Chanier *et al.* 1999). The extension, perpendicular to the Miocene active margin trend, and concurrent with widespread subsidence, was probably related to tectonic erosion inducing gravitational collapse (Chanier *et al.* 1999). However, sub-Waiauan unconformities occur locally in Hawke's Bay and Wairarapa. In Marlborough, strata of Waiauan age are all but absent.

8.4 LATE MIDDLE–LATE MIOCENE

8.4.1 East Coast Basin

The preservation of late Middle Miocene (Waiauan) sediments (Fig. 8.1a) is generally sporadic in the East Coast region, probably because of continued uplift in the west in the early Waiauan, movement on coastal faults the start of rapid subsidence in the latter part of the stage, and eustatic falls in sea level. Sedimentation rates were generally higher than in preceding Cenozoic stages, but with more-localised depocentres with local structural control. Settings appear to have been shallower than at any previous time in the Cenozoic, and locally were probably non-marine by late Waiauan times (Field *et al.* 1997). The Late Miocene (Tongaporutuan–Kapitean) of the region is characterised by rapid subsidence during most of the Tongaporutuan and one or more shallowing events in the Kapitean. In many places the oldest Tongaporutuan sediments preserved are either shallow marine, overlie an unconformity or record an influx of coarse material into deep water. Locally, where an earliest Tongaporutuan unconformity occurs, it is superimposed on earlier (e.g., Waiauan) unconformities (Field *et al.* 1997).

In the far north of the Raukumara Peninsula the late Middle Miocene (Waiauan) portion of the Tolaga Group is generally poorly preserved, and only local outliers of thin Waiauan-aged paralic sediments remain. Commonly, Late Miocene and younger sediments rest unconformably on rocks of early Middle Miocene (Lillburnian) age or older. To the south, preservation improves and late Middle Miocene deposits are thicker, and bathyal siltstone or mudstone, with local intercalations of sandstone, are typical (Field *et al.* 1997). Sediments are thin (110–180 m) in central Raukumara Peninsula, but they thicken to >530 m in the Gisborne area (Strong *et al.* 1993; Field *et al.* 1997).

The early Late Miocene (Tongaporutuan) in the Raukumara Peninsula spans both parts of the Tolaga Group and the overlying Mangaheia Group (Moore *et al.* 1989) that are commonly separated by an angular unconformity (Fig. 8.3a). Late Miocene rocks demonstrate general regression eastwards, with shallow facies overlying older Miocene sediments of more distal facies (Moore *et al.* 1989). The Late Miocene includes shallow-marine, pumiceous, fossiliferous sandstone and limestone, with minor conglomerate and coal, and about 1000 m of siltstone with units of alternating sandstone and mudstone (Chapman-Smith & Grant-Mackie, 1971; Field *et al.* 1997). In the extreme northeast a Late Miocene pebble bed, with common macrofossils and igneous pebbles that rest unconformably on ?Early Miocene siltstone, is overlain by early Late Miocene inner- to mid-shelf sandstone and then thin- to medium-bedded mudstone and sandstone of outer-shelf origin. The succession shallows up-section to inner- to mid-shelf depths and becomes sandier, and locally it has macrofossils and tuffaceous intervals. This history of rapid deepening in the later part of the Tongaporutuan and shallowing in the Kapitean is a common feature of the onshore East Coast Basin. Locally, the early Late Miocene (lower Tongaporutuan) section is about 1100 m thick and is capped at an unconformity by mid- to outer-shelf, probably late Late Miocene (lower Kapitean), sandstone. The early Late Miocene (Tongaporutuan) section appears to thin to the east or southeast, and it contains slumped beds of sandstone that may record contemporaneous tilting, consistent with greater uplift in the east (Gosson, 1986; Field *et al.* 1997). To the south of East Cape, over 700 m of Late Miocene siltstone and sandstone rests unconformably on Late Cretaceous Whangai Formation.

In eastern central Raukumara Peninsula, early Late Miocene (Tongaporutuan) shelly sandstone, up to 600 m thick, lies unconformably on late Middle Miocene (Waiauan) strata in the west, but with a probably conformable basal contact to the east. The unit locally has a basal shelly conglomerate and channel-fill lenses that include rafts of sandy sediment and coquina limestone (Mazengarb *et al.* 1991). It is considered to be of upper-bathyal to outer-shelf origin near the base, passing up-section into mid- to outer-shelf depths. The total Tongaporutuan succession is over 1500 m thick.

To the east, in the Tokomaru Bay–Tolaga Bay area, up to 2500 m of Late Miocene sandstone or alternating sandstone and mudstone, inferred to be turbidites, occurs, with some sandstone units over 500 m thick (Blom, 1982, 1984). An upper bathyal paleobathymetry was inferred, with the main paleocurrent directions to the east and south-southeast. Turbidite content increases to the south, where up to 2200 m of Late Miocene flysch and silty sandstone occurs (Burgreen & Graham, 2014).

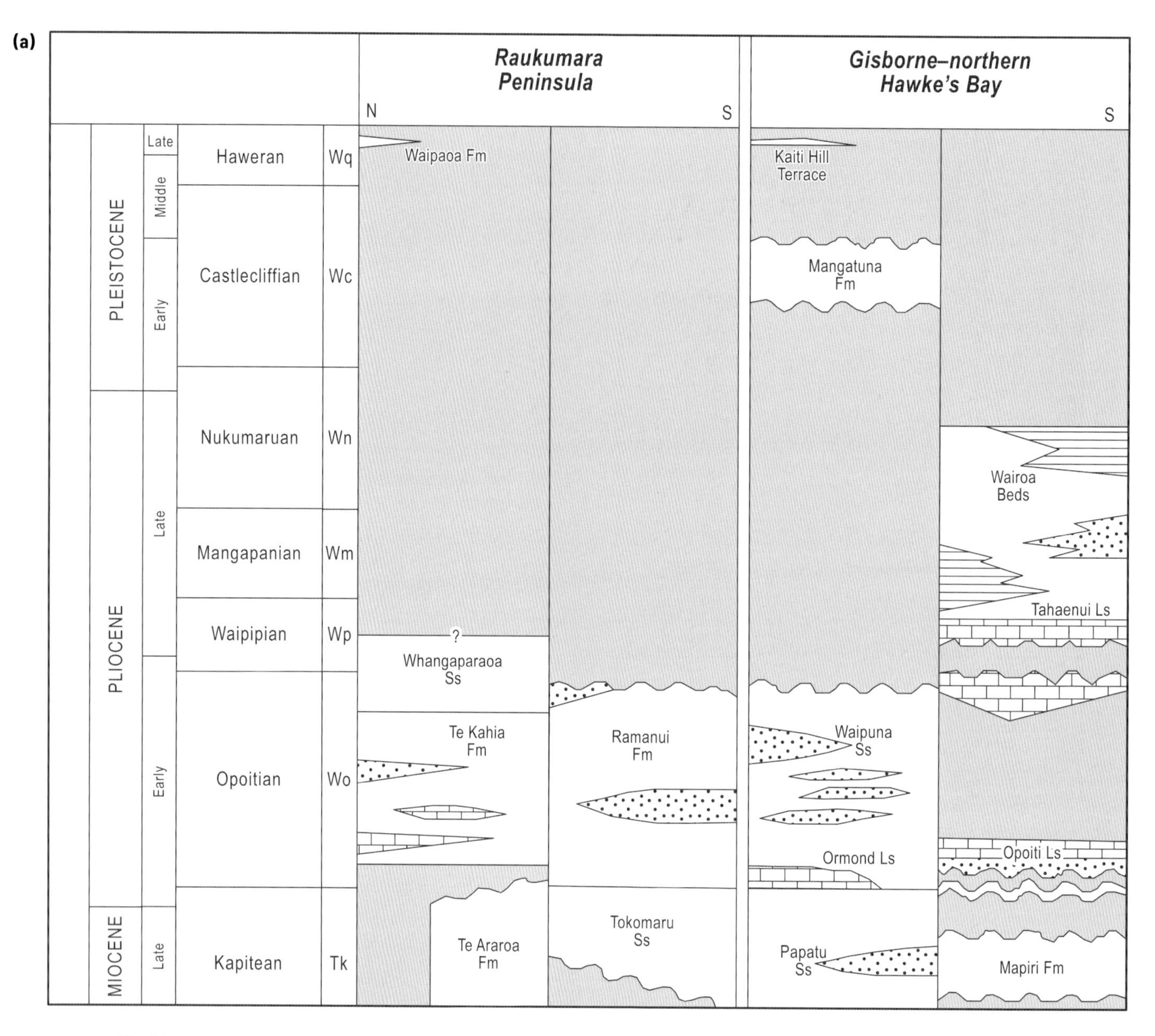

Fig. 8.3. Time-stratigraphic section of Plio–Pleistocene rocks of (a) the Raukumara Peninsula to northern Hawke's Bay, and (b) central Hawke's Bay and northern Wairarapa. For lithologies of named formations, see sections 8.4.1 and 8.6.1. Cng, Conglomerate; Fm, Formation; Gp, Group; Ls, Limestone; Ms, Mudstone; Ss, Sandstone. (Adapted from Field *et al.* 1997.)

In western Raukumara Peninsula up to 4000 m of Late Miocene mudstone and sandstone, with a pebbly base, rests unconformably on Middle Miocene (Lillburnian) mudstone (Kicinski, 1958). These great thicknesses, and probably also their preservation, may have been related to contemporaneous movement on the major axial range faults. West and south of Gisborne, Late Miocene bathyal flysch and mudstone reaches up to 1700 m in thickness. Commonly, there is a thick unit of mid-bathyal sandstone at the base, and locally the succession is capped by Pliocene limestone. Limestone lenses, 10–20 m thick, and deposited at ?mid-shelf depths, occur locally within Tongaporutuan mudstone south of Gisborne and near the Taihape–Napier Road (Beu, 1992).

In northern Hawke's Bay, at Mohaka, the late Middle Miocene (Waiauan) is not preserved, and up to 600 m of outer-shelf to upper-bathyal sandstone

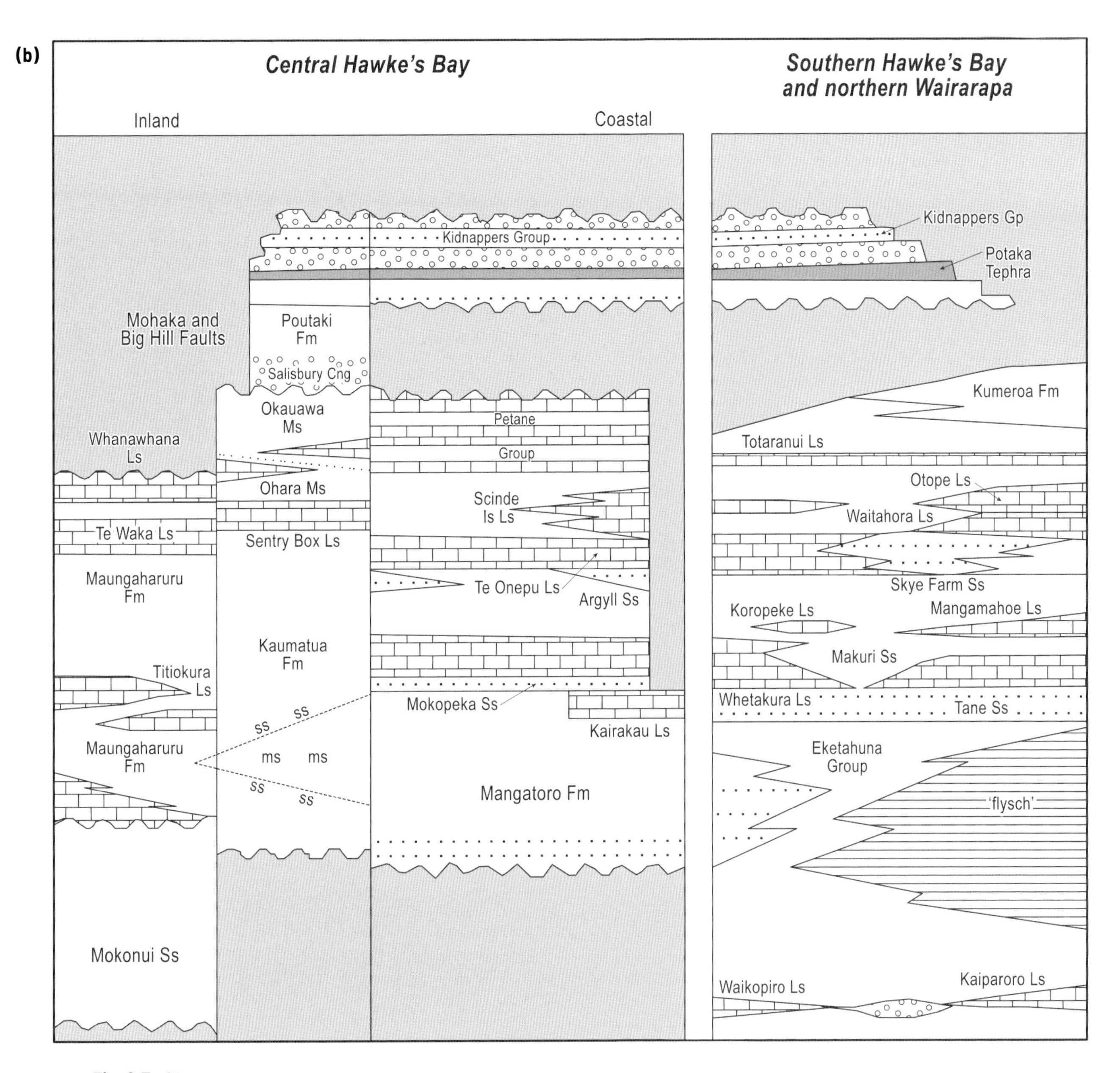

Fig. 8.3. (b)

of early Late Miocene (Tongaporutuan) age rests unconformably on older rocks (Figs 8.1a and 8.3a). The sandstone passes up-section into about 240 m of mid-bathyal flysch, and then into bathyal siltstone, thickening to the north from 850 m to 3000 m. The siltstone is overlain unconformably by late Late Miocene–Early Pliocene (lower Kapitean–Opoitian) sandstone deposited at shelf to upper bathyal depths. Like the underlying siltstone and sandstone, it thickens to the north, from about 300 m to 1000 m. To the east, in the vicinity of Mahia Peninsula, the Late Miocene consists of up to 2000 m of siltstone and mudstone, including tuffaceous intervals (Gosson, 1986; Wright, 1986; Francis, 1993).

In the far northwest of Hawke's Bay, along the Napier–Taupo highway, about 1000 m of late Middle Miocene–early Late Miocene (?Waiauan–Tongaporutuan) shelf mudstone, siltstone and sandstone rests on basement west of the Mohaka Fault (Bland, 2006). The succession shallows and thins to the southwest. A transgressive Late Miocene succession comprising alluvial fan to braidplain conglomerates passing up into late Late Miocene (Kapitean) shallow-marine sandstone, conglomerate, calcareous sandstone and limestone, rests on basement (Browne, 1978).

At the offshore Hawke Bay-l petroleum well, late Middle Miocene (Waiauan) sandstone rests unconformably on Oligocene limestone, and is in turn overlain by 970 m of Late Miocene mainly bathyal mudstone, although episodes of shelf environments occur in places (Scott *et al.* 2004).

In southern Hawke's Bay (Fig. 8.3b), south of Napier, mid-Middle Miocene–early Late Miocene (Lillburnian–Tongaporutuan) deposition occurred mainly in the Makara Basin, which developed on the inner slope of the Hikurangi Trench (van der Lingen & Pettinga, 1980). Basin sediments consist of flysch strata, pebbly mudstones, tuff beds and hemipelagic mudstones, up to 2200 m thick. The strata were inferred to have been deposited by sediment gravity flows, ranging from turbidity currents to fine-grained debris flows. Submarine fans prograded towards the southeast in a basin about 30 km by 20 km, confined by thrust ridges that formed structural highs (van der Lingen & Pettinga, 1980).

At the southern end of the Makara Basin, at Paoanui Point, coarse bioclastic limestone of late Middle Miocene (Waiauan) age overlies a thick unit of mélange (Fig. 8.1a). It contains an unusual microfauna indicating tropical reef conditions, and it seems likely that a structural high associated with the mélange zone reached very shallow depths (van der Lingen & Pettinga, 1980). The overlying Late Miocene succession appears to be over 600 m thick. To the west of Paoanui Point, 220 m of tuffaceous mudstone of ?Tongaporutuan age overlies 70 m of hard limestone, passing southwards into shelf sands resting on basement, suggesting a north- to northeast-trending Tongaporutuan shoreline. Thin limestone of late Late Miocene (Kapitean) age occurs locally (Beu, 1992).

West of Cape Turnagain about 400 m of mid-Middle Miocene (lower Waiauan) bathyal mudstone and alternating sandstone and mudstone overlies an angular unconformity that spans the upper Lillburnian–lower Waiauan interval (Francis, 1989). It marks a significant tectonic event, probably coincident with the Middle Miocene tectonism recorded in the Makara Basin by Pettinga (1982), and may relate to mass flows in northern Wairarapa. Probable lower Waiauan greensand rests unconformably on Oligocene mudstone near Cape Turnagain, and is followed by up to 1300 m of Late Miocene bathyal mudstone to siltstone and fine sandstone (Francis *et al.* 1987; Francis, 1989). Shallowing at Cape Turnagain in the late Late Miocene (upper Tongaporutuan–lower Kapitean) is marked by an upward change to outer-shelf to upper-bathyal greensands and sandstone. To the south, about 400 m of Kapitean coarse sandstone rests directly on Paleocene rocks (Kennett, 1966). Further to the west, beyond the Tinui Fault, a Late Miocene succession of over 2350 m of Tongaporutuan and Kapitean bathyal flysch and mudstone rests with marked angular unconformity on older strata. The rocks comprise turbidites that are inferred to have been derived from a high to the north and east, bounded by the Tinui Fault (Neef & Bottrill, 1992).

In northern Wairarapa (Fig. 8.1b) the late Middle Miocene (Waiauan) succession was deposited largely in shallow-water settings. East of Eketahuna, paralic sandstones and coal measures overlie Mesozoic basement in a transgressive unit that appears to range up into the early Late Miocene. The sea paleoenvironment deepened to the east, where the sediments consist of mid-shelf to upper-bathyal sandstone and mudstone up to 700 m thick (Johnston, 1980). Plant remains and abundant molluscan shells indicate shallow marine conditions adjacent to a vegetated landmass, presumably to the south and west. South of Eketahuna, late Middle Miocene sediments rest unconformably on either older Miocene (Lillburnian) shallow-marine transgressive sandstone (Karamu Sandstone) or Cretaceous basement (Crundwell, 1997). The sandstone fines up-section into mudstone, and deepens from inner- to mid-shelf.

To the southwest, west of Castlepoint, mudstone, siltstone and flysch of Middle Miocene to early Late Miocene age are overlain by about 1200 m of Late Miocene well-bedded shelf sandstone and siltstone, with basal tuffaceous sandstone. The latest Miocene (Kapitean) records at least one and probably two shallowing events, resulting in deposition of paralic to shelf sediments, with thickness changes locally influenced by synsedimentary faulting (Neef, 1984). The effects of synsedimentary faulting can also be seen north of Masterton where eastern and western blocks are separated by strands of the Wairarapa Fault. Over 200 m of terrestrial and shallow-marine late Late Miocene (Kapitean) conglomerate and mudstone is preserved on basement in the western block, and about 1080 m of bathyal Late Miocene in the eastern block (Vella, 1963). The differences between the western and eastern blocks can be interpreted as recording movement on the faults at a time close to the Miocene–Pliocene boundary, with significant lateral offset and juxtaposition of facies and thicknesses (Field *et al.* 1997).

The southernmost exposures in Wairarapa, near Cape Palliser (Figs 8.1b and 8.4), include approximately 200 m of late Middle–early Late Miocene conglomerate (Putangirua Conglomerate) of alluvial fan origin resting unconformably on Torlesse basement rocks (Homer & Moore, 1989; Crundwell, 1997). Pebble imbrication suggests east-facing paleoslopes, but the conglomerates occur on the western flank of the Aorangi Range, suggesting that the range did not form until later (Field *et al.* 1997). The conglomerate grades upwards into 300 m of Late Miocene massive fossiliferous marine sandstone with pebbly horizons (Hurupi Formation), which locally oversteps onto basement. It is succeeded by 450 m of bathyal mudstone (Bells Creek Mudstone), which is areally extensive in southern Wairarapa, cropping out from Palliser Bay to east of Carterton. A thin unit of latest Miocene (Kapitean) limestone, up to 10 m thick, occurs in the northern Aorangi Range. It rests on basement or on a bored surface on Bells

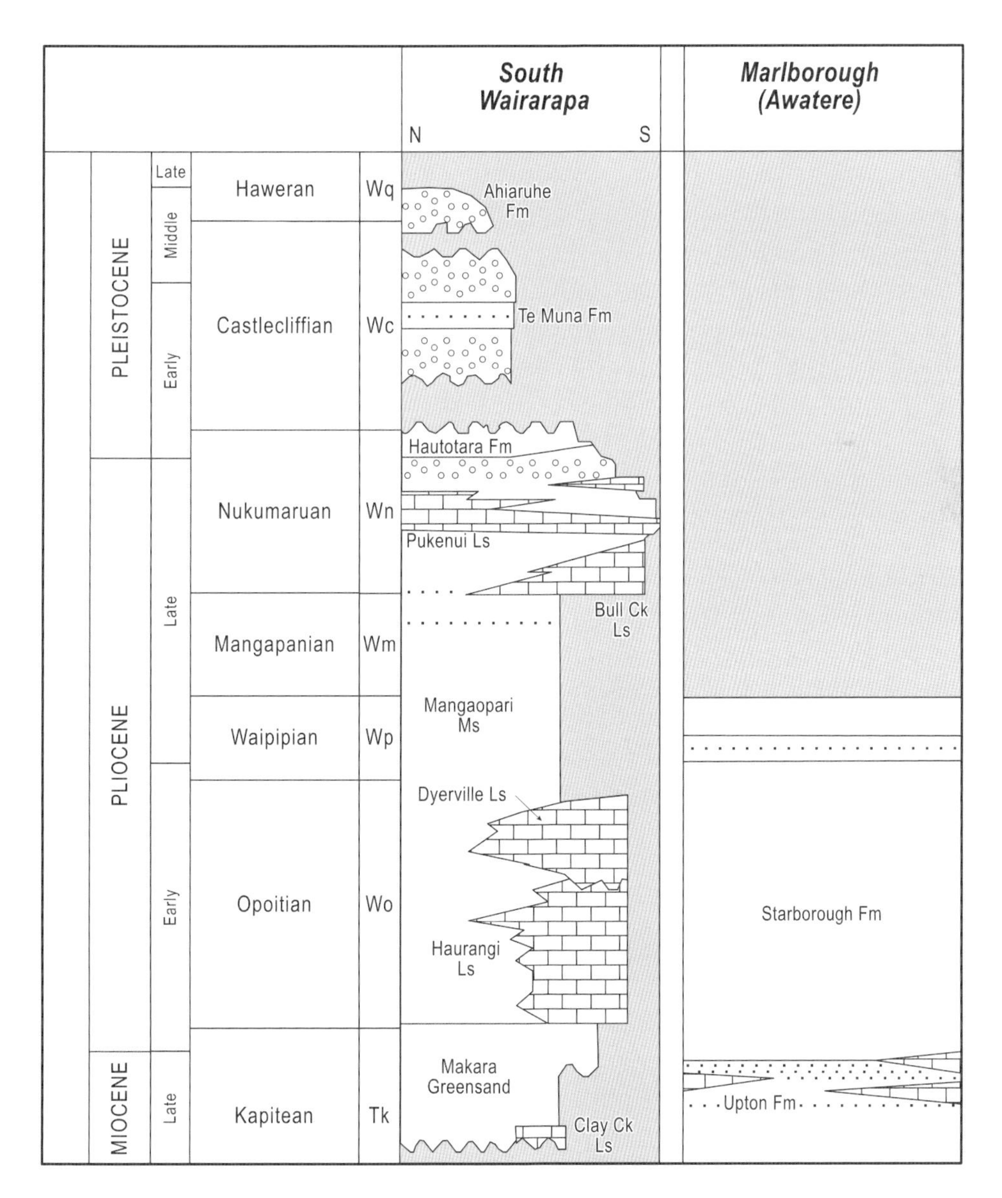

Fig. 8.4. Time-stratigraphic section of Plio–Pleistocene sediments of southern Wairarapa and Marlborough. For lithologies of named formations, see section 8.6.1. Fm, Formation; Ls, Limestone; Ms, Mudstone. (Adapted from Field *et al.* 1997.)

Creek Mudstone. The prominent Makara Greensand (Crundwell, 1997), which appears to span most of the Kapitean, is co-extensive with and rests disconformably on the Bells Creek Mudstone (Field *et al.* 1997).

Early Late Miocene (Tongaporutuan) sediments rest on basement over much of southern Wairarapa, indicating uplift and erosion prior to the lower Tongaporutuan (Eade, 1966). Younger (Kapitean) lithofacies suggest that the Aorangi Range was an island, uplifted by faulting during late Tongaporutuan or early Kapitean times (Vella & Briggs, 1971) and presumably emergent during a lowstand, fringed by shallow marine deposits.

There is little unequivocal record of late Middle Miocene (Waiauan) sedimentation in Marlborough (Fig. 8.4). Up to 60 m of mid-bathyal flysch (Waimana Formation) is present in a discrete fault-bounded block east of Ward. The flysch has been interpreted as slope-basin fill, with paleocurrent flow towards the southeast (Browne, 1992). The base of the flysch is not exposed. A major fault zone separates the Miocene from the underlying Paleocene micrites of the Amuri Limestone, and the entire outcrop may represent an allochthonous block that has moved along the basal shear surface (Field *et al.* 1997).

Late Miocene sedimentation in Marlborough (Fig. 8.1b) is largely confined to areas north of the Waima River, where several discrete basins were filled with a diverse range of marine and non-marine sediments ranging in age from ?late Middle Miocene to Pliocene (Browne, 1995b). They include the Medway and Upton formations (Maxwell, 1990; Roberts & Wilson, 1992) and are assigned to the Awatere Group. The early Late Miocene (early Tongaporutuan) Medway Formation comprises bathyal fine sandy mudstone with interbedded sandstone up to 1600 m thick in the Medway and Awatere rivers area, but, towards the northeast, beds of conglomerate and sandstone appear (Melhuish, 1988; Maxwell, 1990). The conglomerates consist of inner-shelf, matrix- and clast-supported gravels and thin interbedded sandstone forming units up to 390 m thick, overlain by trough cross-bedded sandstone up to 400 m thick (Maxwell, 1990). Paleocurrents indicate paleoflow towards the northeast and northwest (Browne, 1995a), probably in a depositional setting close to an actively rising greywacke-dominated hinterland and deepening towards the north and northeast.

Slightly younger outer-shelf and upper-slope conglomerates interbedded with sandstone and siltstone occur to the north. These rocks form a succession of over 250 m of beds of clast-supported, channelised conglomerates up to 50 m thick, and interbedded fine- to coarse-grained lenticular sandy siltstones up to 100 m thick. The paleogeographic setting inferred was a northeast-sloping, silt-dominated shelf, crossed by a series of channels carrying mass-flow conglomerates feeding the deeper-water canyons of the Late Miocene–Early Pliocene proto-Cook Strait or proto-Hikurangi Trough (Uruski, 1992; Lewis & Barnes, 1999).

The latest Miocene (late Tongaporutuan–early Kapitean) interval, the upper Medway Formation and the Upton Formation, are up to 1700 m thick. These formations include conglomerate units up to 150 m thick at the base but are overlain by very fine sandy siltstone, the dominant lithology, with rare interbedded sandstone and conglomerate. Distinct units consist of apparently massive to horizontally bedded, very fine- to fine-grained sandstones up to 130 m thick with fossiliferous layers and concretionary zones. Siltstones, locally intercalated with layers of tuff or flysch, are more common in the deeper, bathyal, parts of the basin towards the northeast, where they form monotonous sections up to 1500 m thick (Roberts & Wilson, 1992).

At Waihopai River and Ring Creek in northwestern Marlborough (Browne, 1995b) up to 180 m of alluvial fan and fluvio-lacustrine conglomerate and interbedded sandstone and carbonaceous siltstone with paleosols is exposed. These sediments are inferred to be of Late Miocene age. At Ring Creek the alluvial fan conglomerate lithofacies overlies 90 m of well stratified, cross-bedded pebbly sandstone, interpreted as high-energy shoreface sand.

8.4.2 Canterbury, Chatham Rise and Great South basins

In the far north of the Canterbury Basin, near Kaikoura, Middle–Late Miocene calcareous siltstone (Waima Formation) continued to be deposited in outer-neritic to bathyal environments (Fig. 8.4). Localised lenses of mass-flow conglomerate in the

Middle Miocene (Lillburnian–Waiauan) recorded increased uplift in the west.

Further southwest, sedimentation was dominated by inner- to middle-neritic siltstone and sandstone (Mount Brown Formation), locally interbedded with debris-flow conglomerate and limestone (Field *et al.* 1989). The conglomerate units probably reflect active faulting and uplift. The Mount Brown Formation passes basinwards into middle- and outer-neritic fine sandy siltstone (Tokama Siltstone) in the southeast, and outer-neritic to bathyal fine sandy siltstone (Greta Formation) to the northeast (see Figs 6.5 and 6.6).

The Greta Formation is mainly an outer-neritic to bathyal, calcareous, fine to very fine sandy mudstone association. Deposition of the mudstone was interrupted periodically by the influx of inner-shelf sands and conglomerates. Lithofacies represented include coarse-grained debris-flow conglomerate (reflecting local uplift), limestone, interbedded sandstone and siltstone, and massive sandstone and conglomerate (Field *et al.* 1989). The thickness of the Greta Formation (locally over 500 m was deposited in the early Late Miocene (Tongaporutuan)) suggests that a subsiding basin developed in the area. A thin bed of late Middle Miocene (Waiauan: ~12 Ma) tuff at Gore Bay probably correlates with the early phases of Miocene volcanism at Banks Peninsula.

In western central Canterbury the late Middle–Late Miocene is represented by a marginal marine sand–claystone–volcaniclastite sequence, overlain by Late Miocene lacustrine clays (Coalgate Bentonite). Over 150 m of conglomerate at Brechin Burn (Newman & Bradshaw, 1981) is probably Middle Miocene and records increased uplift and erosion of basement rocks at this time. Late Middle Miocene (Waiauan) shallow-marine mudstone and shellbeds following Early–early Middle Miocene non-marine sediments, including a thin coal seam at Lees Valley, were deposited further north and in the east until at least the late Middle Miocene (Wilson, 1956). At Broken River, over 600 m of estuarine siltstone, lignite and sandstone of probable Middle to Late Miocene age passes up-sequence into thick greywacke-derived boulder to pebble conglomerate (Brechin Formation), probably reflecting active faulting. Locally, in the east near Waipara, Waiauan sediments rest unconformably on late Early Miocene (Altonian) deposits, indicating erosion or non-deposition of early Middle Miocene strata.

At Banks Peninsula there is no record of sedimentation between the late Early Miocene (Altonian) Bradley Sandstone and the first volcanics (11 Ma) of the Governors Bay and Lyttelton volcanic groups that mantled the Early Miocene erosion surface on the peninsula and dominated the area until the Pliocene. These Miocene volcanics extended as far west as the Leeston-1 petroleum well and as far south as the Resolution-1 petroleum well (see section 8.7.1). No record of Middle–Late Miocene sedimentation has been preserved.

On the Chatham Islands the rock record in the late Middle–Late Miocene is represented almost entirely by volcanigenic deposits, although thin, lenticular bioclastic limestone is interbedded locally (see section 8.7.1).

In offshore south Canterbury, at the Endeavour-1 petroleum well, Middle–early Late Miocene sediments containing outer- to middle-neritic faunas indicate up-sequence shallowing; and at the Clipper-1 petroleum well, sediments of this age rest unconformably on the middle Early Miocene (Otaian) (Eaton in Crux *et al.* 1984). They record shallowing from upper bathyl through outer- to mid-neritic settings, and include about 40 m of tuffaceous sandstone of Late Miocene age. At the Galleon-1 petroleum well, the late Middle Miocene (Waiauan) rests directly on probable late Early Miocene (Altonian) sediments, and there is no record of the Lillburnian or Clifdenian (see above). Seismic profiles of the late Middle–Late Miocene sediments record widespread channelling down the continental slope as it built out eastwards.

Nannofossil ooze deposition continued into the Late Miocene at DSDP site 594, where a ~2 Ma gap in the middle–upper Tongaporutuan and a ~1 Ma gap at the Miocene–Pliocene boundary were recorded (Nelson *et al.* 1986b). The latter gap probably correlates with a Late Miocene glacio-eustatic sea-level drop and accompanying phase of erosion (Kennett, 1967, 1980; Lewis *et al.* 1985; Hodell & Kennett, 1986). The crest of the Chatham Rise was subject to widespread erosion at this time (Kudrass & von Rad, 1984; Lewis *et al.* 1985).

Most of the western portion of the southern Canterbury Basin was mantled in conglomerate, quartzose sandstone, claystone and lignite (White

Rock Coal Measures and Wedderburn Formation) in the Middle Miocene, reflecting continued regression. In many places (e.g., petroleum well JD George-1, Naseby and the Mackenzie Basin) there is evidence of Late Miocene–Pliocene erosion having removed sediments of Late (and in some cases Middle) Miocene age.

In the Dunedin area, Middle Miocene erosion and volcanism were followed by a minor transgression, giving rise to shallow marine sandstone and grainstone (Waipuna Bay Formation) in the east. In the west, continued erosion produced significant relief leading to local accumulation of non-marine diatomite and oil shales. These are associated with early volcaniclastics of the Dunedin Volcanic Group. The main phase of volcanism was late Middle–early Late Miocene (Lillburnian–Tongaporutuan), and it extended to the Oamaru area (Coombs *et al.* 1986; Reay & Walls, 1994; and see section 8.7.1).

In the Great South Basin, late Middle–Late Miocene sediments are included as part of the Oligocene–Recent Penrod Group (Cook *et al.* 1999). There are very few age data available, because the unit was very poorly sampled in exploration wells. In Tara-1 petroleum well Middle Miocene sediments overlie Late Eocene sediments unconformably, but there are no data from elsewhere. Along the northwestern margin of the Great South Basin, rejuvenation of sediment supply during Miocene to Recent times caused by rapid uplift of the Southern Alps formed a thick (up to 800 m) shelf wedge characterised by southeastwards-prograding clinoform strata. In southeastern parts of the basin the available samples indicate that the sediments are soft foraminiferal oozes (Cook *et al.* 1999).

On Campbell Island, coarse tuffs and volcanic breccia (Shoal Point Formation) of late Cenozoic age are widespread. They range in thickness up to ~200 m, and rest unconformably on the Eocene–Oligocene Tucker Cove Formation (Oliver *et al.* 1950). Cross-bedded deposits occur locally, and lenses of fossils indicate a shallow marine environment. Breccia associated with the tuffs consists mainly of dark scoriaceous fragments and blocks up to 3 m, and fragments of limestone, quartz, flint and schist from older rocks. Although an Early Pliocene age was inferred by Oliver *et al.* (1950) from macrofossils within the tuffs, an early Late Miocene age is probable, because of local interbedding of the Shoal Point Formation with the base of mainly overlying Late Miocene lavas (Weaver & Smith, 1989; and see section 8.7.1).

In Central Otago the Hawkdun Group of late Middle Miocene to Pliocene age is discontinuously exposed, mainly in the margins of tectonic basins (Youngson *et al.* 1998). The lower portion of the group (Wedderburn Formation) comprises predominantly quartzose sediments forming wedges and tongues of predominantly fluvial channel and braidplain deposits of quartzose conglomerate with a variable greywacke and schist content, and interbeds of sandstone, mudstone and rare lignite. Locally the Wedderburn Formation overlies and interfingers with the lacustrine Bannockburn Formation, but towards the basin margins it commonly overlies older strata disconformably or with angular unconformity. The age is mainly Late Miocene, but locally parts of the succession underlie a basalt flow K/Ar dated as 13.4±0.3 Ma, suggesting that it ranges in age down into the mid-Middle Miocene (Youngson *et al.* 1998). The Wedderburn Formation passes gradationally upwards into conglomerate-dominated deposits (Maniototo Conglomerate), consisting of thick beds of greywacke-dominated conglomerate and interbedded loess horizons. The lithofacies is inferred to represent the deposits of a braided river system (Youngson *et al.* 1998). The age is Late Miocene to Early Pliocene.

The Hawkdun Group reflects increased tectonism beginning in the Middle Miocene. The transitional change from quartz-rich basal deposits to lithic conglomerates is characteristic of unroofing successions deposited in areas undergoing tectonic inversion. It marks progressive decrease in recycling of material from pre-existing conglomerates, and increase in erosion of fresh basement detritus from rising adjacent ranges.

8.4.3 Western Southland basins

8.4.3.1 Waiau Basin

Shallow marine sedimentation continued into the Middle Miocene in the east and south of the Waiau Basin, and persisted slightly longer into the late Middle Miocene in the centre of the basin. Mudstone and graded sandstone units still accumulated on the western side. By the end of the Middle Miocene the Blackmount Fault was quiescent, but the Bellmount Fault System further east was at its most active. In the southwest, Middle Miocene sediments had been eroded by uplift of southern Hump Ridge and the Mid Bay High, resulting in an unconformity beneath shallow marine sediments of Late Miocene age at Port Craig. The Late Miocene is marked in the centre of the basin by coarse-grained sandstones (Rowallan Sandstone) and, in the north, in the Monowai Sub-basin, by terrestrial conglomerates (Prospect Formation). At this time the Waiau Basin extended around the northern end of Hump Ridge into the Waitutu Sub-basin, where a different sequence developed. The northern end extended into the Monowai Sub-basin (see below), which thus joined it to the Te Anau area. The Bellmount Fault System, which was probably most active during the Middle Miocene, was a controlling factor in deposition of the Rowallan Sandstone, and now marks the westward limit of the formation (see Figs 6.9 and 6.10).

The Late Miocene Rowallan Sandstone rests unconformably on Early–Middle Miocene rocks in the eastern and southern Waiau Basin. The angularity of the unconformity increases to the east and south, and it is truncated by a local (near top Miocene) unconformity. In the deeper western part of the basin the contact is, however, apparently conformable, with Rowallan Sandstone resting on Waicoe Formation mudstone. The Rowallan Sandstone, which covers much of the central Waiau Basin, consists of at least 500 m of massive, well-sorted, occasionally pebbly, fine- to medium-grained sandstone. It is generally leached and contains a sparse macrofauna. Shellbeds, concretionary horizons and small-scale, 1–2-m-deep, 20–50-m-wide, channels occur throughout. Leaching and bioturbation have largely destroyed diagnostic sedimentary structures. Micro- and macrofauna indicate shallow to very shallow environments with water depths less than 10 m, and are consistent with sedimentary structures such as flaser bedding that suggest very-shallow-marine to intertidal conditions (Turnbull *et al.* 1993).

In the southwestern Waiau Basin, mid-Late Miocene (late Tongaporutuan) sediments overlie an erosion surface, which truncates basement and Eocene–Middle Miocene sediments that are successively younger northwards. It was inferred that at least 3000 m of sediment was removed from the south end of Hump Ridge in the Middle Miocene (Turnbull *et al.* 1993). Post-Oligocene, pre-Late Miocene faulting in the Port Craig area was presumably related to this uplift, and to the continuation of thrusting on the Hump Ridge–Stewart Island Thrust System southwestwards over the northern Solander Basin. The sediments (Port Craig Formation) overlying the unconformity are composed of sandy bioclastic limestone of Late Miocene (Tongaporutuan to Kapitean) age, and represent a shallow-water equivalent of the Rowallan Sandstone. Interbedded lithofacies are bioturbated, calcareous, medium-grained sandstone with hummocky cross-stratification, pebble to boulder sandy conglomerate, calcite-cemented shelly sandstone and sandy shellbeds, and carbonaceous sandstone and lignite. These lithofacies pinch out northwards into the deeper part of the Waiau Basin along the late Early Miocene unconformity, changing laterally to somewhat deeper-water, less calcareous mudstone and sandstone, which grades into the Rowallan Sandstone.

Scarce paleocurrent measurements indicate south-southeast directed currents. The Late Miocene formations include Takitimu Group clasts as well as typical Fiordland debris. The presence of a large non-Fiordland component is reflected in the Monowai Sub-basin, where Carter and Norris (1977) note a dramatic change in provenance from Fiordland to Caples–Haast schist in the mid-Late Miocene.

In the Monowai Sub-basin the late Middle Miocene (Duncraigen Formation) consists of 500–600 m of graded sandstone and mudstone, much of it slump-folded, and a basal muddy carbonaceous member with lenticular conglomerate beds resembling those of the underlying early Middle Miocene Monowai Formation. The microfauna indicates relatively deep-marine, outer-shelf conditions, and an abrupt deepening of the sub-basin in later Middle Miocene times, from the very-shallow-marine delta environment of the Monowai Formation. The basin then abruptly shallowed again in the early Late Miocene, reaching shallow shelf depths in the mid-Late Miocene. Tectonic instability on a southeast-facing slope is indicated by slump folding and intraformational slump breccia horizons in the Duncraigen Formation, and scanty paleocurrent data suggest sediment transport towards the south (Carter & Norris, 1977). The source areas for the formation include Caples Terrane greywacke and semischist, and Fiordland.

8.4.3.2 Te Anau Basin

In the late Middle Miocene (Waiauan–Tongaporutuan) the Te Anau Basin was a dramatic influx of up to 3000 m of greywacke gravel (Prospect Formation) that reached as far south as the Monowai Sub-basin and continued until Early Pleistocene (Hautawan) times (Turnbull *et al.* 1993). The nature of the basal contact of the Prospect Formation, which is marked by a high-amplitude discontinuous seismic reflector, changes across the Te Anau Basin. In the east, adjacent to the Burwood Sub-basin, it is apparently gradational and conformable over the Early–Middle Miocene Takaro Formation (Harrington, 1982). In the central Te Anau Basin up to 100 m of erosional relief has been demonstrated (Turnbull, 1986), while in the west the contact is sharp and apparently conformable. The Upukerora-1 petroleum well in the central Te Anau Basin penetrated 160 m of Prospect Formation conglomerate before intersecting latest Early Miocene mudstone (Carter & Rainey, 1988). The upper contact of the formation is an unconformity, overlain by Pleistocene glacial deposits (McKellar, 1973).

The dominant lithofacies in the Prospect Formation is a planar-bedded, clast- and matrix-supported, sandy conglomerate occurring in thick beds or shallow channels. Trough cross-bedded, clast-supported, sandy conglomerate with lenses and interbeds of sandstone and mudstone with occasional flaser bedding is also common (Turnbull *et al.* 1993). Lignite and carbonaceous mudstone and siltstone occur throughout the unit and have been mined locally. Facies relationships and types suggest the formation is predominantly fluvial, a basin-wide braidplain with subsidiary alluvial fans draining onto it from Fiordland in the west and the Takitimu Mountains in the east. Fanglomerates near the Takitimu Mountains (Carter & Norris, 1977) imply that the streams there were steep and actively eroding. Lignite and carbonaceous mud sequences may represent abandoned channel swamps, or distal overbank deposits. Mudstone units near the base of the formation, some containing marine dinoflagellates, may represent marine-influenced bays adjacent to prograding delta-front gravel lobes (Turnbull *et al.* 1993). Prospect Formation outcrops described from the Monowai Sub-basin by Carter and Norris (1977), and basal Prospect Formation facies in the Forest Burn, have a clear marine influence, with flaser-bedded and rippled siltstone and sandstone and shallow-marine trace fossils.

Paleocurrent indicators suggest flow in a general southerly or southeasterly direction (Harrington, 1982). Clast types vary across the basin, being mainly Caples-derived, but have a minor Fiordland component in the west and south. In the Monowai Sub-basin, clasts are inferred to be derived from the adjacent Takitimu Group (Turnbull *et al.* 1993).

8.4.3.3 Solander and Balleny basins

The Solander and Balleny basins became fully connected during the Early Miocene, although several basement highs remained devoid of Early Miocene sediments and probably contributed debris to adjacent basins. By the end of the Miocene, sediments had submerged all but one basement high in the Balleny Basin. Miocene sediments do not overlap the Hump Ridge–Stewart Island Shelf, and this area is believed to have stayed emergent for much of the Miocene, shedding clastic sediments into the Solander Basin.

Middle–Late Miocene sediments in the offshore portion of the Solander Basin are much more variable than in the onshore regions. Local erosional breaks are common, usually in the form of channels and slump horizons. Disruption increases towards the top of the succession, suggesting an increase in tectonic activity in the Late Miocene. This is confirmed in the northern onshore Waitutu Sub-basin where the latest Miocene succession is overwhelmed by coarse-grained rapidly deposited sediments that persist into the Pliocene.

Middle–Late Miocene mudstone in the onshore Waitutu Sub-basin is a continuation of Lower Miocene marine lithologies (Waicoe Formation), and is typically massive, blue-grey and calcareous. It includes units up to 400 m thick of graded sandstone and mudstone, which become finer-grained and thinner-bedded upwards. Some sandstones are coarse-grained and bioclastic with abundant bathyal macrofossil debris. Microfaunas in this sequence indicate bathyal conditions until the Late Miocene, when the area shallowed to outer-shelf depths (Turnbull *et al.* 1993).

In the offshore Solander-1 petroleum well, mudstones similar to those of the underlying Early Miocene were deposited for much of the Middle Miocene and are overlain by a conglomeratic channel fill or olistostrome of late Middle Miocene age (Clarke, 1986). This is succeeded by more mudstone with scattered limestone beds, reaching into the top Miocene (late Kapitean). In the offshore Parara-1 petroleum well, similar Middle Miocene mudstones include coal fragments and sandstone and limestone beds, but revert to a dominantly bathyal mudstone succession in the Late Miocene. An unconformity in the early Late Miocene is probable, based on the presence of glauconite and the absence of early Late Miocene (Tongaporutuan) sediments (Turnbull *et al.* 1993).

Adjacent to the Solander Fault a westwards-thickening sedimentary wedge within the Middle–Late Miocene section suggests an episode of tensional faulting, or possibly dextral strike-slip motion on the Solander Fault. The Solander Ridge was completely buried during the Middle Miocene, and the top Miocene seismic reflector can be traced into the offshore Waitutu area. Seismic evidence suggests that a maximum sediment thickness of more than 1200 m is reached in the northwest of the offshore Waitutu Sub-basin, thinning to an average thickness of about 400 m (Turnbull *et al.* 1993).

By the end of the Miocene the Balleny Basin was continuous with the Waitutu Sub-basin and the western Solander Basin. Middle–Late Miocene sediments are thickest in the northern part of the basin, with more than 800 m preserved west of the Hauroko Fault. A northwest-trending depocentre containing at least 600 m of sediments is a feature of this northern sector, while in the southern sector of the basin Middle–Late Miocene sediments are relatively thin. The nature of the sediments in the Balleny Basin is difficult to assess because no samples are available.

8.4.4 West Coast basins

The late Middle–Late Miocene history of the West Coast region is characterised by tectonic instability, with intermittent uplift and erosion of rising fault blocks supplying sediment to adjacent subsiding basins. The extreme amount of late Cenozoic basement uplift southeast of the Alpine Fault contrasts with the occurrence of an elongate sedimentary basin to the northwest of the fault in the south of the area, trending parallel to the present coastline and having a maximum depth to basement immediately offshore of over 3500 m (Nathan *et al.* 1986). It has been inferred by Kamp *et al.* (1992a) to be a foreland basin that originated from loading of the Australian plate by the Pacific plate, and to have developed sub-parallel to the Alpine Fault. Other main basins now preserved onshore correspond to the areas of most rapid late Cenozoic subsidence, and include the Murchison Basin and the Grey Valley Trough (see section 8.2) The succession offshore of the present coastline is entirely marine, whereas that in the Murchison Basin and surrounding area is almost wholly non-marine. In the intervening area there is a complex interdigitation between marine and non-marine sediments. Offshore the Western Platform started to subside in Middle Miocene times, with the rate of subsidence increasing southwards. The Cape Foulwind Fault Zone, bounding the northern part of the Western Platform, was active over at least part of its length, with relative subsidence on the western side. Fault control of sedimentation is also marked in the Golden Bay–Tasman Bay area to the north, with major breaks across the Wakamarama and Pepin faults.

The foreland basin in south Westland started to form at about the time that the plate boundary became significantly convergent. Fission track data (Kamp *et al.* 1992a) indicate that basement uplift by thrusting and reverse faulting has inverted the former inner margin of the basin.

Late Middle Miocene sediments are exposed onshore in a narrow, steeply dipping discontinuous strip along the coast north of Milford Sound. In contrast to the apparently continuous succession offshore shown by seismic data, the onshore sequence is cut by unconformities, suggesting intermittent uplift during sedimentation. The late Middle Miocene (late Lillburnian–Waiauan) portion of the Tititira Formation (see Fig. 6.12) has a thickness in places exceeding 1000 m, and rests unconformably on the Oligocene Awarua Limestone (Nathan, 1978a). The lower part of the Tititira Formation comprises a thick coarsening-upwards bathyal succession, passing upwards from hemipelagic mudstone through distal to proximal turbidites, and then into a sequence of granule to boulder conglomerate and interbedded medium- to coarse-grained sandstones and muddy micritic limestone (Nathan, 1978a). The conglomerate layers are commonly matrix-supported, contain clasts up to 15 m in diameter, and are probably debris flows (Aliprantis, 1987). Sparse paleocurrent measurements indicate transport from northeast to southwest, approximately parallel to local Neogene faults. The sediments are largely derived from a granitic source, with a smaller amount of Greenland Group and some material derived from reworking of early Tertiary sediments. No material derived from the present east side of the Alpine Fault has been identified and the original sources now lie well to the south. The stratigraphic top of the Tititira Formation is not seen, as it is everywhere overlain at a thrust contact by the allochthonous Middle Eocene–Late Oligocene Jackson Formation (Sutherland *et al.* 1996).

The Late Miocene (Tongaporutuan–Kapitean) rocks do not crop out onshore, but two deep onshore petroleum drillholes (Harihari-1 and Waiho-1) show an essentially complete succession of Middle–Late Miocene strata, consisting predominantly of marine mudstone up to 1400 m thick containing thick interbeds of granite-derived sandstone and conglomerate, which form prominent, but impersistent, reflectors on seismic profiles. Major increases in sedimentation rate are recorded in the Waiho-1 and Harihari-1 drillholes at between ~15 and 13 Ma (Sutherland, 1996; Sircombe & Kamp, 1998). In an offshore petroleum well, Mikonui-1, drilled 45 km northwest of Hokitika, the Middle–Late Miocene section has thinned to ~200 m, and consists of mudstone resting unconformably on Oligocene limestone (Sircombe & Kamp, 1998). Seismic records for the offshore area and the small onshore area show an almost flat-lying late Cenozoic succession, gradually thinning northwards. The basin is asymmetric in cross-section, and Middle–Late Miocene sediments have a wedge-shaped geometry thickening towards the Alpine Fault (Sutherland, 1996), suggesting that uplift on boundary faults started in Middle Miocene times.

To the north, the Grey Valley Trough continued to subside, and it also expanded southwards in comparison with its Early Miocene extent. A widespread unconformity separates the upper part of the late Early Miocene (Altonian) to Late Miocene (late Tongaporutuan) Stillwater Mudstone from overlying sediments (see Fig. 6.12), which are invariably of shallower facies (Eight Mile Formation). The greater part of the Eight Mile Formation, which is up to 1000 m thick, consists of blue-grey micaceous mudstone and muddy sandstone, with sandstone increasing upwards. However, lithofacies locally include shallow marine greensand, non-marine conglomerate, micaceous quartz sandstone and thin coals, and conglomerate deposited in a coastal to fluvial environment (Suggate & Waight, 1999). The Eight Mile Formation ranges in age from late Late Miocene to Early Pliocene (late Tongaporutuan to Waipipian).

A continuous Middle Miocene–earliest Quaternary succession occurs in the lower Inangahua Valley (see Fig. 6.13). Early Miocene marine beds (Inangahua Formation) contain an upwards-increasing proportion of arkosic sand, and microfaunas indicate rapid shoaling, before passing conformably upwards into a sequence of non-marine sandstone, conglomerate and carbonaceous mudstone, with thick lensoid coal seams near the base (Rotokohu Coal Measures). The top of the Inangahua Formation is dated as late Early Miocene (Altonian) and the Rotokohu Coal Measures range up into the Pliocene. Provenance studies indicate several source areas. Conglomerate boulders are largely Greenland greywacke, presumed to be derived from a rising area to the south and east, and the sand is mainly granite-derived.

In the onshore coastal area north of Greymouth, and up to at least 40 km northeast of Westport, a mid-Miocene (sub-Waiauan) unconformity is strongly developed, and sediments of older Middle Miocene (Clifdenian to Lillburnian) age are entirely missing. Sediments ranging in age from Middle Miocene (latest Waiauan) to early Late Pliocene (Waipipian) are all included in the O'Keefe Formation (Nathan *et al.* 1986; Laird, 1988), which consists predominantly of blue-grey muddy micaceous fine sandstone and siltstone containing scattered bands of calcareous concretions. Interbedded conglomerates and debris-flow deposits are locally found in sections adjacent to the present Paparoa Range (Laird, 1988). In the offshore Haku-1 petroleum well north of Greymouth, a sub-Waiauan unconformity, at which only the upper part of the Lillburnian Stage is missing, is marked by the incoming of mica and a small, but persistent, component of fine sand. The overlying sediments contain shallow-shelf microfaunas of Middle Miocene to Early Pliocene (Waiauan to Opoitian) age.

The sediments immediately above the sub-Waiauan unconformity are invariably a shallow-water facies, commonly glauconitic (Nathan, 1977). Detailed mapping has allowed the unconformity to be traced over a wide area, and has enabled a partial reconstruction of sub-Waiauan geology. Close to the coast a slight angular discordance between the O'Keefe Formation and underlying rocks indicates that the pre-Waiauan beds had been tilted eastwards by 2–4° prior to deposition of the O'Keefe Formation, and that the amount of erosion gradually increases to the northwest. The depth of erosion increases abruptly at the western edge of the Paparoa and Papahaua ranges, where the O'Keefe Formation rests directly on pre-Tertiary basement rocks (granite or gneiss) or Eocene sediments, with the basal sediments containing granitic clasts. This implies major pre-Waiauan uplift of the area immediately east of the Lower Buller (Kongahu) Fault with the erosion of a large (2000+ m) thickness of early Tertiary sediments and exposure of a ridge of granitic rocks. It is probable that uplift also took place immediately east of the offshore Cape Foulwind Fault, as the pre-Middle Miocene beds are tilted on the eastern side but are almost horizontal on the western side.

A similar but apparently complete Middle Miocene–earliest Quaternary succession is seen in the Karamea district to the north (Neef, 1981). Rapid shallowing occurred in early Middle Miocene (Lillburnian) times, the overlying late Middle Miocene–Early Pliocene (Waiauan–Opoitian) sequence consisting predominantly of blue-grey muddy fine sandstone containing scattered near-shelf macrofossils.

The Murchison Basin continued to subside rapidly in Middle and Late Miocene times. Sedimentation of the fluvial Longford Formation and Rappahannock Group, which began in the early Middle Miocene (see section 8.2.5), probably continued into the late Middle and perhaps Late Miocene, although the age of the upper part is unknown. The total preserved thickness of the Longford Formation is 3000 m+. The basal coal seams are of high-volatile bituminous rank, and the whole formation is well indurated, suggesting that perhaps another 3000 m of late Cenozoic sediments was originally deposited, and has since been eroded (Suggate *et al.* 1978). Outside the Murchison Basin and Grey Valley Trough there is only a thin sedimentary mantle.

8.4.5 Taranaki and western offshore Northland basins

The Middle–Late Miocene interval in the Taranaki Basin (upper Wai-iti Group) corresponds to an abrupt increase in the supply of coarse clastic sediment into the basin, beginning in the late Altonian and early Clifdenian. The increased sediment supply resulted in the filling and initial over-spilling of the basin in the eastern foredeep. In general, Middle–Late Miocene strata form a northwestwards-thinning wedge that trends sub-parallel to the paleoslope. In the east, foredeep trough sediments are more than 2000 m thick. The continental margin offlap is depicted on seismic reflection profiles as a series of basinwards-stepping clinoforms. Each clinoform profile represents the successive position of the coeval basin floor, slope and shelf; collectively they illustrate a long-lived pattern of basin floor aggradation, slope progradations, and expansion of shelf areas since at least the beginning of the Middle Miocene (King & Thrasher, 1996).

No Miocene or Pliocene sediments younger than mid-Middle Miocene (Lillburnian) are known onshore in northwest Nelson, but seismic evidence indicates that up to 3000 m of Middle Miocene or younger sediments accumulated offshore locally in Golden Bay and Tasman Bay. Apart from the onshore outcrops mentioned above, the only information on the succession comes from four offshore petroleum exploration holes, all drilled on structural highs, and which may, therefore, have penetrated condensed sequences. The control of sedimentation by subsidence on major faults is clearly shown by the contrast in the thickness on the opposite sides of the Wakamarama Fault. The drillhole successions closely resemble those onshore, consisting of calcareous mudstone near the base, with an upwards-increasing proportion of sand. The upper part of the Miocene succession in Surville-1 drillhole contains thin coal seams, apparently interbedded with shallow marine mudstone and sand, indicating local emergence. The youngest dated Miocene sediments are mid-Middle Miocene (Lillburnian), although it is possible that the uppermost (undated) part of the sequence in Surville-1 may be slightly younger. An unconformity between Middle Miocene sediments and a relatively thin layer of ?Pleistocene shelly sands appears to be a regional feature, indicating uplift and some erosion of the whole northwest Nelson area during the Pliocene (Pilaar & Wakefield, 1978).

Seismic profiles at the southern end of the Taranaki Graben show a widespread, but impersistent, reflector close to the Early–Middle Miocene (Altonian–Clifdenian) boundary, although not always in exactly the same place (Pinchon, 1972). Shelf areas around the southern and southeastern periphery of the basin gradually expanded towards the north and west as the slope prograded during the Late Miocene. Evidence from seismic reflection profiles indicates that uplift on individual anticlines, in some cases to wave base or above sea level, also took place in the southern part of the basin at the same time, leading to erosion of Late Miocene sediments.

Offshore south of the Taranaki Peninsula (Kupe-1 and Kupe South petroleum wells), the initial bathyal foredeep trough had aggraded to shelf depths by the Middle Miocene, and about 200 m of finely interbedded shelf mudstones and siltstones accumulated during the Late Miocene. Similar fine-grained shelf successions occur in wells to the north (Toru-1 and the Kapuni field), but to the southeast, and east of the Manaia Fault, Middle–Late Miocene strata are absent, possibly due to uplift on the fault.

By the late Middle Miocene only a remnant foredeep remained between shelf areas adjacent to the Taranaki Fault. Thin sands were deposited at bathyal depths within this foredeep trough in the Lillburnian–Waiauan, and were overlain by deep-water (slope) muds of Waiauan to early Late Miocene (Tongaporutuan) age. These sediments were unconformably overlain by outer-shelf muds, and by late Late Miocene (late Tongaporutuan) times the foredeep east of the Manaia Fault had apparently been infilled or uplifted to shelf depths.

From the Middle Miocene the area now occupied by the Wanganui Basin was probably exposed basement and may have been actively rising (see Figs 6.3, 6.14 and 8.2), as the southern end of the Taranaki Fault was active in the Late Miocene. Throughout most of the Miocene a very narrow shelf probably existed along the basin's southeastern margin, between the Taranaki Fault scarp and the exposed basement landmass. To the north of the Taranaki Peninsula the Patea–Tongaporutu High formed a narrow, elevated or shallow submarine basement ridge in the Middle Miocene that separated the Taranaki and King Country basins. In wells drilled on the crest of the ridge (Kiore-1 and Uruti-1), Late Miocene strata directly overlie basement, indicating subsidence and submergence of the ridge.

Apart from isolated outliers of Mount Messenger Formation preserved near Tirua Point, all Late Miocene strata have been eroded from the northeast of the Taranaki Basin. Late Middle Miocene (mid-Lillburnian–Waiauan) Mohakatino Formation rocks are exposed in a coastal strip in north Taranaki, and also sub-crop at or near the sea floor immediately offshore. Calcareous sandstones and sandy limestones of similar age are also exposed near Tirua Point. The latter sediments were deposited in bathyal depths, but also contain redeposited, shelf-restricted foraminifera. They were deposited as carbonate slope apron on the eastern flank of the marginal foredeep trough, adjacent to the Patea–Tongaporutu High (Nodder *et al.* 1990).

The Mohakatino Formation (~80 m thick), which consists of sandstones, siltstones and mudstones, with andesitic detritus as a dominant constituent, is exposed along the north Taranaki coastline. It was deposited concurrently with the eruption of the Mohakatino Volcanic Centre, a chain of andesitic, submarine stratovolcanoes in the northeastern Taranaki Basin. The now-buried volcanic edifices are identified from seismic reflection profiles, and from gravity and magnetic anomalies (King & Thrasher, 1996 and references therein).

The volcaniclastic content diminishes up-section, which is partly due to waning of volcanic activity. On seismic reflection profiles, volcaniclastic strata generally have high amplitude reflectivity, and are best identified in units that onlap the flanks of drilled (or inferred) volcanic edifices. They may interfinger with the deposits that make up these structures. The formation has been intersected in several offshore wells, indicating distribution over a large, primarily offshore, area. Significant volcaniclastic content is also present in the Middle–Late Miocene interval in some onshore wells in northernmost Taranaki Peninsula.

The volcaniclastic strata range in age from mid-Middle to Late Miocene (mid-Lillburnian to Kapitean). Exposures along the north Taranaki coast, distal to the eruptive centres, have consistent ages between 14.5 and 14.7 Ma (K/Ar method), which support inception of Mohakatino volcanism in the mid-Middle Miocene (Bergman *et al.* 1992). The most highly volcaniclastic parts of the succession have a restricted late Middle Miocene (Waiauan) age, implying that the main period of volcanism occurred during this time.

Deposition of the volcaniclastic Mohakatino Formation sediments was superimposed on background patterns of clastic sedimentation. In the north Taranaki coastal section, Mohakatino sediments derived from the north at first interfingered with, and were then overwhelmed by, clastic sediments prograding into the trough from the south. Thin volcaniclastic horizons are interspersed throughout the overlying Mount Messenger and lower Urenui formations, indicating that there were intermittent volcanic eruptions at least until the end of the early Late Miocene (mid-Tongaporutuan) (King *et al.* 1993).

The overlying late Middle–Late Miocene (late Lillburnian–Tongaporutuan) Mount Messenger Formation is best developed in the northeast of the onshore part of the basin, and a thickness of about 850 m is exposed along the north Taranaki coast as far north as the Mokau River and in areas immediately inland (King *et al.* 1993). Although mudstone is common as interbeds, sandstones are the predominant lithology over most of the outcrop section, and they are generally fine- to very fine-grained, containing up to 20% mud. Sandstones are commonly normally graded, and individual beds often exhibit Bouma sequences. Fining-upwards depositional cycles are common in lower parts of the formation, and are attributed to submarine fan development and abandonment. These cycles consist primarily of thick-bedded sandstone packages, commonly with basal channelised conglomerates, overlain by thin-bedded sandstone packages, generally totalling a few tens of metres in thickness, topped by a mudstone interval several metres thick (King *et al.* 1994). In several locations interbeds of sandstone and siltstone within the lower Mount Messenger Formation show spectacular intraformational slumping. The high sand content and presence of climbing ripples indicate that sediment supply rates were high. The unit records a rapid influx of coarse clastics into a deep-water trough, and deposition on a series of stacked submarine fans by high-density turbidity currents and fluidised mass flows (King *et al.* 1994).

The entire exposed Mount Messenger Formation (Fig. 8.2) is inferred to represent a succession of (mainly) lowstand systems tract deposits that display an overall progradational stacking pattern (King *et al.* 1994). Lowermost sequences are composed primarily of early lowstand systems tract basin-floor fan deposits, whereas uppermost cycles consist mainly of mid-lowstand slope fan deposits. The cycles represent fourth-order changes in relative base level, probably controlled by fluctuating subsidence and sediment supply. The same general succession is present in nearby offshore petroleum wells.

As stratigraphically underlying thick-bedded sandstones are inferred basin floor deposits, and overlying siltstones of the Urenui Formation are inferred slope deposits, the exposed upper Mount Messenger Formation beds were probably deposited at the foot of the advancing slope (King *et al.* 1993, 1994). The various channelised packages are interleaved and discontinuous. They could represent either a series of relatively small, coalescing slope aprons, or lateral switching of slope aprons caused by the persistent debouching of sediment from a large canyon located upslope.

In the subsurface the Mount Messenger Formation consists primarily of fine- to medium-grained sandstones that grade upwards into interbedded siltstones and minor mudstones. The formation itself fines upwards in most wells, and the contact with the overlying siltstone-dominated Urenui Formation is usually gradational. In the subsurface Taranaki Peninsula the Mount Messenger Formation is well developed in the north but decreases in thickness and sand content towards the south. It is poorly represented or absent in the offshore petroleum wells to the south of the peninsula.

Offshore, Mount Messenger Formation submarine fan sandstones reach a thickness of 300 m, and are distributed on a northeast trend across the centre of the basin. Seismic reflection data show that the Mount Messenger channel complexes in the Maui field consist of entrenched valleys up to 4 km wide within which individual channels meandered (Bussell, 1994). The formation has been encountered in drillholes as far south as the Moki and Maui-4 oil fields.

The Urenui Formation, which stratigraphically overlies deep-water Mount Messenger Formation sandstones, forms a distinctive unit of characteristically fine-grained slope deposits in northeastern onshore regions and in the subsurface of the Taranaki Peninsula. The formation is contiguous with, and a lateral correlative of, upper Manganui Formation slope deposits identified in central and southwestern parts of the basin from seismic reflection and paleobathymetric signatures.

The Urenui Formation is best exposed in cliffs along the north Taranaki coast in a section that is virtually continuous for more than 20 km, and measures about 850 m in stratigraphic thickness (King *et al.* 1993). In outcrop the formation consists mainly of heavily bioturbated mudstones, but it includes sporadic thin tuffaceous layers. It is variably calcareous, and contains scattered shell fragments. Over otherwise massive intervals, bedding is often only discernible from concretion bands, tuff horizons and parting lineations.

At several stratigraphic levels the Urenui Formation coastal section is punctuated by major channel systems, usually between 30 and 60 m deep. They are characterised by deeply incised bases and an infill of coarse-grained lithofacies including thick-bedded and amalgamated sandstones, thin-bedded, wavy-laminated, scoured, carbonaceous sandstones, and debris-flow conglomerates (King *et al.* 1993). Some channels are mud-filled. The exposed channels are lateral equivalents of several broad channels imaged on nearby seismic reflection profiles at approximately the same stratigraphic level. The subsurface channels are each about 1–2 km across, and many are asymmetric in cross-section. The seismic reflection profiles show well-stratified sediments infilling channels, some of which are steep-sided. The sediments thin and onlap the channel margins (King *et al.* 1993). Similar relationships are evident in outcrop. The depositional setting is that of a northwest-facing prograding continental slope that extended across the Taranaki Basin in the Late Miocene from the peninsula region to just north of the South Island.

The conglomerates and sands within otherwise fine-grained Urenui Formation deposits represent sediment pathways interpreted as infilled slope canyon complexes. These canyon complexes acted as slope bypass zones, down which fluvial- and shelf-derived sediments were transported to the basin floor. The base of each of the Urenui Formation canyon complexes represents a type 1 sequence boundary, thought to have been formed by slope failure and retrogradational slumping caused by slope over-steepening, possibly triggered by a drop in relative base level. Each of the sequences infilling the channels is composed mainly of lowstand systems tract deposits. The combined factors of basin subsidence and sediment supply are considered to have been more important in controlling depositional cyclicity throughout the Middle–Late Miocene than any global fluctuations in sea level (King & Thrasher, 1996). The entire progradational system, as exemplified by the exposed Mount Messenger and Urenui formation successions, is inferred to be governed mainly by an over-supply of sediment (King *et al.* 1993).

A lithologically similar unit to the Urenui Formation, the uppermost Miocene–earliest Pliocene Ariki Formation, includes marls encountered in several wells drilled on the northwestern parts of the basin, and takes its name from the petroleum well Ariki-l lying offshore west of Kawhia. The locally developed Ariki Formation is younger and more calcareous than the Urenui Formation, and was deposited in much deeper water in distal northwestern parts of the Taranaki Basin. These thin (70–90 m) marl intervals grade laterally to the south into basin floor mudstones and thin, poorly developed turbidites. The marls form a bold marker horizon on seismic reflection profiles and are used for mapping the base of the Pliocene. This seismic horizon, N2 (Isaac *et al.* 1994) extends northwards into the offshore west Northland Basin, where it also separates the Middle–Late Miocene succession from Pliocene deposits.

The calc-alkaline volcanoes on and west of the Northland Peninsula were extinct by the end of the Early Miocene. Post-Early Miocene uplift of the Northland Peninsula and westward tilting led to deep erosion on land, and the accumulation of a thick, structurally simple, clastic sequence west of the peninsula. This tectonic quiescence indicated by the offshore Middle Miocene–Holocene strata is in marked contrast to the convergent margin tectonism of the Early Miocene. This change is thought to reflect migration of the plate boundary to the east, away from Northland.

In offshore west Northland, the Middle–Late Miocene succession in the Waka Nui-1 petroleum well is represented by 140 m of sediments dominated by very calcareous, slightly silty and pyritic marl with interbedded volcaniclastics, particularly near the base (Milne & Quick, 1999). The age ranges from Lillburnian to Tongaporutuan. The unit, equivalent to the upper part of the Wai-iti Group of the Taranaki Basin, rests unconformably on the Early Miocene volcanic and volcanogenic Waitakere Group, and is in turn overlain unconformably by Early Pliocene sediments. The latest Early Miocene and possibly earliest Middle Miocene (late Altonian and ?Clifdenian) and latest Late Miocene (late Kapitean) are missing (Strong *et al.* 1999). Seismic data show that the Middle–Late Miocene sequence is present throughout western offshore Northland, tapering westwards into the Reinga and New Caledonia basins. It thins locally over Early Miocene volcanoes, and pinches out on the higher flat-topped ones. West of northern Northland the unit pinches out eastwards on the planar surface of the Northland Allochthon. Isopach maps (Isaac *et al.* 1994) show that depocentres were close to the present coast, between the major volcanic massifs. The reflector configuration and seismic facies of the Middle–Late Miocene sequence suggest a marine basin, gradually filled by progradation from an eastern sediment source and from eroding volcanic islands and headlands.

In the far north, calcareous volcaniclastic sandstone with minor bryozoan fragments and glauconite dredged from the sea floor northeast of the Three Kings Islands contains Middle–Late Miocene (late Lillburnian–Kapitean) foraminifera. Conglomeratic limestone from the shelf edge north of Cape Reinga comprises granule- to cobble-size clasts of Cretaceous or Paleogene mudstone in a micrite matrix also with Middle–Late Miocene (late Lillburnian–Kapitean) foraminifera. The dredge samples are all of clasts, probably derived from in situ exposures on the sea floor nearby. Hardground facies indicate low rates of sedimentation, and the limestones are inferred to be present in relatively thin, discontinuous units bounded by unconformities or disconformities (Brook & Thrasher, 1991).

8.4.6 Onshore Northland, South Auckland, King Country and Wanganui basins

The Kaikoura Orogeny caused uplift of much of the onshore Northland and south Auckland regions, marked by extensive block-faulting and tilting during Middle and Late Miocene times. The pattern of block-faulting was largely inherited from earlier, late Mesozoic and early Paleogene trends. As a consequence, sediments of Middle and Late Miocene age are absent from most of Northland, and are entirely absent from the south Auckland region. Only south of Te Kuiti, in the King Country and Wanganui basins, are sediments of this age present (Fig. 8.2).

Although bioclastic limestone and calcareous sandstone of Middle Miocene to Pleistocene age has been dredged offshore from the Cape Reinga–North Cape area (see above), the only outcrops onshore of this age are restricted to a small area at North Cape. Here, several metres of subhorizontal bioclastic limestone (Waikuku Limestone) contain foraminifera indicating a Middle Miocene (Lillburnian to Waiauan) age (Leitch *et al.* 1969; Brook, 1989). The limestone is inferred to overlie folded Early Miocene Parengarenga Group strata unconformably; the top of the limestone is eroded, and is overlain by Late Pleistocene–Holocene dune and swamp deposits.

Southeast of Doubtless Bay an irregular north-sloping erosion surface cut in Tangihua Complex rocks is overlain by weakly lithified fluvial to marginal marine conglomerate, pebbly and carbonaceous sandstones, minor mudstone and lignite, the Mangonui Formation of Brook and Hayward (1989). Conglomerate clasts are mainly from Tangihua Complex rocks, but some are Early Miocene Whangaroa Subgroup andesite. The sediments contain a Late Miocene (Waiauan–Kapitean) palynoflora (Isaac *et al.* 1994). The maximum preserved thickness is between 50 and 100 m.

In the King Country and Wanganui basins the mid-Miocene–earliest Pliocene succession was included in the newly designated Whangamomona Group by Kamp *et al.* (2002, 2004). The group is restricted to the area south of Te Kuiti, and rests unconformably on older units. It is inferred to be a transgressive continental margin wedge prograding northwards into the King Country Basin, which began subsiding in the Middle Miocene. It has a thin transgressive portion (Otunui Formation) and a thick regressive part (Mount Messenger to Matemateaonga formations). The regional Middle Miocene subsidence in the King Country Basin that led to deposition of the Whangamomona Group is expressed along the western margin of the basin by onlap onto the Tongaporutu–Herangi High. This onlap became more marked during the latest Miocene, when the Matemateaonga Formation accumulated in the Wanganui Basin (Fig. 8.2).

The basal unit of the Whangamomona Group is the Middle Miocene (late Clifdenian–Waiauan) Otunui Formation (Kamp *et al.* 2004), representing a transgressive shelf succession. It overlies the Mahoenui Group east of the Ohura Fault, and the Mokau Group west of this fault. The Otunui Formation is 100–200 m thick, the dominant lithologies being fine- to medium-grained, moderately sorted massive to poorly bedded silty fine sandstone and sandy siltstone. Tuffaceous material is locally present, and small broken shell fragments are scattered throughout the sandstone. Sporadic conglomerate channel fills up to 10 m thick and containing boulders up to 40 cm in diameter also occur, becoming more frequent in the upper part of the formation. The basal facies of the Otunui Formation is commonly characterised by the Mangarara Limestone (Henderson & Ongley, 1923), an onlapping shellbed or glauconitic medium- to coarse-grained limestone, which is up to 1.2 m thick and locally shows evidence of traction currents and bioturbation (Gerritsen, 1994). The Otunui Formation extends at least as far south as the Puniwhakau-1 and Tupapakurua-1 petroleum wells, near the western and eastern basin margins respectively of the King Country Basin, where it reaches thicknesses of 140 m and 37 m respectively (St John, 1965; Gerrard, 1971). It probably extends southwards into the northern Wanganui Basin (Kamp *et al.* 2002, fig. 6). To the west of the Puniwhakau-1 petroleum well, the Otunui Formation thins against the Patea–Tongaporutu High, and is not present in the Kiore-1 petroleum well, 17 km to the northwest.

The Otunui Formation passes conformably upwards into the late Middle–early Late Miocene Mount Messenger Formation, which consists of ~60 m of slightly calcareous siltstone interspersed with well-sorted massive sandstone beds interpreted to be deposits of sandy debris flows (Kamp *et al.* 2004). The transition to Mount Messenger Formation reflects rapid late Middle Miocene (mid-Waiauan) subsidence to bathyal depths. The Mount Messenger Formation is overlain conformably by up to 500 m of fine-grained beds of the Late Miocene (late Tongaporutuan) Urenui Formation, consisting mainly of silty to sandy mudstone, locally including thin sandstone layers and shellbeds. In the southwest of the basin, in the Puniwhakau-1 petroleum well, 129 m of Urenui Formation rests disconformably on the Otunui Formation, while further west, in the Kiore-1 petroleum well, 220 m of Urenui Formation rests unconformably on basement on the flanks of the Patea–Tongaporutu High. Late Miocene strata are not present in the southeastern part of the basin.

The overlying Kiore Formation is 500 m thick and lithologically similar to the Urenui Formation. It comprises predominantly sandy siltstone with minor sandstone lithofacies that are punctuated by discontinuous channels infilled with bioclastic material and siliciclastic conglomerate (Vonk *et al.* 2002). It is distinguished by the higher frequency of channels, the slightly coarser texture of the siltstone lithofacies, and the occurrence of interbeds of wavy-bedded, very fine-grained sandstone and siltstone. The Late Miocene (late Tongaporutuan) Urenui and overlying Kiore formations together represent the beginning of regression and the northward progradation of slope deposits (Kamp *et al.* 2004). The deposits overtopped the Patea–Tongaporutu High and formed a narrow shelf and slope along the eastern margin of the Taranaki Basin, where similar slope deposits, but with more prominent channel systems, have been described (King *et al.* 1993; Kamp *et al.* 2002; and see section 8.4.5).

The Matemateaonga Formation (late Late Miocene–Early Pliocene/Kapitean–early Opoitian), which conformably overlies the Kiore Formation in the southwest of the King Country Basin, extends westwards from the Ruahine and Kaimanawa ranges in the east into the eastern Taranaki Peninsula (Fig. 8.2). It also onlaps southwards onto basement, marking the beginning of substantial subsidence in the northern part of the Wanganui Basin, and southward migration of the shoreline. In the eastern Taranaki Peninsula the formation comprises a 1000-m-thick succession characterised by the cyclic repetition of coquina shellbeds, siltstones and sandstones, inferred to have been driven by glacio-eustatic sea-level oscillations (Vonk *et al.* 2002). At least 20 sedimentary cycles are recognised, separated by erosional unconformities. They typically comprise a basal transgressive systems tract, commonly overlain by shellbeds; a highstand systems tract, typically comprising massive siltstone coarsening upwards into sandstone; and a regressive systems tract comprising thick shoreface to inner-shelf sandstones. The magnitude of sea-level change within each cycle is likely to have been of the order of 40–70 m (Vonk *et al.* 2002). Thin carbonaceous or coaly intervals were identified in the lower part of the formation in the southern Kaimanawa Range towards the eastern basin margin, where coal seams up to 0.5 m in thickness are interbedded with shellbeds within a 110-m-thick coarse conglomerate unit (Murphy *et al.* 1994).

In the Wanganui Basin the general absence of strata older than Late Miocene indicates a prolonged period of erosion prior to basin formation. Sedimentation onlapped southwards onto the basement from the King Country Basin in Late Miocene times. Up to 2000 m of Matemateaonga Formation inner mid-shelf sediments are found in the centre of the basin, and paralic sediments in the northern Rangitikei catchment (Thompson *et al.* 1994; Vonk *et al.* 2002).

On the western flank of the basin the Manutahi-1 well penetrated 150 m of conglomerates, sandstones and sub-bituminous coal seams. Correlation with similar sediments in the Puniwhakau-1 petroleum well to the northeast shows that this facies persists for at least 50 km, with coal horizons becoming more prominent towards the south (Murphy *et al.* 1994). It is inferred that this facies association extended eastwards beneath the younger Cenozoic rocks (Thompson *et al.* 1994).

Parts of the Whangamomona Group extended westwards into the eastern Taranaki Basin, overtopping or bypassing the Tongaporutu–Patea High. The Mount Messenger and Urenui formations are exposed in the northern Taranaki coastal section (see section 8.4.5), and the Kiore and Matemateaonga formations crop out to the south in the eastern Taranaki Peninsula (Vonk *et al.* 2002).

8.5 MIOCENE–PLIOCENE BOUNDARY EVENT (KAPITEAN–OPOITIAN BOUNDARY: ~5 MA)

This event is marked by the increased tempo of convergent margin uplift and erosion in response to a change in the Pacific plate relative motion. It is marked by a southward shift of the Australian–Pacific instantaneous pole at ~5 Ma (Cande *et al.* 1995; Sutherland, 1995, 1996). This change is reflected in the outpouring of a huge volume of clastic sediments into most New Zealand basins during the Pliocene–Pleistocene. In areas such as Taranaki and offshore Canterbury it triggered substantial outbuilding of the continental slope and dramatic expansion of shelf areas. In some onshore areas, particularly adjacent to the rising Southern Alps, such as the Moutere Depression and Canterbury, thick, coarse-grained fluvial sediments were deposited extensively.

In the onshore Northland and south Auckland regions, Pliocene and younger rocks, which are predominantly volcanic, rest unconformably on Early Miocene and older strata. However, to the south in the King Country and Wanganui basins the Pliocene and younger sediments rest with apparent conformity on Late Miocene strata.

In the southern portion of the Taranaki Basin, differential uplift of Miocene rocks led to erosion and the development of a pronounced unconformity at the base of Plio–Pleistocene strata. The unconformity surface dips gently to the north, and progressively younger Plio–Pleistocene strata lap southwards onto it. In central parts of the Taranaki Basin there is only local unconformity. In the northwest the Miocene–Pliocene boundary corresponds to the contact between a condensed interval of basin floor carbonates that contains the boundary, and overlying terrigenous clastic deposits. In the northeast the boundary is marked by the contrast between Late Miocene volcaniclastic sediments and their absence in the succeeding Pliocene. The succession appears to be similar in the adjoining offshore west Northland Basin. Here the Pliocene overlies the planed surface of the Northland Allochthon in the north of the basin, and, further south, rests on the eroded surface of Miocene volcanic rocks.

Southwards, the Taranaki Basin merges into the West Coast basins. The northernmost, the Moutere

Depression, is filled with Plio–Pleistocene non-marine gravels that rest unconformably on Early Miocene or older sediments or basement rocks. Elsewhere in the West Coast region the Pliocene saw widespread shallowing and a flood of coarse terrigenous material sourced from the rising Southern Alps on the southeastern side of the Alpine Fault (Pacific plate).

In the East Coast Basin (Fig. 8.3a,b) the Early Pliocene marks the end of Miocene extension, and the inception of a major compressional episode. This is recorded in many parts of the region by Pliocene deposits unconformably overlying tilted and eroded Miocene sediments (Chanier *et al.* 1999). In the Canterbury region, uplift of the Southern Alps, a general increase in folding and faulting, and glacio-eustatic sea-level fluctuation all contributed to the influx of sediment from the west, and the further progradation of the continental shelf wedge to the east. Erosion in the west in the Late Miocene locally removed all 'cover' strata, leaving Torlesse rocks unconformably overlain by Pliocene conglomerate. Further south, in Otago, the increased tempo of tectonism in the Late Miocene–Pliocene resulted in the similar widespread deposition of non-marine conglomerate, locally unconformable on older rocks. The Pliocene of western Southland records the culmination of local regression, as much of the region became fully emergent.

8.6 PLIOCENE–PLEISTOCENE

8.6.1 East Coast Basin

In many parts of onshore East Coast, Pliocene rocks rest unconformably on Miocene or older rocks and the Miocene–Pliocene transition is not preserved. In the Raukumara Peninsula, as far south as Gisborne (Fig. 8.3a), Early Pliocene (Opoitian) rocks, with minor exceptions (see later), are the youngest marine sediments preserved (Beu, 1995). In the northern part of the Raukumara Peninsula and in Marlborough, the Pliocene succession is dominated by clastic sediments. Between Gisborne and southern Wairarapa, however, the Pliocene is characterised by distinctive coarse skeletal carbonates, generally of restricted lateral extent, which grade into or are interbedded on various scales with siliciclastic mudstone or sandstone, often in cyclothemic fashion. These carbonates, which occur repeatedly at several discrete times during the Pliocene, are referred to collectively as the 'Te Aute Limestone lithofacies' (Beu, 1992; Nelson *et al.* 2003; see Plate 16). The limestones were inferred to have developed in protected shallow-water embayments or on submarine 'highs' in a narrow, northeast–southwest-trending seaway or strait extending from the vicinity of modern Cook Strait to north of Gisborne (Beu, 1992).

The latest Miocene–Early Pliocene (Mangaheia Group) in the far northeast of the Raukumara Peninsula between Cape Runaway and East Cape is represented by up to 2000 m of fossiliferous shallow-marine sandstones and sandy mudstones, with minor beds of limestone and pumiceous sandstone, resting unconformably on Early or Middle Miocene rocks (Chapman-Smith & Grant-Mackie, 1971; Beu, 1995; Field *et al.* 1997; Mazengarb & Speden, 2000). Deeper-water sediments occur towards East Cape, where over 300 m of early Opoitian mid-bathyal interbedded fine sandstone and siltstone grade upwards into shallower shelf deposits. To the southwest, inland from Tolaga Bay and Gisborne, latest Miocene–earliest Pliocene (?latest Kapitean–early Opoitian) deposits consist mainly of bathyal to outer-shelf alternating sandstone and mudstone in the lower 1000 m, which coarsen and shallow up-section to inner-shelf sandstone and siltstone, with a few conglomeratic shellbeds and local tuff beds near the top. Opoitian limestone occurs near Gisborne (Francis *et al.* 1987; Beu, 1995), where it forms a basal or near-basal shellbed overlain by up to 1800 m of muddy sandstone, siltstone, flysch and sparse pale-grey tephras, partly of early Late Pliocene (Waipipian) age (Kicinski, 1958; Neef & Botrill, 1992). To the northwest of Mahia Peninsula the Early–early Late Pliocene (Opoitian–Waipipian) strata are mainly sandstones and mudstones, with interbedded discontinuous shelly limestones (Beu, 1995; Mazengarb & Speden, 2000). Further west, near Waikaremoana, about 300 m of Opoitian bathyal sediments consist of alternating siltstone and sandstone, with common beds of redeposited pumice.

Scattered deposits of thin (up to 35 m thick) shallow-marine, fossiliferous, pumiceous, sandy mudstone and sandstone between Cape Runaway and East Cape constitute almost the only known marine deposits of Pleistocene age on the Raukumara Peninsula. They have an age of Oxygen Isotope Stage 9, or approximately 320,000 years BP (Mazengarb & Speden, 2000). Up to 150 m of Pleistocene sediments is present in the vicinity of Gisborne and to the northwest, up the Waipaoa Valley. The sequences unconformably overlie older strata, and have varying lithologies, including fine-grained sand, laminated mud, lignite, tephra, diatomite and gravel. Some of the sediments include freshwater bivalves and other non-marine fossils although, locally, estuarine fossils are present (Mazengarb & Speden, 2000).

In northern Hawke's Bay, near Mahia Peninsula, the Early Pliocene interval is about 1500 m thick and unconformity-bounded, consisting in its lower portion of mudstone and sandstone, which is overlain disconformably by pebbly limestone followed by mudstone and tuff capped by ~500 m of Opoitian coquina and calcarenite. The limestone units (Te Aute lithofacies) are discontinuous and grade laterally into sandstone or sandy limestone (Beu, 1992). West and southwest of Mahia Peninsula, in western Hawke's Bay, the Early Pliocene setting was deeper and forms the upper portion of upper bathyal sandstone (Mokonui Formation), which here is 300 m thick, but which thickens to 1000 m to the north, where it consists mainly of shelf-depth medium- to coarse-grained sandstones including thin limestone beds (Cutten, 1994). To the southwest the Mokonui Formation and overlying limestone rest on a bored hardground, or, locally with angular unconformity, on Late Miocene siltstone. The overlying early Late Pliocene (Waipipian) limestone varies in thickness from 20 m to ~100 m, and oversteps onto Torlesse basement in the far west of Hawke's Bay. Locally it passes upwards into 100 m of brown-weathered sandstone and limestone (Beu, 1992). Outcrops of Early Pliocene mid-shelf sandstone, sandy mudstone and limestone occur in the Ruahine Range. The great thickness, estimated at as much as 1500 m, of shallow-marine strata indicates very rapid subsidence, presumably controlled by the axial range faults (Field *et al.* 1997).

Early Late Pliocene (Waipipian–Mangapanian) successions in northern Hawke's Bay are locally up to 900 m thick and record very high sedimentation rates. North of Mahia Peninsula they consist mainly of poorly bedded to massive silty sandstone with graded sandstone beds (Hornibrook, 1981). Latest Pliocene rocks in northern Hawke's Bay also include small areas of flysch (Hornibrook, 1981; Fleming & Christey, 1989). To the west of Mahia Peninsula, equivalent strata are represented by massive, grey, sandy mudstone or sandstone, and include a 3-m-thick tuff bed. In western Hawke's Bay up to 800 m of Waipipian bathyal mudstone is overlain by over 100 m of Mangapanian probably shelf sands and conglomerates representing a marked shallowing to inner-shelf conditions, in contrast to overlying moderately deep-water Early Pleistocene (Nukumaruan) mudstone (Cutten, 1994).

In southern Hawke's Bay (Fig. 8.3b), south of Napier, Opoitian limestone of the Te Aute lithofacies and Waipipian sandstone (Mokopeka Sandstone) is relatively widespread, reaching a thickness of between 20 and 82 m (Field *et al.* 1997). The limestone also crops out below Te Mata Peak, where it is 30 m thick (Harmsen, 1984), but it thins to the southwest, lensing out into mudstone. The scarp at Te Mata Peak is formed by ~120 m of similar mid-Pliocene (Waipipian) limestone, which is locally planar and trough cross-bedded (Harmsen, 1985). Seventy to 150 m of calcarenite separates this from younger 280-m-thick limestone of the Te Aute lithofacies, which includes up to several tens of metres of calcarenite at its top and is overlain by mudstone (Kelsey *et al.* 1993).

Latest Pliocene (Nukumaruan) rocks are restricted to Hawke's Bay and Wairarapa. The early Nukumaruan includes limestone of Te Aute lithofacies and enclosing siliciclastic rocks. The Te Aute lithofacies differs from older formations of this lithofacies in being paler in colour and even more porous and permeable, presumably because of less diagenesis than the older rocks. The main occurrences are at Napier and to the south, and south of Hastings. These units are overlain by the widespread, cyclothemic mid- to late Nukumaruan rocks of the central Hawke's Bay lowlands. The succession generally has a thick, basal siliciclastic unit, over 1000 m thick in the Taradale-1 petroleum well but thinner and more varied further north, where it contains thick conglomerate units. The basal unit is overlain by four transgressive cyclothems in most areas, beginning with shallow-water bioclastic limestones, which commonly pass upwards into a thin shellbed of large aragonitic bivalves and oysters. A rapid transition to a thick deeper-water mudstone follows (Abbott & Carter, 1994).

A discrete area of Pleistocene (Nukumaruan) rocks, underlain by older Pliocene siliciclastics, lies between the Ruahine Range and Mason Ridge south of Napier (Cashman *et al.* 1992). The basal Nukumaruan limestone consists of a highly varied set of pebbly, barnacle-plate, shellbed lenses, underlain and overlain by bathyal sandy mudstone, interpreted by Beu (1995) to have formed during a glacial lowstand. Overlying deposits consist of mudstones and interbedded shallow-water sandy limestones capped by marine gravels, which are in turn overlain by tephra and pumice layers of Early Pleistocene (Castlecliffian) age (Shane, 1994; Shane *et al.* 1996).

Marginal marine sediments (Kidnappers Group) at Cape Kidnappers are of Early Pleistocene (early to mid-Castlecliffian) age (see Plate 16). The succession consists of 250 m of sandstone, conglomerate, thin fossiliferous estuarine mudstone beds, and white tuffs. Equivalent deposits also occur to the west near Mason Ridge, north of Dannevirke, and near Elsthorpe (Field *et al.* 1997).

In Wairarapa (Fig. 8.3) the latest Miocene–Early Pleistocene rocks have been included in the Onoke Group (Lee & Begg, 2002). Pliocene strata are particularly widespread in western and central Wairarapa, but are present in only limited areas. In the east they occur at Castlepoint, where a 40-m-thick Pleistocene (Nukumaruan) limestone unconformably overlies Early Pliocene (Opoitian) mudstone (Johnston, 1980). At Cape Turnagain Pliocene (Opoitian) cross-bedded coquina limestone and calcareous sandstone lie unconformably on mudstone of similar age (Moore, 1981).

In most places Pliocene rocks conformably overlie Late Miocene strata, but in northern and western Wairarapa they lap unconformably onto basement or rocks of Early Miocene or older age. Pliocene and Quaternary strata lie unconformably on basement west of the Wellington Fault. Several occurrences of marine Early Pliocene (Opoitian) rocks along the axial ranges suggest that an Early Pliocene marine link existed between the East Coast region and the Wanganui Basin to the west. Marine Opoitian siltstone occurs north of Woodville, between the Wellington and Ruahine faults. Neritic Opoitian conglomerate, sandstone and mudstone form an inlier at Makara on the axial ranges near Wellington (Field *et al.* 1997).

Conditions were initially shallow in northern Wairarapa, with deposition of neritic, cross-bedded sandstone (Atea Sandstone), but were deepening during the Opoitian, resulting in deposition of siltstone and bathyal mudstone, flysch and sandstone (Neef, 1984). Further south, near Masterton, remnants of Opoitian pebbly Te Aute lithofacies limestone occur (Morris, 1982; Wells, 1989); and to the east, two limestones of similar age and facies, the lower a lenticular barnacle-dominated coquina reaching up to 100 m thick, and the upper 30 m thick and containing *Phialopecten marwicki*, are associated with greensands on the northern Aorangi Range of southern Wairarapa (Vella & Briggs, 1971). Here, the Pliocene succession thickens up to at least 200 m, and passes into a mainly mudstone succession as the depositional site deepened away from the Aorangi Range (Vella & Briggs, 1971).

Deposition of Late Pliocene (Waipipian) limestone in the Wairarapa appears to have been restricted to the northwest, where it varies in thickness from 7 to 53 m, and locally rests on Opoitian mudstone (Beu, 1992). In southern Wairarapa, along the northern margin of the Aorangi Range, the Waipipian is represented by about 200 m of mudstone (Field *et al.* 1997).

The younger (Mangapanian) limestone of the Te Aute lithofacies is the most geographically restricted of all the Pliocene limestones (Beu, 1992). It occurs principally in the central Wairarapa Raukawa and Puketoi ranges and on the eastern side of the Ruahine and Kaweka ranges, and forms long strike ridges along each side of the postulated Pliocene seaway (Beu, 1992). The western limestone belt consists of two 50-m-thick limestone beds separated by 30 m of brown, weathered mudstone (Beu, 1992). Preservation of this unit is largely confined to the downthrown area between the Mohaka and Ruahine faults, although initial distribution across the present Ruahine and Kaweka ranges is probable. The eastern limestone belt occurs in two blocks separated by a gap of about 30 km. The northern block extends south from the Raukawa Range, west of Te Aute. The southern block comprises the limestone massif of the Puketoi Range and continues south to Eketahuna. The limestone is 39 m thick at its type section at Te Aute (Harmsen, 1985), but it thickens laterally to 140 m where it is composed largely of enormous packets of giant foreset beds (Harmsen, 1985). In the Raukawa Range it is conformably overlain by earliest Pleistocene (early Nukumaruan) mudstone. The limestone thickens to the south in the Puketoi Range, where thicknesses in excess of 300 m have been recorded (Beu, 1992).

Outcrops of Early Pleistocene (Nukumaruan) rocks are extensive and continuous throughout the area from west of Waipukurau in the north to south of Woodville in the south. The stratigraphy consists of a cyclothemic body of shallow-water mudstone, sandstone and limestone, with limestone beds formed as transgressive systems tract shellbeds resting on sequence boundaries. A basal limestone contains common *Zygochlamys delicatula*, a cold-water index fossil. Other, higher, limestone beds are less continuous. East of the Wairarapa Fault Zone, only late Nukumaruan rocks are present, locally resting unconformably on older units (Beu, 1992). An isolated area of early Nukumaruan limestone and minor overlying mudstone lies in the Wairarapa Fault Zone northwest of Masterton. It is an important record of Nukumaruan rocks apparently eroded off a large area of the present Tararua Range.

Nukumaruan rocks of southern Wairarapa (Fig. 8.4) include both limestone and siliciclastic strata. The lowest unit comprises a 50-m-thick wedge of glacial-period shellbed deposited against the northern side of the rising Aorangi Range, passing laterally into sandstone and mudstone, and underlain by 100–120 m of Mangapanian mudstone (Vella & Briggs, 1971; Edwards, 1988; Beu, 1992; Crundwell, 1997). Other limestones are part of the extensive cyclothemic Nukumaruan rocks of southern Wairarapa, and comprise a succession of transgressive cycles consisting of shellbeds, sandstones, and mudstone deposited as the formerly bathyal Wairarapa basin was gradually uplifted into shelf depths at a time of oscillating sea level (Gammon, 1995). Up to four cyclothems are present in the Makara River area, and the resistant lower one or two limestone members crop out extensively along range crests to form most of the conspicuous highlands along the eastern edge of the Wairarapa plains.

Middle Pleistocene (Castlecliffian) marginal marine, lacustrine and alluvial sediments up to several hundred metres in thickness are present in western Wairarapa close to the Tararua Range, where the sediments conformably overlie Late Pliocene–Early Pleistocene strata (Shane, 1991). South and southeast of Masterton, beds of alternating gravel, sand, loess and mud of Early Pleistocene age unconformably overlie Pleistocene limestone, or marine sandstone of Castlecliffian age (Collen & Vella, 1984). Undifferentiated Middle Pleistocene alluvial gravels occur sporadically throughout Wairarapa (Lee & Begg, 2002).

The Pliocene record in Marlborough is sparse (Fig. 8.4). The Opoitian–Waipipian Starborough Formation south of Blenheim consists of more than 350 m of shelf, fine to coarse (sometimes gritty) sandstone and very fine sandy siltstone (Roberts & Wilson, 1992). The only occurrence of Pliocene rocks in southern Marlborough is late Opoitian–early Waipipian fine to medium sandstone and siltstone with interbedded conglomerate just north of the mouth of the Clarence River. These units are lithologically similar to, although much younger than, the Great Marlborough Conglomerate, and are interpreted as outer-shelf to slope debris-flow conglomerate units within the siltstone and sandstone (Browne, 1995b).

Remnants of marine terraces on the Marlborough coast from Cape Campbell to Haumuri Bluff range in age from ~220 to ~60 ka (Late Pleistocene). They are commonly underlain by well-stratified to poorly sorted beach gravel.

8.6.2 Canterbury, Chatham Rise, Great South and Eastern Southland basins

The Pliocene–Pleistocene period saw the continued uplift of the Southern Alps and the development of the present geomorphology. Uplift with active faulting and folding occurred in the west, and the north Canterbury fold and thrust belt developed. Erosion in the west of the Canterbury Basin in the Late Miocene locally removed all Cretaceous–Cenozoic sediments, leaving basement Torlesse rocks overlain by Pliocene conglomerate. Offshore, at DSDP site 594, there is a 1 m.y. hiatus that is not marked by strong lithological change.

Erosion, accompanied by iceberg gouging, occurred on the Chatham Rise, though rafted glacial debris, volcanic ash and, probably, foraminiferal glauconitic sand accumulated locally (Kudrass & von Rad, 1984).

The sedimentation pattern in north Canterbury during the Pliocene–Pleistocene was complex. Deep marine conditions prevailed in much of the northeast (see Fig. 6.5), where massive grey siltstone (Greta Formation) accumulated amidst growing highs of Cretaceous–Cenozoic and Torlesse rock. Grainstone accumulated locally within Greta Formation siltstone in the latest Pliocene–Pleistocene (Mangapanian–Nukumaruan), and coarse mass-flow deposits invaded the area at various times. In the Oaro–Hundalee area south of Kaikoura, and further west, channelised debris-flow conglomerate of Nukumaruan age, the Bourne Conglomerate lithofacies (Browne & Field, 1985; Warren, 1995), occurs within the otherwise fine-grained Greta Formation. The molluscan species in the conglomerate are closely comparable to living bathyal faunas of canyons off the eastern South Island, and the siltstone is interpreted as having been deposited in water depths of 600–800 m (Beu, 1979). The Oaro–Hundalee area formed part of a rapidly subsiding narrow graben developed between subaerial greywacke horsts in the Pliocene–Early Pleistocene and may have been connected with the Kaikoura Canyon or its predecessor, which led eventually into the Hikurangi Trough (Field *et al.* 1989).

Directly south of this area, the basement 'high' of the coastal Hawkswood Range was subjected to major uplift during the Pliocene–Pleistocene. Late Pliocene deep-water mudstones cropping out on the western flanks of the range contain sediment gravity-flow deposits with large rafts of sedimentary rocks, probably reflecting initiation of uplift and emergence. The eastern (seaward) side of the Hawkswood Range is dominated by a Quaternary fan-delta complex consisting largely of gravels transported by debris-flow processes. Delta lobes infilled channels, which transported sediment into the adjacent offshore Conway Trough (Lewis & Ekdale, 1991). At Motunau, north Canterbury, similar channelised subaqueous debris flows of Early Pleistocene (Nukumaruan) age were introduced into the Greta Formation silts from higher in the paleoslope (Lewis, 1976). Lewis suggested a model of retrogressive slumping and the development of channels that served as conduits for the debris flows to pass into deeper-water canyon systems, eventually linking with the Hikurangi Trough (Herzer & Lewis, 1979; Lewis, 1982).

Non-marine to marginal-marine Kowai Formation accumulated further south (Carlson *et al.* 1980), and, in the west, the Pliocene (Opoitian) beds locally dip at progressively lower angles up-section. This suggests that sedimentation occurred on a growing anticline–syncline couplet, and similar relationships are likely elsewhere in north Canterbury. Sediments represent predominantly marginal marine deltaic environments, the conglomerate being proximal to the mouths of rivers, and the sandstone and siltstone–mudstone units recording more-distal interdistributary sites.

In the central Canterbury Basin the crest of the Chatham Rise remained a topographic high largely starved of sediment. On the Chatham Islands (see Fig. 4.2), Late Miocene–Late Pliocene (?Kapitean–Mangapanian) mainly volcanogenic deposits rest with inferred unconformity on older sediments or on basement schist. They include massive extrusive and intrusive rocks of alkaline basaltic and phonolitic composition, volcaniclastic sandstone, tuff and thin and lenticular bioclastic limestone. All are included in the Mairangi Group (Campbell *et al.* 1988, 1993). Thickness of individual volcanic, volcaniclastic and tuff units reaches 150 m, while the limestone bodies range from 2.3 to 26 m in thickness. The tuffaceous rocks are locally richly fossiliferous, with common and diverse micro- and macrofossils. The limestone units are foraminiferal or bryozoan grainstones, locally with a minor component of glauconite and lithics.

An assemblage of Pleistocene (early Nukumaruan) clastic sediments (Karewa Group) disconformably overlies all older rocks of the Chatham Islands (Campbell *et al.* 1988, 1993). The oldest unit (Titirangi Sand), a marine succession of wholly Early Pleistocene age, consists of up to 6 m of well-sorted, quartzo-feldspathic sand, shell grit and shellbeds, and is confined to north-central Chatham Island. The basal layers are dominated by subangular blocks and pebbles of schist, basalt, vein quartz and limestone. The shellbeds are dominated by bivalves. All other Karewa Group units appear to be essentially non-marine and of Late Pleistocene (Haweran) age.

In central Canterbury, volcanism at Banks Peninsula ceased at about 5.8 Ma (Late Miocene) and the composite volcano has been exposed to subaerial erosion since that time. The edges of the cones were probably onlapped by Kowai Formation shallow marine to non-marine sediments during the Pliocene, and clasts of volcanics, probably derived from the eroding cones, occur in probable Pliocene sediments at Leeston-1 petroleum well southwest of Christchurch.

The Kowai Formation is not preserved in inland central Canterbury, but has been penetrated in petroleum wells further east. There it includes up to several hundred metres of Pliocene shellbeds, conglomerate and clay (locally glauconitic and carbonaceous). Offshore, over 300 m of shelly sandy clay with lignite beds occurs at Resolution-1 petroleum well, while at Clipper-1 petroleum well the Pliocene is represented by siltstone, with occasional sandstone beds. The Pliocene at Galleon-1 petroleum well rests on the floor of a submarine canyon. The sediments are dominated by siltstone and sandstone, with shell material becoming more common up-sequence.

The Kowai Formation crops out as far south as the Waitaki Valley in south Canterbury, consisting of Torlesse-derived fluvial sheet conglomerates that interfinger in the east with sandstone, mudstone, lignite and shellbeds. Palynological studies on sparse carbonaceous material show it to be Plio–Pleistocene in age (Morgans *et al.* 1990). Although the typical lithology is conglomerate, deltaic or lacustrine claystone and sandstone also occur. In the east the unit includes Late Pliocene–Pleistocene (Waipipian–Nukumaruan) marine fossils and sand wedges with marine fossils. Foresets of decimetre amplitude in the sand units dip to the northwest, east and southeast, while metre-amplitude foresets in the conglomerate dip to the northwest, suggesting strong shoreward transport in a high-energy coastal environment. Increase in dip in the conglomerate beds down-section may indicate contemporaneous deformation. In the Late Pliocene, considerable vertical displacement occurred during the uplift of the Southern Alps, and cumulative uplift in the west may have reached several kilometres (Field *et al.* 1989).

At DSDP site 594 the Early Pliocene–Recent sediments are represented by 160 m of alternating pelagic nannofossil ooze and hemipelagic nannofossil-bearing clayey silt. They are interpreted as interglacial–glacial cycles, the hemipelagic silty layers representing glacial periods of increased sediment supply from the rising mountains of the central South Island (Griggs *et al.* 1983). Unconformities between Late Miocene and Early Pliocene sediments and between Early and Late Pliocene sediments have been detected biostratigraphically (Nelson *et al.* 1986b).

The oldest Pliocene sediments in the southwest of the Canterbury Basin are immature conglomerates and sandstones referred to as the 'Hawkdun Group' (Youngson *et al.* 1998). This group also includes Early Pleistocene sediments that rest on successively older units from east to west, reflecting increasing pre-Pliocene uplift. The unit occupies much of eastern Central Otago, particularly in the area from Wanaka to Kyeburn (Bishop, 1974a). Most of the Pliocene record in north Otago is missing because of erosion caused by uplift during the Pliocene–Pleistocene.

The widespread Plio–Pleistocene conglomerate lithofacies of Central Otago extends southwards into eastern Southland. It occurs both north and south of Gore and to the west as far as the basement Longwood Range. In the Gore area, basal Waimumu Quartz Gravels and Waikaka Quartz Gravels (Wood, 1956) consist almost entirely of clasts of metamorphic quartz, with thin beds of quartzose sandy clay. They are separated by a hiatus from the underlying Early–Middle Miocene Gore Lignite Measures. The youngest of the three Pliocene gravel formations (Gore Piedmont Gravels) differs from the others mainly in containing up to 60% of weathered schist and greywacke, as well as quartzose clasts. The formation is correlated with the upper Hawkdun Group of Central Otago (Wood, 1956), and is inferred to be a response to increased tectonism and uplift and erosion of local basement horsts. Similar gravel deposits have been mapped throughout the Eastern Southland and Winton basins as far as the Longwood Range.

Oligocene–Recent strata in the Great South Basin were included in the Penrod Group by Cook *et al.* (1999). This unit was poorly sampled in exploration wells. The only evidence for Plio–Pleistocene strata was recovered from the Tara-1 petroleum well, where foraminiferal ooze and chalk of Early Pleistocene (Mangapanian–Nukumaruan) age rest with inferred unconformity on similar sediments of late Middle Miocene (Waiauan) age. These Pleistocene sediments have paleodepths of mid- to outer-shelf, shallowing to mid- to inner-shelf at the top of the sampled interval (Raine *et al.* 1993).

During the Pliocene–Pleistocene there was a period of marked uplift of the South Island, and erosion of Oligocene and Miocene carbonates supplied calcareous detritus to the western margins of the Great South Basin (Turnbull *et al.* 1993). A narrow, progradational shelf wedge, up to 600 m thick, developed. To the east, bathyal conditions persisted, with deposition of approximately 100 m of Late Pliocene and younger pelagic sediments.

8.6.3 Western Southland basins

Pliocene sedimentation in onshore western Southland was restricted to the present-day Te Anau and Waiau topographic basins, and to the onshore Waitutu Sub-basin. The sediments extend throughout the offshore area, and an additional depocentre evolved on the west coast of Fiordland at Five Fingers Peninsula. In the onshore Waitutu Sub-basin, deep marine conditions persisted from the late Miocene. Sediments in offshore areas were probably also deposited in a predominantly deep marine environment, and they form the thickest Pliocene succession in the region.

Pliocene sediments are unconformably overlain by Pleistocene moraine and fluvioglacial gravels in the central Waiau and Te Anau basins, and by raised beach deposits in the southern Waiau Basin and Waitutu Sub-basin. Age control on the Upper Pliocene sequence is poor and only the Five Fingers outcrop is known to be as young as Early Pleistocene (Turnbull *et al.* 1985).

The Pliocene succession in the Waiau Basin (see Fig. 6.10) is represented by the Te Waewae Formation (Turnbull *et al.* 1989), which is restricted almost entirely to the southern part of the basin. The base of the unit is unconformable over a local high in mid-basin, while in the west the lower contact is conformable. The Early Pliocene sediments are up to 400 m thick, thickening offshore, and are probably separated from the Solander Basin sequence by the Mid Bay High (cf. Bishop *et al.* 1992). The Pliocene of the Waiau Basin records the culmination of local regression, as the region became fully emergent after the Early Pliocene.

Pliocene Te Waewae sediments in the southern Waiau Basin include laminated mudstone and sandstone sequences up to 50 m thick, interlayered with massive to trough cross-bedded slightly pebbly sandstone, and shellbeds. An additional facies, as yet undated, but presumed to be Pliocene, is a deeply weathered, sandy, cobble to pebble conglomerate that fills channels eroded into the host Te Waewae Formation in the Rowallan Forest. The conglomerates are metre-bedded, apparently internally massive, locally poorly imbricated, and up to 30 m thick. The Te Waewae Formation contains a macrofauna indicative of inner-shelf depths. The presence of clay pellets and clay laminae, occasional shallow-marine trace fossils such as *Ophiomorpha* and possible crab or shrimp burrows, and flaser and lenticular bedding also suggest sedimentation in a sheltered, intertidal or very-shallow-marine environment. The conglomerate lithofacies may represent infill of proto-Waiau River channels.

Pliocene lithofacies of the central Te Anau Basin Prospect Formation are similar to those of the Miocene, although mudstone, minor lignite and carbonaceous mudstone sequences are less common than in the Miocene part of the unit. Upper Prospect Formation conglomerates have a higher proportion of Fiordland and Takitimu debris in areas adjacent to those basement blocks and are less weathered and deformed. No paleocurrent or paleoenvironmental data are available from this sequence, which is apparently restricted in outcrop to the central parts of the basin, having been removed from the margins by glacial erosion.

The shape of the Te Anau Pliocene basin varies from north to south. In the north it onlaps a rising anticline, which trends southwards down the basin centre. The Moonlight–Hollyford transfer thrust forms its eastern boundary, and the Te Anau Fault the western boundary. Between these two faults, deposition took place around the relict Takaro High, thickening towards Lake Te Anau. The maximum sediment thickness of 2500 m is observed in the south of the basin, in the narrow neck between the Takitimu Mountains and Fiordland, and suggests that faulting in this area remained tensional throughout the Early Pliocene at least (Uruski & Turnbull, 1990).

The sedimentary outlier at the Five Fingers Peninsula (coastal Fiordland) consists of a fault-bounded transgressive sequence a few hundred metres thick. Lithofacies include: locally derived matrix-supported breccia-conglomerate; several fining-upwards conglomerate, sandstone, carbonaceous mudstone and lignite cycles; well-sorted, loose, beach sand; and soft, macrofossiliferous, shallow-marine shelf sand and mud. Macrofaunal, microfaunal and palynological data are consistent with a Plio–Pleistocene age and a change from emergent coastal to shallow (inner- to mid-shelf) conditions of deposition. Turnbull (1985) compared this outlier with Pliocene sediments north of the Alpine Fault (Nathan, 1978b; Nathan *et al.* 1986; Sutherland, 1994).

In the Solander and Balleny basins (see Fig. 6.11), Pliocene sediments were intersected in the offshore Parara-1 well (HIPCO, 1976), but not in Solander-1 (Clarke, 1986). In the Parara-1 well, Pliocene sediments were mostly mudstone with abundant foraminifera and some sandstone towards the base. Seismic data show the Pliocene succession is relatively undeformed, and Early and Late Pliocene units can be differentiated. There is a total thickness of up to 1000 m (Norris & Carter, 1980). Several volcanic centres occur in the Early Pliocene succession, but it is difficult to date the beginning of volcanism.

At the western boundary of the Balleny Basin an early Late Pliocene surface truncates reflectors that are well below the top Miocene reflector, suggesting that the Puysegur Bank region was uplifted before or during the Early Pliocene. Two main depocentres containing more than 500 m of sediments are present, one at the northern end of the Puysegur Bank and the second where the Balleny and Solander basins merge. The general shape of the Early Pliocene Balleny Basin is synclinal in the north, grading into a southwards-facing shelf margin basin in the south. Four mounds, interpreted to be volcanic bodies, are mapped within the succession.

The Early Pliocene Solander Basin is more complex than the Balleny Basin, due mainly to the influence of extensional faulting. Faulting and subsidence in this region extended the area of marine sediment deposition, in contrast to the onshore region, where regression was apparent in the Early Pliocene. The maximum thickness of Early Pliocene sediments is over 1000 m in a small depocentre beside the offshore extension of the West Hump Fault, perhaps indicating continued thrusting and loading. The Early Pliocene Solander Basin appears to have been a shelf and slope basin, merging with the Balleny Basin to the southwest.

Along the eastern margin of the Solander Basin the Hump Ridge–Stewart Island Shelf was subjected to extensional faulting, which produced a graben cross-cutting earlier structures in the same area. The cause of the tensional faulting is unclear, but it could be related to gravitational collapse of the uplifted Stewart Island Shelf, or to back-arc extension related to the Puysegur Trench subduction system.

The Late Pliocene saw a continuation of Early Pliocene paleogeography, with regression continuing onshore and transgression in the present offshore area. Up to 200 m of sediments was deposited in the Foveaux Strait region. The Solander Ridge and the Moonlight Fault System still influenced sedimentation, the former as a compaction ridge and the latter as an active fault zone that breaks the present sea bed. Deposition in the Balleny Basin was minimal, with a maximum of 400 m of sediment at the northern end of Puysegur Bank and less than 200 m over the rest of the basin. The main Late Pliocene–Recent depocentres in the region were between the Parara-1 petroleum well and the Hump Ridge–Stewart Island Shelf, where more than 800 m of sediments was deposited, and at the head of the Solander Trough, where over 600 m of sediment accumulated. The present topography of the offshore region is a broad southwards-plunging trough. This trough is modified by two channel systems, one starting to the northeast of Solander Island and trending southwestwards, and another parallel channel further southeast, starting at the Stewart Island Shelf off Codfish Island. Seismic data show channelling on a similar scale throughout the Pliocene.

Pliocene sediments of the onshore Waitutu Sub-basin are restricted to the area between lakes Hauroko and Poteriteri, and are inferred (Turnbull & Uruski, 1995) to lie beneath an older Miocene sequence, which is thrust over them from the south. The Early Pliocene sediments have subsequently been uplifted along the Hauroko Fault in the latest Pliocene–Pleistocene. This uplift was associated with development of extensive flights of Pleistocene marine terraces on the southern Fiordland coast (Bishop, 1985, 1986; Ward, 1988).

The Pliocene sediments include soft, micaceous mudstone, and a succession of conglomerate, pebbly mudstone, graded sandstone and laminated carbonaceous mudstone up to 700 m thick, which is Late Miocene at the base (Turnbull & Uruski, 1995). The conglomerate varies from massive, thick-bedded, sharp-based sandy clast-supported pebble to boulder conglomerate, to bouldery mudstone. Clasts are predominantly Fiordland-derived with rare greywacke and volcanic pebbles, and abraded macrofossil fragments occur in most beds. Sandstones range from fine to pebbly, and they thin upwards in cycles between, or interrupted by, conglomerates. The mudstone contains a bathyal macrofauna and was the background sediment into which the coarse sediments were mass-emplaced from an active basin margin to the west. Adjacent to the Hauroko Fault and to the east, a slightly younger facies of massive siltstone, macrofossiliferous locally derived breccia and conglomerate, and loose well-sorted bioturbated sand, overlies basement. The conglomerate is a high-energy shoreline deposit; the sand includes isolated, rounded boulders and is interpreted as a beach deposit. These rocks overlie the siltstone, which has outer-shelf microfaunas of Late Miocene to Early Pliocene age (Turnbull *et al.* 1993).

8.6.4 West Coast

The only occurrence of Late Pliocene marine sediments in southern Westland is south of Jackson Bay (see Fig. 6.8). These sediments, of late Opoitian to Waipipian age, are thin (up to 60 m thick), and are composed of probably marine deep-water sand and conglomerate (Sutherland, 1994; Sutherland *et al.* 1996). They rest with angular unconformity on Miocene strata.

From Harihari northwards, petroleum wells record a substantial thickness of Pliocene sediments. Waiho-1, the more southerly hole, penetrated 1800 m of mudstone and minor fine sandstone. Approximately 700 m of mudstone is recorded from Harihari-1, and in Mikonui-1 offshore northwest of Hokitika (Sircombe & Kamp, 1998). Much further north, in the Punakaiki area, combined macro- and micropaleontological evidence suggests marked shallowing to inner- to mid-shelf depths in Early Pliocene (Opoitian) times (Laird, 1988), and, north of Westport, the Early Pliocene portion of the O'Keefe Formation consists of a coarsening-upwards succession of shallow-water sandstone and conglomerate, reaching a thickness in excess of 600 m (Saul, 1994).

In the southern part of the Grey–Inangahua Depression (see Fig. 6.12), both the Late Miocene Callaghans Greensand Member and Rotokohu Coal Measures are conformably overlain by shallow marine sandstone or sandy mudstone (Eight Mile Formation), which ranges in age from latest Miocene to early Late Pliocene (Kapitean to Waipipian). The lithology of the Eight Mile Formation shows marked lateral facies changes, dominated by northward shallowing. Superimposed on this is a gradual regional shallowing until nearshore or estuarine sandstone at the top of the Eight Mile Formation grades upwards into non-marine gravels of the Old Man Group (Bowen, 1967). In the northern part of the Depression, marine mudstone and muddy sandstone, containing abundant shallow-water fossils of early Late Pliocene age, are interbedded with the upper part of the Rotokohu Coal Measures, indicating a brief marine transgression at that time. The coal measures grade upwards into thick conglomerates of the Old Man Group.

The non-marine Old Man Group, of Early Pleistocene (Nukumaruan) age, overlies the Middle Miocene–Early Pliocene succession throughout most of the Grey–Inangahua depression (Suggate & Waight, 1999). The lower part, up to 450 m thick, consists predominantly of boulders of foliated chlorite schist, and minor greywacke and granite interbedded with thin layers of carbonaceous mudstone containing lignified wood. This grades upwards into massive, yellow-brown, weathered conglomerate composed almost entirely of rounded clasts of Torlesse meta-sandstone. Regional stratigraphic evidence indicates that the Grey Valley Trough and other present-day land areas submerged during the Pliocene were buried by a flood of coarse terrigenous material from east of the Alpine Fault, leading to the formation of a widespread gravel plain (similar to the present-day Canterbury Plains). The change was initiated during latest Pliocene times. Although the conglomerate clasts are dominated by Haast Schist and Torlesse sandstone, there is also a proportion of granite and Greenland Group clasts, indicating some uplift of sources west of the Alpine Fault. Gage (1945) recognised that the Old Man Group near Ross contained intercalated glacial sediments (Ross Glaciation), and since then glacial beds have been recognised at several other localities nearby (Bowen, 1967; Suggate & Waight, 1999).

In the southern part of the Murchison Basin (see Fig. 6.13) the lower part of the thick, conglomerate-rich non-marine Rappahannock Group (Cutten, 1979) is correlated with the Middle–?Late Miocene Longford Formation in the north, while the upper part (Devils Knob Formation) is considered equivalent to the Pleistocene Old Man Gravels of the Grey–Inangahua Depression. Like the latter, the Devils Knob Formation contains common pebbles of Alpine schist derived from east of the adjacent Alpine Fault.

The southern end of Tasman Bay merges southwards into the Moutere Depression, a broad half graben approximately 20 km wide, which is filled with locally derived gravels and associated non-marine sediments of Late Miocene to Quaternary age. The older part of the gravel sequence, exposed at places around the margin of the Moutere Depression, is of Late Miocene to Early Pliocene (Kapitean to Opoitian) age (Nathan *et al.* 1986). In the southwest the Early Pliocene rests unconformably on Early Miocene or older sediments, and overlaps westwards onto granite. Near Nelson City, in the northeast, gravels of similar age are locally up to 500 m thick, and rest unconformably on older rocks. The lower part contains clasts of granitic rocks apparently derived from the Separation Point Batholith in the west, whereas the upper part is dominated by clasts of volcaniclastic Permian and Triassic rocks from the eastern side of the Waimea Fault (Johnston, 1979). Geophysical data suggest that the Pliocene gravels are not likely to be much more than 500 m thick (Anderson, 1980), and in most places they directly rest on pre-Tertiary basement rocks.

The younger part of the gravel sequence (Moutere Gravel) consists of weathered, uniform yellow-brown claybound gravel, composed almost entirely of clasts of Torlesse sandstone and semischist. In the southwest there appears to be a gradational contact between the Moutere Gravel and older gravels (Johnston, 1971). Locally the Moutere Gravel laps onto basement granite, and near Nelson there is probably unconformable contact on older gravel (Johnston, 1979). The exact age of the Moutere Gravel is uncertain, but it lies within the Early Pleistocene (Waipipian–Castlecliffian) (Nathan *et al.* 1986), and is a correlative of the Old Man Group. A weathered conglomerate of similar appearance and age, but also containing schist, granite and other igneous and sedimentary clasts, as well as Torlesse type rocks from east of the Alpine Fault, is exposed in the Karamea area and mapped as Old Man Group (Rattenbury *et al.* 1998).

The presence of a vast amount of Torlesse gravel derived from the southeastern side of the Alpine Fault suggests that the Southern Alps started to rise during the Late Pliocene, and that a flood of gravel was transported northwards to the coast. Clasts of lithologies from northwest of the Alpine Fault are locally present but uncommon, indicating that there was little uplift in the northeastern part of the West Coast region at this time.

8.6.5 Taranaki and offshore west Northland basins

The latest Miocene–Pleistocene sedimentary sequence in the Taranaki Basin (Fig. 8.2) is included in the Rotokare Group (King & Thrasher, 1996). Although the distinction between the Rotokare Group and the underlying Miocene Wai-iti Group is partly chronostratigraphic, the nature of the boundary varies across the basin, and in many places there is a clear lithological or tectono-stratigraphic distinction between Miocene and Pliocene deposits. The boundary between the two groups is marked by a prominent seismic reflector, which is traceable throughout the basin.

In the south a pronounced angular unconformity, evident on seismic reflection profiles and formed by differential uplift and erosion, separates the Miocene from the Pliocene. Uplift within the basin was greatest towards the south where progressively older strata were differentially eroded (King & Thrasher, 1996). The unconformity surface dips gently to the north, and progressively younger Plio–Pleistocene strata lap onto it. Offshore, only the latest Miocene (late Kapitean) is missing in the Maui field, but most of the Pliocene–Miocene is missing in southernmost petroleum wells. A facies change occurs across the unconformity, with shelf or marginal marine Rotokare Group strata directly overlying Wai-iti Group shelf and slope mudstones.

At Kapuni, and elsewhere in the Taranaki Peninsula, an additional intra-Early Pliocene (intra-Opoitian) unconformity is inferred to be present, marked by a regionally correlatable conglomerate horizon and a pronounced basinward shift in lithofacies. This introduction of coarse lithologies may record a drop in relative base level, and reflects uplift in the southern hinterland, as well as the final phase of uplift on the Taranaki Fault. In the Northern Graben, uppermost Miocene strata are absent above volcanic edifices, which were areas of positive relief on the sea floor at the time, and a top-Miocene unconformity is locally present elsewhere.

In the northwest of the Taranaki Basin the group boundary corresponds to the lithological contact between a condensed interval of basin floor carbonates containing the Miocene–Pliocene boundary, and overlying terrigenous clastic deposits. In the northeast, volcaniclastic sediments that were deposited over a wide area until virtually the end of the Miocene are absent in the Pliocene.

Plio–Pleistocene successions are thick everywhere except in the southern Taranaki Basin. Beneath the South Taranaki Bight and the southern Taranaki Peninsula the succession is up to 2000 m thick, and east of the buried Patea–Tongaporutu High it increases to 4000 m (Anderton, 1981). In the southern Taranaki Basin, Plio–Pleistocene strata thin southwards and lap onto the Miocene unconformity described above. Plio–Pleistocene accumulations up to 3000 m thick are present in the Central and Northern grabens, and are up to 2200 m thick over the Western Stable Platform.

In the southernmost Taranaki Basin, regional uplift and erosion precluded deposition of the Rotokare Group, apart from a veneer of Pleistocene–Recent sediments. Elsewhere in the basin were several discrete areas of Plio–Pleistocene deposition: Northern and Central grabens, Western Stable Platform, and the Toru Trough, an extension of the South Wanganui Basin into the southern Taranaki Basin across the buried Patea–Tongaporutu High. Within these areas Rotokare Group strata are primarily aggradational or progradational. The overall succession is regressive, the product of an accelerated detrital influx during the Pliocene–Pleistocene from both intra-basinal and hinterland regions of uplift and erosion.

The Plio–Pleistocene succession in the Taranaki Basin is characterised by high sedimentation rates, with undecompacted rates of up to 0.8 m/k.y. (King & Thrasher, 1996). Despite their great thickness, nowhere in the basin is there a complete Plio–Pleistocene succession, mainly because of the effects of relatively high-magnitude eustatic sea-level changes, especially in the last 2 m.y.

The oldest rocks of Pliocene age cropping out in the southern and central Taranaki Peninsula are referred to as the 'Matemateaonga Formation' (Arnold, 1957). It crops out extensively in hill country east of the Taranaki Peninsula, and extends in the subsurface beneath most of the peninsula and areas to the south, reaching a thickness of up to 1400 m in central parts of the Toru Trough. The Formation has a Late Miocene to Early Pliocene (late Tongaporutuan to Opoitian) age, lithologies being predominantly sandstone but including limestone, mudstone, shellbeds and coal. Conglomerate is common in places, especially in areas of basement onlap such as the southeastern corner of the basin.

The entire 1040-m-thick Matemateaonga section was cored in the Manutahi-1 petroleum well, drilled onto the crest of the Patea–Tongaporutu High (Robinson *et al.* 1987). The basal 280 m comprises micaceous and carbonaceous sandstone and siltstone in the lower half, with sandstone, pebbly sandstone and conglomerates dominating the upper portion. Thin coal layers and shell fragments occur sporadically. The remaining 760-m-thick succession consists of repetitive alternations of massive to thick-bedded, fine-grained sandstone beds and massive to medium-bedded siltstone beds. Shell hash layers and cemented horizons are common within the sandstones, generally near their tops. Dispersed pebbles and thin gravel lags are uncommon. Depositional cycles within the formation are evident in cores, and on seismic reflection profiles and wireline logs.

Shelf, marginal marine and terrestrial deposits of the Matemateaonga Formation formed the initial fill within the subsiding Toru Trough. With respect to underlying formations and in regional paleogeographic context the formation represents a significant regressive phase within the basin. However, in general, the formation fines upwards and is internally transgressive; basal strata have terrestrial and coastal affinities, whereas upper parts were mainly deposited in mid- to outer-shelf environments.

Formation thicknesses of up to 1400 m are the result of considerable subsidence accompanied by high rates of sediment supply, the two factors being broadly in balance. Lithological variability and observed cyclicity within the formation indicate that there were numerous intermittent changes in shallow-water environments.

The overlying Tangahoe Formation consists predominantly of fine-grained shelf deposits of early Late Pliocene (Waipipian) age. Locally, proximal

Tangahoe Formation (Fig. 8.2) deposits consist of fine sandstone and mudstone interbedded in varying proportions, but siltstone is the commonest lithology. The formation occurs onshore in the southeast of the Taranaki Peninsula, and offshore in the southwest of the peninsula, where it reaches 200–300 m in thickness. Further north the thickness of subcropping Tangahoe Formation decreases as a result of Late Quaternary tilting and erosion. The Tangahoe Formation was deposited within the central axis of the Toru Trough, where the infilling of the depocentre is recorded in several wells as a bathymetric shallowing from outer-shelf in the upper Matemateaonga Formation, to inner- to mid-shelf in the upper Tangahoe Formation. Undifferentiated Plio–Pleistocene shelf strata overlie the Tangahoe Formation in the western Taranaki Peninsula and in the South Taranaki Bight. Late Pliocene–Pleistocene beds intersected in petroleum wells generally comprise interbedded fine sandstone and mudstone.

The Mangaa Formation is a locally restricted unit consisting of deep-water sandstones of latest Miocene to Pliocene age. Fine- to very fine-grained sandstones intersected in petroleum well Mangaa-1 have a total thickness of nearly 600 m. The formation has a wedge geometry that thins to the northeast (Forder & Sissons, 1992). The formation is thickest along the axis of the Northern Graben, and is controlled to the east and west by synsedimentary faults. Sandstone packages within the Mangaa Formation in Mangaa-1 are well expressed on wireline logs, where they have an appearance indicative of high-density turbidity current or mass-flow deposition on submarine fans. Individual sandstone beds are generally up to 10 m thick. Overall log character is similar to basin floor fan sandstones within the Moki Formation in the south of the basin (de Bock, 1994) and within the exposed basal Mount Messenger Formation along the north Taranaki coast (King *et al.* 1994). The Mangaa Formation has a high-amplitude, sub-parallel seismic reflection character, which is also indicative of submarine fan deposition. Overlying intervals consist of progradational slope reflectors that are sometimes highly disrupted by slumping and channelling (Forder & Sissons, 1992). The Mangaa Formation is interpreted as turbidite sandstones, probably transported from south to north, deposited on the basin floor of the Northern Graben. In Mangaa-1 the formation constitutes the main body of initial sedimentary fill within this part of the Northern Graben in the Early Pliocene (King & Thrasher, 1996).

The Early Pliocene–Recent depositional history of the northern part of the Taranaki Basin is characterised by rapid progradation and aggradation of a continental margin succession that underlies the modern shelf and slope. The sediments are best characterised on seismic reflection profiles, where depositional successions appear as stacked and progressively offlapping (basinward) sigmoidal wedges defined by clinoform-shaped reflectors, from which the sequence (Giant Foresets Formation) takes its name (Pilaar & Wakefield, 1978; Hansen & Kamp, 2002). It consists mainly of siltstones and mudstones with interspersed sandstones, and forms the entire Plio–Pleistocene interval on the Western Stable Platform, extending eastwards across the Northern Graben. Each successive clinoform reflector represents the transient position of the shelf-slope–basin floor profile. Shelf margin offlap was directed towards the west, northwest and north, and over the duration of the Pliocene–Pleistocene the shelf migrated between 8 and 20 km in that quadrant.

The Giant Foresets interval is up to 2200 m thick, and is the thickest stratigraphic unit of any age over much of the platform. Having been deposited over a period of 5 m.y., compared with a total sedimentary record spanning some 70–80 m.y., these deposits also had the highest sedimentation rates on the Western Stable Platform. Facies belts within the formation are highly diachronous. For example, over its southern and central distribution range the Pliocene succession includes thick shelf accumulations, but elsewhere consists mainly of slope and basin floor deposits. Conversely, Pleistocene sediments are mainly of shelf depth everywhere, and extend to the present-day shelf edge, but in northwestern areas they also encompass thick slope accumulations.

Progradation on the Western Stable Platform took place in the form of migrating fan lobes, rather than along a single advancing slope front (Soenandar, 1992). The fan lobes are thought to have been fed by canyons and channels that reached upslope into shelf catchment areas, and some of these are evident on seismic reflection profiles. Those of basal Pliocene age reveal a spectacular contributory network in which sediment was funnelled from the shelf edge onto the New Caledonia Basin abyssal plain. Large volumes of sediment reached the northwestern corner of the basin only in the last 2 m.y., partly as a consequence of the filling and over-spilling of the adjacent Northern Graben (King & Thrasher, 1996). Seismic reflection profiles indicate the presence of large-scale cyclothems, representing eustatic cycles in part (Hansen & Kamp, 2002).

The prominent seismic reflector separating Miocene from younger rocks passes northwards into the N2 regional reflector in offshore west Northland (Isaac *et al.* 1994). Sediments between that reflector and the sea-floor N1 reflector (sequence N1–N2) are correlated with the Plio–Pleistocene Rotokare Group of the Taranaki Basin (King & Thrasher, 1992). This sequence covers most of the area west of the Northland Peninsula, and extends west over the Reinga Basin and into the New Caledonia Basin. These sediments have been penetrated only in the Waka Nui-1 petroleum well west of the Kaipara Harbour, where 50 m of Early Pliocene (early Opoitian) marl rests unconformably on Late Miocene strata (Milne & Quick, 1999; Strong *et al.* 1999).

The Plio–Pleistocene succession thins eastwards towards the present coastline, and further thins over many of the Early Miocene volcanic massifs. It overlies much of the planed surface of the Northland Allochthon at Cape Maria van Diemen, and west of Aupouri in the southern part of the Far North peninsula. It also overlies planed Mount Camel Suspect Terrane basement and Paleogene sequences west of the Three Kings Islands. It generally pinches out as a seismically mappable unit about 10 km southwest of the shore, and is absent through truncation over large areas of the coast northwest of Hokianga Harbour (Isaac *et al.* 1994). The sequence is thickest (800 m) in southwestern and central western Northland, and in the embayments that existed between volcanic headlands (800–1000 m).

In general the reflectors of sequence N1–N2 downlap southwestwards onto the basal reflector N2, onlapping any features of positive relief. Southwesterly prograding clinoforms predominate in the embayment areas between the volcano massifs, and they grade downslope into subhorizontal reflectors. The sediment source area evidently lay to the east, in the region of the present Northland landmass.

The Plio–Pleistocene succession commonly onlaps the Northland Allochthon. Large-scale scour and fill structures are common along the irregular periphery of the allochthon, suggesting that it too was largely eroded to base level by the end of the Miocene, and therefore was not a major source of sediment in Pliocene and Pleistocene times (Isaac *et al.* 1994).

More channels, some of them very large and persistent and passing up into present-day large channel systems, are more apparent in the Plio–Pleistocene sequence than in the underlying sequences. The basin lying off the Kaipara Harbour has particularly well-developed channel complexes, which have persisted to such an extent that the basin has become completely filled, and what was the Kaipara Basin is now the site of a Plio–Pleistocene sediment cone.

The gross geometry of the Middle Miocene–Pleistocene sediment package (equivalent to the upper Wai-iti and Rotokare groups of the Taranaki Basin) is one of a passive margin wedge advancing southwestwards, burying residual relief.

8.6.6 Onshore Northland and South Auckland basins

Pliocene and younger sedimentary rocks are absent from most of the onshore Northland Peninsula, but occur north and south of the Kaipara Harbour, and on the Karikari and adjacent Aupouri peninsulas in the far north (see Fig. 6.16), where they rest unconformably on a highly irregular and commonly weathered erosion surface on Early Miocene and older rocks. In both areas the sedimentary succession (Awhitu Group), which is up to 250 m thick, is dominated by large-scale cross-bedded sandstone, with minor pebbly sandstone, lignite, carbonaceous mudstone and sandstone, and, locally, basal shelly sandstone and mudstone. In the far north of Northland, subaerial dune deposits extend below present sea level and locally overlie shallow marine beds with intercalated lignites, suggesting periodic transgression and regression (Isaac *et al.* 1994). Awhitu Group sediments occur southwest of the Manukau Harbour and extend south of Port Waikato, where they disconformably overlie the Pliocene Kaawa Formation and overstep to rest unconformably on Early Miocene and older rocks. The cross-bedded sandstones are interpreted as subaerial dunes, with associated non-marine to shallow-marine sediments, and range in age from Early Pliocene to Early Pleistocene (Opoitian to Castlecliffian). They are up to 200 m thick (Isaac *et al.* 1994).

South and southwest of Auckland, Pliocene (Opoitian–Waipipian) marine beds of the Kaawa Formation unconformably overlie Early Miocene and older strata, and are overlain disconformably by sedimentary facies of the Awhitu Group and by basaltic rocks of the Ngatutura Volcanics (Briggs *et al.* 1989). The formation crops out southeast of Manukau Harbour and is widely distributed in the subsurface of the Manukau lowland, extending for a few kilometres south of the Waikato River mouth. The lithofacies include mudstone, muddy and shelly sandstone, mollusc-dominated bioclastic conglomerate and lithic conglomerate, with a maximum thickness of 150 m. Bioclastic facies are locally up to tens of metres thick, in both lenticular and laterally persistent beds. The faunas, bed forms and lithofacies indicate deposition in shallow marine and estuarine environments (Beu, 1974; Hollis, 1986).

The Early Pleistocene–Holocene Karioitahi Group comprises moderately consolidated to unconsolidated coastal sand deposits of shallow-marine, beach and dune origins (Isaac *et al.* 1994). It disconformably overlies partly eroded Awhitu Group dunes and unconformably overlies Early Miocene and older rocks around the coast of Northland, Auckland and the Coromandel Peninsula.

Mainly Pliocene–Holocene pumiceous sediments of the Bay of Plenty, Hauraki lowland, Hamilton lowland, lower Waikato and Manakau lowland were included in the Tauranga Group by Kear and Schofield (1978). The sediments are mainly of fluvial, lacustrine and distal ignimbritic origin, and are laterally equivalent to the mainly coastal Kaawa Formation and Awhitu Group deposits further west.

The Tauranga Group is very widespread, dominating surface geology of the Waikato and Waipa lowlands north of Te Kuiti. It unconformably overlies Te Kuiti, Waitemata, Mahoenui and Mohakatino groups, and extends marginally onto Mesozoic basement. It is thickest in the Hamilton Basin, where oil exploration drillholes have recorded thicknesses of up to 600 m (Katz, 1968), but elsewhere it is generally less than 100 m thick. The group consists of a variety of mainly terrestrial sediments dominated by pumiceous and rhyolitic sands, clays and gravels, with interbedded peat.

Palynological studies indicate that the Tauranga Group ranges in age from Pliocene to Holocene, although locally ages as old as Late Miocene (?Tongaporutuan) have been recorded (Mildenhall & Pocknall, 1986). It was deposited in an alluvial environment, strongly influenced by silicic volcanism in the hinterland, tectonic movement, and glacio-eustatic sea-level change and associated change in climate and vegetation (Nelson *et al.* 1988). Tauranga Group sediments have undergone very little deformation and are only gently tilted and warped. They are highly variable in thickness, being thickest in structural depressions bordered by late Miocene faults. Thinner, condensed sequences are present locally on the paleotopographic highs. Paleotopography – and hence sedimentation – was controlled by a series of east- to northeast-trending fault-angle depressions and intervening horsts.

8.6.7 King Country and Wanganui basins

Apart from the uppermost portion of the Whangamomona Group, which extends into the earliest Pliocene (see section 8.4.6), the Pliocene–Pleistocene (late Opoitian–Late Pleistocene) sediments of the King Country and Wanganui basins are represented by the Rangitikei Supergroup. The Rangitikei Supergroup, like the underlying Whangamomona Group, formed as a northwards-prograding continental margin wedge separated by a major flooding surface from the underlying sediments (Kamp *et al.* 2002, 2004). The continental margin comprising the Rangitikei Supergroup advanced northwards on two fronts; one advanced directly northwards from the Southern Alps source through the Wanganui Basin and into southern parts of the King Country Basin, while the other was directed west of the Patea–Tongaporutu High into the Central and Northern grabens of the Taranaki Basin, and ultimately onto the Western Stable Platform (Fig. 8.2).

Deposition of the basal unit, the mid-Pliocene Tangahoe Formation, coincided with the onset of rapid subsidence and regional flooding during the Pliocene in response to changes in eastern North Island subduction geometry (Kamp *et al.* 2002; Naish *et al.* 2005). The base of the Tangahoe Mudstone

(Fig. 8.2) is the most extensive flooding surface in the Middle Miocene–Pleistocene succession in the King Country, Wanganui and eastern Taranaki basins. Across a few metres of section the paleoenvironment in the centre of the basin changes from shoreface or inner-shelf to bathyal water depths, a deepening of about 400 m, which generated a paraconformity marked by glauconitic mudstone (Hayton, 1998). This boundary has been mapped from the Cape Egmont Fault Zone in the Taranaki Basin to the eastern margin of the Wanganui Basin at the Ruahine Range (Kamp *et al.* 2002).

A depocentre, infilled with up to 600 m of bathyal mudstone, evolved in northeastern parts of the Wanganui Basin, while a thinner, regressive wedge of shallow-marine strata prograded from the southwest across the western margin of the depocentre. Sediments of the Tangihoe Formation exposed in cliffs along the south Taranaki coastline were deposited on the southwestern margin of the mid-Pliocene depocentre, and comprise a 270-m-thick, cyclothemic shallow-marine succession, similar to the underlying Matemateaonga Formation. Like this unit, the sediments here consist of a succession of 10–80-m-thick glacio-eustatic sequences, bounded by sharp wave-cut surfaces produced during transgressive shoreface erosion (Naish *et al.* 2005). Each sequence comprises three parts: a deepening-upwards basal shellbed; a gradually shoaling, aggradational siltstone succession; and a strongly progradational, well-sorted shoreline sandstone. The three-fold subdivision corresponds to transgressive, highstand and regressive systems tracts respectively, and represents deposition during a glacio-eustatic sea-level cycle (Naish *et al.* 2005). The shoreline sandstones contain a variety of storm-emplaced sedimentary structures, and represent the rapid and abrupt basinward translation of the shoreline to a storm-dominated, shallow-marine shelf during eustatic sea-level fall. Heavy mineral assemblages from the sandstones indicate derivation from the northwestern South Island (Naish *et al.* 2005).

A similar distribution of paleoenvironments occurs in the interbedded sandstone and mudstone of the early Late Pliocene (middle–late Waipipian) Whenuakura Group, which conformably overlies the Tangahoe Formation. The group ranges in thickness from 420 m in the northwest of the Wanganui Basin near Hawera, where it comprises shallow subtidal to estuarine sediments, to 1050 m of outer-shelf to upper-bathyal sandstone and siltstone in the Waitotara River area 40 km to the east (Murphy *et al.* 1994). The marked difference in thickness and lithofacies is explained by eastward progradation into a rapidly deepening basin.

The incoming of the conformably overlying mid-Late Pliocene (Mangapanian) Paparangi Group marks a regional decrease in water depths, and foraminiferal studies suggest that the basin became temporarily cut off from open-ocean access (McGuire, 1998). The Paparangi Group and the overlying Late Pliocene–Early Pleistocene (late Mangapanian–Nukumaruan) Okiwa Group show a similar paleogeography to the Whenuakura Group, with shallow and marginal marine sediments occurring on the north Wanganui and south Taranaki coastal areas representing the western margin of the Wanganui Basin. The deeper water depocentre lies 80–100 km further east (Naish & Kamp, 1995; Journeaux *et al.* 1996). The two groups, together with the overlying Early Pleistocene (Castlecliffian) deposits, reach a thickness of 2000 m in the axis of the basin, and thin westwards towards the western basin margin (Abbott & Carter, 1994; Saul *et al.* 1999). The succession provides an almost complete Late Pliocene–Early Pleistocene composite record, comprising sands, silts and shellbeds deposited mainly in shoreface and shelf environments. The sedimentary record takes the form of 47 superposed cyclothems of shelf origin, each corresponding to an unconformity-bound stratigraphic sequence, and typically contains a basal transgressive systems tract resting on an eroded surface of marine planation, locally including a mid-cycle shellbed, passing upwards into a highstand systems tract, followed in some cases by a regressive systems tract (Saul *et al.* 1999). Some individual sequences can be traced laterally as far as 70 km across the basin (Abbott & Carter, 1994). The transgressive–regressive cyclothems have been correlated with glacio-eustatic cycles (e.g., Saul *et al.* 1999; Abbot *et al.* 2005).

In areas where Plio–Pleistocene strata are thickest, subsidence rates averaged 0.8 m/k.y. over the last 5 m.y. Basin development also involved regional tilting to the southwest, resulting in a progressive southward migration of depocentres, and uplift and erosion of older deposits to the north (Anderton, 1981). By the Early Pleistocene (Castlecliffian) the depocentre of the Wanganui Basin had shifted to a position offshore south of Wanganui (Anderton, 1981). Continuing southward movement resulted in the Middle Pleistocene submergence of a once extensive river system to form the Marlborough Sounds and, in combination with tectonic downwarp in adjacent basins, the opening of Cook Strait (Lewis *et al.* 1994).

Uplift of the Ruahine Range commenced in the Early Pleistocene (Nukumaruan), the tempo of uplift increasing during Castlecliffian times (Lee & Begg, 2002). In the Wanganui Basin, sediment supply from the uplifted northern basin margin and the axial ranges was supplemented by substantial volumes of rhyolitic pumice from the Taupo Volcanic Zone to the northeast.

8.7 NEOGENE VOLCANISM

During the Neogene, volcanism in the New Zealand region fell into two distinct categories. South of the newly developed plate boundary in the South Island, passive subsidence still continued on the Pacific plate, and volcanism was almost entirely intraplate. By contrast, Neogene volcanism northwest of the plate margin on the Australian plate was dominated by subduction processes.

8.7.1 Eastern and southern New Zealand basins (Pacific plate)

There was no Neogene volcanism east of the axial ranges of the North Island, and none in Marlborough. Rhyolitic tuffs, which occur in the Miocene succession of the eastern North Island, have been attributed to contemporaneous volcanism in the Coromandel Range (Gosson, 1986), which during that time was closer than it is now. Tephras in the East Coast Pleistocene succession of the North Island are attributed to a source in the Taupo Volcanic Zone (Field *et al.* 1997).

Neogene intraplate volcanism occurred south of the new plate boundary in the Canterbury and Great South basins and on the Chatham Rise. Major volcanic centres of alkaline to subalkaline composition were active at Banks Peninsula, Dunedin, Campbell Island and the Auckland Islands. Although compressional tectonics had become well established by Late Miocene times in western Canterbury and Otago, the central volcanoes of Banks Peninsula and Dunedin were on the margin of this zone (Stipp & McDougall, 1968; Weaver & Smith, 1989).

Products of Miocene volcanic activity are abundant in the Canterbury Basin. These are alkaline to subalkaline rocks and include the large composite volcanoes of Banks Peninsula. The first eruptions were largely submarine and are recorded by mildly alkaline basaltic ash and hyaloclastic breccia (Wairiri Volcaniclastite, Sandpit Tuff, Bluff Basalt) interbedded with Middle Miocene greensand and marine sandstone in the Harper Hills district of inland Canterbury. These deposits are overlain by tholeiitic-basalt flows of the Harper Hills Basalt, which have a K/Ar age of 10.5 Ma (Carlson *et al.* 1980). Tholeiitic dikes exposed nearby in the Glentunnel area are thought to be related. Similar tholeiitic-basalt flows containing abundant segregation veins are found in the Oxford district of inland central Canterbury (Oxford Basalt) and have a K/Ar age of 16 Ma. Apart from a nephelinite sill in north Canterbury with a K/Ar age of 15 Ma, these are the only recognised igneous rocks of early Middle Miocene age in Canterbury (Weaver & Smith, 1989).

The largest accumulation of Miocene volcanic rocks along the east coast of the South Island is on Banks Peninsula, southeast of Christchurch City. There are two large composite volcanoes – the older Lyttelton volcano in the northwest and the younger Akaroa volcano in the southeast. The present diameters of the volcanic complexes are 25 and 35 km, respectively, but, on the basis of evidence from boreholes, both extend for some distance beneath the adjacent plains, and they probably had original diameters close to 35 and 50 km, respectively (Sewell, 1988; Weaver *et al.* 1992).

In reality, eruption took place from a large number of individual centres within each complex (Hampton & Cole, 2009), and eruptions can be grouped into three main active periods between 11 and 5.8 Ma. The Lyttelton volcano (11–9.7 Ma) and the Akaroa volcano (9.3–8 Ma) are the major volcanic massifs, with the contemporaneous Mount Herbert Volcanic Group (9.7–8 Ma) occurring between, in the central region of the peninsula. Younger volcanic activity occurred on the eroded outer flanks of the volcanoes and in deep valleys eroded into the volcanoes. These comprise the Church-type lavas (8.1–7.3 Ma) and the Diamond Harbour Volcanic Group (7–5.8 Ma). Detailed descriptions can be found in Sewell (1985), Sewell *et al.* (1992) and Hampton and Cole (2009).

Volcanic activity began at 11 Ma (Barley *et al.* 1988) with the eruption of icelandite, dacite and peraluminous rhyolite (Governors Bay Volcanics) on a basement of Triassic Torlesse rocks and Miocene sediments exposed at the head of the Lyttelton Harbour. They were followed almost immediately by the overlying oldest Lyttelton lavas. They crop out mostly within the eroded centre of the Lyttelton volcano and were buried by the products of diverse subaerial character (Weaver & Smith, 1989) from a number of distinct eruption centres (Hampton & Cole, 2009). All have alkaline to transitional chemistry (Weaver & Smith, 1989).

Activity in the younger Akaroa centre follows a similar pattern, while the contemporaneous but less extensive Mount Herbert Volcanic Group partly buries the boundary zone between the two main eruptive centres. The younger Church Volcanics and Diamond Harbour Volcanic Group occur on the south side of Lyttelton Harbour and on Quail Island, showing that the deep valleys and harbours within the volcano existed by the beginning of the Pliocene.

On the Chatham Islands, with the exception of a thin fossiliferous tuff of late Early Miocene (Altonian) age, no Early or Middle Miocene volcanism is known. Massive to crudely bedded, poorly sorted, coarse pyroclastic breccia of limburgite basalt composition (Rangiauria Breccia), which has an isotopic K/Ar date of 6.1±0.3 Ma (Late Miocene), occurs on Pitt Island and on islands in Pitt Strait. The breccia, which reaches up to 300 m in thickness, unconformably cuts across all older units and is disconformably overlain by elements of the latest Miocene and younger Mairangi and Karewa groups (Campbell *et al.* 1988, 1993). The Mairangi Group, which has an isotopic date of 5.3±0.4 Ma (Late Pliocene–Early Pleistocene), consists mainly of volcanogenic deposits resting unconformably on older successions. The group includes massive extrusive and intrusive rocks of alkaline basaltic and phonolitic composition, volcaniclastic sandstone and tuff, and thin lenticular bioclastic limestone. Thickness of individual volcanic, volcaniclastic and tuff units reaches 150 m.

In south Canterbury, basalts of Pliocene age are represented by two relatively thin sheets in the vicinity of Timaru and Geraldine. Both sheets consist of several individual flows having a range of compositions from tholeiitic to transitional in terms of their normative and modal mineralogy. The Timaru Basalt has a K/Ar age of 2.47±0.37 Ma (Duggan & Reay, 1986).

The term 'Dunedin Volcanic Group' was proposed by Coombs *et al.* (1986) for all late Cenozoic volcanic rocks of eastern and Central Otago. The group includes the Dunedin volcano and all peripheral intrusive bodies, flows and tephra that are present at distances of up to 95 km from the centre of the volcano at Port Chalmers. Rocks of the Dunedin Volcanic Group rest on early to mid-Cenozoic marine strata and are overlain by Late Miocene and younger terrestrial sediments.

Eruption of the Dunedin Volcanic Group probably started in the Early Miocene (upper Otaian) around 21 Ma (Adams, 1981) in the area inland of Oamaru. The main eruption of the Dunedin volcano occurred in Middle Miocene times, although some basalt flows south of Dunedin have Early Miocene K/Ar ages of 21.3 Ma and 22.5 Ma (Bishop, 1994; and Fig. 6.4).

The Dunedin volcano is a major shield that has a diameter of about 25 km and a present-day relief of 700 m. It was constructed during the period 13–10 Ma. The Dunedin volcano for most of its history was subaerial, but the presence of tuff in underlying marine beds may mean that initial eruptions were offshore. Each of the three Main Eruptive Phases consists of a spectrum of compositions from basanite and alkali basalt to phonolite and trachyte (Coombs *et al.* 1986). Mafic compositions predominate, but evolved rocks appeared throughout the history of the volcano.

Most of the outlying vents of the Dunedin Volcanic Group appear to represent short-lived monogenetic eruptions, and most occurrences are remnants of flows or shallow intrusive bodies. K/Ar ages of volcanic rocks from the peripheral vents range from over 20 Ma to slightly less than 10 Ma (Weaver & Smith, 1989). Most rocks are of mafic composition. Basanite is the dominant lithology, and alkali basalt and nephelinite are subordinate. Volcanism took place in a continental intraplate setting that appears to have been mildly extensional.

The late Cenozoic volcanic edifices forming much or part of island groups such as the Auckland Islands, Campbell Island, the Antipodes Islands and the

Chatham Islands are related temporally and provincially to the volcanic shields of Lyttelton, Akaroa and Dunedin on the eastern coast of the South Island. The volcanoes grew on high points of the pre-Cenozoic basement of metasediments and granites, so a deep structural control is inferred for their setting.

The Antipodes Islands consist of dominantly pyroclastic volcanic rocks, together with lava flows and associated minor intrusions. Lavas range from strongly porphyritic ankaramite to near-aphyric glassy rocks, with ages of less than 1 Ma (Weaver & Smith, 1989).

Roughly two-thirds of Campbell Island, which is the most southerly of New Zealand's sub-Antarctic islands, is covered by Late Miocene lavas resting on the Miocene Shoal Point Formation, which is up to 200 m thick and consists of pyroclastic sediments (Morris, 1984). The lavas are predominantly mafic, and both mildly alkaline (nepheline-normative) and subalkaline (hypersthene-normative) compositions occur, together with subsidiary trachyte and rhyolite. Isotopic (K/Ar) ages for whole-rock samples from Campbell Island lavas cluster around 7.4–7.0 Ma, but range from 11.1 to 6.5 Ma. High-level intrusions consisting mainly of gabbro and trachyte appear to be of earlier Miocene age, giving K/Ar dates of 15.0 and 17.0 Ma (Nathan *et al.* 2000).

The Auckland Islands, which lie near the western edge of the Campbell Plateau, consist predominantly of thin, laterally discontinuous lava flows that accumulated to a maximum thickness of 680 m (Wright, 1968). There appear to be two coalesced volcanic shields – the Carnley volcano in the south, and the Ross volcano in the north (Gamble & Adams, 1985). Mafic rocks form about 80% of the lava flows, with intermediate and felsic rocks accounting for the remainder. Locally, lava flows are cut by an intense swarm of dikes, inclined sheets and sills, identical to the extrusive rocks (Wright, 1970). These dikes and sheets decrease in abundance upwards through the volcanic pile and are interpreted as feeders to the lava flows. Gabbroic rocks cropping out in the heart of the Carnley volcano are interpreted as the roof zone of a subvolcanic magma chamber (Gamble & Adams, 1985). The Carnley and Ross rocks are chemically identical, and mafic lavas and intrusions range from mildly alkaline to sub-alkaline (Wright, 1971). The more evolved rocks are quartz-normative trachyte and rhyolite. Adams (1983) reported whole-rock K/Ar age determinations ranging from around 25 to 12 Ma, but the older ages are suspect due to possible contamination from Mesozoic basement granite that crops out on Musgrave Peninsula (Gamble & Adams, 1985). The Auckland Island basalts are, however, likely to be appreciably older than the Campbell Island volcanics.

Solander and Little Solander islands, which lie in the Solander Basin, are the eroded remnants of a major dacitic–andesitic volcano of probable Late Pliocene age (Reay, 1986; Bishop, 1986). They are the only subduction-related volcanic centre on the Pacific plate and they relate to the east-dipping southern section of the current plate boundary (Davey & Smith, 1983; Reay, 1986). Rock types on the islands include agglomerate, tuff with thin lignite beds, andesite and hornblende andesite flows and dikes. The flows are dated at 1.4±0.5 Ma (Adams, in Bishop, 1986) and the lignite as Late Miocene–Early Pleistocene (Mildenhall, in Bishop, 1986). Chemical analyses (Reay, 1986) show the andesites to be high-K, comparable with continental margin calc-alkaline andesites. No associated volcanic deposits are yet known from western Southland Neogene sediments, either on land or offshore. The Solander volcano lies on the crest of the Solander Ridge. Other possible Pliocene igneous bodies occur in the Balleny Basin, mostly on highs west of the offshore portion of the Moonlight Fault System. They are presumed to be eruptive centres, now largely buried by younger Pliocene or Pleistocene sediments.

8.7.2 Western and northern New Zealand basins (Australian plate)

Early–early Middle Miocene igneous activity is unknown in the west coast of the South Island or in the Taranaki Basin, but at this time the Northland Peninsula and adjacent offshore areas were the site of a volcanic arc, comprising marine and subaerial andesitic volcanics forming northwesterly trending twin belts extending for 350 km down the eastern and western sides of the peninsula. Approximately 50 volcanic edifices, from major massifs to small cones, have been identified on marine seismic profiles of the western belt, which is now largely buried (Isaac *et al.* 1994; Herzer, 1995). Western belt complexes consist of basalt, basaltic andesite and two-pyroxene andesite, and rhyolite and dacite are rare. Eastern belt volcanics contain only rare basaltic lithologies, and are dominated by two-pyroxene andesite, hornblende andesite, and hornblende-bearing dacite, with subordinate rhyolite and their intrusive equivalents. Although there are chemical and petrographic differences between the products of the two belts, the geochemical data do not provide any evidence for arc polarity. Compositional differences may reflect differences in crustal thickness, composition and interaction with magma. The Lower Miocene igneous association is calc-alkaline to high-K calc-alkaline in character, and related to a convergent plate regime (Smith *et al.* 1989). Southwest-directed Early Miocene subduction is indicated by the northwest orientation of the twin volcanic belts, and by the regional geology (Isaac *et al.* 1994).

Isotopic ages of the onshore calc-alkaline volcanics are no older than 22–22.5 Ma (earliest Early Miocene), and the volcanics are found on both sides of the peninsula and on eastern offshore islands (Hayward, 1983). Onshore ages are generally no younger than about 16 Ma (latest Early Miocene), the main exception being a north Coromandel Peninsula volcanic complex that remained active until 15 Ma.

Following the cessation of volcanism in Northland at the end of the Early Miocene, the northern Taranaki Basin played host to a series of late Middle–Late Miocene eruptions offshore. Stratovolcanoes of the Mohakatino Volcanic Centre erupted offshore, forming a near-linear north-northeast trend along the axis of the Northern Graben (King & Thrasher, 1996). The volcanoes have been identified from seismic reflection profiles, gravity and magnetic anomalies, and exploration wells that drilled into several volcanic edifices. The volcanic rocks comprise mainly low- to medium-K calc-alkaline andesites, basaltic andesites and subordinate basalts. The main period of extrusive volcanism was about 14–11 Ma, although eruptions continued to at least 7–8 Ma. The belt extends from offshore immediately west of the Taranaki Peninsula to west of the mouth of the Waikato River (King & Thrasher, 1996; Stagpoole, 1997). The Mohakatino volcanics have been generally linked with the offshore Northland volcanics as part of a western volcanic arc system (e.g., Bergman *et al.* 1992; Kear, 1994, 2004). However, although there is some minor overlap in age between the two centres, in general Mohakatino volcanism began after offshore Northland volcanism ceased, and there are differences in physical size and geographic trend between the two groups of edifices.

Herzer (1995) postulated that the two arcs were distinct, and considered that the younger Taranaki arc was part of a volcanic lineament that extended northeastwards to incorporate the Kiwitahi and Coromandel volcanics of the Hauraki Volcanic Region and the submarine Colville Ridge, and that it was sub-parallel to the modern arc. The Taranaki and other volcanoes along the lineament were interpreted to have been part of a separate, northwest-dipping Pacific-facing volcanic arc – the Lau–Colville Ridge. Information from the Coromandel volcanics is not sufficient by itself to indicate an arc trend (Skinner, 1986). The junction between the postulated distinct arcs coincides with an inferred major northeast-trending zone of crustal weakness aligned with and forming an extension of the Waikato Fault (Gage & Kurata, 1996). This lineament also appears to form the boundary between the north- to northeast-trending Late Cretaceous fault systems plus positive negative anomaly belts of the Taranaki and South Island West Coast basins, and the north-northwest- and northwest-trending fault systems plus anomaly belts of the western offshore Northland Basin. By contrast, Stagpoole (1997) proposed a Late Miocene–Pliocene zone of back-arc extension, co-extensive with the Northern Taranaki Graben, and extending north-northeastwards through the Hauraki Gulf (i.e., to the north of Gage and Kurata's proposed lineament).

The Hauraki Volcanic Region (Skinner, 1986), which comprises the onshore Kiwitahi Volcanic Zone and the eastern Coromandel Volcanic Zone, consists dominantly of andesites and dacites, but includes the early examples of the rhyolitic and ignimbritic eruptions that dominated the late Cenozoic volcanic history of the North Island. The Coromandel Volcanic Zone, whose central portion is the Coromandel Peninsula but which stretches from northern offshore islands (including Great Barrier Island) to as far south as Tauranga, evolved more or less continuously throughout the late Cenozoic from late Early Miocene to Pliocene (Adams *et al.* 1994). Between ~18 and 9 Ma, an andesite-dominated volcanic arc was active, and from ~10 Ma ignimbrite eruptions associated with large caldera formation began. A bimodal basalt and basaltic andesite to rhyolite association developed between ~9 and 5.5 Ma, and changed to become entirely basaltic from 4.7 to 4.2 Ma. At about the same time (5.5 Ma), basalt was also erupted on the offshore southern Colville Ridge. During Late Miocene to Early Pliocene times, andesitic to dacitic eruptions continued in the south and southwest of the Coromandel Volcanic Zone. With time, the locus of volcanism moved irregularly eastwards and southwards. Most of the rocks are calc-alkaline and medium-K, but the basaltic rocks are a mixed tholeiitic–high alumina/calc-alkaline suite similar to the volcanics of the Colville Ridge (Skinner, 1986; Adams *et al.* 1994). They range from high-alumina basalts through basaltic andesites and (dominant) andesites to dacites.

The Kiwitahi Volcanics, lying west of, and parallel to, the Coromandel Peninsula and Hauraki Rift, are broadly contemporaneous with the rocks of the Coromandel Volcanic Zone, ranging from 16 Ma to 5.5 Ma (Adams *et al.* 1994). They occupy a belt extending from Waiheke Island south-southeastwards to localities 20–30 km east of Hamilton. As with the Coromandel Volcanic Zone, the rocks have similar andesitic compositions, and there is a southward younging. Both groups of rocks are calc-alkaline and subduction-related, and were probably part of the same volcanic arc until separated by the opening of the Hauraki Rift (post-5.5 Ma). Basaltic rocks also occur, as in the Coromandel Peninsula.

Northwest of Great Barrier Island a north–south elongate, ellipsoidal submarine plateau of igneous rocks, from which flow-banded rhyolite has been dredged, covers nearly 1000 km^2 (Thrasher, 1986). The Mokohinau Islands group and Simpsons Rock 6 km to the south are the only subaerial exposures, and they contain rhyolite and dacite, which give K/Ar ages of 9.0±0.8 Ma. The islands and surrounding sea floor are interpreted as remnants of a Late Miocene–Pliocene rhyolite dome complex, with associated minor flows and dikes of dacite and basalt.

On Great Barrier Island, Middle–Late Miocene Coromandel Group andesitic rocks are unconformably overlain and intruded by Whitianga Group rhyolitic rocks. There are at least two ignimbritic sheets, with an aggregate thickness of at least 200 m, and a number of rhyolite dikes (Skinner, 1986). The ignimbrites overlie 40 m of carbonaceous, lacustrine, volcaniclastic sediments.

Pliocene–Holocene basalt-dominated volcanics of Northland are known collectively as the 'Kerikeri Volcanics' (Edbrooke & Brook, 2009). The volcanic region consists of clusters of monogenetic scoria cones and vents, with associated lava flows. That include minor andesite, alkaline rhyolite and peralkaline rhyolite (Heming, 1980a,b; Ashcroft, 1986), dacite domes and a small rhyolite dome (Smith *et al.* 1993).

Volcanic rocks of the Kaikohe–Bay of Islands area are mainly basaltic, but there is minor andesite and rare alkaline and peralkaline rhyolites. Basalt lavas include hawaiites, highly porphyritic tholeiites and lesser transitional basalts (Ashcroft, 1986). Steep-sided scoria cones are common, with less common small shield volcanoes of overlapping flows. There are several alkaline and peralkaline rhyolite domes, and a similar intrusion at depth may be the heat source for the active geothermal field at Ngawha (Heming, 1980b). Basalts from the area have yielded K/Ar ages in the range 8.1–0.06 Ma (Smith *et al.* 1993).

At Whangarei the distribution of the basaltic volcanoes locally shows a strong correlation with the local fault pattern (Smith *et al.* 1993). K/Ar ages of 1.23±0.05 Ma and 1.45±0.49 Ma have been determined (Isaac *et al.* 1994). The lavas are mainly hawaiites or tholeiites (Ashcroft, 1986), with minor intermediate rock types (Smith, 1989; Smith *et al.* 1993). Flow remnants west and north of Whangarei have K/Ar ages of 6.0 and 4.2 Ma respectively, and basalt from the Vinegar Hill centre has an average age of 2.45 Ma. Other nearby basalts have provided K/Ar ages in the range 0.26–0.33 Ma (Smith *et al.* 1993). Forty kilometres southeast of Whangarei the remnants of a small shield volcano, composed of hawaiites, give K/Ar ages in the range 0.71–0.81 Ma (Smith *et al.* 1993).

Auckland is a much younger field, with about 48 eruption centres present within a north–south elongate ellipse, 21 km by 16 km, centred on Auckland City (Kermode, 1992). The lavas are alkali basalt and basanite, with minor transitional basalt, tholeiitic basalt and nephelinite (Heming & Barnet, 1986). Rangitoto-derived basaltic ash buries volcanic detritus-free beach sand and shell originally ^{14}C-dated at 600 years BP (Brothers & Golson, 1959;

Nichol, 1992). Plagioclase thermoluminescence work suggests the age range of the field may extend back to 141,000 years BP (Phillips, 1989).

The South Auckland Field is mainly within the downfaulted Manukau lowland, but it extends south of the Waikato River and east into the Hunua Ranges. Almost all the South Auckland vents were at least 18 km from the southern margin of the Auckland Field. There are about 44 cones and 30 tuff rings (Rafferty & Heming, 1979). Most of the lavas are subalkalic, with lesser alkalic types (basanites and nephelenites), from two distinct parent magmas (Rafferty & Heming, 1979). K/Ar ages are in the range 0.51–1.56 Ma (Robertson, 1976).

In the Waikato Basin, extinct volcanoes of the Alexandra Volcanic Group (Kear, 1960) are prominent features of the southern part of the basin, exhibiting a northwesterly alignment across the northern part. The volcanoes are of Pliocene and Pleistocene age, with K/Ar ages between 3.79 and 1.80 Ma. The large, low-angle, composite cones of Karioi, Pirongia, Kakepuku, Te Kawa and Tokanui are constructed of thick lava flows and volcanic breccias, with minor dikes, scoria, tuff and lahars (Briggs *et al.* 1989). Tholeiitic rocks (basalts, basaltic andesites and andesites) make up the large cones of Pirongia, Karioi, Kakepuku and Te Kawa, while many small scoria cones and thin lavas are composed of alkaline rocks (basanites, basalts and hawaiites). Geomorphological evidence suggests that the volcanoes become progressively younger to the southeast (Briggs *et al.* 1989).

The Taupo Volcanic Zone in the central North Island is the main focus of young volcanism in New Zealand, comprising the currently active volcanic arc and back-arc basin of the Taupo–Hikurangi arc-trench system (Cole, 1990). Andesitic activity started at ~2 Ma, joined by voluminous rhyolitic (plus minor basaltic and dacitic) activity from ~1.6 Ma (Wilson *et al.* 1995). The Taupo Volcanic Zone, which is up to 60 km wide, stretches for ~300 km in a northeasterly direction from the active Tongariro Volcanic Centre through the hydrothermally active Taupo–Rotorua area into the Bay of Plenty, where White Island is still active. Rhyolite was the dominant magma erupted (mostly as caldera-forming ignimbrite eruptions), andesite is much less abundant, and basalt and dacite are minor in volume.

Aspects of volcanism and its tectonic drivers are very active fields of research and there are many publications. More information may be found in Stern *et al.* (2006), Wilson *et al.* (2009), Mortimer *et al.* (2010), Reyners *et al.* (2011) and Gravley *et al.* (2016), and in references therein.

CHAPTER 9

Event Stratigraphy

9.1 INTRODUCTION

Event stratigraphy seeks to achieve correlation by looking at major physical and biological events in the geological record. The major tectonic events that had a fundamental effect on Cretaceous–Cenozoic basin formation in New Zealand and that resulted in deposition of three major tectono-sedimentary assemblages have been discussed in Chapter 3. They were driven by major changes in plate boundaries and plate motions in the southwest Pacific area. These are first-order events. New Zealand has also been affected by the global-scale events of sea-level change and climate change. Less widespread tectonic events also occur commonly and derive indirectly from plate tectonic changes. To some extent event stratigraphy overlaps with and has been subsumed by sequence stratigraphy, which is closely related to relative sea-level change. Distinguishing tectonic events with local relative sea-level changes from global sea-level events is difficult in tectonically active areas. Unconformities are critical surfaces in this context that may represent local or regional tectonic events, or global events, or some combination of them.

9.2 GLOBAL EVENTS

9.2.1 Cretaceous-Tertiary (K-T) boundary event (65 Ma)

The Cretaceous–Tertiary (K–T) boundary marks a critical time globally (see section 1.1.3). In Zealandia, changes at the K–T boundary are less dramatic than elsewhere, and there are over 20 known K–T boundary sections, most of them on land. Others have been recorded in petroleum prospecting wells or DVDP/ODP drillholes (Hollis, 2003). In most sections studied in detail in New Zealand there is a clear extinction event, at least for some marine microfossils such as foraminifera and nannofossils. The most complete boundary sections so far recognised are found in the northern South Island, where iridium anomalies have also been found in six localities: Woodside Creek, Flaxbourne River, Needles Point, Chancet Rocks, mid-Waipara River and Moody Creek (Brooks *et al.* 1984, 1986; Hollis, 2003; Hollis & Strong, 2003). Shocked quartz, indicative of a meteorite impact event, was also reported from the boundary clay in the Woodside Creek locality (Bohor *et al.* 1987). The iridium-rich horizon at this site was also enriched in elemental carbon (mainly soot), inferred to have been caused by fires triggered by meteorite impact (Wolbach *et al.* 1988). All New Zealand marine K–T sections that have been studied in detail exhibit a significant lithologic and paleoenvironmental change at the boundary.

The northern South Island has some of the best-studied and most complete sections across the K–T boundary in the world, and these represent the only known southern high-latitude transect of the K–T boundary transition from the continental slope to a terrestrial mire (Hollis, 2003). Sections in eastern Marlborough that represent outer-shelf to bathyal paleoenvironments have particularly good stratigraphic preservation across the boundary. One section (Flaxbourne River) has a complete stratigraphic record; two other less well-studied sections (Chancet Rocks and at the rocky point near Wharanui) are possibly complete; and one (Woodside Creek) has a small missing or condensed Early Paleocene interval (Hollis *et al.* 2003a,b). These sections demonstrate that the initial effect of the K–T event was a major reduction in carbonate productivity, associated with the extinction of calcareous plankton and with significant increases in terrigenous clay and biogenic silica content (Hollis *et al.* 2003c). There is a pronounced increase in silica in latest Cretaceous–earliest Paleocene sediments in Marlborough sections generally. The increase in silica has been correlated with an apparent increase in siliceous microfossil abundance that is thought to be a result from enhanced oceanic upwelling associated with climatic deterioration (Hollis *et al.* 2003c).

An apparently complete succession across the K–T boundary also occurs in the middle Waipara River (north Canterbury), in an inferred neritic setting (Hollis & Strong, 2003). Like the more bathyal sections to the north, the K–T boundary in the middle Waipara section coincides with a marked and prolonged decrease in carbonate content and a significant increase in biogenic silica. At this section, and at the non-marine Moody Creek locality near Greymouth on the West Coast, the K–T boundary also exhibits an abrupt change from a diverse flora dominated by gymnosperms (~70%) in the uppermost Cretaceous, to one dominated by a few species of fern (up to 90%) in the earliest Paleocene (Vajda *et al.* 2001, 2003). At Moody Creek a fungal spike also coincides with a boundary iridium anomaly (Vajda & McLoughlin, 2004). The calculated duration of the period of fern-dominance was ~8–20 k.y., until gymnosperms again become dominant (Vajda *et al.* 2004). The fern and fungal spikes, which are recognised globally, are thought to be the result of widespread deforestation due to an 'impact winter' causing low light levels, and/or to massive wildfires resulting from an asteroid impact (Vajda *et al.* 2001). At Moody Creek the K–T boundary lies within a 10-cm-thick coal seam in the upper part of the Rewanui Coal Measure Member of the Paparoa Coal Measures, with no detectable change in lithology across the boundary (Vajda *et al.* 2004).

Although there is no compelling evidence for significant depth changes across the K–T boundary in the bathyal paleoenvironments of coastal Marlborough sections, more-inshore shelf paleoenvironments, such as those represented by sections in the Clarence Valley in inland Marlborough and in the eastern North Island, show evidence of disconformity and relative sea-level fall. In the northern Clarence Valley, sections through the K–T boundary at Mead and Branch streams are incomplete, showing an inferred hiatus of ~50 k.y. years in the earliest Paleocene, probably as a result of erosion following a significant relative sea-level fall (Hollis, 2003; Hollis *et al.* 2003a,b). Elsewhere in the East Coast Basin, in the eastern North Island, the K–T boundary is commonly an erosional surface. At Tora, in southern Wairarapa, the K–T boundary lies within a major channel complex, the deposits in the lower, latest Cretaceous portion suggesting a relative sea-level rise, and those in the upper, Early Paleocene portion indicating an abrupt relative sea-level fall (Laird *et al.* 2003). The paleoenvironment changed from inferred bathyal in the latest Cretaceous to one containing fungal spores and acritarchs, indicating low salinity and suggesting a probably marginal marine environment in the Early Paleocene. Greensand is a dominant lithology of the Early Paleocene channel fill. The uppermost Cretaceous *Manumiella druggii* dinoflagellate zone is missing, wholly or in part, below the erosive base of the Early Paleocene channel system (Laird *et al.* 2003).

Further sections where the K–T boundary has been recognised occur in Hawke's Bay, to the north of Tora, where the boundary commonly lies within the fine-grained Whangai Formation. Dinoflagellate biostratigraphy has been well-established at two localities (Tawanui and the Te Hoe River area), where the boundary is marked by a disconformity (Wilson *et al.* 1989). At both localities the uppermost portion of the latest Cretaceous *M. druggii* zone is absent, suggesting that the upper part of the underlying succession has been removed by erosion. At other localities either there is no clear erosional break, or an erosional break is present but poorly constrained paleontologically (Moore, 1989a). A temporary increase in sand content occurs above the boundary, suggesting a possible lowering of relative sea level (Moore, 1989a; Wilson *et al.* 1989). At Tawanui this inference is further supported by an increase in spore/pollen abundance in immediately overlying sediments.

Elsewhere in New Zealand and the surrounding region there are a number of sections that include the K–T boundary, but they have yet to be studied in detail. These include offshore petroleum exploration wells and ODP/DSDP wells. The boundary section was penetrated at ODP site 1124 on the Hikurangi Plateau north of the Chatham Rise, but unfortunately the relevant portion of the core was not retrieved (Carter, R.M. *et al.* 2004 and references therein). In the Clipper-1, Endeavour-1 and Galleon-1 petroleum exploration wells in the Canterbury Basin the boundary has been located to within a few metres, either within the upper part of the Katiki Formation or at or close to the Katiki Formation–Moeraki Formation contact. It lies within marine mudstone, and the lithologic and paleoenvironmental change is slight. A possible shallowing has been inferred across the K–T boundary in Galleon-1 (Raine, 1994).

Onshore, in eastern Otago, an unconformity close to the K–T boundary has been recognised at Fairfield Quarry (Dunedin) within the lower part of the Abbotsford Formation (McMillan & Wilson, 1997), and less well-examined probably correlative horizons occur north of Dunedin at Hammond Hill and Katiki Beach (McMillan & Wilson, 1997).

Offshore, in the Great South Basin, the interval containing the K–T boundary contains an unconformity of short duration in several petroleum exploration wells (Raine *et al.* 1993). While the K–T boundary shows as a distinct log break in wells, it is not significant in terms of the lithostratigraphy of the basin (Cook *et al.* 1999). The top Cretaceous horizon can be recognised over much of the basin as a prominent, moderate- to high-amplitude and continuous reflector (Cook *et al.* 1999). In the west and northwest of the basin the horizon is traced as an erosional unconformity separating Maastrichtian–Paleocene (Haumurian–Teurian) strata. In the central part of the basin the boundary appears conformable, and to the east it is an onlap and erosional surface. Two potentially complete sedimentary successions inferred to contain the K–T boundary occur on Campbell Island in marginal marine facies, but neither has been investigated in detail.

In the Taranaki Basin a regionally identifiable seismic reflection horizon generally lies close to the paleontologically dated K–T boundary in exploration wells, and it is inferred to separate the Pakawau and Kapuni groups (King & Thrasher, 1996). However, locally the top of the Pakawau Group is marked by an unconformity within lower Paleocene strata (Bal, 1994), and thus, in at least some instances, the reflector lies in the earliest Paleocene. The same bounding reflector can be followed northwards into the offshore west Northland Basin, where it can be traced throughout the basin and as far to the northwest as DSDP site 206 in the New Caledonia Basin (Isaac *et al.* 1994).

9.2.2 Eustatic sea-level changes

It is clear from descriptions in this and the foregoing chapter that the New Zealand Cretaceous–Cenozoic succession contains numerous unconformities, erosion surfaces and sharp facies disjunctions, most of which are associated with changes of base level or relative sea level. Some of these surfaces can be traced throughout the New Zealand region, while others have not been traced beyond local outcrop level. Correlation of surfaces within or between basins is fraught with uncertainty because of commonly poor biostratigraphic control, and mismatches of two or more million years are easy to make.

In active tectonic settings such as the New Zealand region it is extremely difficult to separate the effects of tectonism on depositional systems from other causes, such as eustatic sea-level change, variation of sediment supply or strong ocean currents. Nevertheless, most of the events discussed in Chapter 3 and in the following section (9.3) are clearly tectonic in origin. They are represented either by angular unconformities, suggesting folding and uplift prior to erosion (e.g., the intra-Motuan unconformity at the base of Assemblage 1), or they are breaks in the succession associated with tectonic plate changes in the vicinity of the New Zealand microcontinent (e.g., the unconformities at the base of Assemblages 2 and 3). At least one event (the K–T boundary event) is attributed to an extra-terrestrial cause (see above). Other events appear to be a response to one or all of multiple causes that are nearly coincident in time (e.g., the Early Oligocene unconformities – section 9.3.5).

Of the remaining major events documented in this chapter and Chapter 3, only the late Cenomanian event (~95 Ma) seems to be a potential candidate for at least partial eustatic cause. It is recognisable in every New Zealand basin where mid-Cretaceous sediments are preserved, and is represented by facies disjunction or a concordant erosion surface, except where the contact is with basement. An exception is represented locally by an angular unconformity in the Raukumara Peninsula that suggests at least local uplift and erosion at this time. At the same time there is also half graben formation in the southern Canterbury Basin. Thus a tectonic cause for the break cannot be ruled out. Support for possible eustatic influence is provided by the global sea-level curves of Haq *et al.* (1987) that show a dramatic sea-level fall at 94 Ma followed by a rapid rise.

Relative sea-level changes have been recognised in the sedimentary record of all New Zealand basins, and attempts have been to correlate these with the global sea-level curves of Haq *et al.* (1987). These are generally unreliable, largely because of the poor age control noted above. Attempts to correlate possible eustatic sequence boundaries throughout individual sedimentary basins have also not been highly successful, again mainly because of poor faunal control. One possible successful instance occurs in the Cretaceous of the East Coast Basin. Of five erosional surfaces inferred to be related to relative sea-level falls in the Late Cretaceous Mata River succession (Raukumara Peninsula), two are thought to be related to tectonism, and one, at the base of the Mangaotanean (~92 Ma), can be correlated tentatively with an equivalent surface in Marlborough, 700 km to the southwest (Laird *et al.* 1998). The inferred age of the surface may correlate with the ~90 Ma date of a major eustatic drop in sea level (Haq *et al.* 1987).

The greatest success to date has been with correlating the late Neogene record of the Wanganui Basin with Pliocene and Pleistocene glacial advances and retreats. The ~2-km-thick basin-fill for the last 2.5 Ma comprises 47 superposed cyclothems that have been correlated with successive fifth- (100 ka) and sixth- (41 ka) order glacio-eustatic, sea-level fluctuations on the paleo-New Zealand shelf since Oxygen Isotope Stage 100 (Carter & Naish, 1998 and references therein). Older cyclothemic deposits, also inferred to represent glacio-eustatic sea-level cycles, have been recognised back to the mid-Pliocene (Naish, 2005; Naish & Wilson, 2009; and see section 8.6.7).

Biostratigraphic limitations have hindered inter-basinal correlation of sequence boundaries possibly related to eustatic events. A promising approach is using dinoflagellate biostratigraphy, particularly in the Cretaceous–early Paleogene, to constrain the ages of sedimentary sequences and their bounding surfaces. This approach is still in its infancy, and it has so far been applied successfully only locally (e.g., McMillan & Wilson, 1997; Schiøler *et al.* 2002).

9.3 REGIONAL EVENTS

The events discussed here are restricted to those that are likely to be New Zealand-wide or at least can be correlative through several basins. Most are recognisably associated with tectonic events.

9.3.1 Cenomanian tectono-magmatic event (late Ngaterian: ~97 Ma)

Following the initial half graben phase of deposition in the Albian a regional tectono-magmatic event occurred in Cenomanian (late Ngaterian) times. It is best seen in the northern half of the South Island, where the magmatic component is represented by predominantly alkaline intraplate igneous activity (Laird, 1994). Structurally it is marked by an angular unconformity separating moderately deep marine late Albian–Cenomanian (Urutawan–early Ngaterian) deposits from overlying late Cenomanian (late Ngaterian) non-marine to shallow-marine siliciclastic sediments and volcanic flows. The unconformity is inferred to have resulted from doming and uplift associated with magmatism and half graben formation after the end of subduction at 105±5 Ma (Baker & Seward, 1996; Laird & Bradshaw, 2004).

The volcanic rocks (Gridiron Volcanics and Lookout Volcanics formations, in the Clarence and Awatere valleys respectively) consist mainly of subaerial basalt and trachybasalt lava flows with an ocean island basalt (OIB)-type chemistry that is consistent with magma generation in an extensional tectonic regime (Warner, 1990). Dikes and sills of similar composition to the lavas intrude the underlying Cretaceous strata. The thickest exposed section of volcanic rocks (in the Awatere Valley) exceeds 700 m. A $^{40}Ar/^{39}Ar$ age of 96.1±0.6 Ma has been obtained for a basalt flow in Seymour Stream, central Clarence Valley (Crampton *et al.* 2004b). In the intervening Inland Kaikoura Range a U–Pb age of ~96 Ma has been obtained from the plutonic Tapuaenuku Igneous Complex, the probable source of the volcanics (Baker & Seward, 1996). The Mandamus Complex, 80 km to the south, is petrologically and geochemically similar, has been dated at 97±0.5 Ma (Rb–Sr method) and is consistent with an extensional environment (Weaver & Pankhurst, 1991).

U–Pb SHRIMP ages from zircons of 97±1.5 Ma and 98±1.2 Ma have also been obtained from the calc-alkaline rocks, mainly andesites and garnet-bearing rhyolites of the Mount Somers Volcanics Group of central Canterbury. The group is exposed in the alpine foothills west of Mount Somers and is linked in the subsurface with similar rocks in Banks Peninsula (Tappenden, 2003). The group may be very extensive spatially, as it has also been recognised beneath the Canterbury Plains in the JD George-1 petroleum drillhole. Offshore magnetic data suggest it may also be present to the east and south of Banks Peninsula. The geochemistry suggests high mantle heat flow and resultant crustal anatexis, a setting compatible with an environment of crustal extension (Barley, 1987). The ages of these magmatic rocks in north Canterbury and Marlborough are the same within error, suggesting that they were probably related (Tappenden, 2003).

The igneous province may extend as far north as Cape Palliser in the southeastern North Island, where dike rocks geochemically similar to the Tapuaenuku Igneous Complex occur (Challis, 1960). Alkaline trachyandesites from the Buttress Conglomerate of the southern West Coast of the South Island have yielded a U–Pb age of 97.3±1.1 Ma (Phillips *et al.* 2005), indicating that the magmatism was contemporaneous with that of the eastern South Island and is likely to be related. Although there is no equivalent volcanic or sedimentary record on the northern West Coast of the South Island, histograms of zircon fission track ages for basement rocks show a peak with a mean of 97.7±1.7 Ma, inferred to represent uplift and denudation at that time (Kamp *et al.* 1996).

The geochemistry of the alkaline igneous intraplate rocks of the northeastern South Island is similar to that produced by mantle plumes (Weaver *et al.* 1994; Storey *et al.* 1999). These authors suggested that this plume was regionally widespread, and that it included a voluminous suite of mafic dikes, $^{40}Ar/^{39}Ar$ dated at 107±5 Ma, and slightly younger (102–95 Ma) anorogenic silicic rocks in then adjacent Marie Byrd Land (Storey *et al.* 1999). A mantle plume at ~100 Ma, centred between the present Marie Byrd Land and the future New Zealand, may have weakened the lithosphere and thus controlled partly the locus of rifting. However, the correlative igneous rocks in New Zealand described above are each of limited geographical extent and of short-lived duration, inconsistent with a large-scale plume-head incursion (Tappenden, 2003). The age of the Cenomanian igneous activity in the South Island is somewhat younger than the Antarctic events and clearly post-dates the onset of extension in New Zealand, evidenced by development of metamorphic core complexes and the formation of half grabens during the Albian. The problem is complicated by the fact that isolated volcanism of similar age to the examples given above is also widespread in the Tasman Sea region. Tectonic and magmatic activity of similar age includes rhyolites from DSDP site 207 on the Lord Howe Rise K/Ar dated at 96±1.1 Ma (McDougall & van der Lingen, 1974; age recalculated). Further afield, evidence from southeastern Australia, which was adjacent to the Lord Howe Rise during Cenomanian times, shows a tectono-igneous event of the same age. Alkaline igneous complexes were emplaced at ~98 Ma on the east coast of New South Wales and Tasmania. Tulloch *et al.* (2009b) provide more data and thoroughly examined Late Cretaceous volcanism in the region, including volcanic rocks in Northland that are of similar age. They interpret the magmatism as a precursor to final continental break-up.

9.3.2 Late Cenomanian event (Ngaterian–Arowhanan boundary: ~95 Ma)

The effects of this event are particularly evident in the East Coast Basin (Laird *et al.* 1994). Although strata of Arowhanan age appear to be part of a continuous succession in the central part of the Raukumara Peninsula, in the western and eastern parts sediments of this age rest unconformably on older post-subduction Cretaceous rocks or on basement. In eastern Raukumara, Arowhanan strata rest with angular unconformity on the Ngaterian Waitahaia Formation.

In the Hawke's Bay and Wairarapa regions rocks of Arowhanan age have rarely been recognised. In the only continuously exposed stratigraphic succession involving Arowhanan strata (at Glenburn, in southeast Wairarapa), the base of the Arowhanan, although concordant with underlying strata, is marked by a 12-m-thick conglomerate. The boundary also records an abrupt lithofacies change from mudstone-dominated sediments of Ngaterian age to sandstone-dominated sediments of Arowhanan age (Crampton, 1997).

In Marlborough the Ngaterian–Arowhanan boundary is marked by an erosional break and a major lithofacies change (the boundary between the Wallow and Hapuku groups). There is no angular discordance, but there is a marked regressive shift of facies above the erosion surface, and the break is considered to mark a drop in relative sea level (Laird, 1992, 1996b).

In the Canterbury Basin the Ngaterian–Arowhanan boundary event is not well constrained, because of poor age control, but it is probably recorded by the unconformity at the base of the Clipper Formation in the Clipper-1 petroleum well, and at the base of the Henley Breccia south of Dunedin. The latter accumulated on schistose basement rocks in a half graben formed in response to rapid subsidence on the southeastern side of the Titri Fault (Bishop & Turnbull, 1996). The unconformity is identified as a prominent seismic reflector over much of the Clipper Sub-basin and to the southwest into the area offshore of Dunedin, where it overlies inferred Albian–early Cenomanian strata infilling half grabens (Haskell & Wylie, 1997). The seismic reflector is recognisable as a prominent horizon over most of the Great South Basin, where it is also inferred to be close to the Cenomanian–Turonian (Ngaterian–Arowhanan) boundary (Cook *et al.* 1999). Samples from petroleum wells show little difference between sediments above and below the reflector and they are all non-marine.

Western basins have no rocks of appropriate age to record this event except southwest of Nelson where coarse-grained deposits of the Beebys Conglomerate rest with probable unconformity on older basement (Johnston, 1990).

Although the Cenomanian event is clearly tectonic and magmatic in nature, the cause is unknown but probably is a further pre-cursor to continental separation.

Further evidence of pre-break-up magmatism is provided by new geochronology on mafic dikes in Westland mainly at 88 Ma (Van der Meer *et al.* 2013).

9.3.3 Late Campanian–early Maastrichtian event (late Haumurian: ~70 Ma)

Unlike the events discussed above, this one clearly post-dates continental separation. Although sedimentation appears to have been essentially continuous during Campanian to Maastrichtian (Haumurian) times over much of New Zealand, in western New Zealand and in the eastern South Island, sedimentation was interrupted in the late Haumurian by a tectonic event. On the west coast of the South Island and in the southern Taranaki Basin new sub-basins and half grabens, locally associated with basaltic volcanism, were formed as part of a transtensional rift system. Individual sub-basins were mainly relatively small and elongate parallel to the fault trend, generally about 10–50 km wide and 50–150 km long. Taken together they form a north-northeast-trending linear tectonic zone up to 150 km wide and approximately 800 km long, termed the 'West Coast–Taranaki Rift System' (Laird, 1981, 1994; King & Thrasher, 1996). The new fault-controlled basins were infilled initially with non-marine deposits of Haumurian age that change upwards into shallow marine sediments of latest Haumurian to Early Paleocene age in the north of the region. The tectonic event is seen as recording a dramatic change in the extension fault orientation from a west-northwest–east-southeast direction (Albian to Motuan) to a north-northeast–south-southwest or northeast–southwest direction in latest Cretaceous times.

Dates from basaltic lavas interbedded with near-basal sediments of the Paparoa Coal Measures give K/Ar ages ranging from 68±1.4 Ma to 71±1.3 Ma (Laird, 1996c). The initiation of the rift was therefore likely to be ~70 Ma (early Maastrichtian/late Haumurian). Supporting evidence for uplift and cooling at this time is provided by zircon fission track ages from northwest Nelson, which show a peak at 72.6±0.7 Ma (Kamp *et al*. 1992b), and apatite fission track ages from the coastal region north of Greymouth, which provide a mean of 72 Ma (Seward, 1989). New $^{40}Ar/^{39}Ar$ dating from dikes in Westland shows a distinct period of intrusion of lamprophyre and trachyte dikes between 70 and 68 Ma (Van der Meer *et al.* 2013). Granite pegmatites and enclosing schists in south Westland adjacent to the Alpine Fault give U–Pb monazite/zircon ages of 68±3 Ma and 71±2 Ma respectively, the intrusion and metamorphism inferred to be associated with the period of transtension noted above in the northern West Coast of the South Island (Chamberlain *et al.* 1995; Mortimer & Cooper, 2004).

Age control on the filling of the West Coast–Taranaki Rift System is not well constrained. Initial sedimentation in both the western South Island and the Taranaki Basin was non-marine, and it was restricted to the PM2 pollen zone (Raine, 1984), which is correlated with the upper part of the Haumurian Stage, ranging from late Campanian to late Maastrichtian (~77 Ma–65 Ma). This shows a latest Cretaceous rifting event on the Tasman Sea margin that is separate and younger than onset of sea-floor spreading at ~85 Ma.

The Late Cretaceous Morley Coal Measures of western Southland also fall into the PM2 pollen zone (Warnes, 1990) and were also deposited in graben structures inferred to be contemporaneous with the establishment of the West Coast–Taranaki Rift System (Shearer, 1995).

Faulting at about this time also resulted in localised half graben formation on the eastern side of the South Island, the best known faults being the Birch Fault in north Canterbury, and the Tuapeka and Titri faults in south Otago. The first two are oriented west-northwest to northwest, and the last northeast. The age of the Birch Fault is constrained by the presence of a middle Haumurian (middle–late Campanian: ~75 Ma–72 Ma) dinoflagellate flora near the base of the half graben infill (Roncaglia & Schiøler, 1997). The Otago faults are less well constrained, the maximum age of associated sediments being 'latest Cretaceous' (Bishop & Turnbull, 1996) in the case of the Tuapeka Fault, and Haumurian in the case of the Taratu Coal Measures infilling the half graben formed by the Titri Fault (Bishop, 1994).

In the region of the Clarence Valley in Marlborough a drop in relative sea level in late Haumurian times led to erosion with a dramatic lithofacies change and the local influx of coarse clastic sediments. A drop in sea level has also been suggested as a trigger for the deposition of the micritic limestone Mead Hill Formation, commenced at an estimated date of 72 Ma (Laird, 1992; Laird *et al.* 1994; Strong *et al.* 1995). The relative drop in sea level resulting in the rapid change in lithofacies was possibly related to the ~70 Ma tectonic event that affected much of the rest of the South Island.

In the southeastern North Island, near Cape Palliser, a zircon fission track age of 70±4 Ma on basement rocks adjacent to undeformed, intruding lamprophyre dikes, may reflect resetting of zircon ages by heating associated with dike intrusion at that time (Kamp, 2000). The Red Hill Teschenite, 75 km to the northeast, gave a comparable mean $^{40}Ar/^{39}Ar$ radiometric age of 70.9±2.2 Ma (W.C. McIntosh, pers. comm. 2002). At Tora, southeast Wairarapa, Whangai Formation of late Early and Late Maastrichtian age (~70–65 Ma) includes clasts of older sediments, suggesting active erosion nearby (Laird *et al.* 2003). Moore (1980) inferred activity on the nearby Adams-Tinui Fault and possibly other faults in the Late Cretaceous. No evidence of a late Haumurian event occurs further north in the East Coast Basin, or in the Northland basins.

9.3.3.1 Origin of the event

The north-northeast-trending faults of the West Coast–Taranaki Rift System appear to terminate in the vicinity of the southeastern end of the New Caledonia Basin in the northern Taranaki Basin, and are parallel to transform faults in the Tasman spreading ridge at the southern end. It seems likely that differential movement between the Lord Howe Rise and the remainder of the Pacific plate is responsible. Thus it seems probable that the rift system developed in response to opening (or renewed opening) of the

New Caledonia Basin, with transtensional movement along a rejuvenated fault(s) linking the basin with the Tasman spreading ridge to the south-southwest (Laird, 1994, 1996c; King & Thrasher, 1996). The opening of the New Caledonia Basin may be related to the westward jump of the spreading ridge of the north Tasman Sea between anomalies 32 and 29 (Weissel & Hayes, 1977) at 74–67 Ma (Laird, 1996c; Cooper, 2004). Contemporaneous faulting on the east coast of the South Island may be related, but a mechanism is not clear.

9.3.4 Mid-Eocene event (Porangan-Bortonian: ~45- ~40 Ma)

A number of unconformities and new basins that developed in the mid-Eocene indicate a new phase of extension in western and southern New Zealand. The age of the sediments overlying the unconformities varies, but the older ones range from Porangan (~45 Ma) to Bortonian (~40 Ma).

In southwestern New Zealand, extension formed the Solander Basin, the Solander Trough and a passive margin along the western edge of the Campbell Plateau. Deposition in the Te Anau Basin was initiated in localised fault-bounded sub-basins that were rapidly filled by marine sediments of Bortonian age.

In the southern part of the West Coast of the South Island, sedimentation was essentially continuous from Paleocene through to Late Eocene times. Elsewhere there was a hiatus and deposition began again in Middle Eocene times. The initial sediments were almost always non-marine (Brunner Coal Measures), resting on weathered basement rocks. They immediately precede the deposits of a marine transgression that began in Porangan times in the south, and that progressively inundated the area to the north. In the central and northern areas (Greymouth to Westport) the transgression was coincident with the formation or the reactivation of small fault-bounded basins trending north-northeast to northeast, including the Paparoa Trough, which were separated by areas of low-lying land or shallow sea (Nathan *et al.* 1986). The faults influence the thickness of both coal-measure and marine sediment. The oldest paralic sediments in these basins were commonly of Bortonian age.

The northern West Coast basins merge into the southern Taranaki Basin, where the Brunner Coal Measures equivalent passes northwards into the laterally equivalent Mangahewa Formation. An unconformity of intra-Porangan age is suggested by the sediments, although the range in ages below and above is Heretaungan to Bortonian (>3 m.y.). The basal sediments mark a significant region-wide drop in base level, followed by regional transgression.

No equivalent unconformity is recognised in the East Coast Basin or in much of the Canterbury Basin, but it can be identified in the Chatham Islands as an unconformity overlain by Bortonian sediments. In the southern Canterbury Basin (particularly in onshore Otago) the break is intra-Porangan and Bortonian. In the Great South Basin a hiatus of mainly intra-Porangan age occurs (Cook *et al.* 1999), but on Campbell Island a distinct angular unconformity separates Porangan and Bortonian strata (Hollis *et al.* 1997).

9.3.4.1 Origin of the event

The formation of new basins is attributed to a phase of rifting in the southeast Tasman Sea, southwest of Fiordland, in the Middle Eocene, at ~45–40 Ma (Wood *et al.* 2000). This resulted in new sea-floor spreading and oblique extension, causing a block, which included the Challenger Plateau, the future western South Island and the Resolution Ridge, to begin rifting away from the Campbell Plateau (Sutherland, 1995; Wood *et al.* 2000). The event is marked by a new phase of rifting initiated in western basins (the 'Challenger Rift' of Kamp, 1986).

9.3.5 Intra-Oligocene unconformities: the Marshall Paraconformity

Unconformity and abrupt lithofacies disjunction close to the Eocene–Oligocene boundary appears to represent a New Zealand-wide event, and has been dealt with in Chapter 6. There are, however, other well-documented breaks in sedimentation that occur later in the Oligocene rather than close to the base.

The most widespread intra-Oligocene hiatus surface is particularly evident in the Canterbury Basin. It was named the 'Marshall Paraconformity' by Carter and Landis (1972), with a type locality later defined (Carter *et al.* 1982; Carter, 1985) as the burrowed contact between micritic limestone (Amuri Limestone) and overlying calcareous greensand (Kokoamu Greensand) at Squires Farm in Canterbury.

The surface locally separates strata with a discordant relationship (see Plate 12a) and is an angular unconformity (Lewis, 1992). The amount of Early Oligocene erosion at the Marshall Paraconformity varies laterally, and in places all traces of the underlying Amuri Limestone have been removed (if it had ever been deposited). Where remnants of the underlying limestone are preserved the erosion surface is not smooth but consists of a zone of burrows infilled with the overlying sediment, commonly greensand. In many places in north Canterbury the burrowed surface is phosphatised, and the overlying greensand or limestone contains remnants of the burrowed surface as reworked phosphatised pebbles of Amuri Limestone. In the southern Canterbury Basin near Oamaru the presence, below an unconformity developed on Early Oligocene (Whaingaroan) limestone, of probable paleokarst and burrows enlarged by solution, suggests subaerial exposure prior to infilling with Late Oligocene (Duntroonian) greensand (Lewis & Bellis, 1984; Lewis, 1992). In places where erosion has removed Amuri Limestone and in some places underlying greensand and other units (as in the central Canterbury Basin), wave action and shallow marine currents may have been responsible, since in many Canterbury localities a condensed sequence of Late Oligocene (Duntroonian) age buries the erosion surface (Fulthorpe *et al.* 1996). In other parts of the Canterbury Basin, where more-complete successions have been preserved, multiple surfaces of hiatus or erosion (up to three) occur within the Oligocene succession (Duff, 1975), suggesting episodic sedimentation and erosion.

An erosion surface extends eastwards onto the Chatham Rise and the Chatham Islands, where no sediments of Late Oligocene age have been recorded. It has also been recognised in offshore Canterbury petroleum wells (Fulthorpe *et al.* 1996). Strontium isotopic age estimates suggest that a 2–4 Ma hiatus is associated with onshore outcrops of the unconformity, which is dated at ~32–29 Ma (Fulthorpe *et al.* 1996).

Elsewhere in New Zealand the Marshall Paraconformity has not been recognised in non-marine successions or in some deeper-water sequences, such as in the East Coast Basin. Locally on the West Coast, such as at Punakaiki and Greymouth, poorly dated mid-Oligocene hiatuses, marked by bored surfaces or phosphatic nodules, occur within limestone successions, and may correlate with the Marshall Paraconformity (Nathan, 1978a; Laird, 1988).

To the north, in the Taranaki Basin, there is a widespread unconformity that extends across much of the basin's southeastern and southern margin at that time (King & Thrasher, 1996) and is marked primarily by an absence of Early Oligocene strata. The unconformity generally spans the latest Eocene–Early Oligocene (Runangan–early Whaingaroan), although in some offshore areas it extends into the Late Oligocene (Duntroonian). It is commonly overlain by late Early Oligocene (Whaingaroan) greensand. The pause in sedimentation here is attributed to diminished sediment supply and non-deposition at the peak of transgression, rather than to erosion. In the adjacent western portion of the King Country Basin an unconformity developed in the late Whaingaroan (~32 Ma) in response to uplift and tilting along the margin of the Herangi High caused by mild compression during plate boundary development throughout New Zealand (Nelson, 1986; Nelson *et al.* 1994). In the offshore west Northland Basin a considerable hiatus is recognised in the Oligocene seismic stratigraphic record, and Miocene volcanic rocks commonly rest directly on the basal Oligocene seismic reflector (Isaac *et al.* 1994). Onshore, no hiatuses have been recorded in the Oligocene succession.

Ocean Drilling Program leg 181 penetrated Oligocene and older sediments in two drillholes east of New Zealand (Carter, L. *et al.* 2004; Carter, R.M. *et al.* 2004). At site 1123 there is a substantial hiatus of ~12.5 m.y. that includes the remainder of the Oligocene, subsequent to ~32 Ma. At site 1124 a break at a similar level occurs, with a hiatus of ~5 m.y. The youngest sediments beneath the hiatus are earliest Oligocene (early Whaingaroan) in age, and both hiatuses have been correlated with the Marshall Paraconformity (Carter, R.M. *et al.* 2004).

Early Oligocene unconformities may be a widespread feature throughout the southwest Pacific. Drilling in the north Tasman–Coral Sea regions during leg 21 of the Deep Sea Drilling Project encountered a regional unconformity, which includes part of the Late Eocene through to the middle part of the Oligocene, thus spanning 10–15 m.y. (Kennett *et al.* 1975).

9.3.5.1 Origin of the unconformity(ies)

The Marshall Paraconformity has been equated by Carter, R.M. *et al.* (2004) with the opening of the Tasmanian gateway and the development of the Antarctic Circumpolar Current (ACC), synchronous with the change to carbonate deposition. However, in several instances where the paraconformity has been identified, it appears to be an event separate from, and later than, the earliest Oligocene subsidence. The age range of the event (~32–~29 Ma) coincides with a number of geologic events affecting the New Zealand region. Lewis (1992) and Nicol (1992) inferred that uplift, broad folding and differential erosion had occurred during a mild tectonic event at this time, marking the initiation of a new plate boundary in the New Zealand region that was not fully established until the earliest Miocene. This inference is given weight by the coincidence in timing (~32 Ma) with the uplift and tilting along the margin of the Herangi High separating the King Country and Taranaki basins (Nelson, 1986; Nelson *et al.* 1994; Kamp *et al.* 2004; see section 6.6).

The Marshall Paraconformity is also close to the time of a postulated global mid-Oligocene sea-level fall at ~30 Ma (Haq *et al.* 1987), which could equate with an oxygen isotope excursion at ~32 Ma (Miller *et al.* 1991). As noted earlier in this section the opening of the deep ocean gateway between Tasmania and Antarctica at ~33.5 Ma allowed the development of the ACC through the gap. The New Zealand Plateau at this time lay directly to the east of this gap and was therefore subjected to a strong, eastwards-flowing current system. Thus, much of the erosion and non-deposition that accompanied the development of the Marshall Paraconformity and other Oligocene hiatus surfaces is inferred to have been caused by the ACC. Variations in strength and locus of the current would explain much of the variability in depth of erosion and stratigraphic position of erosion surfaces, and would account for the presence of multiple hiatus surfaces.

9.3.6 Mid-Miocene event (~15–13 Ma)

A major change in tectono-sedimentary patterns occurred during the mid-Miocene throughout most of the New Zealand region and the evidence for this change is outlined in section 8.3. The mid-Middle Miocene event is thought to relate to a change in plate boundary configuration at ~15 Ma, with related changes in convergence rates and vectors. It resulted in a significant change in volcanic style and orientation in the north and west of the North Island. This was also the time of initiation of subduction along the Puysegur Trench southwest of the South Island. Volcanism in the Early Miocene northwest-trending Northland arc had ceased by ~15 Ma, and it was almost immediately followed or overlapped by volcanism in the late Middle–Late Miocene in a new arc in the northern Taranaki Basin inferred to have a north-northeast orientation (Herzer, 1995). Major increases in sedimentation rate that occurred in south Westland at ~15–13 Ma (Sircombe & Kamp, 1998) record the beginning of reverse movement on local faults, considered by Sutherland (1996) to be linked to changes in the Australian–Pacific plate convergence vector.

It is intriguing that there is no significant mid-Miocene event recorded in the sedimentary record over the northern part of the East Coast Basin, which would have been the closest basin to the plate boundary and subduction zone. However, sub-Waiauan unconformities occur locally in Hawke's Bay and Wairarapa. In petroleum well Hawke Bay-1, Waiauan sandstone rests unconformably on Oligocene limestone, and further south, west of Cape Turnagain, lower Waiauan sediments overlie Lillburnian strata with angular unconformity. Locally, in northern Wairarapa, Waiauan paralic sediments directly overlie Mesozoic basement (Field *et al.* 1997). In southern Wairarapa, Waiauan shallow-marine transgressive sandstone rests unconformably on Lillburnian rocks or on Cretaceous basement (Crundwell, 1997). In Marlborough, strata of Waiauan age are all but absent. In short, there appears to be a period of pre-Waiauan uplift followed by a Waiauan transgression that did not reach Marlborough. This pattern may reflect broad arching at the outer edge of the upper plate.

9.3.7 Miocene–Pliocene boundary event (Kapitean–Opoitian boundary: ~5 Ma)

This event triggered the outpouring of a huge volume of clastic sediments into most New Zealand basins during the Pliocene–Pleistocene. In areas such as Taranaki and offshore Canterbury it triggered substantial outbuilding of the continental slope margin and dramatic expansion of shelf areas. In some onshore areas, particularly adjacent to the rising Southern Alps, such as the Moutere Depression and Canterbury, thick, coarse-grained fluvial sediments were deposited extensively.

In onshore Northland and south Auckland regions, Pliocene and younger rocks, which are dominated by volcanics, rest unconformably on Early Miocene and older strata. However, to the south in the King Country and Wanganui basins, the Pliocene and younger sediments rest with apparent conformity on Late Miocene strata.

In the southern portion of the Taranaki Basin a pronounced unconformity formed by differential uplift and erosion separates Plio–Pleistocene strata from those of Miocene age. The unconformity surface dips gently to the north, and progressively younger Plio–Pleistocene strata lap southwards onto it. In central parts of the Taranaki Basin there is only local unconformity, with conformable successions elsewhere. In the northwest the Miocene–Pliocene boundary corresponds to the lithological contact between a condensed interval of basin floor carbonates containing the boundary, and overlying terrigenous clastic deposits. In the northeast the boundary is marked by the contrast between Late Miocene volcaniclastic sediments and their absence in the succeeding Pliocene.

The succession appears to be similar in the adjoining offshore west Northland Basin. Here the Pliocene overlies the planed surface of the Northland Allochthon in the north of the basin, and, further south, it rests on the eroded surface of Miocene volcanic rocks.

The Plio–Pleistocene succession in the Taranaki Basin is characterised by high sedimentation rates and overall transgression. Much of the sequence in the northern and western basin forms part of the Giant Foresets Formation. Shelf-margin offlap was directed towards the west and northwest, and over the duration of the Pliocene–Pleistocene the shelf migrated between 8 and 20 km in that direction.

Southwards the Taranaki Basin merges into the West Coast basins. The northernmost, the Moutere Depression, is filled with Plio–Pleistocene non-marine gravels, which rest unconformably on Early Miocene or older sediments, and in many places on pre-Tertiary basement rocks. An influx of similar coarse-grained non-marine gravels (Old Man Group) occurred in mid-Pliocene times in northern Buller and in the northern part of the Grey–Inangahua Depression, although in these instances the gravels rest conformably on underlying strata. The gravels and the widespread shallowing elsewhere in the West Coast region during the Pliocene were caused by a flood of coarse terrigenous material sourced from the rising Southern Alps on the southeastern side of the Alpine Fault.

In the East Coast Basin the Early Pliocene marks the end of Miocene extension, and the inception of a major compressional episode. This is recorded in many parts of the region by Pliocene deposits unconformably overlying tilted and eroded Miocene sediments (Chanier *et al.* 1999). Although most of the Early Pliocene (Opoitian) sediments are marine, they were commonly deposited in a shelf environment, and shallow upwards.

In the Canterbury region the uplift of the Southern Alps, and general increase in folding and faulting and glacio-eustatic sea-level fluctuation, all contributed to the influx of sediment onto a latest Miocene unconformity in the west, and to the further progradation of the continental shelf wedge. Erosion in the west in the Late Miocene locally removed all 'cover' strata, leaving Torlesse rocks unconformably overlain by Pliocene conglomerate. Further south, in Otago, the increased tempo of tectonism in the Late Miocene–Pliocene resulted in the similar widespread deposition of the non-marine conglomerates of the Hawkdun Group, locally unconformable on older rocks (Youngson *et al.* 1998).

The Pliocene of western Southland records the culmination of local regression, as much of the region became fully emergent after the Early Pliocene. Local basal unconformities occur, particularly in the Waiau Basin.

9.3.7.1 Origin of the event

The event reflects the beginning of an increased tempo of convergent margin uplift and erosion in response to an increase in the rate of tectonism and change in the Pacific plate vector, resulting from changes in position of the Australian–Pacific instantaneous pole at ~5 Ma (Cande *et al.* 1995, 2000; Sutherland, 1995, 1996).

Concluding Remarks

The foregoing chapters allow a comparison of events in a number of sedimentary basins. Some of the basins are small and show patterns of rapid lateral change in thickness and sedimentary facies, usually driven by local tectonics. Though the early chapters are concerned with continental break-up and the separation of Zealandia from Gondwana, the term 'passive margin' has a meaning that differs from that of the typical passive margins related to continental break-up of the type seen on the margins of the Atlantic. Here we see the separation of continental crust within the context of the overall active convergent margin of the Pacific.

The whole story, from continental thinning through break-up to convergence, crust thickening and uplift, is concentrated in a relatively small area and a short period of time. Interpretation is further complicated by a crustal mobility that has changed the relative geographic positions of the sedimentary basins. The Neogene southward migration of the rotation pole between the Australian and Pacific plates illustrates the continuously changing dynamics of the plate boundary and the related evolution of sedimentary basins.

The book incorporates the work of a large number of earth scientists in GNS Science, universities and companies. All have contributed to the present understanding and interpretation, though this understanding, like the continent itself, is still evolving. We hope that the summary will be a useful guide to the last 100 million years of the development of Zealandia and an example of what may happen around the active margins of the Pacific basin. We anticipate that it will stimulate discussion and further research.

References

Note

Single-authored articles are listed in alphabetical order of author, then in chronological order.

Articles with two authors are listed in alphabetical order of first author, then in alphabetical order of second author, then in chronological order.

Articles with three or more authors are treated as '*et al.*' and listed in chronological order, regardless of number or alphabetical order of secondary authors.

Multiple articles by the same author(s) in the same year are listed using unique letter codes (a, b, c, etc.) in order of entry in the text. Thus, order within a year may appear arbitrary here.

Abbott, S.T. and Carter, R.M. 1994. The sequence architecture of mid-Pleistocene (*c*.1.1–0.4 Ma) cyclothems from New Zealand: Facies development during a period of orbital control on sea-level cyclicity. In: P.L. de Boer and D.G. Smith (eds). *Orbital Forcing and Cyclic Sequences*. International Association of Sedimentologists Special Publication 19, pp. 367–394.

Abbott, S.T., Naish, T.R., Carter, R.M. and Pillans, B.J. 2005. Sequence stratigraphy of the Nukumaruan stratotype (Pliocene–Pleistocene, c. 2.08–1.63 Ma), Wanganui Basin, New Zealand. *Journal of the Royal Society of New Zealand* 35: 125–150.

Abreu, V.S. and Anderson, J.B. 1998. Glacial eustasy during the Cenozoic: sequence stratigraphic implications. *AAPG Bulletin* 82: 1385–1400.

Adams, C.J. 1981. Migration of late Cenozoic volcanism in the South Island of New Zealand and the Campbell Plateau. *Nature* 294: 153–155.

Adams, C.J. 1983. Age of the volcanoes and granite basement of the Auckland Islands, Southwest Pacific. *New Zealand Journal of Geology and Geophysics* 26: 227–237.

Adams, C.J. 2004. Rb-Sr age and strontium isotope characteristics of the Greenland Group, Buller Terrane, New Zealand, and correlations at the East Gondwanaland margin. *New Zealand Journal of Geology and Geophysics* 47: 189–200.

Adams, C.J., Harper, C.T. and Laird, M.G. 1975. K-Ar ages of low-grade metasediments of the Greenland and Waiuta Groups in Westland and Buller, New Zealand. *New Zealand Journal of Geology and Geophysics* 18: 39–48.

Adams, C.J., Graham, I.J., Seward, D. and Skinner, D.N.B. 1994. Geochronological and geochemical evolution of late Cenozoic volcanism in the Coromandel Peninsula, New Zealand. *New Zealand Journal of Geology and Geophysics* 37: 359–378.

Adams, C.J., Campbell, H.J. and Griffin, W.L. 2007. Provenance comparisons of Permian to Jurassic tectonostratigraphic terranes in New Zealand: perspectives from detrital zircon age patterns. *Geological Magazine* 144: 701–729.

Adams, C.J., Mortimer, N., Campbell, H.J. and Griffin, W.L. 2009. Age and isotopic characterisation of metasedimentary rocks from the Torlesse Supergroup and Waipapa Group in the central North Island, New Zealand. *New Zealand Journal of Geology and Geophysics* 52: 149–170.

Adams, C.J., Mortimer, N., Campbell, H.J. and Griffin, W.L. 2011. Recognition of the Kaweka Terrane in northern South Island, New Zealand: Preliminary evidence from Rb-Sr metamorphic and U-Pb detrital zircon ages. *New Zealand Journal of Geology and Geophysics* 54: 291–309.

Adams, C.J., Mortimer, N., Campbell, H.J. and Griffin, W.L. 2013a. Detrital zircon geochronology and sandstone provenance of basement Waipapa Terrane (Triassic–Cretaceous) and Cretaceous cover rocks (Northland Allochthon and Houhora Complex) in northern North Island, New Zealand. *Geological Magazine* 150: 89–109.

Adams, C.J., Mortimer, N., Campbell, H.J. and Griffin, W.L. 2013b. The mid-Cretaceous transition from basement to cover within sedimentary rocks of eastern New Zealand: Evidence from detrital zircon age patterns. *Geological Magazine* 150: 455–478.

Adams, C.J., Bradshaw, J.D. and Ireland, T.R. 2013c. Provenance connections between late Neoproterozoic and early Palaeozoic sedimentary basins of the Ross Sea region, Antarctica, south-east Australia and southern Zealandia. *Antarctic Science* 26: 173–182. doi:10.1017/S0954102013000461

Adams, D.P. 1987. Cretaceous and Eocene geology of South Westland. Unpublished MSc thesis, University of Canterbury.

Adamson, T.K. 2008. Structural development of the Dun Mountain Ophiolite Belt, Bryneira Range, western Otago, New Zealand. Unpublished MSc thesis, University of Canterbury.

Aliprantis, M.D. 1987. Tertiary geology of the Paringa region, South Westland, with special attention to a structurally deformed coastal section. Unpublished MSc thesis, University of Canterbury.

Allibone, A.H. and Tulloch, A.J. 2004. Geology of the plutonic basement rocks of Stewart Island, New Zealand. *New Zealand Journal of Geology and Geophysics* 47: 233–256.

Allibone, A.H. and Tulloch, A.J. 2008. Early Cretaceous dextral transpressional deformation within the Median Batholith, Stewart Island, New Zealand. *New Zealand Journal of Geology and Geophysics* 51: 115–134.

Allibone, A.H., Jongens, R., Scott, J.M., Tulloch, A.J., Turnbull, I.M., Cooper, A.F., Powell, N.G., Ladley, E.B., King, R.P. and Rattenbury, M.S. 2009a. Plutonic rocks of the Median Batholith in eastern and central Fiordland, New Zealand: Field relations, geochemistry, correlation and nomenclature. *New Zealand Journal of Geology and Geophysics* 52: 101–148.

Allibone, A.H., Jongens, R., Turnbull, I.M., Milan, L.A., Daczko, N.R., De Paoli, M.C. and Tulloch, A.J. 2009b. Plutonic rocks of western Fiordland, New Zealand: Field relations, geochemistry, correlation and nomenclature. *New Zealand Journal of Geology and Geophysics* 52: 379–415.

Alvarez, L.W., Alvarez, W., Asaro, F. and Michel, H.V. 1980. Extraterrestrial cause for the Cretaceous–Tertiary extinction: Experimental results and theoretical interpretation. *Science* 208: 1095–1108.

Anastas, A.S., Dalrymple, R.W., James, N.P. and Nelson, C.S. 1997. Cross-stratified calcarenites from New Zealand: Subaqueous dunes in a cool-water, Oligo-Miocene seaway. *Sedimentology* 44: 869–891.

Anastas, A.S., James, N.P., Nelson, C.S. and Dalrymple, R.W. 1998. Deposition and textural evolution of cool-water limestones: Outcrop analog for reservoir potential in cross-bedded calcitic reservoirs. *AAPG Bulletin* 82: 160–180.

Anastas, A.S., Dalrymple, R.W., James, N.P. and Nelson, C.S. 2006. Lithofacies and dynamics of a cool-water carbonate seaway: Mid-Tertiary, Te Kuiti Group, New Zealand. In: H.M. Pedley and G. Carannante (eds). *Cool-Water Carbonates: Depositional systems and palaeoenvironmental controls*. Geological Society Special Publication 255, pp. 245–268.

Anderson, H.J. 1980. *Geophysics of the Moutere Depression*. DSIR Geophysics Division Report 159. 41 pp.

Anderton, P.W. 1981. Structure and evolution of the South Wanganui Basin, New Zealand. *New Zealand Journal of Geology and Geophysics* 24: 39–63.

Andrews, P.B. 1963. Stratigraphic nomenclature of the Omihi and Waikari formations, North Canterbury. *New Zealand Journal of Geology and Geophysics* 6: 228–256.

Andrews, P.B. 1968. Patterns of sedimentation during early Otaian (Early Miocene) time in North Canterbury, New Zealand. *New Zealand Journal of Geology and Geophysics* 11: 711–752.

Andrews, P.B. and Ovenshine, A.T. 1975. Terrigenous silt and clay facies: Deposits of the early phase of ocean basin evolution. *Initial Reports of the Deep Sea Drilling Project 29.* US Government Printing Office, Washington, pp. 1049–1063. doi:10.2973/dsdp.proc.29.131.19

Andrews, P.B., Bishop, D.G., Bradshaw, J.D. and Warren, G. 1974. Geology of the Lord Range, central Southern Alps, New Zealand. *New Zealand Journal of Geology and Geophysics* 17: 271–299.

Andrews, P.B., Speden, I.G. and Bradshaw, J.D. 1976. Lithological and paleontological content of the Carboniferous–Jurassic Canterbury Suite, South Island, New Zealand. *New Zealand Journal of Geology and Geophysics* 19: 791–819.

Andrews, P.B., Field, B.D., Browne, G.H. and McLennan, J.M. 1987. Lithostratigraphic nomenclature for the Upper Cretaceous and Tertiary sequence of Central Canterbury, New Zealand. *New Zealand Geological Survey Record* 24. 40 pp.

Arafin, M.S. 1982. Tertiary geology of the Birchwood area. Unpublished BSc (Hons) thesis, University of Otago.

Arnold, H.C. 1957. *The Pliocene Stratigraphy of South Taranaki*. Ministry of Economic Development, Wellington, PR414. 19 pp. + 17 enclosures.

Ashcroft, J. 1986. The Kerikeri Volcanics: A basalt-pantellerite association in Northland. *Royal Society of New Zealand Bulletin* 23: 48–63.

Baker, J. and Seward, D. 1996. Timing of Cretaceous extension and Miocene compression in northeast South Island, New Zealand: Constraints from Rb-Sr and fission-track dating of an igneous pluton. *Tectonics* 15: 976–983.

Bal, A.A. 1994. Cessation of Tasman Sea spreading recorded as a sequence boundary. In: G.J. Van der Lingen, K.M. Swanson and R.J. Muir (eds). *Evolution of the Tasman Sea Basin.* A.A. Balkem, Rotterdam, pp. 105–117.

Bal, A.A. and Lewis, D.W. 1994. A Cretaceous–early Tertiary macrotidal estuarine fluvial succession: Puponga coal measures in Whanganui inlet, onshore Pakawau sub-basin, Northwest Nelson, New Zealand. *New Zealand Journal of Geology and Geophysics* 37: 287–307.

Ballance, P.F. 1974. An inter-arc flysch basin in northern New Zealand: Waitemata Group (Upper Oligocene to Lower Miocene). *Journal of Geology* 82: 439–471.

Ballance, P.F. 1976. Evolution of the Upper Cenozoic Magmatic Arc and plate boundary in northern New Zealand. *Earth and Planetary Science Letters* 28: 356–370.

Ballance, P.F. 1993. The New Zealand Neogene forearc basins. In: P.F. Ballance (ed.). *South Pacific Sedimentary Basins.* Sedimentary Basins of the World 2. Elsevier, Amsterdam, pp. 177–193.

Ballance, P.F. and Campbell, J.D. 1993. The Murihiku arc-related basin of New Zealand (Triassic–Jurassic). In: P.F. Ballance (ed.). *South Pacific Sedimentary Basins.* Sedimentary Basins of the World 2. Elsevier, Amsterdam, pp. 21–33.

Ballance, P.F. and Spörli, K.B. 1979. Northland allochthon. *Journal of the Royal Society of New Zealand* 9: 259–275.

Ballance, P.F., Hayward, B.W. and Wakefield, L.L. 1977. Group nomenclature of the Late Oligocene and early Miocene rocks in Auckland and Northland New Zealand and an Akarana Supergroup. *New Zealand Journal of Geology and Geophysics* 20: 673–686.

Barley, M.E. 1987. Origin and evolution of Mid-Cretaceous, garnet-bearing, intermediate and silicic volcanics from Canterbury, New Zealand. *Journal of Volcanology and Geothermal Research* 32: 247–267.

Barley, M.E., Weaver, S.D. and de Laeter, J.R. 1988. Strontium isotope composition and geochronology of intermediate-silicic volcanics, Mt Somers and Banks Peninsula. *New Zealand Journal of Geology and Geophysics* 31: 179–206.

Barnes, P.M. 1994. Continental extension of the Pacific Plate at the southern termination of the Hikurangi subduction zone: The North Mernoo Fault Zone, offshore New Zealand. *Tectonics* 13: 735–754.

Bassett, K.N. and Orlowski, R. 2004. Pahau Terrane type locality: Fan delta in an accretionary prism trench-slope basin. *New Zealand Journal of Geology and Geophysics* 47: 603–623.

Beggs, J.M. 1978. Geology of the metamorphic basement and Late Cretaceous to Oligocene sedimentary sequence of Campbell Island, southwest Pacific Ocean. *Journal of the Royal Society of New Zealand* 8: 161–177.

Beggs, J.M. 1993. Depositional and tectonic history of the Great South Basin. In: P.F. Ballance (ed.). *South Pacific Sedimentary Basins.* Sedimentary Basins of the World 2. Elsevier, Amsterdam, pp. 365–373.

Beggs, J.M., Challis, G.A. and Cook, R.A. 1990. Basement geology of the Campbell Plateau: Implications for correlation of the Campbell Magnetic Anomaly System. *New Zealand Journal of Geology and Geophysics* 33: 401–404.

Bergman, S.C., Talbot, J.P. and Thompson, P.R. 1992. The Kora Miocene submarine andesite stratovolcano hydrocarbon reservoir, Northern Taranaki Basin, New Zealand. *1991 New Zealand Oil Exploration Conference Proceedings*. Ministry of Commerce, Wellington, pp. 178–206.

Beu, A.G. 1974. Notes from the New Zealand Geological Survey – 8: A Kaawa Creek molluscan fauna from a bore south of Manukau Harbour, Auckland. *New Zealand Journal of Geology and Geophysics* 17: 490–494.

Beu, A.G. 1979. Bathyal Nukumaruan Mollusca from Oaro, southern Marlborough, New Zealand. *New Zealand Journal of Geology and Geophysics* 22: 87–103. doi:10.1080/00288306.1979.10422556

Beu, A.G. 1992. *Late Neogene Limestone of Eastern New Zealand Lithostratigraphy*. New Zealand Geological Survey Paleontological Report 161. 209 pp.

Beu, A.G. 1995. *Pliocene Limestones and their Scallops – Lithostratigraphy, Pectinid Biostratigraphy and Paleogeography of Eastern North Island Late Neogene Limestone*. Monograph 10, Institute of Geological and Nuclear Sciences, Lower Hutt. 243 pp.

Biros, D., Cuevas, R. and Moehl, B. 1995. Well Completion Report, Tithaoa-1, PPL 38318. New Zealand unpublished open-file petroleum report PR2018. Ministry of Commerce, Wellington. 1096 pp. + 13 enclosures.

Bishop, D.G. 1971. *Sheets S1, S3 & pt. S4 Farewell–Collingwood.* Geological Maps of New Zealand 1:63,360. Department of Scientific and Industrial Research, Wellington.

Bishop, D.G. 1974a. Tertiary and Quaternary geology of the Naseby and Kyeburn areas, central Otago. *New Zealand Journal of Geology and Geophysics* 17: 19–39.

Bishop, D.G. 1974b. Stratigraphic, structural, and metamorphic relationships in the Dansey Pass area, Otago, New Zealand. *New Zealand Journal of Geology and Geophysics* 17: 301–335.

Bishop, D.G. 1985. Inferred uplift rates from raised marine surfaces, southern Fiordland, New Zealand. *New Zealand Journal of Geology and Geophysics* 28: 243–251.

Bishop, D.G. 1986. *Sheet B46 Puysegur*. Geological Maps of New Zealand 1:50,000. Department of Scientific and Industrial Research, Wellington. 1 sheet + 36 pp.

Bishop, D.G. 1994. *Geology of the Milton Area*. Institute of Geological and Nuclear Sciences Geological Map 9 1:50,000. Institute of Geological and Nuclear Sciences, Lower Hutt. 1 sheet + 32 pp.

Bishop, D.G. and Laird, M.G. 1976. Stratigraphy and depositional environment of the Kyeburn Formation (Cretaceous), a wedge of coarse terrestrial sediments in Central Otago. *Journal of the Royal Society of New Zealand* 6: 55–71.

Bishop, D.G. and Turnbull, I.M. 1996. *Geology of the Dunedin Area*. Institute of Geological and Nuclear Sciences 1:250,000 QMap 21. Institute of Geological and Nuclear Sciences, Lower Hutt. 1 sheet + 52 pp.

Bishop, D.G., Bradshaw, J.D., Landis, C.A. and Turnbull, I.M. 1976. Lithostratigraphy and structure of the Caples Terrane of the Humboldt Mountains, New Zealand. *New Zealand Journal of Geology and Geophysics* 19: 827–848.

Bishop, D.G., Bradshaw, J.D. and Landis, C.A. 1985. Provisional terrane map of South Island, New Zealand. In: D.G. Howell (ed.). *Tectonostratigraphic Terranes of the Circum-Pacific Region*. Circum-Pacific Council for Minerals and Energy, Houston, pp. 515–521.

Bishop, D.J. 1992. Middle Cretaceous–Tertiary tectonics and seismic interpretation of North Westland and Northwest Nelson, New Zealand. Unpublished PhD thesis, Victoria University of Wellington.

Bishop, D.J., Reay, A., Koons, P.O. and Turnbull, I.M. 1992. Composition and regional significance of Mid Bay and Mason Bay reefs, Foveaux Strait, New Zealand. *New Zealand Journal of Geology and Geophysics* 35: 109–112.

Black, R.D. 1980. Upper Cretaceous and Tertiary geology of Mangatu State Forest, Raukumara Peninsula, New Zealand. *New Zealand Journal of Geology and Geophysics* 23: 293–312.

Bland, K.J. 2006. Analysis of the central Hawke's Bay sector of the Late Neogene forearc basin, Hikurangi Margin, New Zealand. Unpublished PhD thesis, University of Waikato.

Blom, W.M. 1982. Sedimentology of the Tokomaru Formation, Waiapu Subdivision, Raukumara Peninsula. Unpublished MSc thesis, University of Auckland.

Blom, W.M. 1984. Stratigraphy and sedimentology of Tokomaru Formation (Late Miocene–Early Pliocene), eastern Raukumara Peninsula, New Zealand. *New Zealand Journal of Geology and Geophysics* 27: 125–137.

Bohor, B.F., Modreski, P.J. and Foord, E.E. 1987. Shocked quartz in the Cretaceous–Tertiary boundary clays: Evidence for a global distribution. *Science* 236: 705–709.

Bowen, F.E. 1967. Early Pleistocene glacial and associated deposits of the West Coast of the South Island, New Zealand. *New Zealand Journal of Geology and Geophysics* 10: 164–181.

Bradshaw, J.D. 1972. Stratigraphy and structure of the Torlesse Supergroup (Triassic–Jurassic) in the foothills of the Southern Alps near Hawarden (S60–61), Canterbury. *New Zealand Journal of Geology and Geophysics* 15: 71–87.

Bradshaw, J.D. 1973. Allochthonous Mesozoic fossil localities in melange within the Torlesse rocks of North Canterbury. *Journal of the Royal Society of New Zealand* 3: 161–167.

Bradshaw, J.D. 1989. Cretaceous geotectonic pattern in the New Zealand Region. *Tectonics* 8: 803–820.

Bradshaw, J.D. 1993. A review of the Median Tectonic Zone: Terrane boundaries and terrane amalgamation near the Median Tectonic Line. *New Zealand Journal of Geology and Geophysics* 36: 117–125.

Bradshaw, J.D. 1994. Brook Street and Murihiku terranes of New Zealand in the context of a mobile south Pacific margin. *Journal of South American Earth Sciences* 7: 325–332.

Bradshaw, J.D. 2004. Northland Allochthon: An alternative hypothesis of origin. *New Zealand Journal of Geology and Geophysics* 47: 375–382.

Bradshaw, J.D. 2007. The Ross Orogen and Lachlan Fold Belt in Marie Byrd Land, northern Victoria Land and New Zealand: Implication for the tectonic setting of the Lachlan Fold Belt in Antarctica. In: A.K. Cooper and C.R. Raymond (eds). *Proceedings of the 10th ISAES*. USGS Open-File Report 2007-1047. doi:10.3133/of2007-1047.srp059

Bradshaw, J.D., Adams, C.J. and Andrews, P.B. 1981. Carboniferous to Cretaceous on the Pacific margin of Gondwana; the Rangitata phase of New Zealand. In: M.M. Cresswell and P. Vella (eds). *Gondwana Five: Proceedings of the Fifth International Gondwana Symposium*. Balkema, Rotterdam, pp. 217–221.

Bradshaw, J.D., Andrews, P. and Field, D.B. 1983. Swanson Formation and related rocks of Marie Byrd Land and comparison with the Robertson Bay Group of north Victoria Land. In: R.L. Oliver, P.R. James and J.B. Jago (eds). *Antarctic Earth Science*. Australian Academy of Science, Canberra, pp. 274–279.

Bradshaw, J.D., Weaver, S.D. and Muir, R.J. 1996. Mid-Cretaceous oroclinal bending of New Zealand terranes. *New Zealand Journal of Geology and Geophysics* 39: 463–468.

Bradshaw, J.D., Pankhurst, R.J., Weaver, S.D., Storey, B.C., Muir, R.J. and Ireland, T.R. 1997. New Zealand superterranes recognised in Marie Byrd Land and Thurston Island. In: C.A. Ricci (ed.). *The Antarctic Region: Geological evolution and processes*. Terra Antarctica, Siena, pp. 429–436.

Bradshaw, J.D., Weaver, S.D., Pankhurst, R.J. and Storey, B.C. 1998. Segmented oblique convergence: 100 million year history of ridge–trench interaction on the Pacific–Gondwana margin. In: J. Almond, J. Anderson and P. Booth (eds). *Gondwana 10: Event stratigraphy of Gondwana*. Pergamon, London, 36 pp.

Bradshaw, J.D., Gutjahr, M., Weaver, S.D. and Bassett, K.N. 2009. Cambrian intra-oceanic arc accretion to the austral Gondwana margin: Constraints on the location of proto-New Zealand. *Australian Journal of Earth Sciences* 56: 587–594.

Bradshaw, M.A. 1995. Stratigraphy and structure of the Lower Devonian rocks of the Waitahu and Orlando outliers, near Reefton, New Zealand, and their relationship to the Inangahua outlier. *New Zealand Journal of Geology and Geophysics* 38: 81–92.

Bradshaw, M.A. 2000. Base of the Devonian Baton Formation and the question of a pre-Baton tectonic event in the Takaka Terrane, New Zealand. *New Zealand Journal of Geology and Geophysics* 43: 601–610.

Briggs, R.M., Itaya, T., Lowe, D.J. and Keane, A.J. 1989. Ages of the Pliocene–Pleistocene Alexandra and Ngatutura Volcanics, western North Island, New Zealand, and some geological implications. *New Zealand Journal of Geology and Geophysics* 32: 417–427.

Brook, F.J. 1983. Lower Miocene geology of North Kaipara, northern New Zealand, with emphasis on the paleontology. Unpublished PhD thesis, University of Auckland.

Brook, F.J. 1989. *Sheets N1 and N2 North Cape*. Geological Maps of New Zealand 1:63,360. Department of Scientific and Industrial Research, Wellington.

Brook, F.J. and Hayward, B.W. 1989. Geology of autochthonous and allochthonous sequences between Kaitaia and Whangaroa, northern New Zealand. *New Zealand Geological Survey Record* 36. 44 pp.

Brook, F.J. and Thrasher, G.P. 1991. Cretaceous and Cenozoic geology of northernmost New Zealand. *New Zealand Geological Survey Record* 41. 42 pp.

Brook, F.J., Isaac, M.J. and Hayward, B.W. 1988. Geology of autochthonous and allochthonous strata in the Omahuta area, northern New Zealand. *New Zealand Geological Survey Record* 32. 40 pp.

Brooks, R.R., Reeves, R.D., Yang, X-H., Ryan, D.E., Holzbecher, J., Collen, J.D., Neall, V.E. and Lee, J. 1984. Elemental anomalies at the Cretaceous–Tertiary boundary, Woodside Creek, New Zealand. *Science* 226: 539–542.

Brooks, R.R., Strong, C.P., Lee, J., Orth, C.J., Gilmore, J.S., Ryan, D.E. and Holzbecher, J. 1986. Stratigraphic occurrences of iridium anomalies at four Cretaceous/Tertiary boundary sites in New Zealand. *Geology* 14: 727–729.

Brothers, R.N. and Golson, J. 1959. Geological and archaeological interpretation of a section in Rangitoto ash on Motutapu Island, Auckland. *New Zealand Journal of Geology and Geophysics* 2: 569–577.

Browne, G.H. 1978. Wanganui strata of the Mangaohane Plateau, northern Ruahine Range, Taihape. *Tane* 24: 199–210.

Browne, G.H. 1985. Dolomitic concretions in Late Cretaceous sediments at Kaikoura Peninsula, Marlborough. *New Zealand Geological Survey Record* 8: 120–127.

Browne, G.H. 1987. *In situ* and intrusive sandstone in Amuri facies limestone at Te Kaukau Point, southeast Wairarapa, New Zealand. *New Zealand Journal of Geology and Geophysics* 30: 363–374.

Browne, G.H. 1992. A Late Miocene flysch sequence near Ward, Marlborough. *New Zealand Geological Survey Record* 44: 9–14.

Browne, G.H. 1995a. *A Sedimentological Review of Cores from the Mt Messenger Formation at Kaipikari-1 & the Moki Formation at Mystone-1.* Ministry of Economic Development, Wellington, PR2169. 15 pp.

Browne, G.H. 1995b. Sedimentation patterns during the Neogene in Marlborough, New Zealand. *Journal of the Royal Society of New Zealand* 25: 459–483.

Browne, G.H. and Field, B.D. 1985. The lithostratigraphy of Late Cretaceous to Early Pleistocene rocks of northern Canterbury, New Zealand. *New Zealand Geological Survey Record* 6. 63 pp.

Browne, G.H. and Reay, M.B. 1993. The Warder Formation: Cyclic fluvial sedimentation during the Ngaterian (late Albian–Cenomanian) of Marlborough, New Zealand. *New Zealand Journal of Geology and Geophysics* 36: 27–35.

Browne, G.H., Kennedy, E.M., Constable, R.M., Raine, J.I., Crouch, E.M. and Sykes, R. 2008. An outcrop-based study of the economically significant Late Cretaceous Rakopi Formation, northwest Nelson, Taranaki Basin, New Zealand. *New Zealand Journal of Geology and Geophysics* 51: 295–315.

Bryant, I.D. and Bartlett, A.D. 1991. Kapuni 3D reservoir model and reservoir simulation. *1991 New Zealand Oil Exploration Conference Proceedings*. Ministry of Commerce, Wellington.

Burgreen, B. and Graham, S. 2014. Evolution of a deep-water lobe system in the Neogene trench-slope setting of the East Coast Basin, New Zealand: Lobe stratigraphy and architecture in a weakly confined basin configuration. *Marine and Petroleum Geology* 54: 1–22. doi:10.1016/j.marpetgeo.2014.02.011

Bussell, M.R. 1994. Seismic interpretation of the Moki Formation on the Maui 3D survey, Taranaki Basin. *1994 New Zealand Petroleum Conference Proceedings: The post Maui challenge – investment and development opportunities*. Ministry of Commerce, Wellington, pp. 240–255.

Cahill, J.P. 1995. Evolution of the Winton Basin, Southland. *New Zealand Journal of Geology and Geophysics* 38: 245–258.

Campbell, H.J., Andrews, P.B., Beu, A.G., Edwards, A.R., Hornibrook, N. de B., Laird, M.G., Maxwell, P.A. and Watters, W.A. 1988. Cretaceous–Cenozoic lithostratigraphy of the Chatham Islands. *Journal of the Royal Society of New Zealand* 18: 285–308.

Campbell, H.J., Andrews, P.B., Beu, A.G., Maxwell, P.A., Edwards, A.R., Laird, M.G. Hornibrook, N. de B. *et al.* 1993. *Cretaceous–Cenozoic Geology and Biostratigraphy of the Chatham Islands.* Monograph 2, Institute of Geological and Nuclear Sciences, Lower Hutt. 269 pp.

Campbell, H.J., Smale, D., Grapes, R., Hoke, L., Gibson, G.M. and Landis, C.A. 1998. Parapara Group: Permian–Triassic rocks in the Western Province, New Zealand. *New Zealand Journal of Geology and Geophysics* 41: 281–296.

Campbell, H.J., Mortimer, N. and Raine, J.I. 2001. Geology of the Permian Kuriwao Group, Murihiku Terrane, Southland, New Zealand. *New Zealand Journal of Geology and Geophysics* 44: 485–498.

Campbell, H.J., Mortimer, N. and Turnbull, I.M. 2003. Murihiku Supergroup, New Zealand: Redefined. *Journal of the Royal Society of New Zealand* 33: 85–95.

Campbell, K.A., Francis, D.A., Collins, M., Gregory, M.R., Nelson, C.S., Greinert, J. and Aharon, P. 2008. Hydrocarbon seep-carbonates of a Miocene forearc (East Coast Basin), North Island, New Zealand. *Sedimentary Geology* 204: 83–105.

Cande, S.C. and Kent, D.V. 1995. Revised calibration of the geomagnetic polarity time scale for Late Cretaceous and Cenozoic. *Journal of Geophysical Research* 100: 6093–6095.

Cande, S.C. and Stock, J.M. 2004. Pacific–Antarctic–Australia motion and the formation of the Macquarie Plate. *Geophysical Journal International* 157: 399–414.

Cande, S.C., Raymond, C.A., Stock, J. and Haxby, W.F. 1995. Geophysics of the Pitman Fracture Zone and Pacific–Antarctic Plate motions during the Cenozoic. *Science* 270: 947–953.

Cande, S.C., Stock, J.M., Muller, R.D. and Ishihara, T. 2000. Cenozoic motion between East and West Antarctica. *Nature* 404: 145–150.

Cape Roberts Science Team. 2000. Studies from Cape Roberts Project, Ross Sea, Antarctica. Initial report on CRP-3. *Terra Antartica* 7: 1–209.

Carlson, J.R., Grant-Mackie, J.A. and Rodgers, K.A. 1980. Stratigraphy and sedimentology of the Coalgate area, Canterbury, New Zealand. *New Zealand Journal of Geology and Geophysics* 23: 179–192.

Carter, L., Carter, R.M. and McCave, I.N. 2004. Evolution of the sedimentary system beneath the deep Pacific inflow off eastern New Zealand. *Marine Geology* 205: 9–27.

Carter, M. and Rainey, S. 1988. *Well Completion Report, Upukerora-1, PPL38074*. Ministry of Economic Development, Wellington, PR1381. 254 pp. + 9 enclosures.

Carter, R.M. 1985. The mid-Oligocene Marshall Paraconformity, New Zealand: Coincidence with global eustatic sea-level fall or rise? *Journal of Geology* 93: 359–371.

Carter, R.M. 1988a. Post breakup stratigraphy of the Kaikoura Synthem (Cretaceous–Cenozoic), continental margin, southeastern New Zealand. *New Zealand Journal of Geology and Geophysics* 31: 405–429.

Carter, R.M. 1988b. Plate boundary tectonics, global sea-level changes and the development of the eastern South Island continental margin, New Zealand, southwest Pacific. *Marine and Petroleum Geology* 5: 90–107.

Carter, R.M. and Carter, L. 1987. The Bounty Channel system: A 55-million-year-old sediment conduit to the deep sea, southwest Pacific Ocean. *Geo-Marine Letters* 7: 183–190.

Carter, R.M. and Landis, C.A. 1972. Correlative unconformities in southern Australasia. *Nature Physical Sciences* 237: 12–13.

Carter, R.M. and Naish T.R. 1998. A review of Wanganui Basin, New Zealand: Global reference section for shallow marine, Plio–Pleistocene (2.5–0 Ma) cyclostratigraphy. *Sedimentary Geology* 122: 37–52.

Carter, R.M. and Norris, R.J. 1977. *Blackmount, Waiau Basin (field trip guide)*. Geological Society of New Zealand Miscellaneous Series 38C. 31 pp.

Carter, R.M., Landis, C.A., Norris, R.J. and Bishop, D.G. 1974. Suggestions towards a high-level nomenclature for New Zealand rocks. *Journal of the Royal Society of New Zealand* 4: 5–18.

Carter, R.M., Lindqvist, J.K. and Norris, R.J. 1982. Oligocene unconformities and nodular phosphate: Hardground horizons in western Southland and northern West Coast. *Journal of the Royal Society of New Zealand* 12: 11–41.

Carter, R.M., McCave, I.N., Richter, C., Carter, L. *et al*. 1999. *Proceedings of the Ocean Drilling Program, Initial Reports* 181. College Station, Texas. http:odp.tamu.edu/publications/181IR

Carter, R.M., McCave, I.N. and Carter, L. 2004. Fronts, flows, drifts, volcanoes, and the evolution of the southwestern gateway to the Pacific Ocean. *Proceedings of the Ocean Drilling Program, Scientific Reports* 181: 1–111.

Cas, R.A., Landis, C.A. and Fordyce, R.E. 1989. A monogenetic, surtla-type, Surtseyan volcano from the Eocene–Oligocene Waiareka–Deborah Volcanics, Otago, New Zealand: A model. *Bulletin of Volcanology* 51: 281–298.

Cashman, S.M., Kelsey, H.M., Erdman, C.F., Cutten, H.N. and Berryman, K.R. 1992. Strain partitioning between structural domains in the forearc of the Hikurangi Subduction Zone, New Zealand. *Tectonics* 11: 242–257.

Cawood, P.A., Nemchin, A.A., Leverenz, A., Saeed, A. and Ballance, P.F. 1999. U/Pb dating of detrital zircons: Implications for the provenance record of Gondwana margin terranes. *Geological Society of America Bulletin* 111: 1107–1119.

Cawood, P.A., Landis, C.A., Nemchin, A.A. and Hada, S. 2002. Permian fragmentation, accretion and subsequent translation of a low-latitude Tethyan seamount to the high-latitude east Gondwana margin: Evidence from detrital zircon age data. *Geological Magazine* 139: 131–144.

Challis, G.A. 1960. Igneous rocks in the Cape Palliser area. *New Zealand Journal of Geology and Geophysics* 3: 524–542.

Chamberlain, C.P., Zeitler, P.K. and Cooper, A.F. 1995. Geochronologic constraints of the uplift and metamorphism along the Alpine Fault, South Island, New Zealand. *New Zealand Journal of Geology and Geophysics* 38: 515–523.

Chanier, F., Ferriere, J. and Angelier, J. 1999. Extensional deformation across an active margin, relations with subsidence, uplift, and rotations: The Hikurangi subduction, New Zealand. *Tectonics* 18: 862–876.

Chapman-Smith, M. and Grant-Mackie, J.A. 1971. Geology of the Whangaparaoa area, eastern Bay of Plenty. *New Zealand Journal of Geology and Geophysics* 14: 3–38.

Clarke, G. 1986. *Seismic Supervision Report PPL 38206*. Ministry of Economic Development, Wellington, PR1165. 175 pp. + 45 enclosures.

Cluzel, D., Aitchison, J.C. and Picard, C. 2001. Tectonic accretion and underplating of mafic terranes in the Late Eocene intraoceanic fore-arc of New Caledonia (Southwest Pacific): Geodynamic implications. *Tectonophysics* 340: 23–59.

Cluzel, D., Black, P.M., Picard, C. and Nicholson, K.N. 2010. Geochemistry and tectonic setting of Matakaoa Volcanics, East Coast Allochthon, New Zealand: Suprasubduction zone affinity, regional correlations and origin. *Tectonics* 29. doi:10.1029/2009TC002454

Cochrane, P.R. 1988. The stratigraphy and structure of the Early Miocene Mahoenui Group northern Awakino Gorge: Controls on cyclic carbonate-terrigenous sedimentation. Unpublished MSc thesis, University of Waikato.

Cole, J.W. 1990. Structural control and origin of volcanism in the Taupo Volcanic Zone, New Zealand. *Bulletin of Volcanology* 52: 445–459.

Collen, J.D. and Vella, P.P. 1984. Hautotara, Te Muna and Ahiaruhe formations, middle to late Pleistocene, Wairarapa, New Zealand. *Journal of the Royal Society of New Zealand* 14: 297–317.

Cook, R.A., Sutherland, R. and Zhu, H. 1999. *Cretaceous–Cenozoic Geology and Petroleum Systems of the Great South Basin, New Zealand.* Monograph 20, Institute of Geological and Nuclear Sciences, Lower Hutt. 188 pp.

Coombs, D.S., Landis, C.A., Norris, R.J., Sinton, J.M., Borns, D.J. and Craw, D. 1976. The Dun Mountain Ophiolite Belt, New Zealand, its tectonic setting, constitution, and origin, with special reference to the southern portion. *American Journal of Science* 276: 561–603.

Coombs, D.S., Cas, R.A., Kawachi, Y., Landis, C.A., McDonough, W.F. and Reay, A. 1986. Cenozoic volcanism in north, east and Central Otago. *Royal Society of New Zealand Bulletin* 23: 278–312.

Coombs, D.S., Cook, N.J., Kawachi, Y., Johnstone, R.D. and Gibson, I.L. 1996. Park Volcanics, Murihiku Terrane, New Zealand: Petrology, petrochemistry, and tectonic significance. *New Zealand Journal of Geology and Geophysics* 39: 469–492.

Coombs, D.S., Landis, C.A., Hada, S., Ito, M., Roser, B.P., Suzuki, T. and Yoshikura, S. 2000. The Chrystalls Beach-Brighton block, southeast Otago, New Zealand: Petrography, geochemistry, and terrane correlation. *New Zealand Journal of Geology and Geophysics* 43: 355–372.

Cooper, A.F. and Ireland, T.R. 2013. Cretaceous sedimentation and metamorphism of the western Alpine Schist protoliths associated with the Pounamu Ultramafic Belt, Westland, New Zealand. *New Zealand Journal of Geology and Geophysics* 56: 188–199.

Cooper A.F. and Ireland, T.R. 2015. The Pounamu terrane, a new Cretaceous exotic terrane within the Alpine Schist, New Zealand; tectonically emplaced, deformed and metamorphosed during collision of the LIP Hikurangi Plateau with Zealandia. *Gondwana Research* 27: 1255–1269.

Cooper, A.F., Barreiro, B.A., Kimbrough, D.L. and Mattinson, J.M. 1987. Lamprophyre dike intrusion and the age of the Alpine Fault, New Zealand. *Geology* 15: 941–944.

Cooper, R.A. 1989. Early Paleozoic terranes of New Zealand. *Journal of the Royal Society of New Zealand* 19: 73–122.

Cooper, R.A. (ed.) 2004. *The New Zealand Geological Timescale.* Monograph 22, Institute of Geological and Nuclear Sciences, Lower Hutt, 284 pp.

Cooper, R.A. and Tulloch, A.J. 1992. Early Paleozoic terranes in New Zealand and their relationship to the Lachlan Fold Belt. *Tectonophysics* 214: 129–144.

Coote, J.A. 1987. Cenozoic volcanism in the Waiau area, North Canterbury. Unpublished MSc thesis, University of Canterbury.

Cotton, C.A. 1916. The structure and later geological history of New Zealand. *Geological Magazine* 3: 243–249, 314–320.

Cox, S.C. 1991. The Caples/Aspiring terrane boundary – the translation surface of an early nappe structure in the Otago Schist. *New Zealand Journal of Geology and Geophysics* 34: 73–82.

Cox, S.C. and Barrell, D.J. 2007. *Geology of the Aoraki Area.* Institute of Geological and Nuclear Sciences 1:250,000 Geological QMap 15. Institute of Geological and Nuclear Sciences, Lower Hutt. 1 sheet + 71 pp.

Crampton, J.S. 1988. Stratigraphy and structure of the Monkey Face area, Marlborough, New Zealand, with special reference to shallow marine Cretaceous strata. *New Zealand Journal of Geology and Geophysics* 31: 447–470.

Crampton, J.S. 1989. An inferred Motuan sedimentary mélange in southern Hawke's Bay. *New Zealand Geological Survey Record* 40: 1–12.

Crampton, J.S. 1997. *The Cretaceous Stratigraphy of the Southern Hawke's Bay–Wairarapa Region.* Institute of Geological and Nuclear Sciences Science Report 97/08. 96 pp.

Crampton, J.S. and Laird, M.G. 1997. Burnt Creek Formation and Late Cretaceous basin development in Marlborough, New Zealand. *New Zealand Journal of Geology and Geophysics* 40: 199–222.

Crampton, J.S. and Moore, P.R. 1990. Environment of deposition of the Maungataniwha Sandstone (Late Cretaceous), Te Hoe River area, western Hawke's Bay, New Zealand. *New Zealand Journal of Geology and Geophysics* 33: 333–348.

Crampton, J.S., Laird, M.G., Nicol, A., Hollis, C.J., Van Dissen, R.J. 1998. Geology at the northern end of the Clarence Valley, Marlborough: A complete record spanning the Rangitata to Kaikoura Orogenies. In: M. Laird (ed.) Geological Society of New Zealand, New Zealand Geophysical Society 1998 Joint Annual Conference, 30 November–3 December: field trip guides. *Geological Society of New Zealand Miscellaneous Publication* 101B. 35 pp.

Crampton, J.S., Schiøler, P., King, P.R. and Field, B.D. 1999. Marine expression of the complex Late Cretaceous Waipounamu erosion surface in the East Coast region, New Zealand. In: *Geological Society of New Zealand Inc. 1999 Annual Conference, Palmerston North: Programme and abstracts*, p. 34.

Crampton, J.S., Raine, I., Strong, P. and Wilson, G. 2001. Integrated biostratigraphy of the Raukumara Series (Cenomanian–Coniacian) at Mangaotane Stream, Raukumara Peninsula, New Zealand. *New Zealand Journal of Geology and Geophysics* 44: 365–389.

Crampton, J.S., Laird, M.G., Nicol, A., Townsend, D.B. and Van Dissen, R.J. 2003. Palinspastic reconstructions of southeastern Marlborough, New Zealand, for mid-Cretaceous–Eocene times. *New Zealand Journal of Geology and Geophysics* 46: 153–175.

Crampton, J.S., Hollis, C.J., Raine, J.I., Roncaglia, L., Schiøler, P., Strong, C.P. and Wilson, G.J. 2004a. Cretaceous. In: R.A. Cooper (ed.). *The New Zealand Geological Timescale*. Monograph 22, Institute of Geological and Nuclear Sciences, Lower Hutt, pp. 103–122.

Crampton, J.S., Tulloch, A.J., Wilson, G.L., Ramezani, J. and Speden, I.G. 2004b. Definition, age and correlation of the Clarence Series stages in New Zealand (late Early to early Late Cretaceous). *New Zealand Journal of Geology and Geophysics* 47: 1–19.

Craw, D. 1979. Melanges and associated rocks, Livingstone Mountains, Southland, New Zealand. *New Zealand Journal of Geology and Geophysics* 22: 443–454.

Crundwell, M.P. 1997. *Neogene Lithostratigraphy of Southern Wairarapa*. Institute of Geological and Nuclear Sciences Science Report 97/36. 57 pp.

Crux, J.A., Eaton, G.L and Sturrock S.I. (eds) 1984. Biostratigraphy of the Clipper well offshore NZ PPL 38202. New Zealand unpublished open-file petroleum report. Ministry of Commerce, Wellington.

Cutten, H.N. 1979. Rappahannock Group: Late Cenozoic sedimentation and tectonics contemporaneous with Alpine Fault movement. *New Zealand Journal of Geology and Geophysics* 22: 535–553.

Cutten, H.N. 1994. *Geology of the Middle Reaches of the Mohaka River*. Institute of Geological and Nuclear Sciences Geological Map 6. Institute of Geological and Nuclear Sciences, Lower Hutt. 1 sheet + 38 pp.

Daczko, N.R., Klepeis, K.A. and Clarke, G.L. 2001. Evidence of Early Cretaceous collisional-style orogenesis in northern Fiordland, New Zealand and its effects on the evolution of the lower crust. *Journal of Structural Geology* 23: 693–713.

Davey, F.J. and Smith, E.G.C. 1983. The tectonic setting of the Fiordland region, southwest New Zealand. *Geophysical Journal of the Royal Astronomical Society* 72: 23–38.

Davey, F.J., Henyey, T., Holbrook, W.S., Okaya, D., Stern, T.A., Melhuish, A., Henrys, S. *et al*. 1998. Preliminary results from a geophysical study across a modern, continent–continent collisional plate boundary – the Southern Alps, New Zealand. *Tectonophysics* 288: 221–235.

Davy, B.W. 1993. The Bounty Trough – basement structure influences on sedimentary basin evolution. In: P.F. Ballance (ed.). *South Pacific Sedimentary Basins*. Elsevier, Amsterdam, pp. 69–92.

Davy, B.W. 2014. Rotation and offset of the Gondwana convergent margin in the New Zealand region following Cretaceous jamming of Hikurangi Plateau large igneous province subduction. *Tectonics* 33: 1577–1595.

de Bock, J.F. 1994. Moki Formation, a Miocene reservoir sequence, its facies distribution and source in offshore southern Taranaki Basin. *1994 New Zealand Petroleum Conference Proceedings: The post Maui challenge – investment and development opportunities*. Ministry of Commerce, Wellington, pp. 155–167.

de Bock, J.F., Palmer, J.A. and Lock, R.G. 1990. Tariki Sandstone – early Oligocene hydrocarbon reservoir, eastern Taranaki, New Zealand. *1989 New Zealand Oil Exploration Conference Proceedings*. Ministry of Commerce, Wellington, pp. 214–224.

DeCelles, P.G. and Giles, K.A. 1996. Foreland basin systems. *Basin Research* 8: 105–123.

Delteil, J.E. 1992. East Coast Pongaroa transect. Unpublished report, Institute of Geological and Nuclear Sciences, Lower Hutt.

Delteil, J.E., Morgans, H.E.G., Raine, J.I., Field, B.D. and Cutten, H.N.C. 1996. Early Miocene thin-skinned tectonics and wrench faulting in the Pongaroa district, Hikurangi margin, North Island, New Zealand. *New Zealand Journal of Geology and Geophysics* 39: 271–282.

Dibble, R.R. and Suggate, R.P. 1956. Gravity survey for possible oil structures, Kotuku-Ahaura district, North Westland. *New Zealand Journal of Science and Technology. B. General Section* 37: 571–586.

Douglas, B.J. 1986. *Manuherakia Group of Central Otago, New Zealand: Stratigraphy, depositional systems, lignite resource assessment, and exploration models*. New Zealand Energy Research and Development Committee Publication P104. 368 pp.

Duff, S.W. 1975. Early-Mid Oligocene (Whaingaroan-Duntroonian) diastems marked by surfaces of biogenic excavation on the East Coast of the South Island, New Zealand. Unpublished MSc thesis, University of Canterbury.

Duff, S.W. 1985. *Kaitangata Coalfield: Coal resource definition and assessment in the Kaitangata sector.* Mines Division, Ministry of Energy.

Duggan, M.B. and Reay, A. 1986. The Timaru Basalt. *Royal Society of New Zealand Bulletin* 23: 264–277.

Eade, J.V. 1966. Stratigraphy and structure of the Mount Adams area, Eastern Wairarapa. *Transactions of the Royal Society of New Zealand* 4: 103–117.

Eagles, G., Gohl, K. and Larter, R.D. 2004. Life of the Bellingshausen plate. *Geophysical Research Letters* 31. https://doi.org/10.1029/2003GL019127

Eberhart-Phillips, D. and Reyners, M. 1997. Continental subduction and three-dimensional crustal structure: The northern South Island, New Zealand. *Journal of Geophysical Research – Solid Earth* 102(B6): 11843–11861.

Edbrooke, S.W. and Brook, F.J. 2009. *Geology of the Whangarei Area*. Institute of Geological and Nuclear Sciences 1:250,000 Geological QMap 2. Institute of Geological and Nuclear Sciences, Lower Hutt. 1 sheet + 68 pp.

Edbrooke, S.W., Sykes, R. and Pocknall, D.T. 1994. *Geology of the Waikato Coal Measures, Waikato Coal Region, New Zealand.* Monograph 6, Institute of Geological and Nuclear Sciences. Lower Hutt, 236 pp.

Edbrooke S.W., Crouch, E.M., Morgans, H.E.G. and Sykes, R. 1998. Late Eocene–Oligocene Te Kuiti Group at Mount Roskill, Auckland, New Zealand. *New Zealand Journal of Geology and Geophysics* 41: 85–93. doi:10.1080/00288306.1998.9514792

Edwards, A.R. 1988. An integrated biostratigraphy, magnetostratigraphy and oxygen isotope stratigraphy for the Late Neogene of New Zealand. *New Zealand Geological Survey Record* 23. 80 pp.

Els, B.G., Youngson, J.H. and Craw, D. 2003. Blue Spur Conglomerate: Auriferous Late Cretaceous fluvial channel deposits adjacent to normal fault scarps, southeast Otago, New Zealand. *New Zealand Journal of Geology and Geophysics* 46: 123–139.

Evans, R.B. 1989. An outcrop of Waitemata Group strata at Waihou Valley, Hokianga, New Zealand, and implications for Northland stratigraphy. *Royal Society of New Zealand Bulletin* 26: 85–93.

Evans, R.B. 1992. Tertiary tectonic development of Whangaroa district, northeastern Northland, New Zealand. *New Zealand Journal of Geology and Geophysics* 35: 549–559.

Exon, N., Kennett, J., Malone, M., Brinkhuis, H., Chaproniere, G. *et al.* 2002. Drilling reveals climatic consequences of Tasmanian Gateway Opening. *Eos, Transactions American Geophysical Union* 83: 253–259.

Feary, D.A. 1979. Geology of the Urewera Greywacke in Waioeka Gorge, Raukumara Peninsula, New Zealand. *New Zealand Journal of Geology and Geophysics* 22: 693–708.

Ferguson, R.J. 1993. Tectonic controls on basin development, fluvial architecture and coal occurrence in the Paparoa Coal Measures at the Pike River Coalfield, West Coast of the South Island. Unpublished MSc thesis, University of Canterbury.

Field, B.D. 1985. *Mid Tertiary Unconformities in the Mount Somers to Rangitata River Area, Canterbury.* New Zealand Geological Survey Report SL 10. 17 pp.

Field, B.D. and Browne, G.H. 1986. Lithostratigraphy of Cretaceous and Tertiary rocks, southern Canterbury, New Zealand. *New Zealand Geological Survey Record* 14: 1–55.

Field, B.D. and Pocknall, D.T. 1984. The Potts River Tertiary outlier, Canterbury. *New Zealand Geological Survey Record* 3: 52–56.

Field, B.D. and Scott, G. 1993. *Neogene Sediments at Anaurraiti Stream – lower Waiau River, East Coast, North Island.* Geological Society of New Zealand Miscellaneous Publication 79A. 65 pp.

Field, B.D., Browne, G.H., Davy, B., Herzer, R.H., Hoskins, R.H., Raine, J.I., Wilson, G.J., Sewell, R.J., Smale, D. and Watters, W.A. 1989. *Cretaceous and Cenozoic Sedimentary Basins and Geological Evolution of the Canterbury Region, South Island, New Zealand.* New Zealand Geological Survey Basin Studies 2. 94 pp.

Field, B.D., Uruski, C.L., Beu, A.G., Browne, G.H., Crampton, J.S., Funnell, R., Killops, S. *et al.* 1997. *Cretaceous–Cenozoic Geology and Petroleum Systems of the East Coast Region, New Zealand.* Monograph 19, Institute of Geological and Nuclear Sciences, Lower Hutt. 301 pp.

Fleming, C.A. and Christey, G.R. 1989. The synsedimentary slide at Te Awa-awa, Hawkes Bay, New Zealand. In: R.W. Le Maitre (ed.). *Pathways in Geology: Essays in honour of Edwin Sherborn Hills.* Blackwell Scientific, Melbourne, pp. 188–200.

Fleming, C.A., Hornibrook, N.D-B. and Wood, B.L. 1969. Geology of the Clifden Section. In: B.L. Wood, *Geology of Tuatapere Subdivision, Western Southland* (sheets S167, S175). New Zealand Geological Survey Bulletin 79, pp. 71–117.

Flores, R.M. and Sykes, R. 1996. Depositional controls on coal distribution and quality in the Eocene Brunner Coal Measures, Buller Coalfield, South Island, New Zealand. *International Journal of Coal Geology* 29: 291–336.

Ford, P.B., Lee, D.E. and Fischer, P.J. 1999. Early Permian conodonts from the Torlesse and Caples Terranes, New Zealand. *New Zealand Journal of Geology and Geophysics* 42: 79–90.

Forder, S.P. and Sissons, B.A. 1992. The Moki C sands: An example of Mio-Pliocene bathyal fans in the North Taranaki Graben. *1991 New Zealand Oil Exploration Conference Proceedings.* Ministry of Commerce, Wellington, pp. 155–167.

Forster, M.A. and Lister, G.S. 2003. Cretaceous metamorphic core complexes in the Otago Schist, New Zealand. *Australian Journal of Earth Sciences* 50: 181–198.

Forsyth, P.J. 2001. *Geology of the Waitaki Area.* Institute of Geological and Nuclear Sciences 1:250,000 Geological QMap 19. GNS Science, Lower Hutt. 1 sheet + 64 pp.

Forsyth, P.J., Martin, G.M., Campbell, H.J., Simes, J.E. and Nicoll, R.S. 2006. Carboniferous conodonts from Rakaia Terrane, East Otago, New Zealand. *New Zealand Journal of Geology and Geophysics* 49: 329–336.

Francis, D.A. 1989. Outline of the geology of coastal Wairarapa between Cape Turnagain and Honeycomb Rock, with emphasis on stratigraphy and relation to offshore geology, PPL 38318. New Zealand Geological Survey client report 89/33. In: Amoco NZ Exploration Co PPL 38318 East Coast Basin discussion of results 1989 field programme, Appendix 4. New Zealand unpublished open-file petroleum report 1677. Ministry of Commerce, Wellington.

Francis, D.A. 1993. Report on the geology of the Mahia area, northern Hawke’s Bay, adjacent to offshore PPL 38321. New Zealand unpublished open-file petroleum report 1928. Ministry of Commerce, Wellington.

Francis, D.A., Scott, G.H. and Morgans, H.E.G. 1987. An Upper Miocene Oligocene unconformity north of Whangaehu Beach, Southern Hawkes Bay. *New Zealand Geological Survey Record* 20: 60–68.

Fulthorpe, C.S., Carter, R.M., Miller, K.G. and Wilson, J. 1996. Marshall Paraconformity: A mid-Oligocene record of inception of the Antarctic Circumpolar Current and coeval glacio-eustatic lowstand? *Marine and Petroleum Geology* 13: 61–77.

Fyfe, H.E. 1968. *Geology of the Murchison Subdivision.* New Zealand Geological Survey Bulletin, 36. 51 pp.

Gage, M. 1945. The Tertiary and Quaternary geology of Ross, Westland. *Transactions of the Royal Society of New Zealand* 75: 138–159.

Gage, M. 1949. Late Cretaceous and Tertiary geosynclines in Westland, New Zealand. *Transactions of the Royal Society of New Zealand* 77: 325–337.

Gage, M. 1952. *The Greymouth Coalfield.* New Zealand Geological Survey Bulletin 45. 232 pp. + 14 maps.

Gage, M. 1957. *The Geology of Waitaki Subdivision.* New Zealand Geological Survey Bulletin 55. 135 pp. + 4 maps.

Gage, M.S. 1998. A new exploration phase in the Northland Basin of New Zealand. *1998 New Zealand Petroleum Conference Proceedings.* Ministry of Commerce, Wellington, pp. 115–121.

Gage, M.S. and Kurata, Y. 1996. Is there a mid-Cretaceous passive margin sequence beneath the Northland Basin, New Zealand? *1996 New Zealand Petroleum Conference Proceedings*, Volume 1. Ministry of Commerce, Wellington, pp. 1–11.

Gaina, C., Müller, D.R., Royer, J.Y., Stock, J., Hardebeck, J. and Symonds, P. 1998. The tectonic history of the Tasman Sea: A puzzle with 13 pieces. *Journal of Geophysical Research – Solid Earth* 103(B6): 12413–12433.

Gamble, J.A. and Adams, C.J. 1985. Volcanic geology of Carnley volcano, Auckland Islands. *New Zealand Journal of Geology and Geophysics* 28: 43–54.

Gammon, P. 1995. Hautotara Formation, Mangaopari Basin, New Zealand: Record of a cyclothemic Pliocene–Pleistocene marine to non-marine transition. *New Zealand Journal of Geology and Geophysics* 38: 471–481.

German, R.C. 1976. Stratigraphy and sedimentology of the Nile Group (Oligocene) SW Nelson. Unpublished MSc thesis, University of Canterbury.

Gerrard, M.J. 1971. *Tupapakurua-1.* Ministry of Economic Development, Wellington, PR266. 48 pp. + 1 enclosure.

Gerritsen, S.W. 1994. The regional stratigraphy and sedimentology of the Miocene sequence in the Ohura-Taumarunui region. Unpublished MSc thesis, University of Waikato.

Gibbons, M.J. and Herridge, K.J. 1984. *Clipper-1 Canterbury Basin, New Zealand. 1. Biostratigraphy report. 2. Maturity & Source Potential report. PPL38202.* Ministry of Economic Development, Wellington, PR1044. 92 pp. + 1 enclosure.

Gibson, G.M., McDougall, I. and Ireland, T.R. 1988. Age constraints on metamorphism and the development of a metamorphic core complex in Fiordland, southern New Zealand. *Geology* 16: 405–408.

Glennie, K.W. 1959. The graded sediments of the Mahoenui Formation (King Country, North Island). *New Zealand Journal of Geology and Geophysics* 2: 613–621.

Gosson, G.J. 1986. Miocene and Pliocene silicic tuffs in marine sediments of the East Coast Basin, New Zealand. Unpublished PhD thesis, Victoria University of Wellington.

Graham, I.J. and Korsch, R.J. 1990. Age and provenance of granitoid clasts in Moeatoa Conglomerate, Kawhia Syncline, New Zealand. *Journal of the Royal Society of New Zealand* 20: 25–39.

Graham, I.J. and Mortimer, N. 1992. Terrane characterisation and timing of metamorphism in the Otago Schist, New Zealand, using Rb-Sr and K-Ar geochronology. *New Zealand Journal of Geology and Geophysics* 35: 391–401.

Gravley, D.M., Deering, C.D., Leonard, G.S. and Rowland, J.V. 2016. Ignimbrite flare-ups and their drivers: A New Zealand perspective. *Earth-Science Reviews* 162: 65–82.

Gregory, M.R. 1969. Sedimentary features and penecontemporaneous slumping in the Waitemata Group, Whangaparaoa Peninsula, North Auckland, New Zealand. *New Zealand Journal of Geology and Geophysics* 12: 248–282.

Griffith, R.C. 1983. Geology of the Blackmount. Unpublished MSc thesis, University of Otago.

Griggs, G.B., Carter, L., Kennett, J.P. and Carter, R.V. 1983. Late Quaternary marine stratigraphy southeast of New Zealand. *Geological Society of America Bulletin* 94: 791–797.

Grindley G.W. 1980. *Sheet S13 Cobb*. Geological Maps of New Zealand 1:63,360. Department of Scientific and Industrial Research, Wellington.

Grobys, J.W.G., Gohl, K. and Eagles, G. 2008. Quantitative tectonic reconstructions of Zealandia based on crustal thickness estimates. *Geochemistry, Geophysics, Geosystems* 9. doi:10.1029/2007GC001691

Hamilton, R.J., Luyendyk, B.P., Sorlien, C.C. and Bartek, L.R. 2001. Cenozoic tectonics of the Cape Roberts Rift Basin and Transantarctic Mountains Front, Southwestern Ross Sea, Antarctica. *Tectonics* 20: 325–342.

Hampton, S.J. and Cole, J.W. 2009. Lyttelton Volcano, Banks Peninsula, New Zealand: Primary volcanic landforms and eruptive centre identification. *Geomorphology* 104: 284–298.

Hancock, H.J.L., Dickens, G.R., Strong, C.P., Hollis, C.J. and Field, B.D. 2003. Foraminiferal and carbon isotope stratigraphy through the Paleocene–Eocene transition at Dee Stream, Marlborough, New Zealand. *New Zealand Journal of Geology and Geophysics* 46: 1–19.

Hansen, R.J. and Kamp, P.J.J. 2002. Evolution of the Giant Foresets Formation, northern Taranaki Basin, New Zealand. *2002 New Zealand Petroleum Conference Proceedings*. Ministry of Commerce, Wellington, pp. 419–435.

Happy, A.J. 1971. Tertiary geology of the Awakino area, North Taranaki. Unpublished MSc thesis, University of Auckland.

Haq, B.U., Hardenbol, J. and Vail, P.R. 1987. Chronology of fluctuating sea-levels since the Triassic. *Science* 235: 1156–1167.

Harmsen, F.J. 1984. Stratigraphy, depositional history and diagenesis of the Te Aute Group, a Pliocene temperate carbonate-bearing sequence in southern Hawke's Bay, New Zealand. Unpublished PhD thesis, Victoria University of Wellington.

Harmsen, F.J. 1985. Lithostratigraphy of Pliocene strata, central and southern Hawke's Bay, New Zealand. *New Zealand Journal of Geology and Geophysics* 28: 413–433.

Harrington, W.M. 1982. The geology of the Haycocks–Snowdon Forest region, western Southland. Unpublished BSc (Hons) thesis, University of Otago.

Haskell, T. and Wylie, I. 1997. New Zealand's Canterbury basins prospects reviewed in a continental Gondwana setting. *Oil & Gas Journal* 95: 57–62.

Hawkes, P.W., Mound, D.G. and Spicer, P.J. 1985. *Clipper-1 Evaluation Report PPL 38022*. Ministry of Economic Development, Wellington, PR1078. 82 pp. + 5 enclosures.

Hayton, S.H. 1998. Sequence stratigraphic, paleoenvironmental, and chronological analysis of the late Neogene Wanganui River section, Wanganui Basin. Unpublished PhD thesis, University of Waikato.

Hayward, B.W. 1983. *Sheet Q11 Waitakere*. Geological Maps of New Zealand 1:50,000. Department of Scientific and Industrial Research, Wellington.

Hayward, B.W. 1993. The tempestuous 10 million year life of a double arc and intra-arc basin – New Zealand's Northland Basin in the Early Miocene. In: P.F. Ballance (ed.). *South Pacific Sedimentary Basins.* Sedimentary Basins of the World 2. Elsevier, Amsterdam, pp. 113–142.

Hayward, B.W. and Brook, F.J. 1984. Lithostratigraphy of the basal Waitemata Group, Kawau Subgroup (new), Auckland, New Zealand. *New Zealand Journal of Geology and Geophysics* 27: 101–123.

Hayward, B.W. and Smale, D. 1992. Heavy minerals and provenance history of Waitemata Basin sediments (Early Miocene, Northland, New Zealand). *New Zealand Journal of Geology and Geophysics* 35: 223–242.

Hayward, B.W., Brook, F.J. and Isaac, M.J. 1989. Cretaceous to middle Tertiary stratigraphy, paleogeography and tectonic history of Northland, New Zealand. *Royal Society of New Zealand Bulletin* 26: 47–64.

Hematite Petroleum. 1970. *Haku-1 Well Report*. Ministry of Economic Development, Wellington, PR553. 64 pp. + 2 enclosures.

Heming, R.F. 1980a. Patterns of Quaternary basaltic volcanism in the northern North Island, New Zealand. *New Zealand Journal of Geology and Geophysics* 23: 335–344.

Heming, R.F. 1980b. Petrology and geochemistry of quaternary basalts from Northland, New Zealand. *Journal of Volcanology and Geothermal Research* 8: 23–44.

Heming, R.F. and Barnet, P.R. 1986. The petrology and petrochemistry of the Auckland volcanic field. In: I.E.M. Smith (ed.). *Late Cenozoic Volcanism in New Zealand*. Royal Society of New Zealand Bulletin 23, pp. 64–75.

Henderson, J. and Ongley, M. 1923. *The Geology of the Mokau Subdivision*. New Zealand Geological Survey Bulletin 24.

Herzer, R.H. 1995. Seismic stratigraphy of a buried volcanic arc, Northland, New Zealand and implications for Neogene subduction. *Marine and Petroleum Geology* 12: 511–531.

Herzer, R.H. and Lewis, D.W. 1979. Growth and burial of a submarine canyon off Motunau, North Canterbury, New Zealand. *Sedimentary Geology* 24: 69–83.

Herzer, R.H., Chaproniere, G.C.H., Edwards, A.R., Hollis, C.J., Pelletier, B., Raine, J.I. *et al.* 1997. Seismic stratigraphy and structural history of the Reinga Basin and its margins, southern Norfolk Ridge system. *New Zealand Journal of Geology and Geophysics* 40: 425–451.

Herzer, R.H., Sykes, R., Killops, S.D., Funnell, R.H., Burggraf, D.R., Townend, J., Raine, J.I. and Wilson, G.J. 1999. Cretaceous carbonaceous rocks from the Norfolk Ridge system, Southwest Pacific: Implications for regional petroleum potential. *New Zealand Journal of Geology and Geophysics* 42: 57–73.

Hicks, S.R. 1974. *Shallow Seismic Surveys at Wainuiomata, Waikanae and Whangaimoana*. Geophysics Division Report 93. 33 pp.

Higgs, K.E. 2007. *Biostratigraphy and Paleoenvironmental Interpretation of the Cretaceous Sequence at Te Ranga-1, Taranaki Basin, New Zealand*. Ministry of Economic Development, Wellington, PR3663. 57 pp. + 3 enclosures.

HIPCO. 1976. Well Completion Report, Parara-1. New Zealand Geological Survey unpublished open-file petroleum report 673.

HIPCO. 1978. Well Completion Report, Takapu-1. New Zealand Geological Survey unpublished open-file petroleum report 672.

Hodell, D.A. and Kennett, J.P. 1986. Late Miocene–Early Pliocene stratigraphy and paleoceanography of the South Atlantic and southwest Pacific oceans: A synthesis. *Paleoceanography* 1: 285–311.

Hoernle, K., Hauff, F., van den Bogaard, P., Werner, R. and Mortimer, N. 2005. The Hikurangi oceanic plateau: Another large piece of the largest volcanic event on earth. *Geochimica et Cosmochimica Acta* 69(10): A96.

Hoernle, K., White, J., van den Bogaard, P., Hauff, F., Coombs, D., Werner, R., Timm, C., Garbe-Schonberg, D., Reay, A. and Cooper, A. 2006. Lithospheric removal: The cause of widespread Cenozoic intraplate volcanism on Zealandia? *Geochimica et Cosmochimica Acta* 70(18): A256.

Hollis, C.J. 1986. The subsurface Kaawa Formation at Mauku, south Auckland. Unpublished BSc (Hons) thesis, University of Auckland.

Hollis, C.J. 1991. Late Cretaceous to Late Paleocene radiolarian from Marlborough (New Zealand) and DSDP Site 208. Unpublished PhD thesis, University of Auckland.

Hollis, C.J. 2003. The Cretaceous/Tertiary boundary event in New Zealand: Profiling mass extinction. *New Zealand Journal of Geology and Geophysics* 46: 307–321.

Hollis, C.J. and Strong, C.P. 2003. Biostratigraphic review of the Cretaceous/Tertiary boundary transition, mid-Waipara River section, North Canterbury, New Zealand. *New Zealand Journal of Geology and Geophysics* 46: 243–253.

Hollis, C.J., Waghorn, D.B., Strong, C.P. and Crouch, E.M. 1997. *Integrated Paleogene Biostratigraphy of DSDP Site 277 (Leg 29): Foraminifera, calcareous nannofossils, Radiolaria, and palynomorphs*. Institute of Geological and Nuclear Sciences Science Report 97/07. 73 pp. + 14 pp. plates.

Hollis, C.J., Strong, C.P., Rodgers, K.A. and Rogers, K.M. 2003a. Paleoenvironmental changes across the Cretaceous/Tertiary boundary at Flaxbourne River and Woodside Creek, eastern Marlborough, New Zealand. *New Zealand Journal of Geology and Geophysics* 46: 177–197.

Hollis, C.J., Rodgers, K.A., Strong, C.P., Field, B.D. and Rogers, K.M. 2003b. Paleoenvironmental changes across the Cretaceous/Tertiary boundary in the northern Clarence valley, southeastern Marlborough, New Zealand. *New Zealand Journal of Geology and Geophysics* 46: 209–234.

Hollis, C.J., Field, B.D., Rogers, K.M. and Strong, C.P. 2003c. *Stratigraphic, Paleontological, and Geochemical Data from Upper Cretaceous and Lower Paleocene Strata in Southeastern Marlborough, New Zealand*. Institute of Geological and Nuclear Sciences Science Report 03/09. 64 pp.

Hollis, C.J., Dickens, G.R., Field, B.D., Jones, C.M. and Strong, C.P. 2005a. The Paleocene–Eocene transition at Mead Stream, New Zealand: A southern Pacific record of early Cenozoic global change. *Palaeogeography, Palaeoclimatology, Palaeoecology* 215: 313–343.

Hollis, C.J., Field, B.D., Jones, C.M., Strong, C.P., Wilson, G.J. and Dickens, G.R. 2005b. Biostratigraphy and carbon isotope stratigraphy of uppermost Cretaceous–lower Cenozoic Muzzle Group in middle Clarence valley, New Zealand. *Journal of the Royal Society of New Zealand* 35: 345–383.

Hollis, C.J., Lueer, V., Neil, V.H., Manighetti, B. and Scott, G.H. 2005c. Radiolarian-based indicators of oceanographic changes, offshore eastern New Zealand, over the last 500,000 years. *Geological Society of New Zealand 50th Annual Conference: Programme and abstracts*. Geological Society of New Zealand Miscellaneous Publication 119A, pp. 36–37.

Homer, L. and Moore, P. 1989. *Reading the Rocks: A guide to the Wairarapa*. Landscape Publications Ltd, Wellington.

Hornibrook, N. de B. 1981. Globorotalia (planktic foraminiferida) in the Late Pliocene and Early Pleistocene of New Zealand. *New Zealand Journal of Geology and Geophysics* 24: 263–292.

Hornibrook, N. de B., Edwards, A.R., Mildenhall, D.C., Webb, P.N. and Wilson, G.J. 1976. Major displacements in Northland, New Zealand: Micropaleontology and stratigraphy of Waimamaku 1 and 2 wells. *New Zealand Journal of Geology and Geophysics* 19: 233–263.

Hoskins, R.H. 1978. New Zealand Middle Miocene foraminifera: The Waiauan Stage. Unpublished PhD thesis, University of Exeter.

Houghton, B.F. 1981. Lithostratigraphy of the Takitimu Group, central Takitimu Mountains, western Southland. *New Zealand Journal of Geology and Geophysics* 24: 333–348.

Houghton, B.F. and Landis, C.A. 1989. Sedimentation and volcanism in a Permian arc-related basin, southern New Zealand. *Bulletin of Volcanology* 51: 433–450.

Houtz, R., Ewing, J., Ewing, M. and Lonardi, A.G. 1967. Seismic reflection profiles of the New Zealand Plateau. *Journal of Geophysical Research* 72: 4713–4729.

Hunt Petroleum. 1978. *Well Completion Report, Takapu-1 & Takapu-1A*. Ministry of Economic Development, Wellington, PR733. 213 pp.

Hyden, F.M. 1979. Mid-Tertiary temperate shelf limestones, Southland, New Zealand. Unpublished PhD thesis, University of Otago.

Hyden, F.M. 1980. Mass flow deposits on a mid-Tertiary carbonate shelf, southern New Zealand. *Geological Magazine* 117: 409–423.

Isaac, M.J. 1977. Mesozoic geology of the Matawai district, Raukumara Peninsula. Unpublished PhD thesis, University of Auckland.

Isaac, M.J. 1996. *Geology of the Kaitaia Area*. Institute of Geological and Nuclear Sciences 1:250,000 Geological QMap 1. Institute of Geological and Nuclear Sciences, Lower Hutt. 1 sheet + 44 pp.

Isaac, M.J. and Lindqvist, J.K. 1990. *Geology and Lignite Resources of the East Southland Group, New Zealand*. New Zealand Geological Survey Bulletin 101. 202 pp.

Isaac, M.J., Brook, F.J. and Hayward, B.W. 1988. The Cretaceous sequence at Whatuwhiwhi, Northland, and its paleogeographic significance. *New Zealand Geological Survey Record* 26. 32 pp.

Isaac, M.J., Moore, P.R. and Joass, Y.J. 1991. Tahora Formation: The basal facies of a Late Cretaceous transgressive sequence, northeastern New Zealand. *New Zealand Journal of Geology and Geophysics* 34: 227–236.

Isaac, M.J., Herzer, R.H., Brook, F.J. and Hayward, D.W. 1994. *Cretaceous and Cenozoic Sedimentary Basins of Northland, New Zealand*. Monograph 8, Institute of Geological and Nuclear Sciences, Lower Hutt. 203 pp.

Ivany, L.C., Patterson, W.P. and Lohmann, K.C. 2000. Cooler winters as a possible cause of mass extinctions at the Eocene/Oligocene boundary. *Nature* 407: 887–890.

Jamieson, B.S. 1979. The geology of the Takahe Valley area, eastern Fiordland. Unpublished DipSci (Geology) thesis, University of Otago.

Jenkins, D.G. and Jenkins, T.B.H. 1971. First diagnostic Carboniferous fossils from New Zealand. *Nature* 233: 117–118.

Joass, Y.J. 1987. The geology of the Tahora district, southern Raukumara Peninsula, New Zealand. Unpublished MSc thesis, Victoria University of Wellington.

Johnston, M.R. 1971. Pre-hawera geology of the Kaka District, North-West Nelson. *New Zealand Journal of Geology and Geophysics* 14: 82–102.

Johnston, M.R. 1979. Geology of the Nelson urban area 1:25,000. New Zealand Geological Survey Urban Series. New Zealand Department of Scientific and Industrial Research, Wellington. 1 sheet + 52 pp.

Johnston, M.R. 1980. *Geology of the Tinui-Awatoitoi District*. New Zealand Geological Survey Bulletin 94. 62 pp.

Johnston, M.R. 1990. *Geology of the St Arnaud District, Southeast Nelson* (Sheet N29). New Zealand Geological Survey Bulletin 99. 119 pp.

Johnston, M.R., Raine, J.I. and Watters, W.A. 1987. Drumduan Group of East Nelson, New Zealand: Plant-bearing Jurassic arc rocks metamorphosed during terrane interaction. *Journal of the Royal Society of New Zealand* 17: 275–301.

Jongens, R. 2006. Structure of the Buller and Takaka Terrane rocks adjacent to the Anatoki Fault, northwest Nelson, New Zealand. *New Zealand Journal of Geology and Geophysics* 49: 443–461.

Jongens, R., Bradshaw, J.D. and Fowler, A.P. 2003. The Balloon Melange, northwest Nelson: Origin, structure, and emplacement. *New Zealand Journal of Geology and Geophysics* 46: 437–448.

Journeaux, T.D., Kamp, P.J.J. and Naish, T. 1996. Middle Pliocene cyclothems, Mangaweka region, Wanganui basin, New Zealand: A lithostratigraphic framework. *New Zealand Journal of Geology and Geophysics* 39: 135–149.

Jugum, D., Norris, R.H. and Palin, J.M. 2013. Late Jurassic detrital zircons from the Haast Schist and their implications for New Zealand terrane assembly and metamorphism. *New Zealand Journal of Geology and Geophysics* 56: 223–228.

Kamp, P.J.J. 1986. Late Cretaceous–Cenozoic tectonic development of the Southwest Pacific region. *Tectonophysics* 121: 225–251.

Kamp, P.J.J. 2000. Thermochronology of the Torlesse accretionary complex, Wellington region, New Zealand. *Journal of Geophysical Research – Solid Earth* 105(B8): 19253–19272.

Kamp, P.J.J. and Liddell, I.J. 2000. Thermochronology of northern Murihiku Terrane, New Zealand, derived from apatite FT analysis. *Journal of the Geological Society of London* 157: 345–354.

Kamp, P.J.J., Green, P.F. and Tippett, J.M. 1992a. Tectonic architecture of the mountain front-foreland basin transition, South Island, New Zealand, assessed by fission track analysis. *Tectonics* 11: 98–113.

Kamp, P.J.J., Whitehouse, I.W. and Newman, J. 1992b. Thermo-tectonic history and hydrocarbon prospectivity of Greymouth Coalfield, Westland, assessed by apatite fission track analysis. *1991 New Zealand Oil Exploration Conference Proceedings*. Ministry of Commerce, Wellington, pp. 321–335.

Kamp, P.J.J., Webster, K.S. and Nathan, S. 1996. Thermal history analysis by integrated modelling of apatite fission track and vitrinite reflectance data: Application to an inverted basin (Buller Coalfield, New Zealand). *Basin Research* 8: 383–402.

Kamp, P.J.J., Vonk, A.J., Bland, K.J., Griffin, A.G., Hayton, S., Hendy, A.J.W., McIntyre, A.P., Nelson, C.S. and Naish, T. 2002. Megasequence architecture of Taranaki, Wanganui and King Country Basins and Neogene progradation of two continental margin wedges across western New Zealand. *2002 New Zealand Petroleum Conference Proceedings*. Ministry of Economic Development, Wellington, pp. 464–473.

Kamp, P.J.J., Vonk, A.J., Bland, K.J., Hansen, R.J., Hendy, A.J.W., McIntyre, A.P., Ngatai, M., Cartwright, S.J., Hayton, S. and Nelson, C.S. 2004. Neogene stratigraphic architecture and tectonic evolution of Wanganui, King Country, and eastern Taranaki Basins, New Zealand. *New Zealand Journal of Geology and Geophysics* 47: 625–644.

Katz, H.R. 1968. Potential oil formations in New Zealand and their stratigraphic position as related to basin evolution. *New Zealand Journal of Geology and Geophysics* 11: 1077–1135.

Kear, D. 1955. Mesozoic and Lower Tertiary stratigraphy and limestone deposits, Torehina, Coromandel. *New Zealand Journal of Science and Technology. B. General Section* 37: 107–114.

Kear, D. 1960. *Sheet 4-Hamilton*. Geological Maps of New Zealand 1:250,000. Department of Scientific and Industrial Research, Wellington.

Kear, D. 1994. A 'least complex' dynamic model for late Cenozoic volcanism in the North Island, New Zealand. *New Zealand Journal of Geology and Geophysics* 37: 223–236.

Kear, D. 2004. Reassessment of Neogene tectonism and volcanism in North Island, New Zealand. *New Zealand Journal of Geology and Geophysics* 47: 361–374.

Kear, D. and Schofield, J.C. 1965. *Sheet N65 Hamilton*. Geological Maps of New Zealand 1:63,360. Department of Scientific and Industrial Research, Wellington.

Kear, D. and Schofield, J.C. 1978. *Geology of the Ngaruawahia Subdivision*. New Zealand Geological Survey Bulletin 88. 168 pp.

Kelsey H.M., Erdman C.F. and Cashman, S.M. 1993. *Geology of Southern Hawkes Bay from the Maraetotara Plateau and Waipawa Westward to the Wakarara Range and the Ohara Depression*. Institute of Geological and Nuclear Sciences Science Report 93/02. 17 pp.

Kennett, J.P. 1966. Biostratigraphy and paleoecology in upper Miocene–lower Pliocene sections in Wairarapa and southern Hawke's Bay. *Transactions of the Royal Society of New Zealand* (geology) 4: 1–77.

Kennett, J.P. 1967. Recognition and correlation of the Kapitean Stage (Upper Miocence, New Zealand). *New Zealand Journal of Geology and Geophysics* 10: 1051–1063.

Kennett, J.P. 1980. Palaeoceanographic and biogeographic evolution of the Southern Ocean during the Cenozoic, and Cenozoic microfossil datums. *Palaeogeography, Palaeoclimatology, Palaeoecology* 31: 123–152.

Kennett, J.P., Houtz, R.E., Andrews, P.B., Edwards, A.R., Gostin, V.A., Hajós, M., Hampton, M., *et al*. 1975. Cenozoic paleoceanography in the southwest Pacific Ocean, Antarctic glaciation, and the development of the circum-Antarctic current. *Initial Reports of the Deep Sea Drilling Project 29*. US Government Printing Office, Washington, pp. 1155–1169.

Kenny, J.A. 1984. Stratigraphy, sedimentology and structure of the Ihungia decollement, Raukumara Peninsula, North Island, New Zealand. *New Zealand Journal of Geology and Geophysics* 27: 1–19.

Kermode, L.O. 1992. *Geology of the Auckland Urban Area: Sheet R11*. Scale 1:50,000. Institute of Geological and Nuclear Sciences Geological Map 2. Institute of Geological and Nuclear Sciences, Lower Hutt. 1 sheet + 63 pp.

Kicinski, F.M. 1958. Report on correlation of stratigraphic columns from the north east coast of the North Island New Zealand. New Zealand unpublished open-file petroleum report 306. Ministry of Commerce, Wellington.

Killops, S.D., Hollis, C.J., Morgans, H.E.G., Sutherland, R., Field, B.D. and Leckie, D.A. 2000. Paleoceanographic significance of Late Paleocene dysaerobia at the shelf/slope break around New Zealand. *Palaeogeography, Palaeoclimatology, Palaeoecology* 156: 51–70.

Kimbrough, D.L., Mattinson, J.M., Coombs, D.S., Landis, C.A. and Johnston, M.R. 1992. Uranium-lead ages from the Dun Mountain ophiolite belt and Brook Street Terrane, South Island, New Zealand. *Geological Society of America Bulletin* 104(4): 429–443.

Kimbrough, D.L., Tulloch, A.J., Coombs, D.S., Landis, C.A., Johnston, M.R. and Mattinson, J.M. 1994. Uranium-lead zircon ages from the Median Tectonic Zone, New Zealand. *New Zealand Journal of Geology and Geophysics* 37: 393–419.

King, P.R. 1988. *Well Summary Sheets Offshore Taranaki*. New Zealand Geological Survey Report G 127. 18 pp.

King, P.R. 2000. Tectonic reconstructions of New Zealand: 40 Ma to the Present. *New Zealand Journal of Geology and Geophysics* 43: 611–638.

King, P.R. and Robinson, P.H. 1988. An overview of Taranaki region geology, New Zealand. *Energy Exploration & Exploitation* 6. https://doi.org/10.1177/014459878800600304

King, P.R. and Thrasher, G.P. 1992. Post-Eocene development of the Taranaki Basin, New Zealand, convergent overprint of a passive margin. In: J.S. Watkins, F. Zhiquiang and K.J. McMillen (eds). *Geology and Geophysics of Continental Margins*. American Association of Petroleum Geologists Memoir 53, pp. 93–118.

King, P.R. and Thrasher, G.P. 1996. *Cretaceous–Cenozoic Geology and Petroleum Systems of the Taranaki Basin, New Zealand*. Monograph 13, Institute of Geological and Nuclear Sciences, Lower Hutt. 243 pp.

King, P.R., Scott, G.H. and Robinson, P.H. 1993. *Description, Correlation and Depositional History of Miocene Sediments Outcropping Along North Taranaki Coast*. Monograph 5, Institute of Geological and Nuclear Sciences, Lower Hutt. 199 pp.

King, P.R., Browne, G.H. and Slatt, R.M. 1994. Sequence architecture of exposed Late Miocene basin floor fan and channel-levee complexes (Mount Messenger Formation), Taranaki basin, New Zealand. In: P. Weimer, A.H. Bouma and B.F. Perkins (eds). *Submarine Fans and Turbidite Systems: Sequence stratigraphy, reservoir architecture and production characteristics, Gulf of Mexico and international*. SEPM Gulf Coast Section conference 15. Houston, Texas, pp. 177–192.

Kingma, J.T. 1971. *Geology of the Te Aute Subdivision*. New Zealand Geological Survey Bulletin 70.

Knott, S.D. 2001. Gravity-driven crustal shortening in failed rifts. *Journal of the Geological Society of London* 158: 193–196.

Korsch, R.J. and Wellman, H.W. 1988. The geological evolution of New Zealand and the New Zealand region. In: A.E. Nairn, F.G. Stehli and S. Uyeda (eds). *The Oceanic Basins and Margins, Volume 7B: The Pacific Ocean*. Plenum Press, New York, pp. 411–482.

Kudrass, H.R. and von Rad, U. 1984. Underwater television and photography observations, side-scan sonar and acoustic reflectivity measurements of phosphorite-rich areas on the Chatham Rise (New Zealand). *Geologische Jahrbuch* 65: 69–89.

Kula, J., Tulloch, A.J., Spell, T.L. and Wells, M.L. 2007. Two-stage rifting of Zealandia–Australia–Antarctica: Evidence from 40Ar/39Ar thermochronometry of the Sisters shear zone, Stewart Island, New Zealand. *Geology* 35: 411–414.

Kula, J., Tulloch, A.J., Spell, T.L., Wells, M.L. and Zanetti, K.A. 2009. Thermal evolution of the Sisters shear zone, southern New Zealand: Formation of the Great South Basin and onset of Pacific–Antarctic spreading. *Tectonics* 28: 1–22.

Laird, M.G. 1968. The Paparoa Tectonic Zone. *New Zealand Journal of Geology and Geophysics* 11: 435–454.

Laird, M.G. 1981. The Late Mesozoic fragmentation of the New Zealand segment of Gondwana. In: M. Cresswell and P. Vella (eds). *Gondwana Five: Proceedings of the Fifth International Gondwana Symposium*. Balkema, Rotterdam, pp. 311–318.

Laird, M.G. 1982. The Hapuku River/Long Creek stratigraphic drillhole: Immediate report. Unpublished New Zealand Geological Survey report. 3 pp.

Laird, M.G. 1988. *Sheet S37 Punakaiki*. Geological Maps of New Zealand 1:63,360. Department of Scientific and Industrial Research, Wellington.

Laird, M.G. 1991. Sedimentology of the Late Cretaceous lower Mata River sequence, Raukumara Peninsula: Preliminary report. Unpublished DSIR GEO report.

Laird, M.G. 1992. Cretaceous stratigraphy and evolution of the Marlborough segment of the East Coast region. *1991 New Zealand Oil Exploration Conference Proceedings*. Ministry of Commerce, Wellington, pp. 89–100.

Laird, M.G. 1993. Cretaceous continental rifts: New Zealand region. In: P.F. Ballance (ed.). *South Pacific Sedimentary Basins*. Sedimentary Basins of the World 2. Elsevier, Amsterdam, pp. 37–49.

Laird, M.G. 1994. Geological aspects of the opening of the Tasman Sea. In: G.J. van der Lingen, K.M. Swanson and R.J. Muir (eds). *Evolution of the Tasman Sea Basin*. Balkema, Rotterdam, pp. 1–17.

Laird, M.G. 1995. Coarse-grained lacustrine fan-delta deposits (Pororari Group) of the northwestern South Island, New Zealand: Evidence for Mid-Cretaceous rifting. In: A.G. Plint (ed.). *Sedimentary Facies Analysis*. International Association of Sedimentologists Special Publication 22, pp. 197–217.

Laird, M.G. 1996a. Basement/cover relationships and sedimentary environments, western Raukumara Peninsula: An interim report. Unpublished report MGL 1/96.

Laird, M.G. 1996b. Mid and early Late Cretaceous break-up basins of the South Island, New Zealand. In: *Mesozoic Geology of the Eastern Australian Plate*. Geological Society of Australia Inc., Extended Abstracts no. 43, pp. 329–336.

Laird, M.G. 1996c. What plate rearrangement at ~70 Ma initiated the West Coast–Taranaki Rift System? In: *Plate Reconstruction Workshop Proceedings*. Institute of Geological and Nuclear Sciences Science Report 96/35, pp. 15–19.

Laird, M.G. and Bradshaw, J.D. 1996. *A Reassessment of Mid Cretaceous 'Basement/Cover' Relationships and Paleoenvironments, Raukumara Peninsula*. Geological Society of New Zealand Miscellaneous Publication 91A. 107 pp.

Laird, M.G. and Bradshaw, J.D. 2004. The break-up of a long-term relationship: The Cretaceous separation of New Zealand from Gondwana. *Gondwana Research* 7: 273–286.

Laird, M.G. and Hope, J.M. 1968. The Torea Breccia and the Papahaua Overfold. *New Zealand Journal of Geology and Geophysics* 11: 418–434.

Laird, M.G. and Lewis, D.W. 1986. Growth-fault control of mid-Cretaceous shallow marine storm-influenced mass flow sedimentation, northeast South Island, New Zealand. Paper presented at the Sediments Down-under: 12th International Sedimentological Congress, Canberra, Australia.

Laird, M.G. and Nathan, S. 1988. The Eocene Torea Breccia, SW Nelson: Redescription, sedimentology, and regional significance. *New Zealand Geological Survey Record* 35: 104–108.

Laird, M.G. and Schiøler, P. 2005. Is the Late Cretaceous Ward coastal succession allochthonous? *Geological Society of New Zealand 50th Annual Conference, 2005*. Geological Society of New Zealand Miscellaneous Publication 119A, p. 42.

Laird, M.G., Mazengarb, C. and Crampton, J.S. 1994. Mid and late Cretaceous basin development, East Coast region, New Zealand. *1994 New Zealand Petroleum Conference Proceedings: The post Maui challenge – investment and development opportunities.* Ministry of Commerce, Wellington, pp. 79–85.

Laird, M.G., van der Lingen, G.J. and Mazengarb, C. 1998. *Sequence Stratigraphy of a Late Cretaceous Turbidite Succession: Mata River, Raukumara Peninsula.* Geological Society of New Zealand Miscellaneous Publication 101A. 141 pp.

Laird, M.G., Bassett, K.N., Schiøler, P., Morgans, H.E.G., Bradshaw, J.D. and Weaver, S.D. 2003. Paleoenvironmental and tectonic changes across the Cretaceous/Tertiary boundary at Tora, southeast Wairarapa, New Zealand: A link between Marlborough and Hawke's Bay. *New Zealand Journal of Geology and Geophysics* 46: 275–293.

Lamarche, G., Collot, J-Y., Wood, R.A., Sosson, M., Sutherland, R. and Delteil, J. 1997. The Oligocene–Miocene Pacific–Australia plate boundary, south of New Zealand: Evolution from oceanic spreading to strike-slip faulting. *Earth and Planetary Science Letters* 148: 129–139.

Lamb, S.H. and Bibby, H.M. 1989. The last 25 Ma of rotational deformation in part of the New Zealand plate-boundary zone. *Journal of Structural Geology* 11: 473–492.

Lamb, S., Mortimer, N., Smith, E. and Turner, G. 2016. Focusing of relative plate motion at a continental transform fault: Cenozoic dextral displacement >700 km on New Zealand's Alpine Fault, reversing >225 km of Late Cretaceous sinistral motion. *Geochemistry, Geophysics, Geosystems* 17: 1197–1213. doi:10.1002/2015GC006225

Landis, C.A. 1990. *Late Cretaceous Erosion Surface in East Otago – Is it really a peneplain*? (Abstract). Geological Society of New Zealand Miscellaneous Publication 50A. 77 pp.

Landis, C.A. and Coombs, D.S. 1967. Metamorphic belts and orogenesis in southern New Zealand. *Tectonophysics* 4: 501–518.

Landis, C.A., Campbell, H.J., Aslund, T., Cawood, P.A., Douglas, A., Kimbrough, D.L., Pillai, D.D.L., Raine, J.I. and Willsman, A. 1999. Permian–Jurassic strata at Productus Creek, Southland, New Zealand: Implications for terrane dynamics of the eastern Gondwanaland margin. *New Zealand Journal of Geology and Geophysics* 42: 255–278.

Landis, C.A., Campbell, H.J., Begg, J.G., Mildenhall, D.C., Paterson, A.M. and Trewick, S.A. 2008. The Waipounamu Erosion Surface: Questioning the antiquity of the New Zealand land surface and terrestrial fauna and flora. *Geological Magazine* 145: 173–197.

Larson, R.L., Steiner, M.B., Erba, E. and Lancelot, Y. 1992. Paleolatitudes and tectonic reconstructions of the oldest portion of the Pacific plate: A comparative study. *Proceedings of the Ocean Drilling Program* 129: 625–631.

Larter, R.D., Cunningham, A.P., Barker, P.F., Gohl, K. and Nitsche, F.O. 2002. Tectonic evolution of the Pacific margin of Antarctica, 1. Late Cretaceous tectonic reconstructions. *Journal of Geophysical Research – Solid Earth* 107(B12). doi:10.1029/2000JB000052

Lawrence, M.J.F. 1994. Conceptual model for early diagenetic chert and dolomite, Amuri Limestone Group, north-eastern South Island, New Zealand. *Sedimentology* 41: 479–498.

Lawver, L.A. and Gahagan, L.M. 2003. Tectonic reconstructions in the greater Ross Sea region. In: D. Harwood, L. Lacy and R.H Levy (eds). *Future Antarctic Margin Drilling: Developing a science program plan for McMurdo Sound: Report of a workshop, Oxford, UK, April 5–7, 2001.* University of Nebraska-Lincoln, Lincoln, p. 27.

Lawver, L.A., Gahagan, L.M. and Coffin, M.F. 1992. The development of paleoseaways around Antarctica. *American Geophysical Union Antarctic Research Series* 56: 7–30.

Leask, W.L. 1980. Basin analysis of Tertiary strata in Golden Bay, Nelson. Unpublished MSc thesis, Victoria University of Wellington.

Leask, W.L. 1993. Brunner Coal Measures at Golden Bay, Nelson: An Eocene fluvial estuarine deposit. *New Zealand Journal of Geology and Geophysics* 36: 37–50.

Lebrun, J.-F., Lamarche, G., Collot, J-Y. and Delteil, J. 2000. Abrupt strike-slip fault to subduction transition: The Alpine Fault-Puysegur Trench connection, New Zealand. *Tectonics* 19: 688–706. doi:10.1029/2000TC900008

Lee, D.E., Lindqvist, J.K., Beu, A.G., Robinson, J.H., Ayress, M.A., Morgans, H.E.G. and Stein, J.K. 2014. Geological setting and diverse fauna of a Late Oligocene rocky shore ecosystem, Cosy Dell, Southland. *New Zealand Journal of Geology and Geophysics* 57: 195–208. doi:10.1080/00288306.2014.898666

Lee, J.M. 1995. A stratigraphic, biostratigraphic and structural analysis of the geology at Huatokitoki Stream, Glenburn, southern Wairarapa, New Zealand. Unpublished MSc thesis, Victoria University of Wellington.

Lee, J.M. and Begg, J.G. 2002. *Geology of the Wairarapa Area.* Institute of Geological and Nuclear Sciences 1:250,000 Geological QMap 11. Institute of Geological and Nuclear Sciences, Lower Hutt. 1 sheet + 66 pp.

Leitch, E.C., Grant-Mackie, J.A. and Hornibrook, N. de B. 1969. Contributions to the geology of northernmost New Zealand: The mid-Miocene Waikuku Limestone. *Transactions of the Royal Society of New Zealand, Earth Sciences* 7: 21–32.

LeMasurier, W.E. and Landis, C.A. 1996. Mantle-plume activity recorded by low-relief erosion surfaces in West Antarctica and New Zealand. *Geological Society of America Bulletin* 108: 1450–1466.

Lever, H. 2001. An Eocene to early Oligocene unconformity-bounded sequence in the Punakaiki-Westport area, West Coast, South Island, New Zealand. *New Zealand Journal of Geology and Geophysics* 44: 355–363.

Levy, R.H. 1992. The geology of the Limestone-Muzzle Stream area, Clarence Valley, New Zealand. Unpublished MSc thesis, Victoria University of Wellington.

Lewis, D.W. 1976. Subaqueous debris flows of early Pleistocene age at Motunau, North Canterbury, New Zealand. *New Zealand Journal of Geology and Geophysics* 19: 535–567.

Lewis, D.W. 1980. Storm-generated graded beds and debris flow deposits with *Ophiomorpha* in a shallow offshore Oligocene sequence at Nelson, South Island, New Zealand. *New Zealand Journal of Geology and Geophysics* 23: 353–369.

Lewis, D.W. 1982. Channels across continental shelves: Corequisites of canyon-fan systems and potential petroleum conduits. *New Zealand Journal of Geology and Geophysics* 25: 209–225.

Lewis, D.W. 1992. Anatomy of an unconformity on mid-Oligocene Amuri Limestone, Canterbury, New Zealand. *New Zealand Journal of Geology and Geophysics* 35: 463–475.

Lewis, D.W. and Belliss, S.E. 1984. Mid-Tertiary unconformities in the Waitaki subdivision, North Otago. *Journal of the Royal Society of New Zealand* 14: 251–276.

Lewis, D.W. and Ekdale, A.A. 1991. Lithofacies relationships in a late Quaternary gravel and loess fan delta complex, New Zealand. *Palaeogeography, Palaeoclimatology, Palaeoecology* 81: 229–251.

Lewis, D.W., Smale, D. and van der Lingen, G.J. 1979. A sandstone diapir cutting the Amuri Limestone, North Canterbury, New Zealand. *New Zealand Journal of Geology and Geophysics* 22: 295–305.

Lewis, D.W., Laird, M.G. and Powell, R.G. 1980. Debris flow deposits of Early Miocene age, Deadman Stream, Marlborough, New Zealand. *Sedimentary Geology* 27: 83–118.

Lewis, K.B. and Barnes, P.L. 1999. Kaikoura Canyon, New Zealand: Active conduit from near-shore sediment zones to trench-axis channel. *Marine Geology* 162: 39–69.

Lewis, K.B., Bennett, D.J., Herzer, R.H. and Von der Borsch, C.C. 1985. Seismic stratigraphy and structure adjacent to an evolving plate boundary, western Chatham Rise, New Zealand. *Initial Reports of the Deep Sea Drilling Project 90.* US Government Printing Office, Washington, pp. 1325–1337.

Lewis, K.B., Carter, L. and Davey, F.J. 1994. The opening of Cook Strait: Interglacial tidal scour and aligning basins at a subduction to transform plate edge. *Marine Geology* 116: 293–312.

Lillie, A.R. 1953. *The Geology of the Dannevirke Subdivision*. New Zealand Geological Survey Bulletin 46. 156 pp.

Lindqvist, J.K. 1986. Sedimentology of Chalky Island Formation (Oligocene), south Fiordland, New Zealand: Arkosic turbidites and nannofossil pelagites. In: *Abstracts. 12th International Sedimentological Congress*, Canberra, Australia, p. 184.

Lindqvist, J.K. 1990. Mid Cretaceous and Eocene–Oligocene stratigraphy, on-shore Balleny Basin, south Fiordland. In: *Recent Developments in New Zealand Basin Studies*. DSIR Geology and Geophysics, Lower Hutt, pp. 45–47.

Lindqvist, J.K. 1994. Lacustrine stromatolites and oncoids: Manuherikia Group (Miocene), New Zealand. In: J. Bertrand-Safati and C. Monty (eds). *Phanerozoic Stromatolites II*. Kluwer Academic Publishers, Dordrecht, pp. 227–254.

Lindqvist, J.K. 1995. *Wangaloa and Abbotsford Formations: Measly Beach drillhole, South Otago, New Zealand.* Institute of Geological and Nuclear Sciences Science Report 95/12. 44 pp.

Lindqvist, J.K. 1996. *Cretaceous–Cenozoic Sedimentation: East, Central and South Otago*. Field Trip Guide 8. Geological Society of New Zealand Miscellaneous Publication 91B, pp. 1–35.

Lindqvist, J.K. and Douglas B.J. 1987. *Late Cretaceous and Paleocene Fluvial and Shallow Marine Deposits of the Kaitangata Coalfield: Taratu and Wangaloa formations*. Field Trip Guide. Geological Society of New Zealand Miscellaneous Publication 37B, pp. 29–51.

Lister, G.S., Etheridge, M.A. and Symonds, P.A. 1991. Detachment models for the formation of passive continental margins. *Tectonics* 10: 1038–1064.

Little, T.A. and Roberts, A.P. 1997. Distribution and mechanism of Neogene to present-day vertical axis rotations, Pacific-Australian plate boundary zone, South Island, New Zealand. *Journal of Geophysical Research – Solid Earth* 102(B9): 20447–20468.

Luyendyk, B.P. 1995. Hypothesis for Cretaceous rifting of east Gondwana caused by subducted slab capture. *Geology* 23: 373–376.

McCoy-West, A.J., Mortimer, N. and Ireland, T.R. 2014. U-Pb geochronology of Permian plutonic rocks, Longwood Range, New Zealand: Implications for Median Batholith–Brook Street Terrane relations. *New Zealand Journal of Geology and Geophysics* 57: 65–85.

McCulloch, B. 1981. Geology of the Mount Brown Beds. Unpublished MSc thesis, University of Canterbury.

McDougall, I. and van der Lingen, G.J. 1974. Age of the rhyolites of the Lord Howe Rise and the evolution of the southwest Pacific Ocean. *Earth and Planetary Science Letters* 21: 117–126.

McGuire, D.M. 1998. Paleomagnetic stratigraphy and magnetic properties of Pliocene strata, Turakina River, North Island, New Zealand. Unpublished PhD thesis, Victoria University of Wellington.

McKellar, I.C. 1973. *Geology of Te Anau-Manapouri District 1:50,000*. New Zealand Geological Survey Miscellaneous Series Map 4. 1 sheet & booklet.

MacKinnon, T.C. 1983. Origin of the Torlesse terrane and coeval rocks, South Island, New Zealand. *Geological Society of America Bulletin* 94: 967–985.

McLennan, J.M. and Bradshaw, J.D. 1984. Angular unconformity between Oligocene and older Cenozoic rocks at Avoca, Canterbury, New Zealand. *New Zealand Journal of Geology and Geophysics* 27: 299–303.

McMillan, S.G. and Wilson, G.J. 1997. Allostratigraphy of coastal south and east Otago: A stratigraphic framework for interpretation of the Great South Basin, New Zealand. *New Zealand Journal of Geology and Geophysics* 49: 91–107.

McQuillan, H. 1977. Hydrocarbon potential of the North Wanganui Basin, New Zealand. *APEA Journal* 17: 94–104.

Matthews, E.R. 1990. Exploration in the onshore Westland Basin. *1989 New Zealand Oil Exploration Conference Proceedings*. Ministry of Commerce, Wellington, pp. 62–70.

Maxwell, F.A. 1990. Late Miocene to Recent evolution of the Awatere Basin, Medway and Middle Awatere Valleys, Marlborough. Unpublished MSc thesis, Victoria University of Wellington.

Mazengarb, C. 1993. *Cretaceous Stratigraphy of Raukumara Peninsula*. Institute of Geological and Nuclear Sciences Science Report 93/20. 51 pp.

Mazengarb, C. and Harris, D.H. 1994. Cretaceous stratigraphic and structural relations of the Raukumara Peninsula, New Zealand: Stratigraphic patterns associated with the migration of a thrust system. *Annales Tectonicae* 8: 100–118.

Mazengarb, C. and Speden, I.G. 2000. *Geology of the Raukumara Area*. Institute of Geological and Nuclear Sciences 1:250,000 Geological QMap 6. Institute of Geological and Nuclear Sciences, Lower Hutt. 1 sheet + 60 pp.

Mazengarb, C., Francis, D.A. and Moore, P.R. 1991. *Sheet Y16 Tauwhareparae*. Geological Maps of New Zealand 1:50,000. Department of Scientific and Industrial Research, Wellington.

Melhuish, A. 1988. Synsedimentary faulting in the lower Medway River area, Awatere valley, Marlborough. Unpublished BSc (Hons) project, Victoria University of Wellington.

Mildenhall, D.C. and Pocknall, D.T. 1986. *Palynology of the Miocene-Pleistocene Tauranga Group, Ohinewai, South Auckland, New Zealand*. New Zealand Geological Survey Report PAL 120. 22 pp.

Miller, K.G., Wright, J.D. and Fairbanks, R.G. 1991. Unlocking the Ice House: Oligocene–Miocene oxygen isotopes, eustasy, and margin erosion. *Journal of Geophysical Research – Solid Earth and Planets* 96(B4): 6829–6848.

Milne, A. and Quick, R. 1999. Well Completion Report, Waka Nui-1, PEP 38602. New Zealand unpublished open-file petroleum report PR2436. Ministry of Economic Development, Wellington. 922 pp. + 4 enclosures.

Mitchell, M., Craw, D., Landis, C.A. and Frew, R. 2009. Stratigraphy, provenance, and diagenesis of the Cretaceous Horse Range Formation, east Otago. New Zealand. *New Zealand Journal of Geology and Geophysics* 52: 171–183.

Montague, T. 1981. Cretaceous stratigraphy of the Mt. Lookout area and its relation to the proposed regional unconformity at the end of the Rangitata Orogeny. Unpublished MSc thesis, University of Canterbury.

Moore, P.R. 1978a. Geology of western Koranga Valley, Raukumara Peninsula. *New Zealand Journal of Geology and Geophysics* 21: 1–20.

Moore, P.R. 1978b. Petrography of the Late Jurassic–Late Cretaceous rocks from Koranga Valley, Raukumara Peninsula. *New Zealand Journal of Geology and Geophysics* 21: 189–197.

Moore, P.R. 1980. Late Cretaceous–Tertiary stratigraphy, structure, and tectonic history of the area between Whareama and Ngahape, eastern Wairarapa, New Zealand. *New Zealand Journal of Geology and Geophysics* 23: 167–177.

Moore. P.R. 1981. Geology of the Late Tertiary section at Cape Turnagain. *Journal of the Royal Society of New Zealand* 11: 223–230.

Moore, P.R. 1988. Structural divisions of the eastern North Island. *New Zealand Geological Survey Record* 30: 1–24.
Moore, P.R. 1989a. Kirks Breccia: A Late Cretaceous submarine channelised debris flow deposit, Raukumara Peninsula, New Zealand. *Journal of the Royal Society of New Zealand* 19: 195–203.
Moore, P.R. 1989b. Stratigraphy of the Waipawa Black Shale (Paleocene), eastern North Island, New Zealand. *New Zealand Geological Survey Record* 38. 19 pp.
Moore, P.R. and Speden, I.G. 1979. Stratigraphy, structure and inferred environments of deposition of the Early Cretaceous sequence, eastern Wairarapa, New Zealand. *New Zealand Journal of Geology and Geophysics* 22: 417–433.
Moore, P.R. and Speden, I.G. 1984. *The Early Cretaceous (Albian) Sequence of Eastern Wairarapa, New Zealand*. New Zealand Geological Survey Bulletin 97. 98 pp.
Moore, P.R., Isaac, M.J., Mazengarb, C. and Wilson, G.J. 1989. Stratigraphy and structure of Cretaceous (Neocomian–Maastrichtian) sedimentary rocks in the Anini–Okaura Stream area, Urewera National Park, New Zealand. *New Zealand Journal of Geology and Geophysics* 32: 515–526.
Moore, T.A., Li, Z. and Moore, A. 2006. Controls on the formation of an anomalously thick Cretaceous age coal mire. In: S.F. Greb and W.A. DiMichele (eds). *Wetland Through the Ages*. Geological Society of America Special Paper 399, pp. 269–290.
Moore, W.R. 1957. Geology of the Raukokore area, Raukumara Peninsula, North Island. Unpublished PhD thesis, Victoria University of Wellington.
Morgans, H.E. and Raine, J.I. 1988. *Biostratigraphy of Upukerora-1 Onshore Well, Te Anau Basin*. New Zealand Geological Survey Report PAL 132. 23 pp.
Morgans, H.E.G., Wilson, D.C. and Mildenhall, D.C. 1990. The biostratigraphy of the Te Hoe-1 well, northern Hawke's Bay. DSIR Geology and Geophysics Contract Report 1990/12. In: W.A. Dobbie and M.J. Carter (eds). Well Completion Report, Te Hoe-1. New Zealand unpublished open-file petroleum report 1835.
Morgans, H.E.G., Jones, C.M., Crouch, E.M., Field, B.D., Hollis, C.J., Raine, J.I., Strong, C.P. and Wilson, G.J. 2005. *Upper Cretaceous to Eocene Stratigraphy and Sample Collections, Mid-Waipara River Section, North Canterbury*. Institute of Geological and Nuclear Sciences Science Report 03/08. 107 pp.
Morris, J.C. 1987. The stratigraphy of the Amuri Limestone Group, east Marlborough, New Zealand. Unpublished PhD thesis, University of Canterbury.
Morris, J.H. 1982. The geology of Late Tertiary sediments northwest of Masterton, Wairarapa. Unpublished BSc (Hons) thesis, Victoria University of Wellington.
Morris, P.A. 1984. Migration of late Cenozoic volcanism in the South Island of New Zealand and the Campbell Plateau. *Journal of Volcanology and Geothermal Research* 21: 119–148.
Mortimer, N. 1993a. Jurassic tectonic history of the Otago Schist, New Zealand. *Tectonics* 12: 237–244.
Mortimer, N. 1993b. Metamorphic zones, terranes, and Cenozoic faults in the Marlborough Schist, New Zealand. *New Zealand Journal of Geology and Geophysics* 36: 357–368.
Mortimer, N. 2004. New Zealand's geological foundations. *Gondwana Research* 7: 261–272.
Mortimer, N. 2014. The oroclinal bend in the South Island, New Zealand. *Journal of Structural Geology* 64: 32–38.
Mortimer, N. and Cooper, A.F. 2004. U-Pb and Sm-Nd ages from the Alpine Schist, New Zealand. *New Zealand Journal of Geology and Geophysics* 47: 21–28.
Mortimer, N. and Parkinson, D.L. 1996. Hikurangi Plateau: A Cretaceous large igneous province in the southwest Pacific Ocean. *Journal of Geophysical Research – Solid Earth* 101(B1): 687–696.
Mortimer, N. and Roser, B.P. 1992. Geochemical evidence for the position of the Caples–Torlesse boundary in the Otago Schist, New Zealand. *Journal of the Geological Society of London* 149: 967–977.
Mortimer, N., Parkinson, D.L., Raine, J.I., Adams, C.J., Graham, I.J., Oliver, P.J. and Palmer, K. 1995. Ferrar magmatic province rocks discovered in New Zealand: Implications for Mesozoic Gondwana geology. *Geology* 23: 185–188.
Mortimer, N., Tulloch, A.J. and Ireland, T.R. 1997. Basement geology of Taranaki and Wanganui Basins, New Zealand. *New Zealand Journal of Geology and Geophysics* 40: 223–236.
Mortimer, N., Herzer, R.H., Gans, P.B., Parkinson, D.L. and Seward, D. 1998. Basement geology from Three Kings Ridge to West Norfolk Ridge, southwest Pacific Ocean: Evidence from petrology, geochemistry and isotopic dating of dredge samples. *Marine Geology* 148: 135–162.

Mortimer, N., Tulloch, A.J., Spark, R.N., Walker, N.W., Ladley, E., Allibone, A. and Kimbrough, D.L. 1999a. Overview of the Median Batholith, New Zealand: A new interpretation of the geology of the Median Tectonic Zone and adjacent rocks. *Journal of African Earth Sciences* 29: 257–268.

Mortimer, N., Gans, P., Calvert, A. and Walker, N. 1999b. Geology and thermochronometry of the east edge of the Median Batholith (Median Tectonic Zone): A new perspective on Permian to Cretaceous crustal growth of New Zealand. *Island Arc* 8: 404–425.

Mortimer, N., Davey, F.J., Melhuish, A., Yu, J. and Godfrey, N.J. 2002. Geological interpretation of a deep seismic reflection profile across the Eastern Province and Median Batholith, New Zealand: Crustal architecture of an extended Phanerozoic convergent orogen. *New Zealand Journal of Geology and Geophysics* 45: 349–363.

Mortimer, N., Hauff, H.K., Pali, F., Dunlap, J.M., Werner, W.J. and Faure, K. 2006. New constraints on the age and evolution of the Wishbone Ridge, southwest Pacific Cretaceous microplates, and Zealandia–West Antarctica breakup. *Geology* 34: 185–188.

Mortimer, N., Herzer, R., Gans. P.B., Laporte-Magoni, C., Calvert, A.T. and Bosch, D. 2007. Oligocene–Miocene tectonic evolution of the South Fiji Basin and Northland Plateau, SW Pacific Ocean: Evidence from petrology and dating of dredged rocks. *Marine Geology* 237: 1–24.

Mortimer, N., Dunlap, W.J., Palin, J.M., Herzer, R.H., Hauff, F. and Clark, M. 2008. Ultra-fast early Miocene exhumation of Cavalli Seamount, Northland Plateau, Southwest Pacific Ocean. *New Zealand Journal of Geology and Geophysics* 51: 29–42.

Mortimer, N., Raine, J.I. and Cook, R.A. 2009. Correlation of basement rocks from Waka Nui-1 and Awhitu-1, and the Jurassic regional geology of Zealandia. *New Zealand Journal of Geology and Geophysics* 52: 1–10.

Mortimer, N., Gans, P.B., Palin, J.M., Meffre, S., Herzer, R.H. and Skinner, D.N. 2010. Location and migration of Miocene–Quaternary volcanic arcs in the SW Pacific region. *Journal of Volcanology and Geothermal Research* 190: 1–10.

Mortimer, N., Rattenbury M.S., King, P.R., Bland, K.J., Barrell, D.J.A., Bache, F., Begg, J.G. *et al.* 2014. High-level stratigraphic scheme for New Zealand rocks. *New Zealand Journal of Geology and Geophysics* 57: 402–419.

Mound, D.G. and Pratt, D.N. 1984. Interpretation and prospectivity of PPL 38202. Canterbury Basin, New Zealand. New Zealand unpublished open-file petroleum report 1021. New Zealand Department of Trade and Industry, Wellington. 118 pp. + 59 enclosures.

Muir, R.J., Ireland, T.R., Weaver, S.D. and Bradshaw, J.D. 1994. Ion microprobe U-Pb zircon geochronology of granitic magmatism in the Western Province of the South Island, New Zealand. *Chemical Geology* 113: 171–189.

Muir, R.J., Weaver, S.D., Bradshaw, J.D., Eby, G.N. and Evans, J.A. 1995. The Cretaceous Separation Point Batholith, New Zealand: Granitoid magmas formed by melting of mafic lithosphere. *Journal of the Geological Society of London* 152: 689–701.

Muir, R.J., Ireland, T.R., Weaver, S.D. and Bradshaw, J.D. 1996. Ion microprobe dating of Paleozoic granitoids: Devonian magmatism in New Zealand and correlations with Australia and Antarctica. *Chemical Geology* 127: 191–210.

Muir, R.J., Ireland, T.R., Weaver, S.D., Bradshaw, J.D., Waight, T.E., Jongens, R. and Eby, G.N. 1997. SHRIMP U-Pb geochronology of Cretaceous magmatism in northwest Nelson–Westland, South Island, New Zealand. *New Zealand Journal of Geology and Geophysics* 40: 453–463.

Muir, R.J., Ireland, T.R., Weaver, S.D., Bradshaw, J.D., Evans, J.A., Eby, G.N. and Shelley, D. 1998. Geochronology and geochemistry of a Mesozoic magmatic arc system, Fiordland, New Zealand. *Journal of the Geological Society of London* 155: 1037–1052.

Muir, R.J., Bradshaw, J.D., Weaver, S.D. and Laird, M.G. 2000. The influence of basement structure on the evolution of the Taranaki Basin, New Zealand. *Journal of the Geological Society of London* 157: 1179–1185.

Mukasa, S.B. and Dalziel, I.W. 2000. Marie Byrd Land, West Antarctica: Evolution of Gondwana's Pacific margin constrained by zircon U-Pb geochronology and feldspar common-Pb isotopic compositions. *Geological Society of America Bulletin* 112: 611–627.

Münker, C. and Cooper, R.A. 1995. The island arc setting of a New Zealand Cambrian volcano-sedimentary sequence – implications for the evolution of the SW Pacific Gondwana fragments. *Journal of Geology* 103: 687–700.

Münker, C. and Cooper, R.A. 1999. The Cambrian arc complex of the Takaka Terrane, New Zealand: An integrated stratigraphical, paleontological and geochemical approach. *New Zealand Journal of Geology and Geophysics* 42: 415–445.

Murphy, L.S., Leask, W.L. and Collen, J.D. 1994. Source rock potential of the south Wanganui Basin. *1994 New Zealand Petroleum Conference Proceedings: The post Maui challenge – investment and development opportunities*. Ministry of Commerce, Wellington, pp. 95–107.

Naish, T.R. 2005. New Zealand's shallow-marine record of Pliocene–Pleistocene global sea-level and climate change. *Journal of the Royal Society of New Zealand* 35: 1–8.

Naish, T.R. and Kamp, P.J. 1995. Pliocene–Pleistocene marine cyclothems, Wanganui Basin, New Zealand: A lithostratigraphic framework. *New Zealand Journal of Geology and Geophysics* 38: 223–243.

Naish, T.R. and Wilson, G.S. 2009. Constraints on the amplitude of Mid-Pliocene (3.6–2.4 Ma) eustatic sea-level fluctuations from the New Zealand shallow-marine sediment record. *Philosophical Transactions of the Royal Society A*. doi:10.1098/rsta.2008.0223

Naish, T.R., Field, B.D., Zhu, H., Melhuish, A., Carter, R.M., Abbott, S.T., Edwards, S. *et al.* 2005. Integrated outcrop, drill core, borehole and seismic stratigraphic architecture of a cyclothemic, shallow-marine depositional system, Wanganui Basin, New Zealand. *Journal of the Royal Society of New Zealand* 35: 91–122.

Nathan, S. 1974. Stratigraphic nomenclature for the Cretaceous–Lower Quaternary rocks of Buller and North Westland, West Coast, South Island, New Zealand. *New Zealand Journal of Geology and Geophysics* 17: 423–445.

Nathan, S. 1975. *Sheets S23 & S30 Foulwind and Charleston* (1st edition). Geological Maps of New Zealand 1:63,360. Department of Scientific and Industrial Research, Wellington. 1 sheet + 20 pp.

Nathan, S. 1977. Cretaceous and Lower Tertiary stratigraphy of the coastal strip between Buttress Point and Ship Creek, South Westland, New Zealand. *New Zealand Journal of Geology and Geophysics* 20: 615–654.

Nathan, S. 1978a. Upper Cenozoic stratigraphy of South Westland, New Zealand. *New Zealand Journal of Geology and Geophysics* 21: 329–361.

Nathan, S. 1978b. *Sheet S44 Greymouth* (1st edition). Geological Maps of New Zealand 1:63,360. Department of Scientific and Industrial Research, Wellington. 1 sheet + 36 pp.

Nathan, S. 1978c. *Sheet S31 and part S32 Buller–Lyell*. Geological Maps of New Zealand 1:63,360. Department of Scientific and Industrial Research, Wellington.

Nathan, S. 1978d. Tertiary rocks near the White Creek Fault, Upper Buller Gorge, New Zealand. *Journal of the Royal Society of New Zealand* 8: 5–15.

Nathan, S. 1996. *Geology of the Buller Coalfield, 1:50,000*. Institute of Geological and Nuclear Sciences Geological Map 23. Institute of Geological and Nuclear Sciences, Lower Hutt.

Nathan, S., Anderson, H.J., Cook, R.A., Herzer, R.H., Hoskins, R., Raine, I. and Smale, D. 1986. *Cretaceous and Cenozoic Sedimentary Basins of the West Coast Region, South Island, New Zealand*. New Zealand Geological Survey Basin Studies 1. Department of Scientific and Industrial Research, Wellington. 90 pp.

Nathan, S., Thurlow, C., Warnes, P. and Zucchetto, R. 2000. *Geochronology Database for New Zealand Rocks (2nd edition): 1961–1999*. Institute of Geological and Nuclear Sciences Science Report 2000/11. 51 pp.

Nathan, S., Rattenbury, M.S. and Suggate, R.P. 2002. *Geology of the Greymouth Area*. Institute of Geological and Nuclear Sciences 1:250,000 Geological QMap 12. Institute of Geological and Nuclear Sciences, Lower Hutt. 1 sheet + 58 pp.

Neef, G. 1981. Cenozoic stratigraphy and structure of Karamea–Little Wanganui district, Buller, South Island, New Zealand. *New Zealand Journal of Geology and Geophysics* 24: 177–208.

Neef, G. 1984. *Late Cenozoic and Early Quaternary Stratigraphy of the Eketahuna District (N153)*. New Zealand Geological Survey Bulletin 96.

Neef, G. 1992. Turbidite deposition in five Miocene, bathyal formations along an active plate margin, North Island, New Zealand: With notes on styles of deposition at the margins of east coast bathyal basins. *Sedimentary Geology* 78: 111–136.

Neef, G. 1995. Cretaceous and Cenozoic geology east of the Tinui Fault complex in northeastern Wairarapa, New Zealand. *New Zealand Journal of Geology and Geophysics* 38: 375–394.

Neef, G. 1997. Stratigraphy, structural evolution, and tectonics of the northern part of the Tawhero Basin and adjacent areas, northern Wairarapa, North Island, New Zealand. *New Zealand Journal of Geology and Geophysics* 40: 335–358.

Neef, G. and Bottrill, R.S. 1992. The Cenozoic geology of the Gisborne area (1:50,000 metric sheet Y18AB), North Island, New Zealand: With an appendix: clay minerals and quartz, plagioclase, and calcite in some sediments of the Tolaga Group. *New Zealand Journal of Geology and Geophysics* 35: 515–531.

Nelson, C.S. 1978. Stratigraphy and paleontology of the Oligocene Te Kuiti Group, Waitomo County, South Auckland, New Zealand. *New Zealand Journal of Geology and Geophysics* 21: 553–594.

Nelson, C.S. 1986. Lithostratigraphy of Deep Sea Drilling Project Leg 90 drill sites in the Southwest Pacific: An overview. *Initial Reports of the Deep Sea Drilling Project 90.* US Government Printing Office, Washington, 1471–1491.

Nelson, C.S. and Hume, T.M. 1977. Relative intensity of tectonic events revealed by the Tertiary sedimentary record in the North Wanganui Basin and adjacent areas, New Zealand. *New Zealand Journal of Geology and Geophysics* 20: 369–392.

Nelson, C.S., Briggs, R.M. and Kamp, J.J. 1986a. Nature and significance of volcanogenic deposits at the Eocene/Oligocene boundary, hole 593, Challenger Plateau, Tasman Sea. *Initial Reports of the Deep Sea Drilling Project 90.* US Government Printing Office, Washington, pp. 1175–1187.

Nelson, C.S., Hendy, C.H. and Dudley, W.C. 1986b. Quaternary isotope stratigraphy of hole 593, Challenger Plateau, south Tasman Sea: Preliminary observations based on foraminifers and calcareous nannofossils. *Initial Reports of the Deep Sea Drilling Project 90.* US Government Printing Office, Washington, pp. 1413–1424.

Nelson, C.S., Mildenhall, D.C., Todd, A.J. and Pocknall, D.T. 1988. Subsurface stratigraphy, paleoenvironments, palynology, and depositional history of the late Neogene Tauranga Group at Ohinewai, Lower Waikato Lowland, South Auckland, New Zealand. *New Zealand Journal of Geology and Geophysics* 31: 21–40.

Nelson, C.S., Kamp, P.J.J. and Young, H.R. 1994. Sedimentology and petrography of mass-emplaced limestone (Orahiri Limestone) on a Late Oligocene shelf, western North Island, and tectonic implications for eastern margin development of Taranaki Basin. *New Zealand Journal of Geology and Geophysics* 37: 269–285.

Nelson, C.S., Winefield, P.R., Hood, S.D., Caron, V., Pallentin, A. and Kamp, P.J. 2003. Pliocene Te Aute limestones, New Zealand: Expanding concepts for cool-water shelf carbonates. *New Zealand Journal of Geology and Geophysics* 46: 407–424. doi:10.1080/00288306.2003.9515017

Newman, J. 1985. Paleoenvironments, coal properties, and their interrelationships in Paparoa and selected Brunner Coal Measures on the West Coast of the South Island. Unpublished PhD thesis, University of Canterbury.

Newman, J. and Bradshaw, J.D. 1981. Oligocene–Miocene rocks of the Brechin Burn outlier, Waimakariri valley, Canterbury. *New Zealand Journal of Geology and Geophysics* 24: 469–476.

Newman J. and Newman, N. 1992. Tectonic and paleoenvironmental controls on the distribution and properties of Upper Cretaceous coals on the West Coast of the South Island, New Zealand. In: P.J. McCabe and J.T. Parrish (eds). *Controls on the Distribution and Quality of Cretaceous Coals.* Geological Society of America Special Paper 267, pp. 347–368.

Nichol, R. 1992. The eruption history of Rangitoto: Reappraisal of a small New Zealand myth. *Journal of the Royal Society of New Zealand* 22: 159–180.

Nicholson, K.N., Black, P.M. and Picard, C. 2000. Geochemistry and tectonic significance of the Tangihua Ophiolite Complex, New Zealand. *Tectonophysics* 321: 1–15.

Nicholson, K.N., Black, P.M., Picard, C., Cooper, P., Hall, C.M. and Itaya, T. 2007. Alteration, age, and emplacement of the Tangihua Complex Ophiolite, New Zealand. *New Zealand Journal of Geology and Geophysics* 50: 151–164.

Nicol, A. 1992. Tectonic structures developed in Oligocene limestones: Implications for New Zealand plate boundary deformation in North Canterbury. *New Zealand Journal of Geology and Geophysics* 35: 353–362.

Nicol, A. 1993. Haumurian (c. 66–80 Ma) half-graben development and deformation, mid Waipara, North Canterbury, New Zealand. *New Zealand Journal of Geology and Geophysics* 36: 127–130.

Nicol, A. and Campbell, J.K. 1990. Late Cenozoic thrust tectonics, Picton, New Zealand. *New Zealand Journal of Geology and Geophysics* 33: 485–494.

Nicol, A., Mazengarb, C., Chanier, F., Rait, G., Uruski, C. and Wallace, L. 2007. Tectonic evolution of the active Hikurangi subduction margin, New Zealand, since the Oligocene. *Tectonics* 26. https://doi.org/10.1029/2006TC002090

Nodder, S.D., Nelson, C.S. and Kamp, P.J.J. 1990. Mass-emplaced siliciclastic-volcaniclastic-carbonate sediments in Middle Miocene shelf-to-slope environments at Waikawau, northern Taranaki, and some implications for Taranaki Basin development. *New Zealand Journal of Geology and Geophysics* 33: 599–615.

Norris, R.J. and Carter, R.M. 1980. Offshore sedimentary basins at the southern end of the Alpine Fault, New Zealand. In: P.F. Ballance and H.G. Reading (eds): *Sedimentation in Oblique-slip Mobile Zones.* International Association of Sedimentologists Special Publication 4, pp. 237–265.

Norris, R.J. and Carter, R.M. 1982. Fault-bounded blocks and their role in localising sedimentation and deformation adjacent to the Alpine Fault, southern New Zealand. *Tectonophysics* 87: 11–23.

Norris, R.J. and Turnbull, I.M. 1993. Cenozoic basins adjacent to an evolving transform plate boundary, southwest New Zealand. In: P.F. Ballance (ed.). *South Pacific Sedimentary Basins.* Sedimentary Basins of the World 2. Elsevier, Amsterdam, pp. 251–270.

Oliver, P.J., Campbell, J.D. and Speden, I.G. 1982. The stratigraphy of the Torlesse rocks of the Mt Somers area (S81) mid-Canterbury. *Journal of the Royal Society of New Zealand* 12: 243–271.

Oliver, R.L., Finlay, H.J. and Fleming, C.A. 1950. *The Geology of Campbell Island. Cape expedition series.* Department of Scientific and Industrial Research Bulletin 3. 62 pp.

Palmer, J.A. and Andrews, P.B. 1993. Cretaceous–Tertiary sedimentation and implied tectonic controls on the structural evolution of Taranaki Basin, New Zealand. In: P.F. Ballance (ed.). *South Pacific Sedimentary Basins.* Sedimentary Basins of the World 2. Elsevier, Amsterdam, pp. 309–328.

Pankhurst, R.J., Weaver, S.D., Bradshaw, J.D., Storey, B.C. and Ireland, T.R. 1998. Geochronology and geochemistry of pre-Jurassic superterranes in Marie Byrd Land, Antarctica. *Journal of Geophysical Research – Solid Earth* 103(B2): 2529–2547.

Pettinga, J.R. 1982. Upper Cenozoic structural history, coastal southern Hawkes Bay, New Zealand. *New Zealand Journal of Geology and Geophysics* 25: 149–191.

Pettinga, J.R. 1990. Structure, stratigraphy and sedimentology, highest accretionary ridge – Waimarama Beach, Southern Hawke's Bay. In: *Geological Society of New Zealand Annual Conference, November. Conference field trips.* Geological Society of New Zealand Miscellaneous Publication 50b, pp. 39–60.

Pettinga, J.R. and Wise, D.U. 1994. Paleostress adjacent to the Alpine Fault: Broader implications from fault analysis near Nelson, South Island, New Zealand. *Journal of Geophysical Research – Solid Earth* 99(B2): 2727–2736.

Pettinga, J.R., Yetton, M.D., Van Dissen, R.J. and Downes, G. 2001. Earthquake source identification and characterisation for the Canterbury region, South Island, New Zealand. *Bulletin of the New Zealand Society for Earthquake Engineering* 34: 282–317.

Phillips, C.J. 1985. Upper Cretaceous and Tertiary geology of the upper Waitahaia River, Raukumara Peninsula, New Zealand. *New Zealand Journal of Geology and Geophysics* 28: 595–607.

Phillips, C.J., Cooper, A.F., Palin, J.M. and Nathan, S. 2005. Geochronological constraints on Cretaceous–Paleocene volcanism in south Westland, New Zealand. *New Zealand Journal of Geology and Geophysics* 48: 1–14.

Phillips, S. 1989. Aspects of the thermoluminescence dating of Plagioclase feldspars. Unpublished MSc thesis, University of Auckland.

Pick, M.C. 1962. The stratigraphy, structure and economic geology of the Cretaceo-Tertiary rocks of the Waiapu district, New Zealand. Unpublished PhD thesis, University of Bristol. Todd Oil Exploration Co. New Zealand unpublished open-file petroleum report 608. Ministry of Commerce, Wellington.

Pilaar, W.F. and Wakefield, L.L. 1978. Structural and stratigraphic evolution of the Taranaki Basin, offshore North Island, New Zealand. *The APEA Journal* 18: 93–101.

Pinchon, D. 1972. *Adele (Tasman Bay) Marine Seismic Survey*. Ministry of Economic Development, Wellington, PR508. 10 pp. + 25 enclosures.

Pirajno, F. 1979. Geology, geochemistry, and mineralisation of a spilite-keratophyre association in Cretaceous flysch, East Cape area, New Zealand. *New Zealand Journal of Geology and Geophysics* 22: 307–328.

Pocknall, D.T. and Lindqvist, J.K. 1988. Palynology of the Puysegur Group (Mid Cretaceous) at Gulches Peninsula, South Fiordland. Research Notes 1988. *New Zealand Geological Survey Record* 35: 86–93.

Pocknall, D.T. and Turnbull, I.M. 1989. Paleoenvironmental and stratigraphic significance of palynomorphs from Upper Eocene (Kaiatan) Beaumont Coal Measures and Orauea Mudstone, Waiau Basin, western Southland, New Zealand. *New Zealand Journal of Geology and Geophysics* 32: 371–378.

Pound, K.S., Norris, R.J. and Landis, C.A. 2014. Eyre Creek Melange: An accretionary prism shear-zone melange in Caples Terrane rocks, Eyre Creek, northern Southland, New Zealand. *New Zealand Journal of Geology and Geophysics* 57: 1–20.

Prebble, W.M. 1976. The geology of the Kekerengu-Waima River district, northeast Marlborough. Unpublished MSc thesis, Victoria University of Wellington.

Prebble, W.M. 1980. Late Cainozoic sedimentation and tectonics of the East Coast deformed belt, in Marlborough, New Zealand. In: P.F. Ballance and H. Reading (eds). *Sedimentation in Oblique-slip Mobile Zones*. International Association of Sedimentologists Special Publication 4, pp. 217–228.

Price, G.R. 1974. Structural geology of the coastal hills east of Ward. Unpublished BSc (Hons) thesis, University of Canterbury.

Purcell, P.G. 1994. The D'Urville sub-basin, New Zealand. *1994 New Zealand Petroleum Conference Proceedings: The post Maui challenge*. Ministry of Commerce, Wellington, pp. 206–212.

Rafferty, W.J. and Heming, R.F. 1979. Quaternary alkalic and sub-alkalic volcanism in South Auckland, New Zealand. *Contributions to Mineralogy and Petrology* 71: 139–150.

Raine, J.I. 1984. *Outline of a Palynological Zonation of Cretaceous to Paleogene Terrestrial Sediments in the West Coast Region, South Island, New Zealand*. New Zealand Geological Survey Report 109. 82 pp.

Raine, J.I. 1994. Terrestrial K-T boundary studies in New Zealand. *Palaeoaustra* 2: 9–12.

Raine, J.I., Strong, C.P. and Wilson, G.J. 1993. *Biostratigraphic Revision of Petroleum Exploration Wells, Great South Basin, New Zealand*. Institute of Geological and Nuclear Sciences Report 93/32. 146 pp.

Raine, J.I., Beu, A.G., Boyes, A.F., Campbell, H.J., Cooper, R.A., Crampton, J.S., Crundwell, M.P., Hollis, C.J., Morgans, H.E.G. and Mortimer, N. 2015. New Zealand Geological Timescale NZGT 2015/1. *New Zealand Journal of Geology and Geophysics* 58: 398–403.

Rait, G. 1992. Early Miocene thrust tectonics of the Raukumara Peninsula. Unpublished PhD thesis, Victoria University of Wellington.

Rait, G. 2000. Thrust transport directions in the Northland Allochthon, New Zealand. *New Zealand Journal of Geology and Geophysics* 43: 271–288.

Rait, G., Chanier, F. and Waters, D.W. 1991. Landward- and seaward-directed thrusting accompanying the onset of subduction beneath New Zealand. *Geology* 19: 230–233.

Rattenbury, M.S., Cooper, R.A. and Johnston, M.R. 1998. *Geology of the Nelson Area*. Institute of Geological and Nuclear Sciences 1:250,000 Geological QMap 9. Institute of Geological and Nuclear Sciences, Lower Hutt. 1 sheet + 67 pp.

Rattenbury, M.S., Townsend, D.D. and Johnston, M.R. 2006. *Geology of the Kaikoura Area*. Institute of Geological and Nuclear Sciences 1:250,000 Geological QMap 13. Institute of Geological and Nuclear Sciences, Lower Hutt. 1 sheet + 70 pp.

Raymond, G.R. 1985. Paleoenvironmental analysis of the Taratu Formation (upper members) Kaitangata Coalfield. Unpublished MSc thesis, University of Canterbury.

Reay, A. 1986. Andesites from Solander Island. In: I.E.M. Smith (ed.). *Late Cenozoic Volcanism in New Zealand*. Royal Society of New Zealand Bulletin 23, pp. 337–343.

Reay, A. and Sipiera, P.P. 1987. Mantle xenoliths from the New Zealand region. In: P.H. Nixon (ed.). *Mantle Xenoliths*. Wiley, Chichester, pp. 347–358.

Reay, A. and Walls, D.J. 1994. Dunedin Volcanic Group. In: D.G. Bishop (ed.). *Geology of the Milton Area*. Institute of Geological and Nuclear Sciences Geological Map 9 1:50,000. Institute of Geological and Nuclear Sciences, Lower Hutt, pp. 16–20.

Reay, M.B. 1993. *Geology of the Middle Clarence Valley*. Institute of Geological and Nuclear Sciences Geological Map 10. Institute of Geological and Nuclear Sciences, Lower Hutt. 1 sheet +144 pp.

Reay, M.B. and Strong, C.P. 1992. The Branch Sandstone, Clarence Valley, and implications for latest Cretaceous paleoenvironments and geological history of Central Marlborough. *New Zealand Geological Survey Record Research Notes* 44: 43–49.

Retallack, G.J. 1981. Middle Triassic megafossil plants from Long Gully, near Otematata, North Otago, New Zealand. *Journal of the Royal Society of New Zealand* 11: 167–200.

Rey, P.F. and Muller, R.D. 2010. Fragmentation of active continental plate margins owing to the buoyancy of the mantle wedge. *Nature Geoscience Letters* 3: 257–261. doi:10:1038/NGEO825

Reyners, M., Eberhart-Phillips, D. and Bannister, S. 2011. Tracking repeated subduction of the Hikurangi Plateau beneath New Zealand. *Earth and Planetary Science Letters* 311: 165–171.

Richard, S.M., Smith, C.H., Kimbrough, D.L., Fitzgerald, P.G., Luyendyk, B.P. and McWilliams, M.O. 1994. Cooling history of the northern Ford Ranges, Marie Byrd Land, West Antarctica. *Tectonics* 13: 837–857.

Ricketts, B.D., Ballance, P.F., Hayward, B.W. and Mayer, W. 1989. Basal Waitemata Group lithofacies: Rapid subsidence in an early Miocene interarc basin, New Zealand. *Sedimentology* 36: 559–580.

Ridd, M.F. 1964. Succession and structural interpretation of the Whangara-Waimata area, Gisborne, New Zealand. *New Zealand Journal of Geology and Geophysics* 7: 279–298.

Ridd, M.F. 1967. The stratigraphy and structure of the Whangara-Waimate area, Gisborne, New Zealand. Unpublished PhD thesis, University of London.

Riordan, N.K. 2016. A cool-water carbonate seaway in an extensional setting: Oligo-Miocene sedimentology of the Nile Group and Paparoa Trough, western South Island, New Zealand. Unpublished PhD thesis, University of Canterbury.

Riordan, N.K., Reid, C., Bassett, K. and Bradshaw, J. 2014. Reconsidering basin geometries of the West Coast: The influence of the Paparoa Core Complex on Oligocene Rift Systems. *New Zealand Journal of Geology and Geophysics* 57: 170–184.

Ritchie, D.D. 1986. Stratigraphy, structure and geological history of mid-Cretaceous sedimentary rocks across the Torlesse-like and non-Torlesse boundary in the Sawtooth Range, Coverham area, Marlborough. Unpublished MSc thesis, University of Canterbury.

Roberts, A.P. and Wilson, G.S. 1992. Stratigraphy of the Awatere Group, Marlborough, New Zealand. *Journal of the Royal Society of New Zealand* 22: 187–204.

Robertson, D.J. 1976. A paleomagnetic study of volcanic rocks in the South Auckland area. Unpublished MSc thesis, University of Auckland.

Robinson, P.H., Morris, B.D. and Scott, G.H. 1987. *Lithologic Log and Micropaleontology of Manutahi-1 Core, Onshore South Taranaki*. New Zealand Geological Survey Report PAL 121. 47 pp.

Roncaglia, L. and Schiøler, P. 1997. *Dinoflagellate Biostratigraphy of Piripauan-Haumurian Sections in Southern Marlborough and Northern Canterbury, New Zealand*. Institute of Geological and Nuclear Sciences Science Report 97/09. 50 pp.

Roser, B.P. and Korsch, R.J. 1999. Geochemical characterization, evolution and source of a Mesozoic accretionary wedge: The Torlesse terrane, New Zealand. *Geological Magazine* 136: 493–512.

Sagar, M.W. and Palin, J.M. 2011. Emplacement, metamorphism, deformation and affiliation of mid-Cretaceous orthogneiss from the Paparoa Metamorphic Core Complex lower plate, Charleston, New Zealand. *New Zealand Journal of Geology and Geophysics* 54: 273–289.

Saul, G. 1994. The basin development and deformation associated with the Kongahu (Lower Buller) fault zone over the last 12 Ma, Mokihinui River, West Coast, South Island, New Zealand. *Journal of the Royal Society of New Zealand* 24: 277–288.

Saul, G., Naish, T.R., Abbott, S.T. and Carter, R.M. 1999. Sedimentary cyclicity in the marine Pliocene–Pleistocene of the Wanganui basin (New Zealand): Sequence stratigraphic motifs characteristic of the past 2.5 m.y. *Geological Society of America Bulletin* 111: 524–537.

Schellart, W.P. 2007. North-eastward subduction followed by slab detachment to explain ophiolite obduction and Early Miocene volcanism in Northland, New Zealand. *Terra Nova* 19: 211–218.

Schellart, W.P., Lister, G.S. and Toy V.G. 2006. A Late Cretaceous and Cenozoic reconstruction of the Southwest Pacific region: Tectonics controlled by subduction and slab rollback processes. *Earth-Science Reviews* 76: 191–233.

Schiøler, P. and Wilson, G.J. 1998. Dinoflagellate biostratigraphy of the middle Coniacian–lower Campanian (Upper Cretaceous) in south Marlborough, New Zealand. *Micropaleontology* 44: 313–349.

Schiøler, P., Crampton, J.S. and Laird, M.G. 2002. Palynofacies and sea-level changes in the Middle Coniacian–Late Campanian (Late Cretaceous) of the East Coast Basin, New Zealand. *Palaeogeography, Palaeoclimatology, Palaeoecology* 188: 101–125.

Schiøler, P., Rogers, K.M., Sykes, R., Hollis, C.J., Ilg, B.R., Meadows, D., Roncaglia, L. and Uruski, C.I. 2010. Palynofacies, organic geochemistry and depositional environment of the Tartan Formation (Late Paleocene), a potential source rock in the Great South Basin, New Zealand. *Marine and Petroleum Geology* 27: 351–369.

Schulte, D.O., Ring, U., Thomson, S.N., Glodny, J. and Carrad, H. 2014. Two stage development of the Paparoa Metamorphic Core Complex, West Coast, South Island, New Zealand: Hot continental extension precedes sea floor spreading by ~25 m.y. *Lithosphere* 6: 177–194.

Scott, G.H., King, P.R. and Crundwell, M.P. 2004. Recognition and interpretation of depositional units in a late Neogene progradational shelf margin complex, Taranaki Basin, New Zealand: Foraminiferal data compared with seismic facies and wireline logs. *Sedimentary Geology* 164: 55–74.

Scott, J.M. 2013. A review of the location and significance of the boundary between the Western Province and Eastern Province, New Zealand. *New Zealand Journal of Geology and Geophysics* 56: 276–293.

Scott, J.M. and Cooper, A.F. 2006. Early Cretaceous extensional exhumation of the lower crust of a magmatic arc: Evidence from the Mount Irene Shear Zone, Fiordland, New Zealand. *Tectonics* 25: TC3018. 15 pp.

Scott, J.M., Cooper, A.F., Tulloch, A.J. and Spell, T.L. 2011. Crustal thickening of the Early Cretaceous paleo-Pacific Gondwana margin. *Gondwana Research* 20: 380–394.

Seilacher, A. 1967. Bathymetry of trace fossils. *Marine Geology* 5: 413–428.

Seward, D. 1989. Cenozoic basin histories determined by fission-track dating of basement granites, South Island, New Zealand. *Chemical Geology* 79: 31–48.

Sewell, R.J. 1985. The volcanic geology and geochemistry of central Banks Peninsula and relationships to Lyttelton and Akaroa volcanoes. Unpublished PhD thesis, University of Canterbury.

Sewell, R.J. 1988. Late Miocene volcanic stratigraphy of central Banks Peninsula, Canterbury, New Zealand. *New Zealand Journal of Geology and Geophysics* 31: 41–64.

Sewell, R.J. and Nathan, S. 1987. Geochemistry of Late Cretaceous and Early Tertiary basalts from South Westland. *New Zealand Geological Survey Record* 18: 87–94.

Sewell, R.J., Weaver, S.D. and Reay, M.B. 1992. *Geology of Banks Peninsula*. Scale 1:100,000. Institute of Geological and Nuclear Sciences Geological Map 3. Institute of Geological and Nuclear Sciences, Lower Hutt.

Shane, P.A. 1991. Remobilised silicic tuffs in middle Pleistocene fluvial sediments, southern North Island, New Zealand. *New Zealand Journal of Geology and Geophysics* 34: 489–499. doi:10.1080/00288306.1991.9514485

Shane, P.A. 1994. A widespread, early Pleistocene tephra (Potaka tephra, 1 Ma) in New Zealand: Character, distribution, and implications. *New Zealand Journal of Geology and Geophysics* 37: 25–35.

Shane, P.A., Alloway, B., Black, T. and Westgate, J. 1996. Isothermal plateau fission-track ages of tephra beds in an early-middle Pleistocene marine and terrestrial sequence, Cape Kidnappers, New Zealand. *Quaternary International* 34–36: 49–53.

Shearer, J.C. 1992. Sedimentology, coal chemistry and petrography of the Cretaceous Morley Coal Measures and the Eocene Beaumont Coal Measures, Ohai Coalfield, South Island, New Zealand. Unpublished PhD thesis, University of Canterbury.

Shearer, J.C. 1995. Tectonic controls on styles of sediment accumulation in the Late Cretaceous Morley Coal Measures of Ohai Coalfield, New Zealand. *Cretaceous Research* 16: 367–384.

Shell BP Todd. 1984. *Drilling Completion Report, Clipper-1. Offshore Canterbury, South Island, New Zealand. PPL 38202.* Ministry of Economic Development, Wellington, PR1036. 909 pp. + 22 enclosures.

Sherwood, A.M., Lindqvist, J.K., Newman, J. and Sykes, R. 1992. Depositional controls on Cretaceous coals and coal measures in New Zealand. In: P.J. McCabe and J.T. Parrish (eds). *Controls on the Distribution and Quality of Cretaceous Coals* (Vol. 267). Colorado, Geological Society of America, pp. 325–346.

Sikumbang, N. 1978. Miocene regressive strata, Dunedin district. Unpublished MSc thesis, University of Otago.

Silberling, N.J., Nichols, K.M., Bradshaw, J.D. and Blome, C.D. 1988. Limestone and chert in tectonic blocks from the Esk Head subterrane, South Island, New Zealand. *Geological Society of America Bulletin* 100: 1213–1223.

Simms, B. and Nelson, C.S. 1998. *Regional Stratigraphy of the Early Miocene Mahoenui Group, Western North Island, New Zealand*. Geological Society of New Zealand Miscellaneous Publication 101A. 204 pp.

Sircombe, K.N. and Kamp, P.J. 1998. The South Westland Basin: Seismic stratigraphy, basin geometry and evolution of a foreland basin within the Southern Alps collision zone, New Zealand. *Tectonophysics* 300: 359–387.

Sivell, W.J. and McCulloch, M.T. 2000. Reassessment of the origin of the Dun Mountain Ophiolite, New Zealand: Nd-isotopic and geochemical evolution of magma suites. *New Zealand Journal of Geology and Geophysics* 43: 133–146.

Skinner, D.N. 1986. Neogene volcanism of the Hauraki Volcanic Region. *Royal Society of New Zealand Bulletin* 23: 21–47.

Smale, D. 1980. *Petrology of Some Tertiary Sandstones from Murchison*. New Zealand Geological Survey Report G 35. 6 pp.

Smale, D. 1993. *Heavy Minerals in Cretaceous Sediments of East Coast, North Island and Marlborough*. Institute of Geological and Nuclear Sciences Science Report 93/05. 29 pp.

Smale, D. and Laird, M.G. 1995. Relation of heavy mineral populations to stratigraphy of Cretaceous formations in Marlborough, New Zealand. *New Zealand Journal of Geology and Geophysics* 38: 211–222.

Smith, I.E.M. 1989. New Zealand intraplate volcanism: North Island. In: R.W. Johnson, J. Knutson and S.R. Taylor (eds). *Intraplate Volcanism in Eastern Australia and New Zealand*. Cambridge University Press, Cambridge, pp. 157–162.

Smith, I.E., Ruddock, R.S. and Day, R.A. 1989. Miocene arc-type volcanic/plutonic complexes of the Northland Peninsula, New Zealand. *Royal Society of New Zealand Bulletin* 26: 205–213.

Smith, I.E., Okada, T., Itaya, T. and Black, P.M. 1993. Age relationships and tectonic implications of late Cenozoic basaltic volcanism in Northland, New Zealand. *New Zealand Journal of Geology and Geophysics* 36: 385–394.

Soenandar, H.B. 1992. Seismic stratigraphy of the Giant Foresets Formation, offshore North Taranaki western platform. *1991 New Zealand Oil Exploration Conference Proceedings*. Ministry of Commerce, Wellington, pp. 207–233.

Speden, I.G. 1975. *Cretaceous Stratigraphy of Raukumara Peninsula. Part 1. Cretaceous stratigraphy of Koranga (parts N87 and N88). Part 2. Geology of the Lower Waimana and Waiotahi Valleys (part N78).* New Zealand Geological Survey Bulletin 91. 70 pp.

Speden, I.G. 1976. Geology of Mt Taitai, Tapuaeroa Valley, Raukumara Peninsula. *New Zealand Journal of Geology and Geophysics* 19: 71–119.

Spell, T.L., McDougall, I. and Tulloch, A.J. 2000. Thermochronologic constraints on the breakup of the Pacific Gondwana margin: The Paparoa metamorphic core complex, South Island, New Zealand. *Tectonics* 19: 433–451.

Spörli, K.B. 1978. Mesozoic tectonics, North Island, New Zealand. *Geological Society of America Bulletin* 89: 415–425.

Spörli, K.B. and Aita, Y. 1992. Tectonic significance of Late Cretaceous radiolaria from the obducted Matakaoa Volcanics, East Cape, North Island, New Zealand. *Geoscience Reports of Shizuoka University* 20: 115–133.

Spörli, K.B. and Ballance, P.F. 1989. Mesozoic–Cenozoic ocean floor/continent interaction and terrane configuration, southwest Pacific area around New Zealand. In: Z. Ben-Avraham (ed.). *The Evolution of the Pacific Ocean Margins*. Oxford Monographs on Geology and Geophysics (Book 8). Oxford University Press, New York, pp. 176–190.

Spörli, K.B. and Lillie, A.R. 1974. Geology of the Torlesse Supergroup in the northern Ben Ohau Range, Canterbury. *New Zealand Journal of Geology and Geophysics* 17: 115–141.

Spörli, K.B., Takemuri, A. and Hori, R.S. 2007. *The Oceanic Permian Boundary Sequence at Arrow Rocks (Orautemanu) Northland, New Zealand.* Monograph 24, Institute of Geological and Nuclear Sciences, Lower Hutt. 229 pp.

Sprott, A. 1997. Aspects of the Albany Conglomerate, Waitemata Group, North Auckland. Unpublished MSc thesis, University of Auckland.

Stagpoole, V. 1997. A geophysical study of the northern Taranaki Basin, New Zealand. Unpublished PhD thesis, Victoria University of Wellington.

Stainton P.W. and Gibson, G.W. 1964. The geology of Central Taranaki. Compilation report, Shell, BP and Todd Oil Services Ltd New Zealand. Unpublished.

Stark, C.J. 1996. Interpretation of Paleocene fluvial sediments from the Upper Pakawau and Kapuni Groups, Pakawau Sub-Basin, north-west Nelson. Unpublished MSc thesis, University of Canterbury.

St John, D.H. 1965. *Puniwhakau-1 Exploration Well Resumé*. Ministry of Economic Development, Wellington, PR455. 142 pp. + 2 enclosures.

Stern, T.A. and Davey, F.J. 1989. Crustal structure and origin of basins formed behind the Hikurangi subduction zone, New Zealand. In: R.A. Price (ed.). *Origin and Evolution of Sedimentary Basins and Their Energy and Mineral Resources.* Geophysical Monograph 48. American Geophysical Union, Washington, pp. 73–85.

Stern, T.A., Stratford, W.R. and Salmon, M.L. 2006. Subduction evolution and mantle dynamics at a continental margin: Central North Island, New Zealand. *Reviews of Geophysics* 44: RG4002.

Stilwell, J.D., Consoli, C.P., Sutherland, R., Salisbury, S., Rich, T.H., Vickers-Rich, P.A., Currie, P.J. and Wilson, G.J. 2006. Dinosaur sanctuary on the Chatham Islands, Southwest Pacific: First record of theropods from the K-T boundary Takatika Grit. *Palaeogeography, Palaeoclimatology, Palaeoecology* 230: 243–250.

Stipp, J.J. and McDougall, I. 1968. Geochronology of Banks Peninsula volcanoes, New Zealand. *New Zealand Journal of Geology and Geophysics* 11: 1239–1260.

Stoneley, R. 1968. A lower Tertiary décollement of the East Coast, North Island, New Zealand. *New Zealand Journal of Geology and Geophysics* 11: 128–156.

Storey, B.C., Leat, P.T., Weaver, S.D., Pankhurst, R.J., Bradshaw, J.D. and Kelley, S. 1999. Mantle plumes and Antarctica–New Zealand rifting: Evidence from mid-Cretaceous mafic dykes. *Journal of the Geological Society, London* 156: 659–671.

Strogen, D.P., Bland, K.J., Nicol, A. and King, P.R. 2014. Paleogeography of the Taranaki Basin region during the latest Eocene–Early Miocene and implications for the 'total drowning' of Zealandia. *New Zealand Journal of Geology and Geophysics* 57: 110–127. doi:10.1080/00288306.2014.901231

Strogen, D.P., Seebeck, H., Nicol, A. and King, P.R. 2017. Two-phase Cretaceous–Paleocene rifting in the Taranaki Basin region, New Zealand; implications for Gondwana break-up. *Journal of the Geological Society of London* 174: 929–946. doi:org/10.1144/jgs2016-160

Strong, C.P. 1976. Notes from the New Zealand Geological Survey – 9: Cretaceous foraminifera from the Matakaoa Volcanic Group. *New Zealand Journal of Geology and Geophysics* 19: 140–143.

Strong, C.P. 1979. Late Cretaceous foraminifera from Kahuitara Tuff, Pitt Island, New Zealand. *New Zealand Journal of Geology and Geophysics* 22: 593–611.

Strong, C.P. 1980. Early Paleogene foraminifera from Matakaoa Volcanic Group (Note). *New Zealand Journal of Geology and Geophysics* 23: 267–272.

Strong, C.P. and Beggs, J.M. 1990. Late Cretaceous–early Paleogene stratigraphic sequence in Marlborough and possible offshore seismic equivalent. *1989 New Zealand Oil Exploration Conference Proceedings*. Ministry of Commerce, Wellington, pp. 173–180.

Strong, C.P., Scott, G.H. and Morgans, H.E.G. 1993. *Biostratigraphic Review of Selected Drillholes, Hawkes Bay Gisborne Area*. Institute of Geological and Nuclear Sciences Science Report 93/19. 133 pp.

Strong, C.P., Hollis, C.J. and Wilson, G.J. 1995. Foraminiferal, radiolarian, and dinoflagellate biostratigraphy of Late Cretaceous to Middle Eocene pelagic sediments (Muzzle Group), Mead Stream, Marlborough, New Zealand. *New Zealand Journal of Geology and Geophysics* 38: 171–209.

Strong, C.P., Mildenhall, D.C., Raine J.I., Wilson, G.J. and Edwards, A.R. 1999. *Biostratigraphy of Waka Nui-1 Offshore Petroleum Exploration Well, Northland Basin, New Zealand*. Institute of Geological and Nuclear Sciences Client Report 1999/123. 52 pp.

Suggate, R.P. 1984. *Sheet M29 AC Mangles Valley*. Geological Maps of New Zealand 1:50,000. Department of Scientific and Industrial Research, Wellington. 1 sheet + booklet.

Suggate, R.P. and Waight, T.E. 1999. *Geology of the Kumara-Moana Area, 1:50,000*. Institute of Geological and Nuclear Sciences Geological Map 24. Institute of Geological and Nuclear Sciences, Lower Hutt. 1 sheet + 24 pp.

Suggate, R.P., Stevens, G.R. and Te Punga, M.T. 1978. *The Geology of New Zealand*. Government Printer, Wellington. Two volumes, 820 pp.

Suggate, R.P., Kamp, P.J.J., Whitehouse, I.W.S. and Newman, J. 2000. Letters to the Editor. *New Zealand Journal of Geology and Geophysics* 43: 651–654.

Sutherland, R. 1994. Displacement since the Pliocene along the southern section of the Alpine Fault, New Zealand. *Geology* 22: 327–330.

Sutherland, R. 1995. The Australia–Pacific boundary and Cenozoic plate motions in the SW Pacific: Some constraints from Geosat data. *Tectonics* 14: 819–831.

Sutherland, R. 1996. Transpressional development of the Australia–Pacific boundary through southern South Island, New Zealand: Constraints from Miocene–Pliocene sediments, Waiho-1 borehole, South Westland. *New Zealand Journal of Geology and Geophysics* 39: 251–264.

Sutherland, R. 1999. Basement geology and tectonic development of the greater New Zealand region: An interpretation from regional magnetic data. *Tectonophysics* 308: 341–362.

Sutherland, R. and Hollis, C.J. 2001. Cretaceous demise of the Moa plate and strike-slip motion at the Gondwana margin. *Geology* 29: 279–282.

Sutherland, R. and Melhuish, A. 2000. Formation and evolution of the Solander Basin, southwestern South Island, New Zealand, controlled by a major fault in continental crust and mantle. *Tectonics* 19: 44–61.

Sutherland, R., Hollis, C.J., Nathan, S., Strong, C.P. and Wilson, G.J. 1996. Age of Jackson Formation proves Late Cenozoic allochthony in South Westland, New Zealand. *New Zealand Journal of Geology and Geophysics* 39: 559–563.

Sutherland, R., King, P.R. and Wood, R.A. 2001. Tectonic evolution of Cretaceous rift basins in south-eastern Australia and New Zealand: Implications for exploration risk assessment. *Eastern Australasian Basins Symposium, 25–28 November 2001*. Melbourne, Victoria, pp. 3–13.

Sykes, R. 1985. Paleoenvironmental and tectonic controls on coal measure characteristics, Ohai Coalfield, Southland. Unpublished MSc thesis, University of Canterbury.

Sykes, R. 1988. *The Morley Coal Measures, Ohai Coalfield, Southland*. New Zealand Energy Research and Development Committee Report 170. 85 pp.

Tappenden, V.E. 2003. Magmatic response to the evolving New Zealand margin of Gondwana during the mid-late Cretaceous. Unpublished PhD thesis, University of Canterbury.

Tappenden, V., Hoernle, K., Weaver, S.W. and Ireland, T.R. 2002. Slab detachment during the initial stages of mid-Cretaceous rifting of New Zealand from Gondwana. Programme and abstracts. Gondwana 11: Correlations and Connections, 25–30 August 2002, University of Canterbury, Christchurch.

Thompson, N.K., Bassett, K.N. and Reid, C.M. 2014. The effect of volcanism on cool-water carbonate facies during maximum inundation of Zealandia in the Waitaki–Oamaru region. *New Zealand Journal of Geology and Geophysics* 57: 149–169. doi:10.1080/00288306.2014.904385

Thompson, T.L., Leask, W.L. and May, B.T. 1994. Petroleum potential of the South Wanganui Basin. *1994 New Zealand Petroleum Conference Proceedings: The post Maui challenge – investment and development opportunities*. Ministry of Commerce, Wellington, pp. 108–127.

Thrasher, G.P. 1986. Basement structure and sediment thickness beneath the continental shelf of the Hauraki Gulf and offshore Coromandel region, New Zealand. *New Zealand Journal of Geology and Geophysics* 29: 41–50.

Thrasher, G.P. 1992. Late Cretaceous geology of Taranaki Basin, New Zealand. Unpublished PhD thesis, Victoria University of Wellington.

Thrasher, G.P., Suggate, R.P. and Funnell, R.H. 1996. The Kotoku Anticline, West Coast, South Island. *1996 New Zealand Petroleum Conference Proceedings*. Ministry of Commerce, Wellington, pp. 71–75.

Titheridge, D.G. 1977. Stratigraphy and sedimentology of the upper Pakawau and lower Westhaven Groups (Upper Cretaceous–Oligocene), NW Nelson. Unpublished MSc thesis, University of Canterbury.

Titheridge, D.G. 1993. The influence of half-graben syn-depositional tilting on thickness variation and seam splitting in the Brunner Coal Measures, New Zealand. *Sedimentary Geology* 87: 195–213.

Topping, R.M. 1978. Foraminifera from the Mahoenui Group, North Wanganui Basin. Unpublished PhD thesis, University of Auckland.

Toy, V.G. and Spörli, K.B. 2008. Stratigraphic and structural evidence for an accretionary precursor to the Northland Allochthon: Mt Camel Terrane, northernmost New Zealand. *New Zealand Journal of Geology and Geophysics* 51: 331–347.

Trewick, S.A. and Bland, K.J. 2012. Fire and slice: Palaeogeography for biogeography at New Zealand's North Island/South Island juncture. *Journal of the Royal Society of New Zealand* 42: 153–183.

Tulloch, A.J. and Kimbrough, D.L. 1989. The Paparoa metamorphic core complex, New Zealand – Cretaceous extension associated with fragmentation of the Pacific margin of Gondwana. *Tectonics* 8: 1217–1234.

Tulloch, A.J. and Kimbrough, D.L. 2003. *Paired Plutonic Belts in Convergent Margins and the Development of High Sr/Y: Peninsular Ranges Batholith of Baja-California and Median Batholith of New Zealand*. Geological Society of America Special Paper 374, pp. 275–295.

Tulloch, A.J., Kimbrough, D.L. and Waight, T.E. 1994. The French Creek Granite, North Westland, New Zealand: Late Cretaceous A-type plutonism on the Tasman passive margin. In: G.J. van der Lingen, K.M. Swanson and R.J. Muir (eds). *Evolution of the Tasman Sea Basin.* Balkema, Rotterdam, pp. 65–66.

Tulloch, A.J., Ramezani, J., Kimbrough, D.L., Faure, K. and Allibone, A.H. 2009a. U-Pb geochronology of mid-Paleozoic plutonism in western New Zealand: Implications for S-type granite generation and growth of the east Gondwana margin. *Geological Society of America Bulletin* 121: 1236–1261.

Tulloch, A.J., Ramezani, J., Mortimer, N., Mortensen, J., van den Bogaard, P. and Maas, R. 2009b. Cretaceous felsic volcanism in New Zealand and the Lord Howe Rise (Zealandia) as a precursor to final Gondwana break-up. *Geological Society of London, Special Publication* 321: 89–118.

Turnbull, I.M. 1979. Stratigraphy and sedimentology of the Caples terrane of the Thomson Mountains, northern Southland, New Zealand. *New Zealand Journal of Geology and Geophysics* 22: 555–574.

Turnbull, I.M. 1985. *Sheet D42AC & part sheet D43 Te Anau Downs*. Geological Maps of New Zealand 1:50,000. Department of Scientific and Industrial Research, Wellington. 1 sheet + 31 pp.

Turnbull, I.M. 1986. *Sheet D42BD & part sheet D43 Snowdon*. Geological Maps of New Zealand 1:50,000. Department of Scientific and Industrial Research, Wellington. 1 sheet + booklet.

Turnbull, I.M. 1993. Cretaceous and Cenozoic stratigraphic columns from the western Southland region. New Zealand unpublished open-file petroleum report PR4200. Ministry of Economic Development, Wellington. 24 pp. + 2 enclosures.

Turnbull, I.M. 2000. *Geology of the Wakatipu Area*. Institute of Geological and Nuclear Sciences 1:250,000 Geological QMap 18. Institute of Geological and Nuclear Sciences, Lower Hutt. 1 sheet + 72 pp.

Turnbull, I.M. and Uruski, C.I. 1993. *Cretaceous and Cenozoic Sedimentary Basins of Western Southland, South Island, New Zealand.* Monograph 1, Institute of Geological and Nuclear Sciences, Lower Hutt. 86 pp. + 3 enclosures.

Turnbull, I.M. and Uruski, C.I. 1995. *Geology of the Monowai-Waitutu Area*. Sheets C46 and part C45, 1:50,000. Institute of Geological and Nuclear Sciences Geological Map 19. Institute of Geological and Nuclear Sciences, Lower Hutt.

Turnbull, I.M., Barry, J.M., Carter, R.M. and Norris, R.J. 1975. The Bobs Cove Beds and their relationship to the Moonlight Fault Zone. *Journal of the Royal Society of New Zealand* 5: 355–394.

Turnbull, I.M., Lindqvist, J.K., Mildenhall, D.C., Hornibrook, N. de B., Beu, A.G. and Mildenhall, D.C. 1985. Stratigraphy and paleontology of Pliocene–Pleistocene sediments on Five Fingers Peninsula, Dusky Sound, Fiordland. *New Zealand Journal of Geology and Geophysics* 28: 217–231. doi:10.1080/00288306.1985.10422221

Turnbull, I.M., Lindqvist, J.K., Norris, R.J., Carter, R.M., Cave, M.P., Sykes, R. and Hyden, F.M. 1989. Lithostratigraphic nomenclature of the Cretaceous and Tertiary sedimentary rocks of Western Southland, New Zealand. *New Zealand Geological Survey Record* 31. 55 pp.

Turnbull, I.M., Uruski, C.I., Anderson, H.J., Lindqvist, J.K., Scott, G.H., Morgans, H.E.G., Hoskins, R.H. *et al.* 1993. *Cretaceous and Cenozoic Sedimentary Basins of Western Southland, South Island, New Zealand*. Monograph 1, Institute of Geological and Nuclear Sciences, Lower Hutt. 86 pp.

Turnbull, R.E., Tulloch, A.J. and Ramezani, J. 2013. Zetland Diorite, Karamea Batholith, west Nelson: Field relationships, geochemistry and geochronology demonstrate links to the Carboniferous Tobin Suite. *New Zealand Journal of Geology and Geophysics* 56: 83–99.

Turner, G.M., Michalk, D. and Little, T.A. 2012. Paleomagnetic constraints on Cenozoic deformation along the northwest margin of the Pacific–Australian plate boundary zone through New Zealand. *Tectonics* 31. doi:10.1029/2011TC002931

Uruski, C.I. 1992. Seismic evidence for dextral wrench faulting on the Moonlight Fault System. *New Zealand Geological Survey Record* 44: 69–75.

Uruski, C.I. and Baillie, P. 2002. Petroleum systems of the deepwater Taranaki Basin, New Zealand. *2002 New Zealand Petroleum Conference Proceedings*. Ministry of Economic Development, Wellington, pp. 402–407.

Uruski, C.I. and Turnbull, I.M. 1990. Stratigraphy and structural evolution of the west Southland sedimentary basins. *1989 New Zealand Oil Exploration Conference Proceedings*. Ministry of Commerce, Wellington, pp. 225–240.

Uruski, C. and Wood, R. 1991. A new look at the New Caledonia Basin, an extension of the Taranaki Basin, offshore North Island, New Zealand. *Marine and Petroleum Geology* 8: 379–391.

Uruski, C.I., Cook, R.A., Herzer, R.H. and Isaac, M.J. 2004. Petroleum geology of the Northland Sector of the Greater Taranaki Basin. *2004 New Zealand Petroleum Conference Proceedings*. Ministry of Economic Development, Wellington, pp. 1–10.

Vajda, V. and McLoughlin, S. 2004. Fungal proliferation at the Cretaceous–Tertiary Boundary. *Science* 303: 1489.

Vajda, V., Raine, J.I. and Hollis, C.J. 2001. Indication of global deforestation at the Cretaceous–Tertiary boundary by New Zealand fern spike. *Science* 294: 1700–1702.

Vajda, V, Ocampo, A. and Buffetaut, E. 2003. Unmasking the KT catastrophe; evidence from flora, fauna and geochemistry. In: C. Cockell (ed.). Abstracts. Workshop on Biological Processes Associated with Impact Events – Cambridge (United Kingdom), Impact Programme, European Science Foundation, 57.

Vajda, V., Raine, J.I., Hollis, C.J. and Strong, C.P. 2004. Global effects of the Chicxulub Impact on terrestrial vegetation – review of the palynological record from New Zealand Cretaceous/Tertiary boundary. In: H. Dypvik, M.J. Burchell and P. Claeys (eds). *Cratering in Marine Environments and on Ice. Impact Studies*. Springer, Berlin, Heidelberg, pp. 57–74.

Van den Heuvel, H.B. 1960. The geology of the flat point area, eastern Wairarapa. *New Zealand Journal of Geology and Geophysics* 3: 309–320. doi:10.1080/00288306.1960.10423603

van der Lingen, G.J. 1973. The Lord Howe Rise rhyolites. *Initial Reports of the Deep Sea Drilling Project*, DFSD proceedings, pp. 523–540.

van der Lingen, G.J. 1982. Development of the North Island subduction system, New Zealand. In: J.K. Leggett (ed.). *Trench-Forearc Geology*. Geological Society of London Special Publication 10, pp. 259–272.

van der Lingen, G.J. and Pettinga, J.R. 1980. The Makara Basin: A Miocene slope-basin along the New Zealand sector of the Australian-Pacific obliquely convergent plate boundary. In: P.F. Ballance and H.G. Reading (eds). *Sedimentation in Oblique-slip Mobile Zones*. International Association of Sedimentologists Special Publication 4, pp. 191–215.

van der Lingen, G.J., Smale, D. and Lewis, D.W. 1978. Alteration of pelagic chalk below a paleokarst surface, Oxford, South Island, New Zealand. *Sedimentary Geology* 21: 46–66.

Van der Meer, Q.H., Scott, J.M., Waight, T.E., Sudo, M., Schersten, A., Cooper, A.F. and Spell, T.L. 2013. Magmatism during Gondwana break-up: New geochronological data from Westland, New Zealand. *New Zealand Journal of Geology and Geophysics* 56: 229–242.

Van der Meer, Q.H., Waight, T.E., Whitehouse, M.J. and Andersen, T. 2017. Age and petrogenetic constraints on the lower glassy ignimbrite of the Mount Somers Volcanic Group, New Zealand. *New Zealand Journal of Geology and Geophysics* 60: 209–219.

Veevers, J.J., Powell, C.M. and Roots, S.R. 1991. Review of sea-floor spreading around Australia. I. Synthesis of the patterns of spreading. *Australian Journal of Earth Sciences* 38: 373–389.

Vella, P. 1963. Plio-Pleistocene cyclothems, Wairarapa, New Zealand. *Transactions of the Royal Society of New Zealand* 2: 15–50.

Vella, P. and Briggs, W.M. 1971. Lithostratigraphic names, Upper Miocene to Lower Pleistocene, northern Aorangi Range, Wairarapa. *New Zealand Journal of Geology and Geophysics* 14: 253–274.

Vickery, S. and Lamb, S. 1995. Large tectonic rotations since the Early Miocene in a convergent plate boundary zone, South Island, New Zealand. *Earth and Planetary Science Letters* 136: 44–59.

Vonk, A.J. and Nelson, C.S. 1998. *Stratigraphic Re-evaluation of the Early Miocene Mokau Group across North Wanganui Basin*. Geological Society of New Zealand Miscellaneous Publication 101A. 236 pp.

Vonk, A.J., Kamp, P.J. and Hendy, A.J.W. 2002. Outcrop to subcrop correlations of late Miocene–Pliocene strata, eastern Taranaki Peninsula. *2002 New Zealand Petroleum Conference Proceedings*. Ministry of Economic Development, Wellington, pp. 234–255.

Waight, T.E., Weaver, S.D., Ireland, T.R., Maas, R., Muir, R.J. and Shelley, D. 1997. Field characteristics, petrography and geochronology of the Hohonu Batholith and the adjacent Granite Hill Complex, North Westland, New Zealand. *New Zealand Journal of Geology and Geophysics* 40: 1–17.

Waight, T.E., Weaver, S.D., Muir, R.J., Maas. R. and Eby, G.N. 1998a. The Hohonu Batholith of North Westland, New Zealand: Granitoid compositions controlled by source H_2O contents and generated during tectonic transition. *Contributions to Mineralogy and Petrology* 130: 225–239.

Waight, T.E., Weaver, S.D. and Muir, R.J. 1998b. Mid-Cretaceous granitic magmatism during the transition from subduction to extension in southern New Zealand: A chemical and tectonic synthesis. *Lithos* 45: 469–482.

Walcott, R.I. 1978. Present tectonics and late Cenozoic evolution of New Zealand. *Geophysical Journal of the Royal Astronomical Society* 52: 137–164.

Walcott, R.I. 1984a. The kinematics of the plate boundary zone through New Zealand: A comparison of short- and long-term deformations. *Geophysical Journal International* 79: 613–633.

Walcott, R.I. 1984b. Reconstructions of the New Zealand region for the Neogene. *Paleogeography, Paleogeography, Paleoecology* 46: 217–231.

Walcott, R.I. 1998. Modes of oblique compression: Late Cenozoic tectonics of the South Island of New Zealand. *Reviews of Geophysics* 36: 1–26.

Wandres, A.M. and Bradshaw, J.D. 2005. New Zealand tectonostratigraphy and implications from conglomeratic rocks for the configuration of the SW Pacific of Gondwana. In: A.M.P. Vaughan, P. Leat and R.J. Pankhurst (eds). *Terrane Processes at the Margins of Gondwana*. Geological Society Special Publication 246, pp. 179–206.

Wandres, A.M., Bradshaw, J.D., Weaver, S.D., Maas, R., Ireland, T.R. and Eby, G.N. 2004a. Provenance analysis using conglomerate clast lithologies: A case study from the Pahau terrane of New Zealand. *Sedimentary Geology* 167: 57–89.

Wandres, A.M., Bradshaw, J.D., Weaver, S.D., Maas, R., Ireland, T.R. and Eby, G.N. 2004b. Provenance of the sedimentary Rakaia sub-terrane, Torlesse Terrane, South Island, New Zealand: The use of igneous clast compositions to define the source. *Sedimentary Geology* 168: 193–226.

Ward, C.M. 1988. Marine terraces of the Waitutu district and their relation to the late Cenozoic tectonics of the southern Fiordland region, New Zealand. *Journal of the Royal Society of New Zealand* 18: 1–28.

Ward, D.M. and Lewis, D.W. 1975. Paleoenvironmental implications of storm-scoured, ichnofossiliferous mid-Tertiary limestones, Waihao district, south Canterbury, New Zealand. *New Zealand Journal of Geology and Geophysics* 18: 881–908.

Ward, S.D. 1997. Lithostratigraphy, palynostratigraphy and basin analysis of the late Cretaceous to early Tertiary Paparoa Group, Greymouth Coalfield, New Zealand. Unpublished PhD thesis, University of Canterbury.

Warner, T.L. 1990. The extrusive igneous rocks of the Marlborough region and their geological significance. Unpublished MSc thesis, University of Canterbury.

Warnes, M.D. 1990. *The Palynology of the Morley Coal Measures, Ohai Coalfield*. Resource Information Report, Energy and Resources Division, Ministry of Commerce, Wellington. 27 pp.

Warren, G. 1995. *Geology of the Parnassus Area*. Sheets O32 & part N32, 1:50,000. Institute of Geological and Nuclear Sciences Geological Map 18. Institute of Geological and Nuclear Sciences, Lower Hutt. 1 sheet + 36 pp.

Warren, G. and Speden, I.G. 1978. *The Piripauan and Haumurian Stratotypes (Mata Series, Upper Cretaceous) and Correlative Sequences in the Haumuri Bluff District, South Marlborough*. New Zealand Geological Survey Bulletin 92. 60 pp.

Waterhouse, B.C. and White, P.J. 1994. *Geology of the Raglan-Kawhia Area*. 1:50,000. Institute of Geological and Nuclear Sciences Geological Map 13. Institute of Geological and Nuclear Sciences, Lower Hutt. 1 sheet + 48 pp.

Watters, W.A. 1982. *Petrographic Notes on Conglomerate Pebbles from the Mangles and Longford Formations, Murchison District.* New Zealand Geological Survey Report G 65. 31 pp.

Weaver, S.D. and Pankhurst, R.J. 1991. A precise Rb-Sr age for the Mandamus Igneous Complex, North Canterbury, and regional tectonic implications. *New Zealand Journal of Geology and Geophysics* 34: 341–345.

Weaver, S.D. and Smith, I.E.M. 1989. New Zealand intraplate volcanism. In: R.W. Johnson, J. Knutson and S.R. Taylor (eds). *Intraplate Volcanism in Eastern Australia and New Zealand.* Cambridge University Press, Cambridge, pp. 157–188.

Weaver, S.D., Adams, C.J., Pankhurst, R.J. and Gibson, I.L. 1992. Granites of Edward VII Peninsula, Marie Byrd Land: Anorogenic magmatism related to Antarctic–New Zealand rifting. In: P.E. Brown and B.W. Chappell (eds). *The Second Hutton Symposium on the Origin of Granites and Related Rocks: Proceedings of a symposium held at the Australian Academy of Science, Canberra, 23–28 September 1991.* Geological Society of America, Boulder, pp. 281–290.

Weaver, S.D., Storey, B.C., Pankhurst, R.J., Mukasa, S.B., DiVenere, V.J. and Bradshaw, J.D. 1994. Antarctica–New Zealand rifting and Marie Byrd Land lithospheric magmatism linked to ridge subduction and mantle plume activity. *Geology* 22: 811–814.

Webb, P.N. 1971. New Zealand Late Cretaceous (Haumurian) foraminifera and stratigraphy: A summary. *New Zealand Journal of Geology and Geophysics* 14: 795–828.

Weissel, J.K. and Hayes, D.E. 1977. Evolution of the Tasman Sea reappraised. *Earth and Planetary Science Letters* 36: 77–84.

Weissel, J.K., Hayes, D.E. and Herron, E.M. 1977. Plate tectonics synthesis: The displacements between Australia, New Zealand and Antarctica since the Late Cretaceous. *Marine Geology* 25: 231–277.

Wellman, H.W. 1959. Divisions of the New Zealand Cretaceous. *Transactions of the Royal Society of New Zealand* 87: 99–163.

Wells, P.E. 1989. Late Neogene vertical tectonic movements in Western Wairarapa, New Zealand. Unpublished PhD thesis, Victoria University of Wellington.

Whattam, S.A., Malpas, J.G., Ali, J.R., Smith, I.E. and Lo, C-H. 2004. Origin of the Northland Ophiolite, northern New Zealand: Discussion of new data and reassessment of the model. *New Zealand Journal of Geology and Geophysics* 47: 383–389.

Whattam, S.A., Malpas, J., Ali, J.R., Lo, C-H. and Smith, I.E. 2005. Formation and emplacement of the Northland Ophiolite, northern New Zealand, SW Pacific tectonic implications. *Journal of the Geological Society of London* 162: 225–241.

White, P.J. and Waterhouse, B.C. 1993. Lithostratigraphy of the Te Kuiti Group: A revision. *New Zealand Journal of Geology and Geophysics* 36: 255–266.

Wilgus, C.K., Hastings, B.S., Posamentier, H., Van Wagoner, J., Ross, C.A. and Kendall, C.G.St.C. (eds) 1988. *Sea-Level Changes: An integrated approach.* SEPM Special Publication 42.

Williams, J.G. 1978. Eglinton Volcanics: stratigraphy, petrography, and metamorphism. *New Zealand Journal of Geology and Geophysics* 21: 713–732.

Williams, J.G. and Smith, I.E.M. 1979. Geochemical evidence for paired arcs in the Permian volcanics of southern New Zealand. *Contributions to Mineralogy and Petrology* 68: 285–291.

Willis, I. 1965. Stratigraphy and structure of the Devonian strata at Baton River, New Zealand. *New Zealand Journal of Geology and Geophysics* 8: 35–48.

Wilson, B.T. 1994. Sedimentology of the Miocene succession (coastal section), eastern Taranaki Basin margin: Sequence stratigraphic interpretation. Unpublished MSc thesis, University of Waikato.

Wilson, C.J.N., Houghton, B.F., McWilliams, M.O., Lanphere, M.A., Weaver, S.D. and Briggs, R.M. 1995. Volcanic and structural evolution of Taupo Volcanic Zone, New Zealand: A review. *Journal of Volcanology and Geothermal Research* 68: 1–28.

Wilson, C.J.N., Gravley, D.M., Leonard, G.S. and Rowland, J.V. 2009. Volcanism in the central Taupo Volcanic Zone: Tempo, styles and controls. In: T. Thordarson, S. Self, G. Larson, S.K. Rowlands and A. Huskuldsson (eds). *Studies in Volcanology: The legacy of George Walker.* Special Publication of IAVCEI, Volume 2. Geological Society of London, pp. 225–247.

Wilson, D.D. 1956. The late Cretaceous and early Tertiary transgression in South Island New Zealand. *New Zealand Journal of Science and Technology. B. General Section* 37(5): 610–622.

Wilson, D.D. 1963. *Geology of the Waipara Subdivision* (Amberley and Motunau sheets S68 & 69). New Zealand Geological Survey Bulletin 64. 1 sheet & 122 pp.

Wilson, G.J. 1976. Notes from the New Zealand Geological Survey – 9. Late Cretaceous (Senonian) dinoflagellate cysts from the Kahuitara Tuff, Chatham Islands. *New Zealand Journal of Geology and Geophysics* 19: 127–132. doi:10.1080/00288306.1976.10423553

Wilson, G.J. 1991. *Revised Ages for Some Northland Cretaceous Localities*. GNS internal report, Lower Hutt. 1 p.

Wilson, G.J., Morgans, H.E.and Moore, P.R. 1989. Cretaceous–Tertiary boundary at Tawanui, southern Hawkes Bay, New Zealand. *New Zealand Geological Survey Record* 40: 29–40.

Wilson, G.J., Schiøler, P., Hiller, N. and Jones, C.M. 2005. Age and provenance of Cretaceous marine reptiles from the South Island and Chatham Islands, New Zealand. *New Zealand Journal of Geology and Geophysics* 48: 377–387.

Wizevich, M.C. 1994. Sedimentary evolution of the onshore Pakawau Subbasin: Rift sediments of the Taranaki Basin deposited during Tasman Sea spreading. In: G.J. van der Lingen, K.M. Swanson and R.J. Muir (eds). *Evolution of the Tasman Sea Basin*. Balkema, Rotterdam, pp. 83–104.

Wolbach, W.S., Gilmour, I., Anders, E., Orth, C.J. and Brooks, R.R. 1988. Global fire at the Cretaceous–Tertiary boundary. *Nature* 334: 665–669.

Wood, B.L. 1956. *The Geology of Gore Subdivision*. New Zealand Geological Survey Bulletin 53. 128 pp.

Wood, B.L. 1966. *Sheet 24 Invercargill*. Geological Maps of New Zealand 1:250,000. Department of Scientific and Industrial Research, Wellington.

Wood, B.L. 1969. *Geology of Tuatapere Subdivision, Western Southland* (sheets S167, S175). New Zealand Geological Survey Bulletin 79. 161 pp.

Wood, R.A. and Ingham, C.E. 1989. *Chatham Islands Refraction Survey*. New Zealand Geological Survey Report G 48. 16 pp.

Wood, R.A., Andrews, P.B., Herzer, R.H. *et al.* 1989. *Cretaceous and Cenozoic Geology of the Chatham Rise Region, South Island, New Zealand*. New Zealand Geological Survey Basin Studies 3. New Zealand Geological Survey, Lower Hutt.

Wood, R.A., Lamarche, G., Herzer, R., Delteil, J. and Davy, B. 1996. Paleogene seafloor spreading in the southeast Tasman Sea. *Tectonics* 15: 966–975.

Wood, R.A., Herzer, R.H., Sutherland, R. and Melhuish, A. 2000. Cretaceous–Tertiary tectonic history of the Fiordland margin, New Zealand. *New Zealand Journal of Geology and Geophysics* 43: 289–302.

Wright, I.C. 1986. Paleomagnetic studies of the Late Miocene Mangapoike River section, northern Hawke's Bay, New Zealand. Unpublished PhD thesis, Victoria University of Wellington.

Wright, J.B. 1968. Contributions to the volcanic succession and petrology of the Auckland Islands, New Zealand, III: Minor intrusives on the Ross Volcano. *Transactions of the Royal Society of New Zealand* 6: 1–11.

Wright, J.B. 1970. Contributions to the volcanic succession and petrology of the Auckland Islands, New Zealand, IV: Chemical analyses from the lower half of the Ross Volcano. *Transactions of the Royal Society of New Zealand* 8: 109–115.

Wright, J.B. 1971. Contributions to the volcanic succession and petrology of the Auckland Islands, New Zealand, V: Chemical analyses from upper parts of the Ross Volcano, including the minor intrusions. *Journal of the Royal Society of New Zealand* 1: 175–183.

Youngson, J.H., Craw, D., Landis, C.A. and Schmitt, K.R. 1998. Redefinition and interpretation of late Miocene–Pleistocene terrestrial stratigraphy, Central Otago, New Zealand. *New Zealand Journal of Geology and Geophysics* 41: 51–68.

Location Index

Bold page numbers indicate maps and diagrams. Colour photographs have individual plate numbers. The term *passim* indicates that topics are mentioned discontinuously in a page range.

Akaroa 227, 229
Antarctic Peninsula 40
Antarctica 20, 21, 41, 42, 88, 140, 157
  in Gondwana break-up 48–49
Antipodes Islands 37, 228–29
Aorangi Range 214, 215
Auckland area 43, 152
  south Auckland 28, 31, 36, 38, 153, 187, 192, 209, 211, 225, 232, 243
  volcanism 38, 231–32
Auckland Islands 227, 228–29
Aupouri Peninsula 224
Australia 40
  in Gondwana break-up 21, 22, 48, 85–86
  Cenomanian event 21, 237
Awakino area 153, 185, 191
Awakino Gorge 137, 190
Awatere River/Valley *Plates 4b, 5b*, 72, 73, 74, 83–84, 174, 198, 237

Banks Peninsula 23, 33, 84, 105, 175, 199, 216, 227, 237
Bay of Islands 231
Bay of Plenty 38, 59–60, 225, 232
Benneydale 137
Blackmount (Southland) 126, 178
Blenheim area 215
Blue Mountain Stream (Marlborough) 138
Boatmans Harbour (Oamaru) *Plate 11b*, 139
Branch Stream (Marlborough) 234
Broken River (Canterbury) 105, 199
Brynderwyn Hills (Northland) 136
Buller district *Plates 5a, 6a & b, 13a & b*, 192
Buller Gorge 63, 65, 148
Buller River and Valley 129, **131**

Campbell Island 41, 94, 101, 117, 120, 122, 143, 200, 227, 228–29, 235, 240
Canterbury region **101**, **102**
  Jurassic 47, 48
  Cretaceous 21, 66, 67, 92, 94, 97, 100, 104, 239
  Paleocene 23, 25, 105, 121
  Eocene 25–26, 120–21
  Oligocene 29, 120, 141–43, 155, 156–57, 240–41
  Miocene 32–33, 56, 163, 174–75, 198–200, 227
  Pliocene–Pleistocene 37, 211, 212, 215–17, 228, 242–43
  volcanism 29, 37, 84, 121, 228, 237
Cape Foulwind 130–31
Cape Kidnappers *Plate 16b*, 214
Cape Maria van Diemen 224
Cape Palliser 84, 115, 197, 237, 239
Cape Reinga 188, 209
Cape Turnagain 196, 214
Castle Hill (Canterbury) *Plate 11a*, 155
Castlepoint area *Plate 14a*, 83, 169, 196, 214
central North Island 28, 46, 51, 164, 232
Chalky Island (Southland) 127
Chancet Rocks 233–34
Charleston (Westland) 130, 149
Chatham Islands **61**, 97, 105, 115
  Cretaceous 54–55, **59**, 61–62, 76, 94
  Paleogene (Paleocene, Eocene, Oligocene) 25, 26, 117, 120, 121, 138–39, 142, 240, 241
  Neogene (Miocene–Pliocene) 32, 37, 175, 199, 216, 228
  volcanism 138–39, 228, 229
Cheviot 32, 174
Clarence Valley *Plate 15a*, 70, 71–72, 74, 80–81, 84, 97, 104, 107, 116, 119, 138, 155, 234, 237, 239
Clifden area 178
Cook Strait 38, 227
Coromandel Peninsula 27, 36, 152, 187, 230–31
Coromandel Range 227
Coverham (Marlborough) 71–72, 73, 74, 81, 84

Dannevirke 172
Doubtless Bay 209
Dunedin area 33, 106, 122, 175, 200, 227, 235, 238

East Cape 193, 212
Eglinton Valley 27, 126, 146
Eketahuna area 196
Elsthorpe 214

Fairfield Quarry (Dunedin) 235
Fiordland
  Cambrian 41
  Cretaceous 42, 51, 66, 87, 109–10
  Paleogene (Paleocene, Eocene, Oligocene) 29–30, 147, 157

Miocene 33, 34, 164
Pliocene–Pleistocene 218–19
Five Fingers Peninsula 218–19
Flat Point (Wairarapa) 173
Flaxbourne River 233–34
Forest Burn (Southland) 202
Foveaux Strait 219

Geraldine area 228
Gisborne area *Plates 7a & b*, *8a*, 169, 193, 194, **194**, 212–13
Glenburn (Wairarapa) 74, 75, 81, 103–04, 238
Glentunnel area (Canterbury) 227
Golden Bay 133, 203, 205
Goldie Hill (Clifden area) 178
Gore area 217
Gore Bay **101**, 199
Governors Bay 33, 34, 199
Great Barrier Island 36, 230, 231
Grey Valley **129**, 163
Greymouth area 27, 35, 131, 234, 241

Hamilton area 153, 189, 225, 231
Hapuku River/Valley *Plate 8b*, 72, 74
Harper Hills (Canterbury) 227
Hastings area 169, 172
Hauhungaroa Range 137, 189
Hauraki Gulf 230
Hauraki Plain/lowland 38, 225
Hawai River 59–60
Hawera area 226
Hawke's Bay
Cretaceous 21, **59**, 71, 73, 91, 238
K–T boundary 234–35
Paleogene (Paleocene, Eocene, Oligocene) **90–91**, 107, 118
Miocene 56, 169–72, **170**, 192, 194–96
Pliocene–Pleistocene 37, 38, **194–95**, 213–14
Hawkswood Range 216
Helmet Hill (Clifden area) 178
Herangi Range 137, 189, 191
Hicks Bay 83
Hinemahanga Rocks (Hawke's Bay) 75
Hohonu Range 182
Hokianga 78, 188
Hokitika 65, 130
Hollyford Valley 27
Hump Ridge 30, 34, 124, 126, 145, 180–81, 201
Huntly–Rotowaro area 137
Hunua Ranges 137, 187, 232
Hurunui River *Plate 4a*, 117, 142

Inangahua 34, 204

Jackson Bay 220

Kaikohe 79, 231
Kaikoura area *Plates 12a*, *14b*, 104, 198, 216
Kaikoura Range 84, 156, 237
Kaimanawa Range 210–11
Kaipara 27, 36, 95, 115, 135, 187–88, 224
Kaitangata 24
Kaiwhata Stream (Wairarapa) 115
Kaketu Range 137
Kapuni 221
Karamea area 35, 139, 148, 205, 221
Karikari Peninsula 69, 99, 224
Katiki beach 106
Kaweka Range 214
Kawhia area 31, 65, 115, 152, 153
Kekerengu Valley 81, 97
King Country 137, 191
Kokopumatara Stream (Gisborne district) *Plate 8a*, 60, 71, 73
Koranga *Plate 7a & b*, 71
Kyeburn (Otago) 33, 77

Lake Hauroko 34, 37
Lake Te Anau 126
Lake Waikaremoana 32, 168, 169
Little Wanganui River 148
Longwood Range 30, 144, 176, 178, 217
Lyttelton 33, 199, 227, 228, 229

Mahia Peninsula 75, 167, 168, 172, 195, 212, 213
Makara (Wairarapa) 169, 214, 215
Manawatu Gorge 37
Mangaotane Stream 80
Manukau Harbour and lowland 224–25, 232
Marie Byrd Land 40, 41, 49, 237
Marlborough 44, 47, 51, 62, 99, 159–60
Cretaceous 21, 22, 48, **59**, 69–70, 71–72, 73–74, 80–81, 83–84, 91–92, 94, 96–97, 100, 236, 238, 239
K–T boundary 234
Late Cretaceous–Paleocene 23, 104
Paleogene (Paleocene, Eocene, Oligocene) 25, 29, 54, **90–91**, 107, 116, 119, 138, 140, 155
Miocene 32, 140, 162, 163, 168, **170–71**, 173–74, 198
Pliocene–Pleistocene 37, **197**, 212, 215
Pleistocene–Holocene 164
Marlborough Sounds 227
Mason Ridge (Hawke's Bay) 214
Masterton area 173, 196–97, 214, 215
Mata River 80
Matawai *Plates 7a & b*, *8a*, 60, 167
Maui field 133, 151, 185, 207, 221
Mead Stream (Marlborough) 107, 119, 234
Medway River 198
Milford Sound 203

Milton 176
Moeraki beach 106
Mohaka (Hawke's Bay) 172, 194–95
Moki oil field 207
Mokohinau Islands 231
Moody Creek (West Coast) 233–34
Mororimu Stream (Marlborough) 73, 74
Motu Falls 71, 100
Motunau (Canterbury) 142, 216
Mount Aorangi 75
Mount Somers 67, 155
Mount Taitai 75
Mount Wharekia 75
Moutohora (Bay of Plenty) 60
Murchison area 35, 130

Napier 172, 213
Naseby 142
Needles Point 233–34
Nelson
  Cretaceous 78, 239
  Paleogene (Paleocene, Eocene, Oligocene) 27, 114, 130, 132, 149, 150–51, 157
  Neogene (Miocene, Pliocene) 34–35, 182, 184, 185, 205–06, 221
  Pleistocene–Holocene 164
New England 40, 42
Ngawha geothermal field 231
North Cape 82–83, 155, 188, 192
North Island, central 28, 46, 51, 164, 232
Northland 46, 47, 99, 166–67
  Cretaceous 20, 22, 23, 43
    Clarence Series rocks 82–83
  Paleogene (Paleocene–Oligocene) 23, 31
  Miocene 192, 209
  Pliocene–Pleistocene 211, 243
  marine successions 62, 69, 78–79, 95, 152
  terranes 46, 47
  volcanism 192, 237
Northland Peninsula **135**, **136**, **159**
  Cretaceous 95
  Paleogene (Paleocene, Eocene, Oligocene) 25, 27, 28, 35, 137–38
  Miocene 35–36, 158–59, 161–62, 192, 208, 229–31
  Pliocene–Pleistocene 38, 224

Oamaru area 29, 33, 120, 122, 139, 142, 155, 200, 241
Ohai area (Southland) 126, 144, 178
Oponae (Bay of Plenty) 70–71, 72, 79
Otago 44, 45, 47
  Cretaceous 22, 48, 51, 54, 66–67, 77, 106, 239
    K–T boundary 235
  Cenozoic rocks **98**, 228
  Paleogene (Paleocene, Eocene, Oligocene) 24, 94, 116, 120, 122, 123, 139, 142, 143, 157, 240
  Neogene (Miocene–Pliocene) 33, 37, 143, 175–76, 200, 212, 217, 227, 228, 243
Ouse Stream (Marlborough) 71, 81
Owen Valley 183
Oxford (Canterbury) 155, 227

Paoanui Point 196
Papahaua Range 131, 192, 205
Paparoa Range 35, 63–65, 131, 182, 192, 204–05
Parapara Peak, Takaka 42
Paringa (Westland) 65, 78, 110, 128
Penk River *Plates 4b, 5b,* 72
Picton 150
Piopio 190
Pirongia Mountain 188
Pitt Island 61, **61**, 76, 77, 78, 82, 97, 105, 115, 121, 138, 228
Pongaroa (Wairarapa) 172, 173
Port Craig 126, 201
Port Waikato 224
Puketoi Range 214–15
Punakaiki area 34, 130, 148, 149, 163, 184, 220, 241
Puysegur Point 127

Quail Island 228
Queensland 42

Raglan area 153, 154
Rangitikei 211
Rangitoto Range 189
Raukumara Peninsula 167
  Cretaceous 20–21, 48, 51, 54, 55, 58–61, **58**, 69–71, 72–73, 74–75, 79–80, 83, 91, 94, 95–96, 97, 100, 236, 238
  Paleogene **90–91**
    Oligocene 141
  Miocene 32, 162, 168–169, **170**, 193–94
  Pliocene–Pleistocene **194**, 212, 213
Red Island (Hawke's Bay) 75
Reefton 65, 130
Rimutaka Range 46
Ring Creek (Marlborough) 198
Ross area (West Coast) 220
Ross Sea 140
Rowallan Forest 218
Ruahine Range 37, 38, 210, 213, 214, 226, 227
Ruakawa Range 214–15
Ruatoria 80, 169

Seymour Stream (Marlborough) 74
Shag Point (Otago) 77, 97, 101, 175
Silverdale (Northland) 78, 166
Simpsons Rock 231
The Sisters islands 76
Solander Islands 37, 181, 219, 229

South Taranaki Bight 222
Southern Alps 32, 36–37, 56, 164, 211, 212, 215, 217, 221, 243
Southland
Permian 43
Cretaceous 54, 239
Paleogene (Paleocene, Eocene, Oligocene) 28, 29, 56, 116, 143–48
Neogene (Miocene–Pliocene) 33–34, 56, 192, 200–02, 243
Pliocene–Pleistocene 37, 212, 217, 218
basins 123–27, **124–25**, 140, 144–48, 200–02
Stewart Island 42, 44, 181

Taieri River 77
Takaka area 184
Takitimu Mountains 30, 33, 34, 126, 145, 146, 176, 202
Tapuaeroa Valley 75, 83
Taranaki 56, 87–88, **134**, 165
volcanism in 35, 36, 192, 206–207
Taranaki Bight 88
Taranaki Peninsula 133, 150, 184, 185, 206–07, 210–11, 221–23
Tararua Range 38, 215
Tasman Bay 203, 205, 221
Tasman Sea 240
Cretaceous 22–23, 39, 49–50, 54–55, 85–86, 89, 108, 237, 240
Paleocene–Eocene 25, 56, 117, 156, 240
Miocene 164
Tasmania 237, 242
Taumarunui 36, 153, 189
Tawanui (Hawke's Bay) 107, 234–35
Te Anau area 177
Te Hoe River area 234–35
Te Kiwikiwi Hill 83
Te Kuiti area 31, 153, 190, 209
Te Mata Peak 213
Te Waewae Bay 37, 178
Three Kings Islands 136, 209, 224
Timaru area 228
Tinui (Wairarapa) 73, 172, 173
Tirua Point 206
Tokomaru Bay area 193
Tolaga Bay area 193, 212
Tora (Wairarapa) 71, 100, 103, 234, 239

Victoria Land (Antarctica) 40, 41, 157
Victoria Range (West Coast) 182
Vinegar Hill (Northland) 231

Waiau River area 126, 178
Waiheke Island 231
Waihopai River 198
Waikaremoana 212
Waikato 225
Waipa lowlands 225
Waipaoa Valley 213
Waipara area (Canterbury) 199
Waipara River 105, 121, 233–34
Waipukurau area 215
Wairarapa 51, 62, 167, 192
Cretaceous 21, 23, 58, **59**, 69, 71, 73, 74, 75, 81–82, 83, 91, 95, 96, 97, 99, 100, 115, 238, 239
K–T boundary 234
Paleogene (Paleocene, Eocene, Oligocene) 23, 25, **90–91**, 103, 118, 119–20, 140
Miocene 32, 56, 163, 168, 169–73, **170–71**, 196–98
Pliocene–Pleistocene **194–95**, **197**, 214–15
Wairoa 169
Waitaki Valley 217
Waitomo 190
Waitotara River area 226
Wakatipu area 178
West Coast (South Island) 37, 165
Cretaceous 21, 22, 54, 63–65, 77–78, 84, 237, 238–39
Paleogene (Paleocene, Eocene, Oligocene) 24–28 *passim*, 27, 28, 31, 56, 110, 116, 148–49, 240
Neogene (Miocene–Pliocene) 182–184, 192, 203–05, 243
see also Westland
Westland
Cretaceous and Cenozoic rocks **111**
mid-Cretaceous 48
Late Cretaceous–Paleocene 22, 56, 110–11, 116, 238, 239
Paleogene (Paleocene, Eocene, Oligocene) 117, 128, 139, 157
Miocene 34, 163–64, 192, 203–04
Pliocene–Pleistocene 219
Westport 27, 131, 139, 220
Whakatane 46, 70
Whangai Range 71, 73, 99
Whangarei area 38, 137, 152, 187, 231
Whangaroa Harbour 69, 78–79, 95
Whatuwhiwhi (Karikari) 69
White Island 232
Whitecliffs (Buller) 148
Woodside Creek (Marlborough) 119, 138, 233–34
Woodville area 214, 215

General Index

Bold page numbers indicate maps and diagrams. Colour photographs have individual plate numbers. Geological time units and main general topic entries are in **bold**. The term *passim* indicates that topics are mentioned discontinuously in a page range. For modern geographic place names, see the Location Index.

Abbey Formation 128
Abbotsford Formation 122, 235
Adams–Tinui Fault 172, 173, 239
Adare Trough 157
Albany Conglomerate 187
Albian Age *Plate 8a*, 47, 51, 54–55
  Albian Unconformity 51, 54–55
  basin evolution 20–21, 58–63, 69–76
  igneous rocks 82–83
  see also Clarence Series
Albian Basin 21
Alexandra Volcanic Group 232
Alma Group 155
Alpine Fault **19**, 50, 87, 126, 157, 163, 164, 177, 183, 212, 220–21
  proto-Alpine Fault 88, 144
Altonian Stage 168, 169, 172, 174–78, 182–83, 185–86, 191
Amuri Limestone *Plates 12a, 15b*, 25–26, 29, 107, 118–21, 142, 143, 155, 173, 174, 198, 240–41
Anatoki Fault 40
Annick Group 126–27
Antarctic Circumpolar Current (ACC) 29, 140–41, 242
Aotea Sandstone 153
Aptian Age 51, 54, 58–60, 70
Arahura-1 well 112
Ararimu-1 well 190
Ariki Formation 208
Ariki-1 well 108, 113, 117, 134, 208
Arnott Basalt 110, 116, 139
Ashley Mudstone 121, 122, 142
Atea Sandstone 214
Australian plate 87, 164
Australian–Pacific instantaneous pole 211, 243
Australian–Pacific plate boundary 28, 31, 56, 87, **88**, 156, 157, 159, 163, 165, 192
Awapapa Limestone *Plate 16a*
Awapoko facies 78
Awarua Limestone 182, 203
Awatere Fault 162
Awatere Group 198
Awhea Formation 103
Awhitu Group 224–25

Balleny Basin 26–27, 30, 34, 37, 66, 123–24, **125**, 127, 144, 147–48, 177, 180–82, 202–03, 219, 229
Balleny Group 128
Bannockburn Formation 176, 200
Barque prospect 116
basement terranes *Plate 1*, 40–47, 57
basin evolution
  Cretaceous 21, 57–74, 85–86
  Paleocene 23–25
  Eocene 25–28
  Oligocene 28–31
  Miocene 32–36
  Pliocene–Pleistocene 37–38
Beaumont Formation 126
Beebys Conglomerate 65, 238
Bellmount Fault 145, 201
Bells Creek Mudstone 197–98
Benioff zone 160
Birch Fault 239
Blackmount Fault 124, 145, 201
Blackmount Formation 145
Blackmount Sub-basin 126, 145
Blue Spur Conglomerate 106
Bluff Basalt 227
Bluff Sandstone Formation *Plate 8b*, 74, 84
Borland Formation 179, 180
Bounty Channel System 123
Bounty Trough 26, 49, 76, 85, 95
Bourne Conglomerate 216
Bradley Sandstone 199
Branch Sandstone 104
Brechin Formation 175, 199
Brighton Formation 106
Broken River Formation 104, 105
Brook Street Terrane 43–44, 48
Brothers Basalt 155
Brunner Coal Measures *Plate 10a & b*, 27, 112, 117, 128, 130, 131, 240
Buckland Granite 49
Buller Coalfield *Plate 10a*, 129–31
Buller Gorge Basin 65
Buller Terrane 20, 40–41
Burmeister Formation 184
Burnside Mudstone 122
Burnt Creek Formation 81, 96

Burwood Sub-basin 34, 145–46, 177, 179–80
Buttress Point Conglomerate 84, 237

Callaghans Greensand Member 220
Cambrian period 41
Campanian Age 22, 97
late Campanian–early Maastrichtian event 22, 238–40
Campanian–Paleocene Succession 98–115
Campbell Plateau 20, 23, 41, 87, 94–95, 101, 120, 157, 240
Canterbury Basin **92**, 100–01, 106
Cretaceous 22, 23, 95, 236, 238
K–T boundary 235
Paleogene (Paleocene, Eocene, Oligocene) 24, 25, 29, 54, 117, 118, 119, 120, 122, 138, 140, 141–42, 155, 240–41
Neogene (Miocene–Pliocene) 32, 173, 174–76, 198–200, 215–17
sedimentary deposition *Plate 9*, **67**, 77, 89, 92, 94, 141–42, 198–200
volcanism in 115–16, 138, 155, 227–28
Canterbury Bight High 29
Cape Egmont Fault Zone 57, 226
Cape Foulwind Fault Zone 116, 203, 205
Caples Terrane 33, 43, 45, 47–48, 77, 183, 202
Caples–Haast schist 201
Carboniferous period 41, 42, 45, 46
Carnley volcano 229
Caroline Basin 67
Castle Hill Fault 106
Cavalli core complex 166
Caversham Formation 175
Cenomanian Age 83–84, 236–38
basin evolution 21, 62–63, 65–67, 72–75, 78–81
see also Clarence Series
Cenozoic Era 50, **98**, **101**, **102**, 123, **129**, **131**, **136**
Cretaceous–Cenozoic megasequence **53**, 54
tectonism 72, 156–64
see also individual periods and epochs
Central Graben (Taranaki Basin) 222, 225
Challenger Plateau 22, 24, 87, 94, 155, 240
Challenger Rift System 87–88
Champagne Formation 72
Charteris Bay Sandstone 105
Chatham Rise
Cretaceous 20–22, **20**, 61–62, 76, 82, 92, 94–95
Late Cretaceous–Paleocene 105–06, 117
Paleogene (Paleocene, Eocene, Oligocene) 23–24, 25–26, 29, 120, 121, 140, 142, 174, 241
Neogene 227
Miocene 32–33, 159, 162, 174–75, 199
Pliocene 37, 215, 216
Pleistocene–Holocene 164
Chatham Schist 105
Chatton Formation 143, 177–78
Church Volcanics 228
Clarence Fault 162
Clarence Series 62–77, 82–84
Clarendon Sandstone 176
Clent Hills Group 48, 62
Clifden Subgroup 178
climate changes 23, 25, 140–41, 223, 225
Clipper Basin 24, 67, 77, 106
Sub-basin 238
Clipper Formation 77, 238
Clipper-1 well 97, 100, 106, 107, 116, 120, 122, 142, 143, 175, 217, 235, 238
Coalgate Bentonite 199
Cobden Limestone 183
Colville Ridge 159, 230, 231
Colville Trench 164
Coniacian Age 21–22, 55–56, 79, 80–81, 84, 89–92, 94–98
Coniacian–Maastrichtian basin evolution 21–23
Conway Formation 104–05
Cookson Volcanics 155, 173
Coromandel Group 231
Coromandel Volcanic Zone 230–31
Coromandel Volcanics 230
Coverham Group 72, 73
Cretaceous period *Plates 4b, 5a & b, 6a & b, 7a, 8a*, 51–56, **52**, 236, 237–39
Cretaceous–Cenozoic megasequence **53**, 54
Late Cretaceous–Paleogene phase 89–114
New Zealand region formation 17–23, **20**, 39–50, **50**, 94
sedimentary basin evolution 57–76, **58–59**, **62**, **64**, **68**, **78**
see also Albian, Aptian, Campanian, Cenomanian, Coniacian, Maastrichtian, Santonian, Turonian ages; Clarence Series, Raukumara Series
Cretaceous–Cenozoic Project (CCP) 11, 12, 14, 15
Cretaceous–Tertiary (K–T) boundary 23, 99, 233–36

Danseys Pass Fault 66
Deborah Volcanics 122, 155
Deep Sea Drilling Project (DSDP) 13, 241
site 206 27, 113, 135–36, 235
site 207 101, 135–36
site 208 135–36
site 275 101
site 277 120
site 593 155
Devils Knob Formation 220–21
Devonian period 40–41, 42
Diamond Harbour Volcanic Group 227
Dowling Bay Limestone 176
Drumduan Terrane 42

Dun Mountain Ophiolite 44
Duncraigen Formation 201–02
Dunedin Volcanic Group 33, 176, 200, 228
Dunollie Coal Measure Member 112
Dunstan Formation 176
Dunton Formation 180
Dunton Sub-basin 146, 179
d'Urville Sub-basin 114

Earl Mountains Sandstone 127
East Coast Allochthon 32, 56, 59, 62, 69, 74, 95, 97, 99, 118, 141, 159, 162, 166–68
East Coast Basin
 Cretaceous 20–22, 51, 54, 55–58, 69–75, 79–84, 86, 89, 91–92, 95, 103, 236, 238
  K–T boundary 234
 Paleogene (Paleocene, Eocene, Oligocene) 23, 25, 29, **90–91**, 107, 116, 117, 118, 140, 155
 Neogene (Miocene–Pliocene) 168–74, 192, 193–98, 243
 Pliocene–Pleistocene 37, 212–15, 243
 structural sub-belts 62, **62**, 95–97
East Southland Group 143
Eastern Mobile Belt 160
Eastern Province 17, 20, 40, 43–47, 57
Eastern Southland Basin 24, 176, 177
Eight Mile Formation 204, 220
Emerald Basin 28, 89
Endeavour High 29, 32–33, 142, 175
Endeavour-1 well 97, 100–101, 106, 107, 116, 122, 175, 199, 235
Eocene Epoch 51
 basin evolution 25–28, 56, 88
 marine transgression 25–28, 118–19, 123, 128, 130–37
 sedimentation 118–38, 240
 tectonism 124, 126
 volcanism in 138–39
Eocene–Oligocene boundary event 139–44
Esk Head Mélange *Plate 4a*, 45, 46
eustatic sea-level changes 29, 52, 199, 212, 222, 225, 226, 233, 235–36, 239, 242–43
extinctions 14, 23, 140

Farewell Formation 114, 117, 132–33
Feary Greensand 141
Fells Greensand 119
Flags Creek Thrust 163
Forest Hill Formation 177–78
fossils and traces 46, 96, 101, 104, 105–06, 110, 120, 121, 152, 154, 169, 180, 218–19
 bivalves 14, 44, 46, 47, 70, 71, 80, 107
 dinoflagellates 77, 81, 96, 99, 105, 107, 110
 foraminifera 105, 107, 110, 116, 138, 141, 155
 gastropods 107
 marine reptiles 105
 mollusca 110, 116, 216
 plants 43, 46, 77
 pollen 133
 theropods 105–06
French Creek Granite 95

Galleon-1 well 97, 100–101, 106, 116, 120, 122, 175, 199, 235
Garden Cove Formation 101
Gentle Annie Formation 71, 73
geological time scale 18
Giant Foresets Formation 223, 243
Glen Massey Sandstone 153
Glenburn Formation 75, 81–82, 97, 115
Golden Bay Group 41
Goldlight Coal Measure Member 112
Gondwana 17, **17**, 39–50, 54, 89–90
Goodwood Limestone 175–76
Gore Lignite Measures 143, 177, 217
Gore Piedmont Gravels 217
Governors Bay Volcanics 227
Grasseed Volcanics 119, 138
Great Marlborough Conglomerate *Plate 15a & b*, 32, 174
Great South Basin 41, 49, **67**, **68**, 85, 101, 115–16, 163
 Cretaceous 54, 56, 63, 67, 77, 89, 92–95, 98, 106, 107, 238
  Campanian–Maastrictian 22, 115–16
  K–T boundary 235
 Paleogene (Paleocene, Eocene, Oligocene) 23–26 *passim*, 54, 108, 118, 120, 122–23, 140, 143, 240
 Miocene 33, 164, 175, 176, 200
 Pliocene–Pleistocene 37, 217–18
Grebe Shear Zone 49
Greenland Group 41, 111, 204, 220
Greta Formation 199, 216
Greville Basin 110
Grey Valley Trough 34–35, 110, 111, 130, 182–83, 204, 220
Grey–Inangahua Depression 164, 182, 220, 221, 243
Greymouth Coalfield 111–12, 117, 129, 130
Gridiron Volcanics Formation 74, 237

Haast Schist 44–45, 48, 66, 220
Haerenga Supergroup 52
Haku-1 well 184, 204
Hamilton Basin 225
Hampden Formation 122
Hapuku Group 80
Harihari-1 well 182, 204, 220
Harper Hills Basalt 227
Haumurian–Teurian succession 98–115
Hauraki Rift 231
Hauraki Volcanic Region 230

Hauroko Fault 30, 34, 124, 128, 145, 147, 180–81, 203, 219–20
Hautere Sub-basin 30, 66, 127–28, 147, 180
Havre Trough 159
Hawkdun Group 200, 217, 243
Hawke Bay-1 well 196
Hawks Crag Breccia *Plate 6a & b*, 63, 65
Haycocks Formation 180
Hematite Petroleum 184
Henley Basin 85
Henley Breccia 77, 106, 238
Herangi High 28, 31, 152, 154, 156, 189, 241, 242
Herbert Formation 106
Herring Formation 99
Hikurangi margin 159–60, 162, 164
Hikurangi Plateau 20, **20**, 39, 48, 50, 85, 100, 140, 160, 235
Hikurangi subduction system 166
Hikurangi Trench 196
Hikurangi Trough 216
Hogburn Formation 123
Hohonu Conglomerate Member 182
Hohonu-1 well 183
Hoiho Group/Sequence 67, 77, 98
Hoiho-1 wells 67, 143
Hokianga Basin 186
Hollyford Fault 30, 146, 218
Homebush Sandstone 121
Hope Fault 162
Houhora Complex 47, 69
Hukerenui Mudstone 79, 95
Hump Ridge Formation 126, 144–45
Hump Ridge–Mid Bay High 34
Hump Ridge–Stewart Island High 27
Hump Ridge–Stewart Island Shelf 34, 180, 202, 219
Hump Ridge–Stewart Island Thrust System 34, 201
Hunua Facies 47
Hurunui High 23–24, 25, 104
Hurupi Formation 197

igneous rocks
  pre-mid-Cretaceous 40, 46
  mid-Cretaceous (Korangan; Clarence Series) 60, 82–84
  Late Cretaceous (Cenomanian–Maastrichtian) 21, 84, 115–16
  Paleocene 115–16
  Oligocene 154–55
Inangahua Depression 182
Inangahua Formation 182–83
Iron Creek Greensand 121
Iselin Bank 86
Island Sandstone 130

Jackson Formation 164, 204
JD George-1 well 84, 200, 237
Jurassic period *Plate 2a*, 41–45, 47–48, 51, 62, 65, 133

Kaawa Formation 225
Kahuitara Tuff 76, 97, 105, 115
Kaiata facies 148
Kaiata Formation 130–32, 146, 148
Kaikoura Canyon 216
Kaikoura Orogeny 56, 141, 156–63, 209
Kaikoura Sequence/Synthem 54
Kaimiro Formation 133
Kaipara Basin 38, 186, 224
Kaipuke Siltstone 185
Kaitangata Coalfield 106
Kaiwhata Limestone 120
Kakahu Siltstone 122
Kakepuku volcano 232
Kamo Coal Measures 136, 137
Kapitean–Opoitian boundary event 211–12, 242–43
Kapuni field 206
Kapuni Group 114, 115, 132–34, 133–34, 235
Karamea Batholith 41
Karamea Limestone *Plate 13a & b*
Karaumu Sandstone 196
Karekare Formation *Plate 8a*, 71, 72–73, 79, 80
Karewa Group 216, 228
Karioi volcano 232
Karioitahi Group 225
Katiki Formation 97–98, 100–01, 106, 107, 235
Kauru Formation 122
Kawau Subgroup 188
Kaweka Terrane *Plate 4a*, 43, 46, 47, 57
Kekenodon Group 143
Kerikeri Volcanics 231
Kermadec subduction zone 159
Kidnappers Group *Plate 16b*, 214
King Country Basin 28, 31, 35, 36, 38, 118, 136, 150, 153, 185, **189**, 190, 191, 192, 209–11, 211, 225–26, 241, 243
Kiore Formation 210, 211
Kiore-1 well 190, 210
Kirks Breccia 99–100
Kiwitahi Volcanic Zone 230
Kiwitahi Volcanics 230, 231
Knife and Steel Formation 181
Koiterangi Hill (Westland) 111
Kokoamu Greensand 142, 240
Kongahu Fault 131
Koranga Formation *Plate 7a*, 59–60, 70
Kotuku Dome 111
Kowai Formation 216–17
Kowai-1 well 105
K–T boundary *see* Cretaceous–Tertiary (K–T) boundary
Kupe South wells 24, 206

Kupe-1 well 206
Kuriwao Group 43
Kyeburn Basin 66, 85
Kyeburn Formation 66–67, 77

Lachlan Fold Belt (Australia) 40
Lagoon Stream Formation 104
Laing Formation 122
Lau–Colville Ridge 192, 230
Leeston-1 well 199
Letham Shear Zone 48
Lillburnian Stage 169, 173–74, 182
Little Totara Sand 130
Livingstone Fault 33, 44
Loburn Formation 105
London Hill Fault 100
Longford Formation 183, 205, 220
Longwood High 30
Lookout Volcanics Formation 74, 174, 237
Lord Howe Rise 21, 22, 23, 86, 87, 101–03, 135–36, 139, 239
Lower Awakino Limestone 190
Lower Buller Fault 205
Lower Limestone (Amuri) 119
Lower Marl (Amuri) 119
Lyttelton volcano 227, 228

Maastrichtian Age 21–23, 93, 99–106, 108–12, 115–16, 121, 138
Mackenzie Basin 200
Macnamara Formation 127, 147
Macquarie Ridge Complex 157
Magazine Point Formation 150
magmatism 20, 42, 46, 48–50, 154, 162, 236–38
Mahoenui Formation 190
Mahoenui Group 188, 189, 190–91, 210, 225
Mairangi Group 216, 228
Maitai Terrane 43, 44, 183
Makara Basin 172–73, 196
Makara Greensand 198
Mako Formation 123
Manaia Sub-basin 24, 112–13, 149
Mandamus Igneous Complex 84, 237
Mangaa Formation 223
Mangaa-1 well 223
Mangaheia Group 193
Mangahewa Formation 132, 133, 240
Mangakahia Complex 69, 78, 95, 99, 103, 166–67
Mangakotuku Formation 152–53
Manganui Formation 184–86, 191, 208
Mangapa Mudstone 136, 138
Mangapurupuru Group 71
Mangarara Limestone 210
Mangles Formation 182, 183
Mangonui Formation 209
Maniototo Conglomerate 200
Manuherikia Group 176
Manurewa Formation 103
Manutahi-1 well 211, 222
marine transgression 52
  mid–Late Cretaceous 22–23, 54–56, 89–103, 109–10, 113–14
  Paleocene 23–25, 56
  Eocene 25–28, 118–19, 123, 128, 128–34, 130–37, 136, 139
  Oligocene 28–31, 54, 56, 139, 140, 141–47, 149–50
  Miocene 33–36
Marlborough Fault 32, 173
Marshall Paraconformity/Unconformity 29, 54, 120, 142, 173–74, 240–42
Maruia Basin 161, 163, 164
Maryville Coal Measures 190
Mata River succession 236
Matakaoa Volcanics 74, 83, 166, 167
Matemateaonga Formation 210–11, 222, 226
Matiri Formation 149, 183
Maui Supergroup 52
Maungataniwha Sandstone 95–96
Mawhera Formation 130
McIvor Formation 178
McKee Formation 133
Mead Hill Formation 99, 104, 107, 116, 239
Mead Stream succession 107
Median Tectonic Zone (Median Batholith) 17–18, 40, 42, 44, 46, 49, 57
Medway Formation 198
Meremere Subgroup 188–89
Mernoo Bank 26
Mesozoic Era *Plate 2*, **20**, 39, 41–42, 44, 45, 47, 49, 133, 161, 176
  see also Cretaceous, Jurassic, Permian, Triassic periods
Mid Bay High 34, 37, 201, 218
Mikonui-1 well 182, 204, 220
Milburn Limestone 176
Miocene Epoch 47, **170–71**
  Waitakian–Clifdenian stages 56
  basin evolution 32–36
  sedimentation *Plate 14a & b*, 143–47, 149–54, 165–212, 242–43
  subsidence 32, 34–36, 163, 168–69, 177, 181–84, 186–87, 192–93, 203, 205–08
  tectonism 141, 156–58, 161–227
  volcanism 34, 35, 36, 151, 186, 187, 199, 200, 206–07, 208–09, 227–31
Miocene–Pliocene boundary event 211–12, 242–43
Moa Group 114, 115, 132–34, 149, 151
Moa-1B well 42
Moanui Formation 79, 81
Moeraki Formation 106, 107, 122, 235

Mohaka Fault 195, 214
Mohakatino Formation 206, 207
Mohakatino Group 191, 225
Mohakatino Volcanic Centre 206–07, 230
Mokau Formation 185, 191
Mokau Group 190–91, 210
Moki Formation 185–86, 223
Mokoiwi Formation 75, 83, 167
Mokonui Formation 213
Mokopeka Sandstone 213
Momotu Supergroup 52
Monowai Formation 34, 179, 201
Monowai Sub-basin 34, 145, 147, 177, 179, 180, 201
Moonlight Fault 30, 33, 87, 128, 140, 144, 145, 147, 157, 177, 180, 219, 229
Moonlight Sea 33
Moonlight–Hollyford transfer thrust 218
Morgan Volcanics 112
Morley Coal Measures 239
Morley Formation 109
Morrinsville Facies 47
Motatau Complex 137–38, 166–67
Motuan Stage 51, 58
Motukaraka facies 78
Mount Brown Formation 174, 199
Mount Camel Suspect Terrane 47, 69, 155, 224
Mount Herbert Volcanic Group 227, 228
Mount Messenger Formation 206, 207–08, 210, 211, 223
Mount Somers Volcanics Group 48, 67, 84, 237
Moutere Depression 56, 211, 211–12, 221, 243
Moutere Gravels 37, 221
Mungaroa Limestone 120
Murchison Basin 31, 34–35, 56, 132, 148–49, 161, 163, 164, 182, 183–84, 205, 220
Murchison Mountains High 30
Murihiku Supergroup 65, 114, 115
Murihiku Terrane *Plate 2a*, 43–44, 48, 133, 137

Neogene period 54, 87, 163, 165–232
  volcanism 227–32
  see also Miocene Epoch; Pliocene Epoch
Nessing Greensand 142
New Brighton Conglomerate 109
New Caledonia Basin 25, 27–28, 49, 65, 85, 86, 87, 108, 114, 115, 135, 139, 186, 209, 224, 235, 239–40
New Caledonia Trough 22, 113, 117
New Zealand region/landmass **19**
  formation pre mid-Cretaceous 39–50
  mid-Cretaceous (Albian–Cenomanian) 20–21
  Late Cretaceous **20**, 94
  Paleogene (Paleocene, Eocene, Oligocene) 56, 139, 141, 242
Ngahape Igneous Complex 115
Ngatoro Group 149–50, 151, 186
Ngatutura Volcanics 225
Ngaumu Mudstone 173
Niagara Sandstone Member 183
Niagara-1 and 2 wells 183
Nightcaps Group 126, 128
Nile Group 182
Norfolk Ridge 22, 87
North Cape Formation 113–14
Northern Graben (Taranaki Basin) 221–22, 223–24, 225, 230
Northland Allochthon 28, 35, 47, 50, 54, 56, 62, 69, 78–79, **78**, 82–83, 95, 99, 103, 108, 136, 137–38, 152, 154, 158, **158**, **159**, 161–62, 166–67, 186–88, 209, 211, 224, 243
Northland Basin **135**, 166–67
  Cretaceous 22, 56, 65, 78, 82–83, 114
    K–T boundary 235
  Paleogene (Paleocene, Eocene, Oligocene) 27, 28, 115, 117, 118, 134–36, 139, 151–52, 154, 241
  Neogene (Miocene–Pliocene) 184, 186–88, 208–09, 211, 230, 243
Northland Ophiolite 154, 166
Northland Peninsula **135**, **136**, **159**
  Cretaceous 95
  Paleogene (Paleocene, Eocene, Oligocene) 25, 27, 28, 35, 137–38
  Miocene 35–36, 158–159, 161–62, 192, 208, 229–31
  Pliocene–Pleistocene 38, 224

Oamaru Volcanics 120
Ocean Drilling Program (ODP)
  site 1121 101
  site 1123 140, 241
  site 1124 100, 140, 241
Ohai Basin 86
Ohai Coalfield 109
Ohai Depression 26–27, 177, 178
Ohai Group 109
Ohura Fault 36, 190, 191, 210
O'Keefe Formation 204–05
Okiwa Group 226
Old Man Group 37, 220–21, 243
Oligocene Epoch *Plates 11a, 12a & b, 13a & b*
  basin evolution 28–31, 56
  marine transgression 54, 56
  plate movement 87
  sedimentation 118–30, 134–55, 240–42
  tectonism 156–58, 162
  volcanism 151, 154–55
Omahuta Sandstone 138
Omihi Formation 142, 174
Onemama Formation 151–52
Onoke Group 214
Oponae Mélange 70, 72
Orahiri Limestone 153–54
Orauea Formation mudstone 126, 144, 146

Ordovician period 41, 42
Otago Schist 123
Otaraoa Formation 149–50, 151
Otaua Basin 188
Otekaike Limestone *Plate 12b*, 142, 175
Otitia Basalt 128, 139
Otorohanga Limestone 154
Ototara Limestone 120, 122
Otumotu Formation 65
Otunui Formation 191, 210
Ouse Fault 81
Owhena Formation 96
Oxford Basalt 227

Pacific plate 28, 39, 49, **50**, 86–87, 158–60, 165, 211
  see also Australian–Pacific plate boundary
Pacific–Antarctic Ridge 86, 117
Pahau Siltstone 174
Pahau Terrane/Subterrane *Plates 4b*, *7a & b*, 20–21, 45–46, 47, 48, 57, 59, 60, 71–72, 73
Pakaha Group 98
Pakawau Basin/Sub-basin 24, 86, 110, 112–14, 113
Pakawau Group 112–14, 235
Pakihi Supergroup 52
Paleocene Epoch 98–108, 107–08
  basin evolution 23–25
  cessation of sea-floor spreading 56, 86–87, 116–17
  sedimentation in 118–23, 128–29, 132–34
  volcanism in 116, 138–39
Paleogene period 55–56, **90–91**
  see also Eocene, Oligocene, Paleocene epochs
Paleozoic Era 17, 40–42
  see also Cambrian, Carboniferous, Ordovician, Silurian periods
Paparangi Group 226
Paparoa Basin 86, 148–49
Paparoa Coal Measures 108, 111–12, 129, 234, 239
Paparoa Metamorphic Core Complex *Plates 5a*, *6a & b*, 22, 49, 63, **64**
Paparoa Tectonic Zone 111, 128, 132, 148, 192
Paparoa Trough 24, 27, 31, 110, 111, 112, 128–31, 130, 131, 148, 182, 240
Parara Fault 34
Parara Sub-basin 30, 128, 147, 180
Parara-1 well 123, 128, 147, 203, 219
Parengarenga Basin 188
Parengarenga Group 209
Parnell Grit 187
Patea–Tongaporutu High 36, 190, 191, 206, 210, 222, 225
Paton Formation 96–97, 99, 100
Penrod Group 143, 164, 176, 217
Pepin Fault 203
Permian period *Plate 2b*, 17, 35, 39, 41–48
petroleum exploration **67**
Phoenix plate 39, 48, 49
Phoenix-Pacific Ridge 39, 86
Picton Fault 87
Pike River Coalfield 112, 117
Pike River detachment zone 49
Pirongia volcano 232
plate boundary evolution **17**, 26, 28–29, 31–32, 35, 37, 39, 49–50, **50**, 56, 85–87, **88**, 154, 156–57, 159–65
Pleistocene Epoch 194–95, **197**
  basin evolution 33, 37–38
  sedimentation 211–27
  tectonism 164
Pliocene Epoch 194–95, **197**
  basin evolution 36–38, 161
  sedimentation 56, 211–26
  tectonism 164, 166
  uplifting 219
  volcanism 219, 229, 231
Pliocene–Recent epochs 157, 231–32
Pomahaka Formation 143
Pororari Basin 85
Pororari Group 55, 63, **64**, 65, 77–78, 110, 111
Port Craig Formation 201
Prospect Formation 34, 180, 192, 201, 202, 218
Pukemuri Siltstone 118
Punakitere Sandstone 78–79, 95
Puniwhakau-1 well 190, 210, 211
Putangirua Conglomerate 197
Puysegur Bank 219
Puysegur Group 66
Puysegur Trench 157, 192, 219

Quaternary period 37, 38
  see also Pleistocene Epoch

Rahu Suite granitoids 42
Rakaia Subterrane *Plate 2b*, 45–46, 47–48, 57
Rakiura Group 120, 122–23
Rakiura-1 well 122
Rakopi Formation 113
Rangiauria Breccia 228
Rangiawhia Volcanics 47
Rangitata Orogeny 17, 47–48
Rangitikei Supergroup 225
Rangitoto High 152
Rapahoe Group 130
Rappahannock Group gravels 37, 183, 205, 220–21
Raukumara Series 75, 77–82, 84
Red Bluff Tuff 117, 121, 138
Red Hill Teschenite 115, 239
Reefton Basin 27, 130
Reinga Basin 27, 115, 117, 135, 139, 151–52, 209, 224
Resolution Ridge 87, 157, 240
Resolution-1 well 105, 120, 121, 175, 199, 217
Rewanui Coal Measure Member 112, 234

rifting 22–23, 54–55, 63, 76, 86–89, 108–09, 114, 238–39, 240
Rip Volcanics 83
Ross Glaciation 220
Ross volcano (Auckland Islands) 229
Rotokare Group 221, 222, 224
Rotokohu Coal Measures 182, 204, 220
Rowallan Sandstone 201
Ruahine Fault 214
Ruatangata Sandstone 136, 137, 152
Ruatoria Group 74

Saddle Hill Siltstone 106
Sand Hill Formation 126
Sandfly Formation 126–27
Sandpit Tuff 227
Santonian Age 21, 22, 55, 89–92, 94–101
Scargill Siltstone 174
sea-floor spreading 21–22, 89–95, 98–99, 101–03
cessation 25, 56, 117
sea-level changes *see* eustatic sea-level changes
sedimentation 52, **189**
Cretaceous 21–23
Clarence Series **62**, 63–76, **64**
Raukumara Series 77–82
Haumurian–Teurian 98–115
Paleogene (Paleocene, Eocene, Oligocene) 25, 26, 27, 118–38, 141–54, 240
Miocene 32–36, 143–47, 149–54, 165–212
Pliocene–Pleistocene 36–38, **197**, 211–27
Seek Cove Formation 66
Seismic Unit IIB 76
Separation Point Batholith 18, 42, 221
Separation Point Granite 183
Seymour Group 81
Shag Point Group 67, 77
Shoal Point Formation 200, 229
Silurian period 41
Solander Basin
Cretaceous 66
Eocene–Oligocene 26–27, 30, 123–124, 127–28, 147, 148, 240
Miocene 34, 177, 180–81, 202–03
Pliocene 37, 219, 229
Solander Fault 34
Solander Ridge 27, 30, 33, 124, 147–48, 219, 229
Solander Trough 26, 181, 240
Solander volcano 229
Solander-1 well 181, 203
South Auckland Basin 28, 31, 116, 118, 136–37, 139, 152–53, 188
South Auckland Field 232
South Fiji Basin 157, 159, 192
South Mernoo graben 76
South Tasman Rise 140–41
South Wanganui Basin 222
Southburn Sand 175
southeast Tasman oceanic crust (STOC) 86–87
Southern Volcanics basalt 115
Southland Shelf 30
Split Rock Formation 72, 74
Springhill Formation 71, 73, 82, 83, 96
Spyglass Formation *Plate 12a*, 173–74
Stanton Conglomerate Formation 104
Stewart Island Shelf 219
Stillwater Mudstone 163, 183, 204
subsidence
Late Cretaceous–Paleocene 21–24, 55, 89, 94–95, 98–99, 105, 112
Eocene 25–28, 122
Oligocene 28–31, 89, 139, 140, 149–50
Miocene 32, 34–36, 163, 168–69, 177, 181–84, 186–87, 192–93, 203, 205–08
Pliocene–Pleistocene 37
superterranes 40
Surville-1 well 185, 205
Swinburn Foundation 123

Tahora Formation 95–96
Taimana Formation 151, 184
Taipa Mudstone 137–38
Taitai Sandstone Member 75
Takaha Limestone 149
Takaha Terrane 18, 41–42
Takapu-1A well 106, 176
Takaro Formation 180, 202
Takaro High 218
Takatika Grit 105
Takiritini Formation 172
Takitimu Group 201, 202
Takitimu Mountains block 30
Tane Member 114, 117
Tangahoe Formation 222–23, 225–26
Tangaroa Formation 134
Tangaroa-1 well 41, 42
Tangaruhe Formation 96
Tangihua Complex 69, 78, 82–83, 154–55, 158, 161, 166–67, 209
Taniwha Formation 55, 65, 113
Tapuaenuku Igneous Complex 74, 84
Tapuwaeroa Formation 97, 99
Tara-1 well 24, 33, 67, 122, 143, 200, 217
Taradale-1 well 213
Tarakohe Mudstone 184, 185
Taranaki Basin **93**
Cretaceous 22, 56, 62, 63, 65, 85, 89, 93, 108, 112–15, 238–39
K–T boundary 235
Paleogene (Paleocene, Eocene, Oligocene) 24–25, 27–28, 31, 88, 132–34, 139, 149–51, 240, 241

Neogene (Miocene–Pliocene) 35, 160–61, 164, 184–86, 191, 192, 205–08, 210, 211–12, 225–26, 230
Pliocene–Pleistocene 37–38, 221–24, 242–43
Taranaki Fault 24, 28, 31, 35, 65, 115, 133, 150, 151, 154, 160–61, 184–85, 206, 221
Taranaki Graben 35, 57, 206
Taranaki Peninsula 133, 150, 184, 185, 206–07, 210–11, 221–23
Taranaki shelf 41, 42
Tarata Thrust Zone 161
Taratu Coal Measures 239
Taratu Formation 106
Tariki Sandstone Member 150
Tartan Formation 108
Tasmanian deep ocean gateway 140
Tatu-1 well 190
Taumarunui Formation 189–90
Taumatamaire Formation 189–90
Tauperikaka Coal Measures 110
Taupo Volcanic Zone 38, 162, 164, 227, 232
Taupo–Hikurangi arc-trench system 232
Tauranga Group 225
Te Akatea Siltstone 153
Te Anau Basin **125**
Eocene 26–27, 123–24, 126–27, 240
Oligocene 30, 144, 145–47
Miocene 33–34, 176–77, 179–80, 192, 202
Pliocene–Pleistocene 37, 218
Te Anau Fault 218
Te Aute Limestone lithofacies *Plate 16a*, 37, 212, 213, 214
Te Hoe-1 well 169
Te Kawa volcano 232
Te Kuiti Basin 152–53
Te Kuiti Group 136–37, 151–54, 186–87, 188–89, 190, 191, 225
Te Ranga-1 well 24, 65, 113, 115, 133
Te Rata unit 167
Te Waewae Formation 218
Te Wera Formation *Plate 7b*, 59, 70–71, 72, 83
Te Whanga Limestone 120, 142
tectonism 50, 51–56, **53**
before mid-Cretaceous 39–50
mid-Cretaceous 54–55
Late Cretaceous–early Paleocene 55–56, 85–88, 98, 100, 108, 110
Paleogene (Paleocene, Eocene, Oligocene) 124, 126, 140, 150, 156–58
Miocene 141
tectono-sedimentary assemblages 52, 54–56
Teredo Limestone 119, 120
terranes *see* basement terranes
Three Kings Rise 157–58, 159
Tikihore Formation 75, 80, 81, 99, 167
Tikorangi Formation 149, 150–51, 160
Timaru Basalt 228
Tinui Fault 196
Tinui Group 95
Tioriori Group 105, 121, 138–39
Tiropahi Limestone 149
Titihaoa-1 well 172
Titirangi Sand 216
Tititira Formation 164, 182, 203–04
Titri Fault 37, 77, 238, 239
Tokakoriri Formation 110, 116, 128
Tokama Siltstone 175
Tokanui volcano 232
Tokerau Facies 47
Tolaga Group 168, 169, 193
Tonga Trench 159
Tonga–Kermadec arc 159
Tongaporutuan mudstone 194
Tongaporutu–Herangi High 210
Tongaporutu–Patea High 211
Tongariro Volcanic Centre 232
Torlesse Supergroup 44, 48, 60, 66–67, 70, 71, 73, 96, 140, 141, 220, 221, 227
Torlesse Terrane *Plate 2b*, 43, 45–46, 47, 55, 59, 72, 174
Toroa-1 well 24, 122
Toru Trough 37, 222, 223
Toru-1 well 206
transgression *see* marine transgression
Triassic period *Plates 2*, *11a*, 41, 43, 44, 45–46, 133, 176
Tuapeka Fault 106, 239
Tucker Cove Formation 122, 123, 200
Tunnel Burn Formation 180
Tupapakurua-1 well 153, 154, 190
Tupou Complex 54, 69, 79, 166–67
Tupuaenuku Igneous Complex 237
Tupuangi Formation 76
Turi Formation 114, 115, 117, 133–34
Turonian Age 66, 78–79, 80, 81
Turret Peaks Formation 146–47, 180
Tutaki Member 183–84

uplift (general) 21–22, 32–38, 49, 154, 165–66, 215–17
Upper Limestone (Amuri) 119, 138
Upper Marl (Amuri) 119, 138
Upton Formation 198
Upukerora-1 well 180, 202
Urenui Formation 207, 208, 210, 211
Urewera Group 59–60, 79

Van der Linden fracture zone 158
Vening Meinesz fracture zone 158, **158**
Victoriella Limestone 142

View Hill Volcanics 121, 138
volcanism
Cretaceous 21, 22, 82–84, 115–16
Paleogene
Paleocene–Eocene 25–26, 116, 121, 138–39
Oligocene 29, 142, 154–55
Neogene (Miocene–Pliocene) 33, 34, 35, 36, 151, 186, 187, 192, 199, 200, 206–07, 208–09, 216–17, 227–32
Pliocene–Pleistocene 37, 38, 219
Waiareka Volcanics *Plate 11b*, 122, 139
Waiari Formation 99
Waiau Basin 26–27, 29–30, 33–34, 37, 109, 123–24, 124–26, **125**, 144–45, 176–77, 178–79, 200–02, 218, 243
Waiau Formation *Plate 15a*
Waicoe Formation mudstone 33–34, 144–46, 177, 179, 180, 181, 201, 202
Waihao Greensand 122
Waihemo Fault 37, 66–67
Waihere Bay Group 97
Waiho-1 well 182, 183, 192, 204, 220
Waihoaka Formation 144
Wai-iti Group 151, 184, 186, 221
Waikaka Quartz Gravels 217
Waikaraka Mudstone 79
Waikari Formation 174
Waikato Basin 232
Waikato Coal Measures 137, 152, 153
Waikato Coalfield 24
Waikato Fault 230
Waikuku Limestone 188, 209
Waima Formation *Plate 14b*, 174, 198
Waimamaku-2 well 103, 115
Waimana Formation 198
Waimana Sandstone 70, 72
Waimea Fault 183, 221
Waimumu Quartz Gravels 217
Wainui-1 well 113
Waipapa Formation 152
Waipapa Terrane 43, 45, 46–47, 48, 69, 136–37, 152, 187
Waipara Greensand 105, 121
Waipawa 'Black Shale' facies 25, 103, 107–08, 117
Waipawa Formation 107, 116, 118, 167
Waipoua Basin 38
Waipoua Plain 186
Waipounamu erosion surface 94
Waipuna Bay Formation 200
Wairarapa Fault 172, 196, 215
Wairata Sandstone 79
Wairau Fault 162, 163, 164
Wairio Formation 109
Wairiri Volcaniclastite 227
Waitahaia Formation 71, 73, 80, 167, 238
Waitakere Group 209
Waitakere Limestone 149
Waitakere Volcanic Arc 151
Waitemata Basin 36, 187
Waitemata Group 36, 186, 187, 188–89, 225
Waitomo Sandstone 154
Waitui Sandstone 184
Waitutu Sub-basin 30, 33–34, 66, 124, 127–28, 147, 148, 177, 180, 181–82, 201, 202–03, 218, 219
Waka Nui-1 well 25, 65, 108, 114, 115, 135, 186, 209, 224
Waka Supergroup 52
Wakamarama Fault 113, 203, 205
Wallow Group 73–74
Wangaloa Formation 106–07
Wanganui basins 28, 36–37, 38, 136, 164, **189**, 206, 209–11, 211, 225–27, 236, 243
Wanstead Formation 107, 116, 118
Warder Formation 73–74
Weber Formation 118, 141, 168
Wedderburn Formation 175, 176, 200
Weka Pass Stone 173, 174
Wellington Fault 214
Welsh Formation 163, 184
West Coast Basin 89, 93, 139–40
West Coast–Taranaki Rift 22, 86, 108–09, 111, 128–29, 238–40
West Hump Fault 219
West Norfolk Ridge 42, 78
Western Fiordland Orthogneiss 48
Western Province 17, 20, 40–42, 48–49, 57, 164
Western Stable Platform (Taranaki Basin) 37–38, 160, 222, 223–24
Whaingaroa Siltstone 153
Whakamarino Formation 169
Whakapohai Sandstone 110
Whakataki Formation *Plate 14a*, 172
Whales Back Limestone 173
Whangai facies 22, 23, 92, 106–07
Whangai Formation 79, 95–96, 99–103, 163, 167, 193
Whangamomona Group 209–10, 211, 225
Whangarei Limestone 151–52
Whangaroa Subgroup 209
Whenuakura Group 226
White Creek Fault 132
White Rock Coal Measures 176, 199–200
Whitianga Group 231
Wickliffe Formation 101, 107
Winton Basin 26, 29, 30, 33, 109, **109**, 123, 144, 176–78
Winton Hill Formation 144

Zealandia **17**, 20, 22, 39, **40**, 48–50, 52, 54, 86–87, 89, 94